D0142561

Electronics:
Circuits and Devices

THIRD EDITION

Electronics: Circuits and Devices

THIRD EDITION

Ralph J. Smith

Department of Electrical Engineering, Stanford University

J O H N   W I L E Y   &   S O N S

New York · Chichester · Brisbane · Toronto

WITHDRAWN
OESTERLE LIBRARY, NCC
NAPERVILLE, IL 60540

Copyright © 1973, 1980, and 1987 by John Wiley & Sons, Inc.

All rights reserved. Published simultaneously in Canada.

Reproduction or translation of any part of this work beyond that
permitted by Sections 107 or 108 of the 1976 United States
Copyright Act without the permission of the copyright owner is
unlawful. Requests for permission or further information should
be addressed to the Permissions Department, John Wiley &
Sons, Inc.

Library of Congress Cataloging-in-Publication Data

Smith, Ralph Judson.
    Electronics: circuits and devices.

    Includes index.
    1. Electronics. I. Title.

TK7816.S6   1987   621.381   86-28910
ISBN 0-471-84446-2

Printed in the United States of America

16 15 14 13 12 11

Printed and bound by R.R. Donnelley & Sons, Inc.

621.381
Smbe
3 ed

Preface

This is a book on electronics for engineering students. The pervasive nature of electronics is illustrated by the modern automobile. Conceived as an assemblage of pistons, levers, and gears, the automobile has become an electronic machine with its combustion controller, engine diagnostics, traction and anti-skid systems, and trip computer. Similar examples can be found in every branch of engineering involving instrumentation, control, or data processing. Electronics has become an essential subject for engineers.

What are the special characteristics of engineers? They want to understand; for them, explanations of abstract concepts should be derived from basic principles and be completely satisfying. They have background in physics and skill in mathematics that can and should be used and developed. They are problem solvers who need practice in applying new concepts to real problems, in design as well as analysis. They are willing to work on difficult material if they are convinced of its career value. I kept their needs in mind while preparing this edition.

This text is an introduction to the broad field of electronics at the sophomore or junior level. The prerequisites are freshman calculus and electrical physics. It focuses on the most important circuits and devices and their applications. With an accompanying laboratory, it provides an integrated treatment of principles, practical applications, problem solving, and design techniques.

The choice of topics represents a compromise. The objective is to take beginning students to the point where they can make effective use of modern integrated circuits in the design of simple digital and analog systems. To this end, the operation and application of gates, flip-flops, memories, converters, and op amps in integrated circuit form are discussed in detail. However, to exploit packaged units effectively, students must understand the physical behavior of semiconductor devices and how they can be modeled, and they must master basic concepts such as circuit analysis, frequency response, and feedback. To meet the needs of non-electrical engineers, the number of concepts and variety of devices have been limited so that a satisfying level of quantitative understanding can be reached in a one-term course.

APPROACH

The guidelines followed in the planning included:

- The discussion of circuits is limited to those techniques needed to handle the more important device applications, and these are introduced only as they are needed.
- Ideal diodes and amplifiers (op amps) and their use in signal processing are introduced early to provide motivation through interesting applications and opportunity for lab experiments.
- Application of diodes and transistors is first to digital devices where dc concepts and simple device models are adequate. Analog applications requiring ac circuit theory and more sophisticated models are postponed until later.
- In the digital section, the treatment of logic gates, counters, registers, and memory devices is directed toward ultimate application in information, control, and microcomputer systems.
- In the analog section, after discussion of digital and analog applications of op amps, the emphasis shifts to the operation and design of amplifiers: large-signal, small-signal, multistage, and feedback amplifiers, with special attention to device modelling and frequency response.

CONTENT

Chapters 1, 2, and 3 provide a basic understanding of electrical quantities, circuit laws, and signal processing. Network theorems and applications are in terms of direct currents, but the properties of exponential and sinusoidal signals are described mathematically. Chapter 3 starts with a description of signals and waveforms; then ideal amplifiers and diodes are introduced to illustrate the application of basic concepts to electronic circuits that are widely used in signal processing.

The next three chapters develop a quantitative understanding of electron behavior, semiconductor physics, and device characteristics. Electron motion is described in Chapter 4, and the cathode-ray oscilloscope is introduced here for use in accompanying lab work. Chapter 5 focuses on the semiconductor diode; conduction, doping, and junction phenomena are used in deriving the external characteristics and a simple mathematical model. Chapter 6 focuses on transistors and their integrated circuit forms; the explanations of physical phenomena lead to quantitative descriptions of external characteristics.

Chapters 7, 8, and 9 provide an introduction to digital electronics. Chapter 7 uses simple models of diodes and transistors to explain the operation of electronic switches, logic gates, and flip-flops. In Chapter 8 the emphasis is on binary representation, the analysis and synthesis of logic circuits, and their application in registers, counters, and memories. Chapter 9 provides a first-level treatment of the microprocessor: basic concepts, computer architecture, programming, and the application of practical microprocessors to information processing.

Chapter 10 represents a sideways step to derive convenient methods for handling steady-state ac circuit problems. Practice is provided in series and parallel circuit analysis and the design of frequency selective circuits. This chapter stands alone and could follow Chapter 3 in the Circuits and Signals section.

The last five chapters provide an introduction to linear electronics. Chapter 11 describes practical operational amplifiers and shows how inexpensive IC op amps can be used in the design of practical amplifiers, buffers, integrators, converters, regulators, oscillators, or analog computers. Chapter 12 treats amplifiers in general and large-signal discrete-transistor amplifiers in particular; bias design and efficiency calculations for audiofrequency amplifiers are included. In Chapter 13, small-signal models of varying levels of sophistication are derived for field-effect and bipolar junction transistors. The models are applied to the analysis and design of small-signal amplifiers in Chapter 14, and the virtues of positive and negative feedback are explored in Chapter 15.

A feature of this book is the emphasis on operational amplifiers, digital devices, and microprocessors. The microprocessor is more than just a new device with a host of diverse applications. Its computing power, small size, and low cost have completely changed our way of thinking. It has altered our philosophy of design; to solve an information processing problem we now take a mass-produced "brain" with a flexible "memory" and "instruct" it to solve our problem. Fortunately, the microprocessor is an ideal subject for study; it integrates our knowledge of registers, counters, memories, and logic units. Mastery requires no difficult-to-understand concepts or hard-to-acquire mathematics. Students who understand microprocessors have gained a general understanding of digital systems as well as a powerful tool for solving engineering problems. Teachers have the satisfaction of knowing that they have made a significant contribution to their students' education. (Students report that their experience with microprocessors makes them more employable.)

LEARNING AIDS

Electronics is a difficult subject; successful students must master abstract concepts and sophisticated techniques. The following features have been included to aid the student:

- There are 112 completely worked-out **Examples** to illustrate important new concepts as they are developed and to serve as models of problem solution.
- There are 75 elementary **Practice Problems** with answers to allow students to test their understanding of key points as they are presented.

- **Key statements** in the text are set off, in color, to emphasize their importance.
- The **two-color illustrations** provide extra clarity and make it possible to focus attention on the significant details in a complicated circuit or graph.
- The chapter **Summaries** have been carefully prepared to represent a distillation of essential concepts, mathematical relations, and analytical techniques.
- At the end of each chapter, the **Review Questions** are primarily for the student's use in testing understanding of the new ideas and terminology.
- The **Exercises** vary in difficulty, but, in general, they are straightforward applications of the new principles. Answers to Selected Exercises are included in the Appendix.
- The **Problems** are more involved and may require extending a concept, making simplifying assumptions, or putting ideas together in a design. Hints from the teacher may be required.

ORGANIZATION

The entire book could be covered in about sixty 50-minute lectures. By selecting material to emphasize particular objectives, a variety of one-semester courses (of 45 lectures) is possible. If there is a preceding circuits course, Chapters 1, 2, and 10 can be omitted and approximately 36 lectures would be appropriate for the strictly electronics portion. The first 11 chapters would fit a course of about 30 lectures emphasizing digital electronics.

An accompanying laboratory is essential, and the text material has been planned to support a series of 8 to 12 lab experiments to improve students' grasp of abstract concepts. More emphasis could be placed on microprocessors by adding current material on a specific processor and supplementing it with design practice and laboratory experience.

The *Instructor's Manual*, available from the publisher to teachers adopting the book for class use, contains complete solutions to the *Exercises* and the *Problems*. Also included are suggested lecture and reading schedules for courses of different length and emphasis, along with a brief description of each chapter from the teacher's viewpoint.

ACKNOWLEDGMENTS

I acknowledge with pleasure the assistance of many persons in this revision. The manuscript was read critically by Allen Nussbaum of the University of Minnesota, Wallace H. McDonald of the U.S. Merchant Marine Academy, Carleton M. Brown of the University of Maine, Russell A. Hannen of Wright State University, Thomas C. Jannett of the University of Alabama at Birmingham, and Bruce P. Johnson of the University of Nevada, Reno. Although all their suggestions could not be incorporated, this third edition is greatly improved as a result of their comments. I especially thank Kim Kihlstrom of Westmont College for her insight into student difficulties, her helpful suggestions, and her valuable contributions to the *Instructor's Manual*.

Mary Forkner of Publication Alternatives skillfully supervised the production of this book. Her grasp of the entire production process, her devotion to keeping the project running smoothly, and her ability to provide an ideal relationship between author and publisher are greatly appreciated.

For careful organizing of diverse materials, for patient proofreading, and for cheerful assistance in the details of publication for the eleventh time, I again thank my wife Louise.

Stanford, California   Ralph J. Smith

Contents

Analog Electronics

Electronics:
Circuits and Devices

THIRD EDITION

1

Electrical Quantities

Introduction

Definitions and Laws

Circuit Elements

The history of electronics starts with the discovery of cathode rays and continues into tomorrow. It is a fascinating story of individual contributions by mathematicians, physicists, engineers, and inventors and of the painstaking building of a solid technology.

While Hittorf and Crookes were studying cathode rays (1869), Maxwell was developing his mathematical theory of electromagnetic radiation. Soon after Edison observed electronic conduction in a vacuum (1883), Hertz demonstrated (1888) the existence of the radio waves predicted by Maxwell. At the time of J. J. Thomson's measurement of e/m (1897), Marconi was becoming interested in wireless and succeeded in spanning the Atlantic (1901). While Einstein was generalizing from the photoelectric effect, Fleming was inventing the first electron tube (1904), a sensitive diode detector utilizing the Edison effect. DeForest's invention of the triode (1906) made it possible to amplify signals electronically and led to Armstrong's sensitive regenerative detector (1912) and the important oscillator.

Zworykin's invention of the picture tube (1924) and the photomultiplier (1939) opened new areas of electronics. Watson-Watts' idea for radio detection and ranging, or *radar,* developed rapidly under the pressure of World War II. Postwar demands and increased knowledge of semiconductor physics led to the invention of the transistor by Shockley, Bardeen, and Brattain (1947) and the silicon solar cell by Pearson (1954). Kilby's integrated circuit (1958) permitted building active and passive circuit elements simultaneously and Noyce (1958) provided the technique for fabricating a complete network containing many semiconductor devices on a single monolithic chip. Einstein had pointed out (1917) the possibility of stimulated emission of radiation, and Townes used this concept (1953) to achieve *maser* operation. Townes and Schawlow (1958) described a device that could extend the maser principle into the optical region, and Maiman succeeded in doing so (1960) with his pulsed ruby *laser*.

Eckert and Mauchly started the computer revolution (1946) with ENIAC, their 18,000-vacuum tube electronic digital computer; Hoff (1969) matched the computational ability of ENIAC with his microprocessor on a single silicon chip. The availability of tiny, cheap, powerful computers in the 1970s revolutionized computation, communication, information processing, and control.

INTRODUCTION

The electrical engineer is primarily concerned with phenomena involving electric charges, particularly forces between charges and energy exchanges between charges. Electronics is the branch of electrical engineering in which the behavior of a device is determined by or explained in terms of the behavior of electrons in electric and magnetic fields. In some situations *energy* is the important quantity, and in others energy is merely a means of conveying *information*.

In an observation satellite, for example, the useful output from the solar cells is electrical energy that is converted into various forms for powering control and communication systems. In a subsystem that converts data to digital form, performs preliminary calculations, and sends the results back to earth, energy is merely the means for processing information.

Forces and Fields

Electric charges are defined by the forces they exert on one another; experimentally, the forces are found to depend on the magnitudes of the charges, their relative positions, and their velocities. Forces due to the position of charges are called *electric forces,* and those due to the velocity of charges are called *magnetic forces*. All electric and magnetic phenomena of interest in electrical engineering can be explained in terms of the forces between charges.

In a television picture tube, the electrodes are designed to provide accelerating and deflecting *electric fields* to control the path of electrons in forming the picture. In a computer memory bank, minute magnetic "bubbles" become magnetized by pulses of current and then "remember" the digital information in the form of *magnetic fields*. In a motor, the combination of steel armature and copper winding is shaped to create an intense magnetic field through which the conductors pass in developing motor torque. Fields are characteristically distributed throughout a region and must be defined in terms of two or three dimensions.

Circuits

In contrast to fields, the behavior of a *circuit* can be completely described in terms of a single dimension, the position along the path constituting the circuit. In an electric circuit, the variables of interest are the voltage and current at various points along the circuit. In circuits in which voltages and currents are constant (not changing with time), the currents are limited by resistances. In the case of a battery being charged by a generator, 10 meters of copper wire may provide a certain resistance to limit the current flow. The same resistance could be provided by 10 centimeters of resistance wire, or by 1 centimeter of resistance carbon, or by 1 millimeter of semiconductor.

When the dimensions of a component are unimportant and the total effect can be considered to be concentrated at a point or "lumped," the component can be represented by a *lumped parameter*. In contrast, the behavior of 10 meters of copper wire as an antenna is dependent on its dimensions and the way in which voltage and current are distributed along it; an antenna must be represented by *distributed parameters*. In this book, we are concerned with lumped parameter circuits only.

Devices

Circuits are important in guiding energy within *devices* and also to and from devices that are combined into *systems*. Electrical devices perform such functions as generation, amplification, modulation, and detection of signals. For example, at an AM radio broadcasting station a modulator changes the amplitude of the transmitted wave in accordance with a musical note to produce amplitude modulation. In a microcomputer, logic gates, counters, and registers are organized to receive, process, and store digital signals. A *transducer* is a device that converts energy or information from one form to another; a microphone is a transducer that converts the acoustical energy in an input sound wave into the electrical energy of an output current. A cathode-ray tube converts the information stored in a solid-state memory to a graphic display.

Systems

Systems incorporate circuits and devices to accomplish desired results. A communication system includes a microphone transducer, an oscillator to provide a high-frequency carrier for efficient radiation, a modulator to superimpose the sound signal on the carrier, an antenna to radiate the electromagnetic wave into space, a receiving antenna, a detector for separating the desired signal from the carrier, various amplifiers and power supplies, and a loudspeaker to transduce electrical current into a replica of the original acoustic signal. A space-vehicle guidance system includes a transducer to convert a desired heading into an electrical signal, an error detector to compare the actual heading with the desired heading, an amplifier to magnify the difference, an actuator to energize the precise control jets, a sensor to determine the actual heading, and a feedback loop to permit the necessary comparison.

Models

Circuits are important for another reason: frequently it is advantageous to represent a device or an entire system by a *circuit model*. Assume that the 10 meters of copper wire mentioned previously is wound into the form of a multiturn coil and that a voltage of variable frequency is applied. If the ratio of applied voltage V to resulting current I is measured as a function of frequency, the observations will be as shown

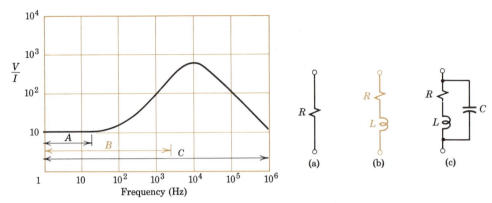

Figure 1.1   Coil characteristics and circuit models.

in Fig. 1.1. Over region A, the coil can be represented by a single lumped parameter (resistance R); in other words, the results obtained from the circuit of Fig. 1.1a are approximately the same as the results obtained from the actual coil of wire. Similarly, the behavior of the coil in region B can be represented by the circuit in Fig. 1.1b which contains another lumped parameter (inductance L). To represent the coil over a wide range of frequencies, an additional parameter (capacitance C) is necessary, and the circuit model of Fig. 1.1c is used.

The technique of representing, approximately, a complicated physical device by a relatively simple model is an important part of electrical engineering. In this book we use circuit models to represent such devices as transistors and amplifiers. One advantage of such a model is that it is amenable to mathematical analysis. Representing a transistor by a circuit model (Fig. 1.2) permits the use of well-known circuit laws in predicting the behavior of the actual transistor in an amplifier circuit.

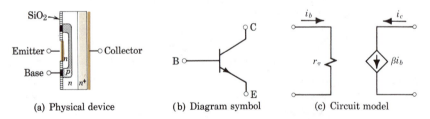

(a) Physical device   (b) Diagram symbol   (c) Circuit model

Figure 1.2   Transistor representation.

A Preview

This book has four parts. In this chapter we start our study of circuits and signals by defining the important electrical quantities and describing circuit elements in terms of physical phenomena. Then we establish the basic principles and techniques needed in predicting the behavior of electronic circuits. In Chapter 3 we add ideal diodes and amplifiers to our repertoire of circuit elements, and we learn to analyze and to design simple circuits with many practical applications.

In the second part we learn about electronic devices. Chapter 4 describes the motion of electrons in electric and magnetic fields and the operation and application of cathode-ray tubes. In Chapter 5 we study the physical principles of electron motion in semiconductors and see how diodes operate. Then we look at the fabrication and performance of various transistors in discrete and integrated-circuit form.

With this background, we are ready to study digital systems. We learn how diodes and transistors are used in logic gates and memory elements and see how practical digital devices can be used in data processing. In Chapter 9 we combine the digital devices we have studied into a basic microprocessor and learn to program a practical microcomputer.

The last part is devoted to analog systems. In Chapter 10 we study ac circuits and develop the necessary analytical tools. Then, in Chapter 11, we take a closer look at operational amplifiers and see how to design amplifiers, converters, regulators, and signal generators with inexpensive "op amps." The next three chapters discuss the design and operation of large-signal and small-signal amplifiers. Then we study the use of feedback to improve amplifier performance. Throughout the book we emphasize basic principles and useful techniques that find frequent application in electronic engineering.

DEFINITIONS AND LAWS

In beginning the study of circuits, we must first define the important circuit quantities and adopt a standard set of units, symbols, and abbreviations. Much of this material is a review of basic physics, but it deserves careful attention because it constitutes the "language" in which ideas are presented, concepts formed, and conclusions stated.

Next we take a new look at three laws based on early experiments conducted on resistors, inductors, and capacitors. Using these experimental results and modeling techniques, we invent idealized circuit components with highly desirable characteristics and then see how they behave in circuits and how they transform energy.

The International System of Units

In engineering, we must be able to describe physical phenomena quantitatively in terms that will mean the same to everyone. We need a standard set of units consistent among themselves and reproducible any place in the world. In electrical engineering, we use the SI (System International) system in which the *meter* is the unit of length, the *kilogram* the unit of mass, and the *second* the unit of time. Another basic quantity is temperature, which in the SI system is measured in *kelvins*. To define electrical quantities, an additional unit is needed; taking the *ampere* as the unit of electric current satisfies this requirement. The *candela* is needed to define illumination quantities. All quantities encountered in this book can be defined in terms of the six units displayed in Table 1-1 on page 6. When data are specified in other units, they are first converted to SI units and then substituted in the applicable equations. Three conversions frequently needed are: 1 meter = 39.37 inches, 1 kilogram = 2.205 pounds (mass), and 1 newton = 0.2248 pound (force). Other useful conversion factors are given in the appendix.

Table 1-1   Basic Quantities

Quantity	Symbol	Unit	Abbreviation
Length	l	meter	m
Mass	m	kilogram	kg
Time	t	second	s
Temperature	τ	kelvin	K
Current	i	ampere	A
Luminous intensity	I	candela	cd

Definitions

For quantitative work in circuits, we need to define the quantities displayed in Table 1-2. These are probably familiar from your previous study, but a brief review here may be helpful.

Force. A force is a push or a pull. A force of 1 newton is required to cause a mass of 1 kilogram to change its velocity at a rate of 1 meter per second per second ($f = ma$). In this text we are concerned primarily with electric and magnetic forces.

Energy. Energy is the ability to do work. An object requiring a force of 1 newton to hold it against the force of gravity (i.e., an object "weighing" 1 newton) receives 1 joule of potential energy when it is raised 1 meter (PE $= mgh$). A mass of 1 kilogram moving with a velocity of 1 meter per second possesses $\frac{1}{2}$ joule of kinetic energy (KE $= \frac{1}{2}mu^2$).

Power. Power measures the rate at which energy is transformed. The transformation of 1 joule of energy in 1 second represents an average power of 1 watt. In general, instantaneous power p and average power P are defined by

$$p = \frac{dw}{dt} \quad \text{and} \quad P = \frac{W}{T} \tag{1-1}$$

Charge. The quantity of electricity q is electric charge, a concept useful in explaining physical phenomena. Charge is said to be "conservative" in that it can be neither created nor destroyed. It is said to be "quantized" because the charge on 1 electron (1.602×10^{-19} C) is the smallest amount of charge that can exist. The coulomb can be defined as the charge on 6.24×10^{18} electrons, or as the charge experiencing a force of 1 newton in an electric field of 1 volt per meter, or as the charge transferred in 1 second by a current of 1 ampere ($q = it$).

Current. Electric field effects are due to the presence of charges; magnetic field effects are due to the motion of charges. The current through an area A is defined by the electric charge passing through the area per unit of time. In a current of 1 ampere, charge is being transferred at the rate of 1 coulomb per second. In general, the charges may be positive and negative, moving through the area in both directions. The current is the *net* rate of flow of *positive* charges, a scalar quantity. In the specific case of

Table 1-2 Important Derived Quantities

Quantity	Symbol[†]	Definition	Unit	Abbreviation	(Alternative)
Force	f	push or pull	newton	N	$(kg \cdot m/s^2)$
Energy	w	ability to do work	joule	J	$(N \cdot m)$
Power	p	energy/unit of time	watt	W	(J/s)
Charge	q	quantity of electricity	coulomb	C	$(A \cdot s)$
Current	i	rate of flow of charge	ampere	A	(C/s)
Voltage	v	energy/unit charge	volt	V	(W/A)
Electric field strength	$\mathcal{E}$	force/unit charge	volt/meter	V/m	(N/C)
Magnetic flux density	B	force/unit charge momentum	tesla	T	(Wb/m^2)
Magnetic flux	ϕ	integral of magnetic flux density	weber	Wb	$(T \cdot m^2)$

[†] In general, but not always, we use lower-case letters for time varying quantities and capital letters for constant or average values. B is an exception.

positive charges moving to the right and negative charges to the left, the net effect of both actions is positive charge moving to the right; the instantaneous current to the right is given by the equation

$$i = \frac{dq}{dt} = +\frac{dq^+}{dt} + \frac{dq^-}{dt} \qquad (1\text{-}2)$$

In a neon light, for example, positive ions moving to the right and negative electrons moving to the left contribute to the current flowing to the right.

Voltage. The energy-transfer capability of a flow of electric charge is determined by the electric potential difference or voltage through which the charge moves. A charge of 1 coulomb receives or delivers an energy of 1 joule in moving through a voltage of 1 volt or, in general, instantaneous voltage is defined by

$$v = \frac{dw}{dq} \qquad (1\text{-}3)$$

Voltage measures the energy carried by each unit of charge. The function of an energy source such as an automobile battery is to add energy to the current; a 12-V battery adds twice as much energy per unit charge as a 6-V battery.

Practice Problem 1-1

A battery operated radio requires a current of 0.15 A at 12 V. Calculate the power required and the energy consumed in 2 hours of operation; express the results in SI units.

Answers: 1.8 W; 12,960 J.

Electric Field Strength. The "field" is a convenient concept in calculating electric and magnetic forces. Around a charge we visualize a region of influence called an "electric field." The electric field strength $\mathcal{E}$, a vector, is defined by the magnitude and direction of the force **f** on a unit positive charge in the field. In vector notation the defining equation is

$$\mathbf{f} = q\mathcal{E} \tag{1-4}$$

where magnitude $\mathcal{E}$ could be measured in newtons per coulomb. However, bearing in mind the definitions of energy and voltage, we note that

$$\frac{\text{force}}{\text{charge}} = \frac{\text{force} \times \text{distance}}{\text{charge} \times \text{distance}} = \frac{\text{energy}}{\text{charge} \times \text{distance}} = \frac{\text{voltage}}{\text{distance}}$$

and electric field strength in newtons per coulomb is just equal and opposite to the *voltage gradient*[†] or

$$\mathcal{E} = -\frac{dv}{dl} \quad \text{in volts per meter} \tag{1-5}$$

The minus sign indicates that the force is directed away from the higher potential.

Magnetic Flux Density. Around a moving charge or current we visualize a region of influence called a "magnetic field." In a bar magnet the current consists of spinning electrons in the atoms of iron; the effect of this current on the spinning electrons of an unmagnetized piece of iron results in the familiar force of attraction. The intensity of the magnetic effect is determined by the magnetic flux density **B**, a vector defined by the magnitude and direction of the force **f** exerted on a charge q moving in the field with velocity **u**. In vector notation (see Fig. 4.3) the defining equation is

$$\mathbf{f} = q\mathbf{u} \times \mathbf{B} \tag{1-6}$$

A force of 1 newton is experienced by a charge of 1 coulomb moving with a velocity of 1 meter per second normal to a magnetic flux density of 1 tesla.

Magnetic Flux. Historically, magnetic fields were first described in terms of *lines of force* or *flux*. The flux lines (so called because of their similarity to flow lines in a moving fluid) are convenient abstractions that can be visualized in the familiar iron-filing patterns. Magnetic flux ϕ (phi) in webers is a total quantity obtained by integrating magnetic flux density over area **A**. The defining equation is

$$\phi = \int \mathbf{B} \cdot d\mathbf{A} \tag{1-7}$$

where the dot product indicates the contributions of B *normal* to dA. Because of this background, magnetic flux density is frequently considered as the derived unit and expressed in webers per square meter. In this text, however, we consider magnitude B in teslas as the primary unit.

[†]In the vicinity of a radio transmitter, the radiated field may have a strength of a millivolt per meter; in other words, one meter of antenna may develop a millivolt of signal voltage that can be fed into a receiver for amplification.

Electrical Power and Energy

A common problem in electric circuits is to predict the power and energy transformations in terms of the expected currents and voltages. Since, by definition, $v = dw/dq$ and $i = dq/dt$, instantaneous power is

$$p = \frac{dw}{dt} = \frac{dw}{dq}\frac{dq}{dt} = vi \tag{1-8}$$

Therefore, total energy is

$$w = \int p \, dt = \int vi \, dt \tag{1-9}$$

EXAMPLE 1

The "electron gun" of a cathode-ray tube provides a beam of high-velocity electrons.

(a) If the electrons are accelerated through a potential difference of 20,000 V over a distance of 4 cm (Fig. 1.3), calculate the average field strength.

(b) Calculate the power supplied to a beam of 50 million billion electrons per second.

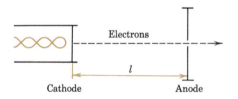

Cathode   Anode

Figure 1.3   Current and power.

(a) By definition, $\varepsilon = -dv/dl$ or

$$\varepsilon_{av} = \frac{\Delta v}{\Delta l} = \frac{20,000}{0.04} = 5 \times 10^5 \text{ V/m}$$

(b) By definition, $i = dq/dt$ or

$$i = \frac{\text{charge}}{\text{electron}} \times \frac{\text{electrons}}{\text{second}}$$

$$= 1.6 \times 10^{-19} \times 50 \times 10^6 \times 10^9 = 0.008 \text{ A}$$

By Eq. 1-8, the power is

$$p = vi = 2 \times 10^4 \times 8 \times 10^{-3} = 160 \text{ W}$$

Note: In calculations, enter all values in SI units. Only in the results are units shown.

Experimental Laws

In contrast to the foregoing arbitrary definitions, there are three useful laws that were formulated (about 200 years ago) from experimentally observed facts. Our present understanding of electrical phenomena is much more sophisticated and these laws are readily derived from basic theory. At this stage in our treatment of circuits, however, it is preferable to consider these relationships as they would be revealed in experiments on real devices in any laboratory. From observations on the behavior of real devices we shall derive idealized models of circuit elements. Then we shall establish the rules governing the behavior of such elements when they are combined in simple or complicated circuits.

Resistance

In circuit analysis, we are interested in the relationship between the voltage across a device and the current flowing through it. Consider an experiment (Fig. 1.4a) in which a generator is used to supply a varying current i to a copper rod. The current and resulting voltage v are observed on an oscilloscope, and voltage is plotted as a function of current (Fig. 1.4c). If all other factors (such as temperature) are held constant, the voltage is observed to be approximately proportional to the current. The experimental relation for this *resistor* can be expressed by the equation

$$v \cong Ri \qquad (1\text{-}10)$$

where R is a constant of proportionality. Where v is in volts and i in amperes, Eq. 1-10 defines the *resistance R* in *ohms* (abbreviated Ω) and is called *Ohm's law* in honor of Georg Ohm, the German physicist whose original experiments led to this simple relation. The same relation can be expressed by the equation

$$i \cong Gv \qquad (1\text{-}11)$$

which defines the *conductance G* in siemens (S) in honor of the German inventor Werner von Siemens. Numerically, G in siemens[†] is just equal to $1/R$ where R is in ohms.

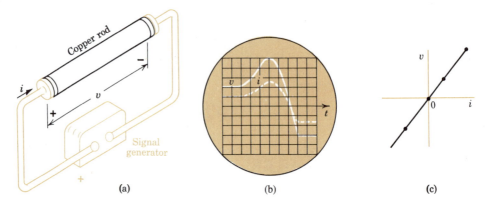

(a)	(b)	(c)

Figure 1.4   Experimental determination of resistor characteristics.

We now know that a metallic conductor such as copper contains many relatively free electrons. The application of a voltage creates an electric field that tends to accelerate these *conduction* electrons, and the resulting motion is superimposed on the random thermal motion of the electrons at, say, room temperature. Electrons are accelerated by the field, collide with copper atoms, and give up their energy. They are accelerated again, gaining energy from the electric field, collide again, and give up their energy. Superimposed on the random motion due to thermal energy, there is an average net directed motion or *drift* due to the applied electric field. The speed of drift is found to be directly proportional to the applied electric field. In a given conducting

---

[†] In the United States, conductance was formerly designated in mhos ($\mho$).

element, therefore, the rate of flow of electric charge or the current is directly proportional to the electric field, which, in turn, is directly proportional to the applied voltage. Ohm's law can be derived directly from a consideration of this concept of conduction.

Capacitance

Now consider an experiment (Fig. 1.5a) in which a copper rod is cut and reshaped into two flat plates separated by air as an insulator. Available, but not shown, are a voltmeter, an ammeter, and an electroscope to measure voltage, current, and charge. When a voltage is applied as shown, it is observed that positive charge appears on the left-hand plate and negative on the right. If the generator is disconnected, the charge persists. Such a device that stores charge is called a *capacitor*. If the charge on the capacitor is measured accurately, it is observed to be approximately proportional to the voltage applied or

$$q \cong Cv \tag{1-12}$$

where C is a constant of proportionality called *capacitance* and measured in *farads* (abbreviated F) in honor of the ingenious English experimenter, Michael Faraday.

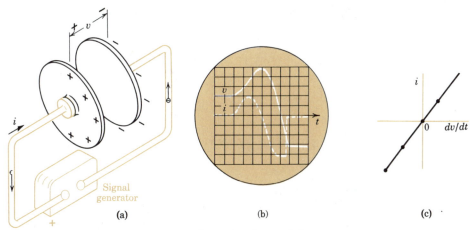

Figure 1.5 Experimental determination of capacitor characteristics.

As long as the voltage is constant, the charge is maintained and no current flows. If the voltage is changing with time (Fig. 1.5b), the current must flow to provide the necessary change in charge. This current is observed to be approximately proportional to the *rate of change* of voltage. The experimental relation can be expressed by the equation

$$i \cong C \frac{dv}{dt} \tag{1-13}$$

where C has the same value as determined previously. Note that Eq. 1-1 can be derived from Eq. 1-12 since $i = dq/dt$ by definition.

Practice Problem 1-2

A sinusoidal voltage $v = 2 \sin 100t$ V is applied across a resistor of 5 Ω.

(a) Use Ohm's law to predict the current that flows.
(b) What is the conductance of this resistor?
(c) If the same voltage is applied to a 0.0005-F capacitor, what current flows?

Answers: (a) $0.4 \sin 100t$ A; (b) 0.2 S; (c) $0.1 \cos 100t$ A.

Inductance

If, instead, the copper rod is drawn out and formed into the configuration of Fig. 1.6a, an entirely different result is obtained. Now the generator is used to supply

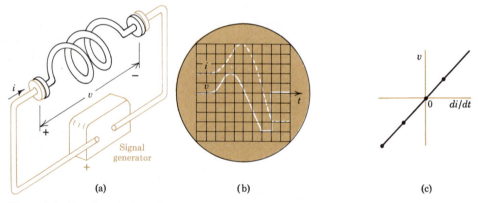

(a) (b) (c)

Figure 1.6 Experimental determination of inductor characteristics.

current to a multiturn coil. It is observed that only a small voltage is required to maintain an appreciable steady current, but a relatively large voltage is required to produce a rapidly changing current. As indicated in Fig. 1.6c, the voltage is observed to be approximately proportional to the *rate of change* of current. The experimental relation can be expressed by the equation

$$v \cong L\frac{di}{dt} \tag{1-14}$$

where L is a constant of proportionality called *inductance*[†] and measured in *henrys* (abbreviated H) in honor of the American inventor and experimenter, Joseph Henry.

This important relation can be derived by using our knowledge of magnetic fields. We know that charge in motion produces a magnetic field and the multiturn coil

---

[†] More properly called "self-inductance"; *mutual inductance M* is defined by the equation $v_{12} = M \, di_2/dt$ where v_{12} is the voltage induced in coil 1 due to changing current i_2 in a second coil.

is an efficient configuration for concentrating the effect of all the moving charges in the conductor. In the region in and around the coil, the magnetic flux density is high for a given current. But we also know (and this was Henry's great discovery) that a changing magnetic field *induces* a voltage in any coil that it links and the induced voltage is directly proportional to the rate of change of the magnetic field. If the magnetic field is due to the current in the coil itself, the induced voltage always tends to oppose the change in current that produces it. The voltage required of the generator is determined primarily by this voltage of *self-induction* and is approximately proportional to the time rate of change of magnetic flux density. For the coil of Fig. 1.6, magnetic flux density is proportional to current and Eq. 1-14 is confirmed.

EXAMPLE 2

A current varies as a function of time as shown by the colored line in Fig. 1.7.

(a) Determine and plot the voltage produced by this current in a 30-Ω resistor.

(b) Determine and plot the voltage produced by this current in a coil of 0.5 Ω resistance and 4 H inductance.

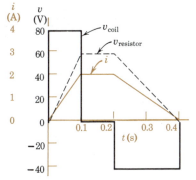

Figure 1.7 Current-voltage relations in a resistor and an inductor.

(a) Assuming that this is a linear resistor whose behavior is defined by $v = Ri$, the voltage is directly proportional to the current. For $i = 2$ A, $v = 30 \times 2 = 60$ V; the voltage as a function of time is as shown by the dashed line.

(b) For the maximum current of 2 A, the voltage due to the coil resistance is

$$v_R = Ri = 0.5 \times 2 = 1 \text{ V}$$

The voltage due to the coil inductance is

$$0 < t < 0.1 \quad v_L = L\frac{di}{dt} = 4\frac{2}{0.1} = 80 \text{ V}$$

$$0.1 < t < 0.2 \quad v_L = L\frac{di}{dt} = 4\frac{0}{0.1} = 0$$

$$0.2 < t < 0.4 \quad v_L = L\frac{di}{dt} = 4\frac{-2}{0.2} = -40 \text{ V}$$

Later we shall learn how to combine these two effects. For many purposes, however, satisfactory results can be obtained by neglecting v_R where v_L is comparatively large and considering v_R where v_L is comparatively small. On this basis the voltage across the coil is as shown by the solid line.

Decimal Notation

To specify the *value* of a measurable quantity, we must state the *unit* and also a *number*. We shall use the SI units, but the range of numbers encountered in electrical engineering practice is so great that a special notation has been adopted for convenience. For example, the power of a public utility system may be millions of watts, whereas the power received from a satellite transmitter may be millionths of a watt. The engineer describes the former in *mega*watts and the latter in *micro*watts.

Table 1-3   Standard Decimal Prefixes

Multiplier	Prefix	Abbreviation	Pronunciation
10^{12}	tera	T	těr′ à
10^{9}	giga	G	jĭ′ gà
10^{6}	mega	M	měg′ à
10^{3}	kilo	k	kĭl′ ð
10^{2}	hecto	h	hěk′ tð
10^{1}	deka	da	děk′ à
10^{-1}	deci	d	děs′ ĭ
10^{-2}	centi	c	sěn′ tĭ
10^{-3}	milli	m	mĭl′ ĭ
10^{-6}	micro	μ	mī′ krð
10^{-9}	nano	n	năn′ ð
10^{-12}	pico	p	pē′ cð
10^{-15}	femto	f	fěm′ tð
10^{-18}	atto	a	ăt′ tð

The notation is based on the decimal system that uses powers of 10. The standard prefixes in Table 1-3 are widely used to designate multiples and submultiples of the fundamental units. Thus 20 MW is read "20 megawatts" and is equal to 20×10^{6} or 20 million watts; and 20 μW is read "20 microwatts" and is equal to 20×10^{-6} or 20 millionths of a watt. In Example 3, note that all numerical quantities are first converted to SI units before being substituted in the applicable equations.

EXAMPLE 3

A current varies as a function of time as shown in Fig. 1.8. Predict and plot the voltage produced during the first 3 ms by this current flowing in an initially uncharged 1-μF capacitor.

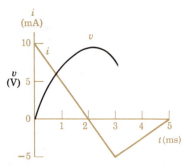

Figure 1.8 Current-voltage relations in a capacitor.

For an ideal capacitor, $i = C \, dv/dt$.

$$\therefore v = \frac{1}{C} \int_0^t i \, dt + V_0$$

For $0 < t < 3$ ms, $V_0 = 0$ and the slope of the current curve is

$$m = \frac{\Delta i}{\Delta t} = \frac{-0.01}{0.002} = -5 \text{ A/s}$$

As a function of time, the current is

$$i = I_0 - mt = 0.01 - 5t$$

$$\therefore v = \frac{1}{10^{-6}} \int_0^t (0.01 - 5t) \, dt + 0$$

$$= 10^6 (0.01t - 2.5t^2), \text{ a parabola}$$

At $t = 3$ ms, for example,

$$v = 10^6 [0.01 \times 0.003 - 2.5(0.003)^2] = 7.5 \text{ V}$$

Check: We note that the slope of the voltage curve is proportional to the current; where i is large, zero, or negative, dv/dt has a corresponding value.

CIRCUIT ELEMENTS

An electric circuit consists of a set of interconnected components. Components carrying the same current are said to be connected in *series*. A simple series circuit might include a battery, a switch, a lamp, and the connecting wires. Components subjected to the same voltage are said to be connected in *parallel*. The various lights and appliances in a residence constitute a parallel circuit. The terms *circuit* and *network* are sometimes used interchangeably, but usually network implies a more complicated interconnection such as the communication network provided by the telephone company. There is also an implication of generality when we refer to network analysis or network theorems.

Circuit Components

From our standpoint, the distinguishing feature of a circuit component is that its behavior is described in terms of a voltage-current relation at the terminals of the component. This relation, called the "*v-i* characteristic," may be obtained analytically by using field theory in which the geometry of the associated electric or magnetic field is important, or it may be obtained experimentally by using measurements at the terminals. Once the *v-i* characteristic is known, the behavior of the component in combination with other components can be determined by the powerful methods of *circuit theory,* the central topic of this part of the book.

Using hypothetical experiments and applying elementary field concepts, we obtained *v-i* characteristics for a resistor, a capacitor, and an inductor. In writing mathematical expressions for the *v-i* characteristics (Eqs. 1-10, 1-13, and 1-14) we neglected the distributed nature of the fields involved and we assumed linearity. But we know that no real physical device is exactly linear. For example, the resistance of the incandescent lamp of Fig. 1.9a increases rapidly with current, and the inductance of the coil of Fig. 1.9b is greatly affected by the value of average current. Therefore we shall invent three ideal circuit components, or *models,* that *are* linear.

There are good reasons for doing this. Some real components can, within limits, be represented by such lumped linear models in many problems. Other real components can be represented accurately by combinations of two or more ideal com-

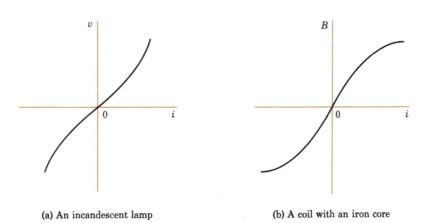

(a) An incandescent lamp   (b) A coil with an iron core

Figure 1.9   Characteristics of real circuit components.

ponents. Also, by using the mathematical methods applicable to linear models we can derive general results of great value in predicting the behavior of complicated devices and systems of devices.

Circuit Element Definitions

The three linear models are shown symbolically in Fig. 1.10 along with their defining equations and a set of alternative equations.[†] In a real resistor, the resistance is a function of current; in the ideal unit, resistance R (in ohms) is a constant. A real coil or inductor exhibits the property of inductance, but every real inductor also exhibits the property of resistance. The corresponding linear model is called an "ideal inductor" or, since it exhibits only a single property, it is called an "inductance element," or simply an "inductance" where L is in henrys. Similarly, an ideal capacitor is called a "capacitance" where C is in farads. The three linear models—resistance, inductance, and capacitance—are referred to as *circuit elements*; with these elements as building blocks, we can devise an infinite number of circuits.

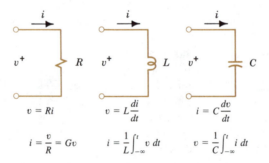

$$v = Ri \qquad v = L\frac{di}{dt} \qquad i = C\frac{dv}{dt}$$

$$i = \frac{v}{R} = Gv \qquad i = \frac{1}{L}\int_{-\infty}^{t} v\, dt \qquad v = \frac{1}{C}\int_{-\infty}^{t} i\, dt$$

Figure 1.10   Circuit elements.

Energy Storage in Linear Elements

We can obtain valuable insights into the behavior of real circuit components by considering the energy transformations that occur in the corresponding linear models. Recalling that energy $w = \int vi\, dt$, Eq. 1-9, we see that, once the v-i characteristic is defined, the energy storage or dissipation property is determined.

Inductance. Where $v = L\, di/dt$ and $i = 0$ at $t = 0$, the energy stored is

$$w_L = \int_0^T L\frac{di}{dt} i\, dt = \int_0^I Li\, di = \tfrac{1}{2}LI^2 \tag{1-15}$$

[†] The plus sign above the "v" indicates that the upper terminal is the positive reference direction of potential. The arrow under the "i" indicates that flow to the right is the positive reference direction of current. These conventions are discussed in detail in Chapter 2.

The total energy input to an inductance is directly proportional to the square of the final current. The constant of proportionality is $L/2$; in fact, this expression for energy could have been used to define inductance.

Inductance is a measure of the ability of a device to store energy in the form of moving charge or in the form of a magnetic field.

The equation reveals that the energy is stored rather than dissipated. If the current is increased from zero to some finite value and then decreased to zero, the upper limit of the integration becomes zero and the net energy input is zero; the energy input was stored in the field and then returned to the circuit.

Capacitance. Where $i = C\, dv/dt$ and $v = 0$ at $t = 0$,

$$w_C = \int_0^T vC\frac{dv}{dt}\, dt = \int_0^V Cv\, dv = \tfrac{1}{2}CV^2 \tag{1-16}$$

Can *you* interpret this equation? In words, the total energy input to a capacitance is directly proportional to the square of the final voltage. The constant of proportionality is $C/2$.

Capacitance is a measure of the ability of a device to store energy in the form of separated charge or in the form of an electric field.

The energy is stored rather than dissipated; if the final voltage is zero, the energy stored in the field is returned to the circuit.

Practice Problem 1-3

At a current of 2 A, a well-designed coil stores 0.5 J of energy with negligible losses. Represent the coil by a circuit element.

Answer: $L = 0.25$ H.

Energy Dissipation in Linear Elements

The expressions for stored energy remind us of the expressions for kinetic energy of a moving mass and potential energy of a stretched spring. Each of these elements (and there are others in thermal and chemical systems) has the ability to store energy and then return it to the circuit or system. If we make a similar analysis of an electrical resistance, the results are quite different.

Resistance. Where $v = Ri$ and $i = 0$ at $t = 0$,

$$w_R = \int_0^T Ri\, i\, dt = \int_0^T Ri^2\, dt \tag{1-17}$$

To evaluate the total energy supplied, we must know current i as a function of time t. For the special case where the current is constant or $i = I$,

$$w_R = \int_0^T RI^2 \, dt = RI^2 T \tag{1-18}$$

In general, for finite values of i (either positive or negative) and finite values of t, the energy supplied to the resistor is finite and positive. There is no possibility of controlling the current in such a way as to return any energy to the circuit; the energy has been *dissipated*. In a real resistor, the dissipated energy appears in the form of heat; in describing the behavior of the linear model we say that the energy has been dissipated in an *irreversible transformation,* irreversible because there is no way of heating an ordinary resistor and obtaining electric energy.

The *rate* of dissipation of energy or power is a useful characteristic of resistive elements. By definition, $p = dw/dt$; therefore,

$$p_R = \frac{dw_R}{dt} = Ri^2 \tag{1-19}$$

The power dissipation in a resistance is directly proportional to the square of the current. The constant of proportionality is R and this expression for power is frequently used to define resistance.

Resistance is a measure of the ability of a device to dissipate power irreversibly.

Sometimes it is more convenient to work with the voltage across a resistance. Because $i = v/R$, we can express the power dissipated as

$$p_R = Ri^2 = R\frac{v^2}{R^2} = \frac{v^2}{R} = Gv^2 \tag{1-20}$$

Practice Problem 1-4

When a 120-V, 60-W light bulb is turned on, the initial current is 4 A. Represent the light bulb by ideal circuit components for the "cold" and the "operating" conditions.

Answers: $R_c = 30 \ \Omega$; $R_{op} = 240 \ \Omega$.

Continuity of Stored Energy

We know that a massive body tends to oppose rapid changes in velocity; we say that such a body has *inertia*. In solving problems in dynamics we take advantage of the fact that the velocity of a mass cannot be changed instantaneously; for example, the velocity just after a force is applied must be the same as that just before the force is applied. Similar conclusions can be drawn about electric circuit quantities by applying the fundamental principle that in any physical system the energy must be a continuous function of time.

Since power is the time rate of change of energy ($p = dw/dt$), an instantaneous change in energy would require an infinite power. But the existence of an infinite power is contrary to our concept of a physical system, and we require that the energy stored in any element of a system, real or ideal, be a continuous function of time. Recalling that the energy of a moving mass is $\frac{1}{2}Mu^2$, we conclude that the velocity u cannot change instantaneously.

Following the same reasoning, we note that the energy stored in an inductance is $\frac{1}{2}Li^2$ and, therefore,

The current in an inductance cannot change instantaneously.

Since the energy stored in a capacitance is $\frac{1}{2}Cv^2$,

The voltage across a capacitance cannot change instantaneously.

These concepts are particularly useful in predicting the behavior of a circuit just after an abrupt change, such as closing a switch. Note that in considering energy changes there is no limitation on the rapidity with which inductance voltage or capacitance current can change. Also, because resistance does not store energy, there is no limitation on the rapidity with which resistance voltage and current changes can occur.

EXAMPLE 4

The ignition coil of an automobile can be represented by a series combination of inductance L and resistance R (Fig. 1.11). A source of voltage V, the battery, is applied through contacts or points represented by switch S.

(a) With the switch open, the current i is zero. What is the current just after the switch is closed?

(b) After a finite time the current has reached a value of 1 A and the switch is suddenly opened. What is the current just after the switch is opened?

(a) Since the current in the inductance cannot change instantaneously, the current just after closing the switch must be zero (but starting to increase rapidly for proper operation of the ignition system).

(b) The current just after the switch is opened is still 1 A because it takes time to dissipate the energy stored in the inductance ($\frac{1}{2}Li^2$). Some of the energy goes to ionize the air between the switch contacts and an arc is formed; the points are "burned."

In practice, a "condenser" (a capacitor) is connected across the switch to minimize arcing by absorbing energy from the coil. The condenser stops the current quickly so that di/dt is large and a large $L\,di/dt$ voltage is passed to the spark plugs by transformer action.

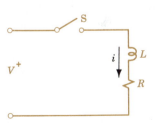

Figure 1.11 Ignition coil.

Energy Sources and Reversible Transformations

Energy can be stored in inductance or capacitance and dissipated in resistance. These are called *passive* circuit elements in contrast to *active* elements that can serve as sources of electrical energy. The many different forms of electrical "generators" include the rotating dynamo, the chemical storage battery, the solar cell, the fuel cell, and the thermocouple. In each of these real devices the electrical energy is "generated" by the conversion of some other form of energy. The conversion represents a *reversible transformation* in that the conversion process can go either way, in contrast to the situation in a resistor. To illustrate: in a dynamo operating as a "generator," mechanical energy is converted into electrical energy; but electrical input to a typical dynamo will cause it to operate as a "motor" producing mechanical energy.

Although every generator supplies both voltage and current ($w = \int vi\,dt$), it is desirable to distinguish between two general classes of devices. If the voltage output is relatively independent of the circuit to which it is connected (as in the battery of Fig. 1.12a), the device is called a "voltage generator." If over wide ranges the output current tends to be independent of the connected circuit (as in the transistor of Fig. 1.12b), it is treated as a "current generator."

An ideal voltage generator in which the output voltage is completely independent of the current is called an "ideal voltage source," or simply a "voltage source." The

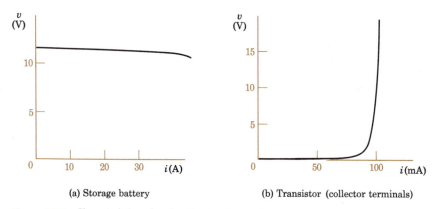

(a) Storage battery   (b) Transistor (collector terminals)

Figure 1.12   Characteristics of real voltage and current generators.

output voltage is specified, usually as a function of time, and it is unaffected by changes in the circuit to which it is connected. At low currents, an automobile battery supplies a nearly constant voltage and may be represented by a 12-V voltage source. At high currents, however, the output voltage is appreciably less than 12 V.[†] The difference is due to the fact that the energy-conversion process is not perfect; part of the chemical energy is converted to heat in an irreversible process. This result suggests the possibility of representing a real electrical generator by a linear model consisting of a pure voltage source and a resistance, the resistance accounting for the irreversible energy transformations.

[†] How is this fact evidenced when an automobile is started with its headlights on?

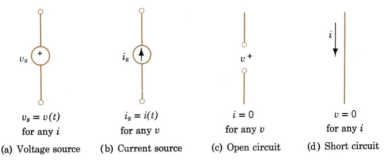

$v_s = v(t)$ $\qquad$ $i_s = i(t)$ $\qquad$ $i = 0$ $\qquad$ $v = 0$
for any i $\qquad$ for any v $\qquad$ for any v $\qquad$ for any i

(a) Voltage source $\quad$ (b) Current source $\quad$ (c) Open circuit $\quad$ (d) Short circuit

Figure 1.13 Other circuit elements.

The circuit symbol and the defining equation for a voltage source are shown in Fig. 1.13a. The corresponding ideal current generator or "current source" is shown in Fig. 1.13b. Note that here the output current is specified, usually as a function of time, and it is independent of the voltage across the source. As we shall see, a combination of a current source and a resistance also may be used to represent a real generator in predicting the behavior of real circuits. There is a finite limit to the power supplied by such a combination, whereas an ideal voltage or current source can supply, or absorb, an unlimited amount of power.

Two additional circuit elements need to be defined before we begin circuit analysis. The *open circuit* shown symbolically in Fig. 1.13c is characterized by finite voltage and zero current; the *short circuit* is characterized by finite current and zero voltage. (To what real components do open and short circuits correspond?) While other circuit elements are conceivable (and, in fact, will be desirable later on), the passive elements R, L, and C, the active sources $v(t)$ and $i(t)$, and the open and short circuits constitute a set that will be sufficient to describe a large number of useful circuits.

SUMMARY

- The important electrical quantities, their definitions, and their SI units are summarized in Table 1-2 on page 7.
  In electrical terms, instantaneous power and energy are expressed by

$$p = \frac{dw}{dt} = vi \quad \text{and} \quad w = \int p\,dt = \int vi\,dt$$

- The approximate behavior of real circuit components is used to define the five passive and two active circuit elements displayed in Table 1-4 on page 22.
  Voltage and current sources represent reversible energy transformations.

Table 1-4   Electrical Circuit Elements

Element	Unit	Symbol	Characteristic
Resistance (Conductance)	ohm (siemen)	$i \xrightarrow{} R$ $+v$ (G)	$v = Ri$ $(i = Gv)$
Inductance	henry	$i \xrightarrow{} L$ $+v$	$v = L\dfrac{di}{dt}$ $i = \dfrac{1}{L}\displaystyle\int_{-\infty}^{t} v\,dt$
Capacitance	farad	$i \xrightarrow{} C$ $+v$	$i = C\dfrac{dv}{dt}$ $v = \dfrac{1}{C}\displaystyle\int_{-\infty}^{t} i\,dt$
Short circuit		$i \xrightarrow{}$	$v = 0$ for any i
Open circuit		$-\!\!\circ\; +v \;\circ\!\!-$	$i = 0$ for any v
Voltage source	volt	$\ominus$ v_s	$v = v_s$ for any i
Current source	ampere	$\ominus$ i_s	$i = i_s$ for any v

- Resistance is a measure of the ability of a device to dissipate energy irreversibly ($p_R = Ri^2$).
  Inductance is a measure of the ability of a device to store energy in a magnetic field ($w_L = \frac{1}{2}Li^2$).
  Capacitance is a measure of the ability of a device to store energy in an electric field ($w_C = \frac{1}{2}Cv^2$).
- Energy stored in an element must be a continuous function of time.
  The current in an inductance cannot change instantaneously.
  The voltage on a capacitance cannot change instantaneously.

REVIEW QUESTIONS

(These questions are primarily for use by the student in checking his or her familiarity with the ideas in the text.)

1. Without referring to the text, list with their symbols and units the six quantities basic to the SI system.
2. Without referring to the text, define "force," "energy," and "power," and give their symbols and units.
3. Without referring to the text, define "current," "voltage," "electric field strength," and "magnetic flux density," and give their symbols and units.
4. Some automobile manufacturers are considering going to an 18-V battery. What specific advantages would this have?
5. Explain the minus sign in Eq. 1-5.
6. Given two points a and b in a field of known strength ε, how could voltage v_{ab} be calculated?

7. Explain the difference between an "electric signal" that moves along a wire at nearly the speed of light and an "electric charge" that drifts along at only a fraction of 1 m/s.

8. Define "resistance," "inductance," and "capacitance" in terms of voltage-current characteristics and also in terms of energy transformation characteristics.

9. Draw a curve representing current as a function of time and sketch on the same graph curves of the corresponding voltage across inductance and capacitance.

10. How can a current flow "through" a capacitor containing an insulator?

11. Once the effect of an electric field has been represented by a circuit element, is the geometry of the field significant?

12. Draw a curve representing voltage as a function of time and sketch on the same graph curves of the corresponding current in an inductance and a capacitance.

13. List three physical components not mentioned in the text that have the ability to store energy.

14. The concept of continuity of stored energy places what restrictions on the circuit behavior of inductances and capacitances?

15. Give three examples of reversible transformations involving electrical energy.

16. Explain, in terms of source characteristics, the dimming of the house lights when the refrigerator motor starts.

17. Distinguish between an "inductor" and an "inductance"; give an example of each.

EXERCISES

(These exercises are intended to be straightforward applications of the concepts of this chapter.)

1. A woman weighing 110 lb runs up a 12-ft (vertical distance) flight of stairs in 3 s.
   (a) Express in newtons the force such a woman would exert on a bathroom scale and in kilograms the "reading" of the scale.
   (b) Express in joules and in kilowatt-hours the net work done in climbing the stairs.
   (c) Express in watts the average power expended in climbing the stairs.

2. A glass tube with a cross-sectional area of 10^{-2} m^2 contains an ionized gas. The densities are 10^{15} positive ions/m^3 and 10^{11} free electrons/m^3. Under the influence of an applied voltage, the positive ions are moving with an average axial velocity of 5×10^3 m/s. At that same point the axial velocity of the electrons is 2000 times as great. Calculate the electric current.

3. A laser printer produces circular dots 125 μm in diameter. The exposure required is 1.5 μJ/cm^2 of printed area. Calculate the dot writing rate for a 5-mW helium-neon laser. If a typical letter requires 10 dots, about how long would it take to print a page of text in this book?

4. A 24-W automobile lamp operates at 12 V. Calculate the current, the number of electrons passing through the filament per second, and the total charge transferred in 1 minute.

5. A voltage of 4000 V exists across the 2-cm insulating space between two parallel conducting plates. An electron of mass $m = 9.1 \times 10^{-31}$ kg is introduced into the space.
   (a) Calculate the electric field strength in the space.
   (b) Calculate the force on the electron and the resulting acceleration.

6. In a picture-tube deflection system, electrons moving at a velocity of 39,000 km/s enter a magnetic field of flux density 0.2 T.
   (a) Predict the force exerted on the electrons.
   (b) If flux is from left to right on the screen, what is the direction of the resulting deflection? (See Fig. 4.3 for the definition of $\mathbf{u} \times \mathbf{B}$.)

7. A calculator battery supplies a current of 0.002 A at a voltage of 9 V. Calculate the rate of flow of charge and the energy supplied per second.

8. In a TV picture tube, the accelerating voltage is 25 kV.
   (a) Express in joules the energy gained by an electron moving through this potential.
   (b) If the power delivered to the fluorescent screen by the high-velocity electrons is to be 2 W, calculate the necessary beam current.

9. Predict the total energy available from a 9-V radio battery rated at 0.6 A·h and the total operating time at a current of 0.018 A.

10. In Fig. 1.14, the conductance of the resistor is 0.02 S. For $v = 15$ V, what current flows through the resistor?

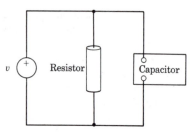

Figure 1.14

11. The voltage in Fig. 1.14 is $v(t) = 2000t$ V where t is in seconds. For time $t = 5$ ms, predict the current through the 4-Ω resistor and through the 300-μF capacitor.

12. The signals from Voyager 2, describing Uranus, traveled for 2 hr, 44 min, and 50 s at the speed of light to deliver 1 millionth of a billionth of a watt. Express the power and the distance traveled (in miles) in decimal notation.

13. In Fig. 1.15, current $i(t) = I_m \sin \omega t$.
  (a) Write an expression for the voltage across the resistor of R ohms.
  (b) Write an expression for the voltage across the inductor of L henrys.

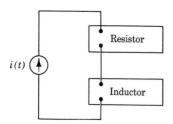

Figure 1.15

14. The current in Fig. 1.15 is $i(t) = 10t$ A where t is in seconds. For time $t = 5$ ms, predict the voltages across the 120-Ω resistor and across the 3-H inductor.

15. The current shown in Fig. 1.16 flows through a 5-H inductor and an initially uncharged 2-μF capacitor. Plot to scale graphs of v_L and v_C.

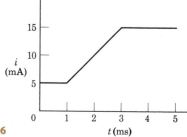

Figure 1.16

16. The voltage shown in Fig. 1.17 is applied to a 2-μF capacitor and a 0.5-H inductor (no initial current). Plot to scale graphs of i_C and i_L.

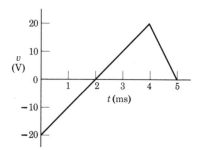

Figure 1.17

17. (a) Calculate the energy stored in an inductance $L = 10$ H at time $t = 2$ ms by the waveform of Fig. 1.16.
  (b) Calculate the energy stored in a capacitance $C = 100$ μF at time $t = 3$ ms by the waveform of Fig. 1.17.

18. For the current of Fig. 1.16 flowing in a 2-H inductance, calculate the voltage across the inductance and the power into the inductance at $t = 2$ ms. Repeat for $t = 4$ ms.

19. For the voltage of Fig. 1.17 across a 20-μF capacitance, calculate the current and the power into the capacitance at $t = 1$ ms. Repeat for $t = 4.5$ ms.

20. The current source in the circuit of Fig. 1.15 provides an output defined by $i = 5t$ A where t is in seconds. For time $t = 2$ s, calculate:
  (a) The power dissipated in the 15-Ω resistor.
  (b) The power supplied to the 10-H inductor.
  (c) The power delivered by the source.
  (d) The total energy dissipated in the resistor.
  (e) The total energy stored in the inductor.

21. The voltage source in the circuit of Fig. 1.14 provides an output defined by $v = 200t$ V where t is in seconds. For time $t = 0.5$ s, calculate:
  (a) The power dissipated in the 200-Ω resistor.

(b) The power supplied to the 1000-μF capacitor.

(c) The power delivered by the source.

(d) The total energy dissipated in the resistor.

(e) The total energy stored in the capacitor.

22. In Fig. 1.18, capacitance C_1 is charged to voltage V and C_2 is uncharged. Immediately after closing switch S,

(a) What are V_{C1} and V_{C2}?

(b) What is current i?

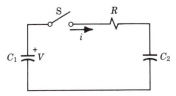

Figure 1.18

23. In Fig. 1.19, capacitor C is "charged up" by closing switch S_1. After equilibrium is reached, switch S_2 is closed.

(a) Just *before* closing S_2, what is the voltage across C? The current in inductor L?

(b) Just *after* closing S_2, what is the voltage across C? The current i_L? The current i_2?

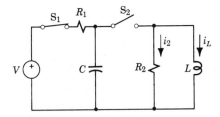

Figure 1.19

PROBLEMS

(These problems require extending the basic concepts of this chapter to new situations.)

1. A "Q-switching" carbon dioxide laser capable of producing 50 W of continuous-wave power will produce 50 kW of pulsed power in bursts 150 ns long at a rate of 400 bursts per second. Calculate the average power output in the pulsed mode and the energy per burst.

2. Show that the following mathematical models describe a resistance, capacitance, or inductance.

(a) $v = kq$

(b) $v = k\, dq/dt$

(c) $v = k\, d^2q/dt^2$

3. As we shall see (in Chapter 3), the circuit shown in Fig. 1.20 can be used to smooth out or "filter" variations in input current i so that a fairly constant voltage appears across load resistor R. For good filtering, the capacitor should be able to store 10 times as much energy as is dissipated in one cycle by R. If $R = 10$ kΩ and variations in i occur 60 times per s, specify the value of C.

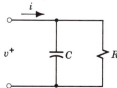

Figure 1.20

4. Laboratory tests on a coil yield the following data: a change in current from $+10$ mA to -10 mA in 50 μs results in an average voltage of 2 V; a steady current of 10 mA requires a power input of 0.2 W. Devise a linear circuit model to represent the coil.

5. Plot as a function of time the power to the inductor of Exercise 15 and the energy stored in the inductor.

6. Plot as a function of time the power to the capacitor of Exercise 16 and the energy stored in the capacitor.

7. Devise a linear model incorporating a current source to represent the solar battery whose output characteristics are shown in Fig. 1.21. Specify numerical values valid for currents up to 40 mA.

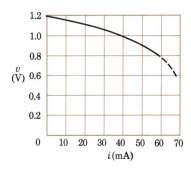

Figure 1.21

2

Circuit Principles

Circuit Laws

Network Theorems

Nonlinear Networks

The emphasis so far has been on the behavior of individual components and the ideal circuit elements that serve as models. Now we are ready to consider combinations of elements into circuits, using two well-known experimental laws. Our first objective is to learn how to formulate and solve the equations that will enable us to analyze a given electronic circuit or to design a circuit to meet specifications.

Circuits of considerable complexity or generality are called *networks* and circuit principles capable of general application are called *network theorems*. Some of these theorems are useful in reducing complicated networks to simple ones. Other theorems enable us to draw general conclusions about network behavior. This chapter focuses on networks consisting of resistances and steady (dc) sources; later the theorems will be extended to include networks containing other elements and other sources.

We start with *linear* networks, i.e., combinations of components that can be represented by the ideal circuit elements R, L, and C, and ideal energy sources V and I. However, most electronic devices and many practical components are nonlinear. Therefore, we also consider how linear methods can be used in analyzing some nonlinear networks, and we learn some new techniques applicable to nonlinear devices.

CIRCUIT LAWS

The foundations of network theory were laid about 150 years ago by Gustav Kirchhoff, a German university professor, whose careful experiments resulted in the laws that bear his name. In the following discussion, a *branch* is part of a circuit with two terminals to which connections can be made, a *node* is the point where two or more branches come together, and a *loop* is a closed path formed by connecting branches.

Kirchhoff's Current Law

To repeat Kirchhoff's experiments in the laboratory, we could arrange to measure the currents in a number of conductors or "leads" soldered together (Fig. 2.1a). In every case, we would find that the sum of the currents flowing into the common point (a node) at any instant is just equal to the sum of the currents flowing out. (Because a current is a flow of charge, if the sum of the currents were *not* equal to zero it would mean that electric charge could pile up at a point!)

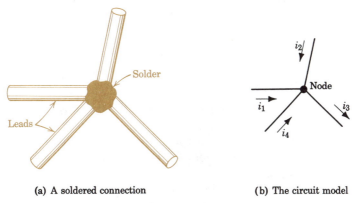

(a) A soldered connection   (b) The circuit model

Figure 2.1   Kirchhoff's current experiment.

In Fig. 2.1b, a circuit model is used to represent an actual connection. The arrows define the *reference* direction for positive current where current is defined by the motion of positive charges. The quantity "i_1" specifies the *magnitude* of the current and its algebraic *sign* with respect to the reference direction. If i_1 has a value of "+5 A," the effect is as if positive charge is moving toward the node at the rate of 5 C/s. If $i_1 = +5$ A flows in a metallic conductor in which charge is transported in the form of negative electrons, the electrons are actually moving away from the node but the effect is the same. If i_2 has a value of "−3 A," positive charge is in effect moving away from the node. In a practical situation the direction of current flow is easy to determine; if an ammeter reads "upscale," current is flowing into the meter terminal marked + and through the meter to the terminal marked −.

By Kirchhoff's current law, the algebraic sum of the currents into a node at any instant is zero.

It is sometimes convenient to abbreviate this statement, and write "$\Sigma i = 0$," where the Greek letter sigma stands for "summation." As applied to Fig. 2.1b,

$$\Sigma i = 0 = i_1 + i_2 - i_3 + i_4 \tag{2-1}$$

Obviously some of the currents may be negative. (See Example 1 on page 28.)

EXAMPLE 1

In Fig. 2.2, if $i_1 = +5$ A, $i_2 = -3$ A, and $i_4 = +2$ A, what is the value of i_3?

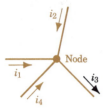

Figure 2.2 Current calculation.

By Kirchhoff's law, the current into the node is

$$\Sigma i = 0 = i_1 + i_2 - i_3 + i_4$$

$$\Sigma i = 0 = +5 - 3 - i_3 + 2$$

or

$$i_3 = +5 - 3 + 2 = +4 \text{ A}$$

Note: We could just as well say that the algebraic sum of the currents *leaving* a node is zero and write

$$\Sigma i_{out} = 0 = -i_1 - i_2 + i_3 - i_4$$

which yields $i_3 = +4$ A as before.

Kirchhoff's Voltage Law

The current law was originally formulated on the basis of experimental data. The same result can be obtained from the principle of conservation of charge and the definition of current. Kirchhoff's voltage law also was based on experiment, but the same result can be obtained from the principle of conservation of energy and the definition of voltage.

To repeat Kirchhoff's observations about voltages, we could set up an electrical circuit and arrange to measure the voltages across a number of components that form a closed path. Only a portion of the circuit is shown in Fig. 2.3, but the combination

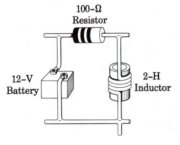

Figure 2.3 A voltage experiment.

of a battery, a resistor, an inductor, and the associated leads forms the desired closed path. In every case, we would find that the voltages around the loop, when properly combined, add up to zero.

The circuit model of Fig. 2.4 is more convenient to work with. There the voltages are labeled and the + signs define the *reference* direction for positive voltage or potential difference. The quantity v_S specifies the *magnitude* of the source voltage and its algebraic *sign* with respect to the reference. If v_S has a value of $+12$ V, the voltage of node b with respect to node a is positive. Since, by definition, voltage is energy per unit charge, a positive charge of 1 C moving from node a to b gains 12 J of energy from the voltage source.

We define the voltage across the resistance as v_R, arbitrarily placing the $+$ sign near the b node. If v_R has a value of $+5$ V, a positive charge of 1 C moving from b to c loses 5 J of energy; this energy is removed from the circuit and dissipated in the resistance R. If v_L, for example, has a value of -7 V, node d is at a higher voltage than c and a positive charge of 1 C moving from c to d gains 7 J of energy. A current flowing from c to d could receive energy previously stored in the magnetic field of inductance L.

> *By Kirchhoff's voltage law, the algebraic sum of the voltages around a loop at any instant is zero.*

As an abbreviation, we may write "$\Sigma v = 0$." As applied to Fig. 2.4,

$$\Sigma v = 0 = v_{ba} + v_{cb} + v_{dc} + v_{ad} \qquad (2\text{-}2)$$

where v_{ba} means "the voltage of b with respect to a"; if v_{ba} is positive, node b is at a higher potential than node a. In applying the law, the loop should be traversed in one continuous direction, starting at one point and returning to the same point. Starting from node a in this case,

$$\Sigma v = 0 = +v_S - v_R - v_L + 0 \qquad (2\text{-}3)$$

A minus sign is affixed to the v_R term because the plus sign on the diagram indicates that node b is nominally at a higher potential than c or $v_{cb} = -v_R$. There is no voltage across the ideal conductor joining d and a.

EXAMPLE 2

In Fig. 2.4, if $v_S = +12$ V and $v_R = +5$ V, what is the value of v_L? Is energy being stored in or received from the inductance?

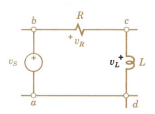

Figure 2.4   Voltage calculation.

By Kirchhoff's voltage law,

$$\Sigma v = 0 = v_S - v_R - v_L + 0$$

$$\Sigma v = 0 = +12 - 5 - v_L + 0$$

or

$$v_L = +12 - 5 = +7 \text{ V}$$

The loop can be traversed in either direction, starting at any point. Going counterclockwise, starting at node c,

$$\Sigma v = 0 = + v_R - v_S + 0 + v_L$$

In this case, a positive charge moving from c to d gives up energy; energy is being stored in the magnetic field of inductance L.

Application of Kirchhoff's Laws

To illustrate the application of Kirchhoff's laws in solving electric circuits, consider the circuit shown in Fig. 2.5. In this case, the source voltages and the resistances are given and the element currents and voltages are to be determined.

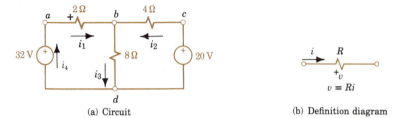

(a) Circuit   (b) Definition diagram

Figure 2.5 Application of Kirchhoff's laws.

The first step is to label the unknown currents arbitrarily; as drawn, the arrow to the right indicates the *reference direction* of i_1. If the value of i_1 is calculated and found to be positive, current i_1 actually flows to the right and an ammeter inserted between node a and the 2-Ω resistor with the + terminal at a would read *upscale* or positive. If the value of i_1 is found to be negative, current i_1 actually flows to the left (the ammeter inserted as previously would read *downscale* or negative).

Applying Kirchhoff's current law to node a,

$$\Sigma i_a = 0 = +i_4 - i_1$$

or

$$i_4 = i_1$$

From this we conclude that the current is the same at every point in a series circuit. Since $i_4 \equiv i_1$, it is not another unknown and will be considered no further. Note that i_2 is the current in the 20-V source as well as in the 4-Ω resistance.

The next step is to define the unknown element voltages in terms of the arbitrarily assumed currents. The diagram of Fig. 2.5b defines the *v-i* relation for a resistance; Ohm's law holds and $v = Ri$ if the current reference arrow enters the resistance at the terminal designated +. In flowing from a to b in Fig. 2.5a, the positive charges constituting current i_1 lose energy in the 2-Ω resistance; this loss in energy indicates that the potential of a is higher than that of b, or $v_{ab} = Ri_{ab} = +2i_1$. As a reminder, the left-hand terminal of the 2-Ω resistance is marked with a + to indicate the polarity of the element voltage *in terms of the assumed current direction*. Following the same analysis, the upper end of the 8-Ω resistance and the right-hand end of the 4-Ω resistance could be marked + in accordance with the following *element equations:*

$$v_{ab} = +2i_1 \quad \text{or} \quad v_{ba} = -2i_1$$

$$v_{bd} = +8i_3 \quad \text{or} \quad v_{db} = -8i_3 \tag{2-4}$$

$$v_{cb} = +4i_2 \quad \text{or} \quad v_{bc} = -4i_2$$

Additional relations are obtained in the form of *connection equations.* Applying Kirchhoff's current law to node b,

$$\Sigma i_b = 0 = +i_1 + i_2 - i_3 \tag{2-5}$$

Since we are summing the currents *into* node b, i_3 is shown with a minus sign. Applying the current law to node d,

$$\Sigma i_d = 0 = -i_1 - i_2 + i_3$$

Note that this equation is *not independent;* it contributes no new information.

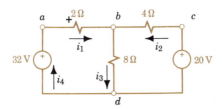

Figure 2.6   Writing circuit equations.

Applying the voltage law to the left-hand loop *abda* in Fig. 2.6, and considering each element in turn,

$$\Sigma v = 0 = v_{ba} + v_{db} + v_{ad} \tag{2-6}$$

For the right-hand loop *bcdb*,

$$\Sigma v = 0 = v_{cb} + v_{dc} + v_{bd} \tag{2-7}$$

Kirchhoff's voltage law applies to any closed path; for the outside loop *abcda*,

$$\Sigma v = 0 = v_{ba} + v_{cb} + v_{dc} + v_{ad}$$

But this equation is just the sum of Eqs. 2-6 and 2-7 and no new information is obtained. Although this equation is not independent, it can be valuable in checking.

It is always possible to write as many independent equations as there are unknowns. For a circuit with six unknowns (three voltages and three currents), we have written six equations (three element and three connection equations). The three element equations, Eqs. 2-4, express the *v-i* characteristics of individual elements; the three connection equations, Eqs. 2-5, 2-6, and 2-7, describe how the elements are connected in the circuit. Other equations could be written but they would contribute no new information. If the currents are of primary interest, the unknown voltages may be eliminated by substituting Eqs. 2-4 into Eqs. 2-6 and 2-7, which become

$$\Sigma v_{abda} = 0 = -2i_1 - 8i_3 + 32$$

$$\Sigma v_{bcdb} = 0 = +4i_2 - 20 + 8i_3$$

When these are rewritten along with Eq. 2-5, we have

$$i_1 + i_2 - i_3 = 0 \tag{2-8}$$

$$+2i_1 \qquad + 8i_3 = 32 \tag{2-9}$$

$$+ 4i_2 + 8i_3 = 20 \tag{2-10}$$

To evaluate the unknowns, these three equations are to be solved simultaneously. Some commonly employed methods are illustrated in the solution of this problem.

Solution by Determinants

The method of determinants is valuable because it is systematic and general; it can be used to solve complicated problems or to prove general theorems. The first step is to write the equations in the *standard form* of Eqs. 2-8, 2-9, and 2-10 with the constant terms on the right and the corresponding current terms aligned on the left.

A determinant is an array of numbers or symbols; the array of the coefficients of the current terms is a *third-order* determinant

$$D = \begin{vmatrix} 1 & 1 & -1 \\ 2 & 0 & 8 \\ 0 & 4 & 8 \end{vmatrix} \tag{2-11}$$

The value of a determinant of second order is

$$\begin{vmatrix} a_1 & a_2 \\ b_1 & b_2 \end{vmatrix} = a_1 b_2 - a_2 b_1 \tag{2-12}$$

The value of a determinant of third order is

$$\begin{vmatrix} a_1 & a_2 & a_3 \\ b_1 & b_2 & b_3 \\ c_1 & c_2 & c_3 \end{vmatrix} = +a_1 b_2 c_3 + a_2 b_3 c_1 + a_3 b_1 c_2 - a_1 b_3 c_2 - a_2 b_1 c_3 - a_3 b_2 c_1 \tag{2-13}$$

A simple rule for second- and third-order determinants is to take the products "from upper left to lower right" and subtract the products "from upper right to lower left." Determinants of higher order can be evaluated by the process called "expansion by minors."[†]

Repeating the first two columns for clarity and using the rule of Eq. 2-13,

$$D = \begin{vmatrix} 1 & 1 & -1 \\ 2 & 0 & 8 \\ 0 & 4 & 8 \end{vmatrix} \begin{matrix} 1 & 1 \\ 2 & 0 \\ 0 & 4 \end{matrix} = \begin{aligned} &+(1)(0)(8) + (1)(8)(0) + (-1)(2)(4) \\ &-(1)(8)(4) - (1)(2)(8) - (-1)(0)(0) \end{aligned}$$

$$= 0 + 0 - 8 - 32 - 16 - 0 = -56$$

To solve for i_3, for example, we form a new determinant D_3 by replacing the coefficients of i_3 with the constant terms on the right-hand side of the equations in the standard form:

$$D_3 = \begin{vmatrix} 1 & 1 & 0 \\ 2 & 0 & 32 \\ 0 & 4 & 20 \end{vmatrix} = 0 + 0 + 0 - 128 - 40 - 0 = -168 \tag{2-14}$$

By Cramer's rule for the solution of equations by determinants,

$$i_n = \frac{D_n}{D} \qquad \text{or} \qquad i_3 = \frac{D_3}{D} = \frac{-168}{-56} = 3 \text{ A}$$

Currents i_1 and i_2 could be obtained in a similar manner.

[†] See Gere and Weaver: *Matrix Algebra for Engineers*, Brooks/Cole, Monterey, CA, 1983, or any college algebra textbook.

Practice Problem 2-1

Evaluate the determinant D_1 derived from Eqs. 2-8, 2-9, and 2-10 and solve for current i_1.

Answers: -224; 4 A.

Substitution Method

When the number of unknowns is not large, the method of *substitution* is sometimes more convenient than the method of determinants. In this method, one of the equations is solved for one of the unknowns and the result is substituted in the other equations, thus eliminating one unknown. Let us use this method to find current i_2 and the voltage across the 4–Ω resistance. Solving Eq. 2-8 for i_3 yields

$$i_3 = i_1 + i_2$$

Substituting in Eqs. 2-9 and 2-10 yields

$$2i_1 + 8(i_1 + i_2) = 32$$
$$4i_2 + 8(i_1 + i_2) = 20$$

which can be rewritten as

$$(2 + 8)i_1 + \qquad 8i_2 = 32 \qquad\qquad (2\text{-}15)$$
$$8i_1 + (8 + 4)i_2 = 20 \qquad\qquad (2\text{-}16)$$

One of the two unknowns has been eliminated. Solving Eq. 2-15 for i_1 yields

$$i_1 = \frac{32 - 8i_2}{10}$$

We can eliminate a second unknown by substituting this expression for i_1 in Eq. 2-16 to obtain

$$8\frac{32 - 8i_2}{10} + 12i_2 = 20 \qquad \text{or} \qquad 256 - 64i_2 + 120i_2 = 200$$

Solving this equation yields

$$i_2 = \frac{-56}{56} = -1 \text{ A}$$

and the voltage across the 4-Ω resistance is

$$v_{cb} = 4i_2 = 4(-1) = -4 \text{ V}$$

The interpretation of the minus sign is that in reality node b is at a higher potential than c, and positive current flows to the right in the 4-Ω resistance.

Loop-Current Method

The reduction in the number of unknowns and in the number of simultaneous equations achieved by substitution can be obtained automatically by an ingenious approach to circuit analysis. Instead of thinking in terms of *branches* in analyzing a circuit, we now think in terms of *loops*, closed paths formed by connecting branches. We identify enough loops so that every branch is included at least once. Around each loop, we assume that a *loop current* flows, independent of currents in other loops.

In Fig. 2.7, a *loop current* I_1 is assumed to circulate around loop *abda*, and another loop current I_2 is assumed to circulate around loop *cbdc*. By inspection, the branch current i_1 is equal to loop current I_1, and i_2 is equal to I_2; but branch current i_3 is the *sum* of loop currents I_1 and I_2.

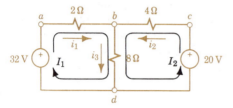

Figure 2.7   Loop currents.

By applying Kirchhoff's voltage law to the two loops we obtain

$$\Sigma v_{abda} = 0 = -2I_1 - 8(I_1 + I_2) + 32$$

$$\Sigma v_{cbdc} = 0 = -4I_2 - 8(I_1 + I_2) + 20$$

which can be rewritten as

$$(2 + 8)I_1 \qquad + 8I_2 = 32 \qquad\qquad (2\text{-}17)$$

$$8I_1 + (8 + 4)I_2 = 20 \qquad\qquad (2\text{-}18)$$

These equations are similar to Eqs. 2-15 and 2-16, but the physical interpretation is different. Equation 2-17 says: "The source voltage in loop 1 is equal to the sum of two voltage drops. The first, $(2 + 8)I_1$, is the product of loop current I_1 and the sum of all the resistances in loop 1. The second, $8I_2$, is the product of loop current I_2 and the sum of all resistances that are common to loops 1 and 2." The *self-resistance* of loop 1 is $2 + 8 = 10\ \Omega$; the *mutual resistance* between the two loops is $8\ \Omega$. A similar statement could be made by using Eq. 2-18. To obtain I_1, multiply Eq. 2-17 by 3 and Eq. 2-18 by (-2) and add the resulting equations. Then

$$
\begin{array}{rcr}
30I_1 + 24I_2 & = & 96 \\
-16I_1 - 24I_2 & = & -40 \\
\hline
14I_1 \qquad\quad & = & 56
\end{array}
\qquad \text{or} \qquad I_1 = 4\ \text{A} = i_1
$$

A loop current is an abstract but very useful concept. By defining loop currents we automatically reduce the number of unknown currents; here $i_3 = I_1 + I_2$ and we need to solve only two equations. If only certain currents or voltages are required, a judicious choice of loops will reduce the amount of work. (See Example 3.)

Practice Problem 2-2

Choose loops so that the 8–Ω resistance in Fig. 2.7 appears in only one loop. Evaluate the self- and mutual resistances of each loop and calculate i_3.

Answers: $R_s = 6\ \Omega$ and 10 Ω or 6 Ω and 12 Ω; $R_m = 2\ \Omega$ or 4 Ω depending on choice; $i_3 = 3$ A.

Checking

In solving complicated circuits there are many opportunities for mistakes and a reliable check is essential. This is particularly true when the solution is obtained in a mechanical way, as when using determinants. A new equation such as that obtained by writing Kirchhoff's voltage law around the outside loop of Fig. 2.7 provides a good check. With a little experience you will be able to write (by inspection, without first writing the element equations)

$$\Sigma v = 0 = -2i_1 + 4i_2 - 20 + 32$$

Substituting $i_1 = 4$ A and $i_2 = -1$ A,

$$\Sigma v = 0 = -2(4) + 4(-1) - 20 + 32 = 0$$

and the values of i_1 and i_2 are checked.

In the foregoing discussion and in Example 3 below, the loop currents are defined so that all currents in a mutual resistance flow in the same direction. If this is not possible or not convenient, some terms in Eqs. 2-17, 2-18, 2-19, and 2-20 will carry minus signs. (See Exercise 2-12b.)

EXAMPLE 3

Given the circuit of Fig. 2.8, calculate the current in the 10-Ω resistance using loop currents.

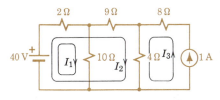

Figure 2.8 Loop-current analysis.

1. Assume loop currents so that the desired current is I_1 by definition and $I_3 = 1$ A as given.
2. Applying Kirchhoff's voltage law to two loops,

$$\Sigma v = 0 = 40 - 2(I_1 + I_2) - 10I_1$$

$$\Sigma v = 0 = 40 - 2(I_1 + I_2) - 9I_2 - 4(I_2 + 1)$$

In terms of self- and mutual resistance,

$$(10 + 2)I_1 + \qquad\qquad 2I_2 \qquad\qquad = 40 \qquad (2\text{-}19)$$

$$2I_1 + (2 + 9 + 4)I_2 + 4 \times 1 = 40 \qquad (2\text{-}20)$$

From Eq. 2-20, $I_2 = (36 - 2I_1)/15$.

Substituting into Eq. 2-19,

$$12I_1 + (72 - 4I_1)/15 = 40$$

or

$$I_1 = \frac{600 - 72}{180 - 4} = \frac{528}{176} = 3 \text{ A}$$

Node-Voltage Method

The wise selection of loop currents can significantly reduce the number of simultaneous equations to be solved in a given problem. In the preceding example, the number of equations was reduced from three to two by using loop currents. Would it be possible to get all the important information in a single equation with a single unknown? In this particular case, the answer is "yes" if we choose as the unknown the voltage of node b with respect to a properly chosen reference. In some practical devices many components are connected to a metal "chassis," which in turn is "grounded" to the earth. Such a ground, often shown as a common lead at the bottom of a circuit diagram, is a convenient reference. In this problem the greatest simplification will result if we choose node d; now the voltage of any node is understood to be with respect to node d, in other words, the voltage of node d is zero. (See Fig. 2.9.)

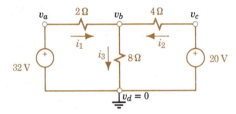

Figure 2.9   Node voltages.

Next, apply Kirchhoff's current law to each independent node. Here there is only one independent node (since the voltages of nodes a and c are fixed by the voltage sources), so the sum of the currents into node b is

$$\Sigma i_b = 0 = i_1 + i_2 - i_3 \tag{2-21}$$

The current i_1 to node b from node a is the current in the 2-Ω resistance. It is equal to the difference in potential between nodes a and b divided by the resistance between a and b, or

$$i_1 = \frac{v_{ab}}{R_{ab}} = \frac{v_a - v_b}{R_{ab}} \tag{2-22}$$

Similarly,

$$i_2 = \frac{v_c - v_b}{R_{cb}} \quad \text{and} \quad i_3 = \frac{v_b - 0}{R_{bd}} \tag{2-23}$$

where all voltages are in reference to node d. In this particular problem, v_a is given as $+32$ V because the voltage source holds its positive terminal at 32 V "above ground." Voltage v_c is given as $+20$ V.

Finally, bearing in mind Eq. 2-21 and the concepts represented by Eqs. 2-22 and 2-23, we write one equation with a single unknown,

$$\Sigma i_b = 0 = \frac{32 - v_b}{2} + \frac{20 - v_b}{4} - \frac{v_b - 0}{8} \tag{2-24}$$

Multiplying through by 8,

$$128 - 4v_b + 40 - 2v_b - v_b = 0$$

Therefore, the unknown voltage at independent node b is

$$v_b = \frac{168}{7} = 24 \text{ V}$$

By inspection of Fig. 2.9 we can write

$$i_3 = \frac{24}{8} = 3 \text{ A}, \qquad i_1 = \frac{32 - 24}{2} = 4 \text{ A}, \qquad \text{and} \qquad i_2 = \frac{20 - 24}{4} = -1 \text{ A}$$

which agree with the results previously obtained. As illustrated in Example 4, the node-voltage approach can be applied to more complicated circuits.

EXAMPLE 4

Given the circuit of Fig. 2.10, calculate the current in the 10-Ω resistance using node voltages.

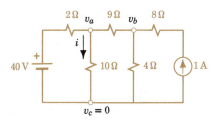

Figure 2.10 Node-voltage analysis.

1. Choose node c (one end of the branch of interest) as the reference node.
2. Apply Kirchhoff's current law to independent nodes a and b.

$$\Sigma i_a = 0 = \frac{40 - v_a}{2} + \frac{v_b - v_a}{9} - \frac{v_a - 0}{10}$$

$$\Sigma i_b = 0 = \frac{v_a - v_b}{9} + 1 - \frac{v_b - 0}{4}$$

Multiplying through by 90 and 36, respectively,

$$1800 - 45v_a + 10v_b - 10v_a - 9v_a = 0$$

$$4v_a - 4v_b + 36 - 9v_b = 0$$

Collecting terms,

$$64v_a - 10v_b = 1800$$

$$4v_a - 13v_b = -36$$

3. Solving by determinants,

$$v_a = \frac{D_a}{D} = \frac{\begin{vmatrix} 1800 & -10 \\ -36 & -13 \end{vmatrix}}{\begin{vmatrix} 64 & -10 \\ 4 & -13 \end{vmatrix}} = \frac{-23,400 - 360}{-832 + 40} = 30 \text{ V}$$

$$\therefore i_{10} = \frac{v_a}{10} = 3 \text{ A}$$

At this point you should be familiar with electrical quantities (their definitions, units, and symbols), you should understand the distinction between real circuit components and idealized circuit elements, you should know the voltage-current characteristics of active and passive circuit elements, and you should be able to apply Kirchhoff's laws to obtain circuit equations. In engineering, the solution of equations is usually less demanding than the formulation, and it is proper to place primary emphasis on the formulation of equations.

Practical electronic circuits may be quite complicated, however, and the solution of the system of equations obtained for multiloop circuits containing electronic devices will be tedious and time consuming. Furthermore, the mathematics may obscure the critical relationships that point the way to effective designs. Fortunately, over the years engineers have discovered some general principles that allow us to reduce complicated networks to simple ones and, in other ways, simplify the process of circuit analysis or design. Let us now take advantage of their work.

NETWORK THEOREMS

A common problem in circuit analysis is to determine the response of a two-terminal circuit to an input signal. In Fig. 2.11a there is one "port" at which excitation is applied and response is measured. If the desired response is measured across R_2 as in Fig. 2.11b, this circuit becomes a *three-terminal circuit* (one terminal is common to input and output) or a *two-port circuit*, or simply a *two-port*. Transistors, transformers, and transmission lines are two-ports; typically there are an input port and a separate output port. The multiwinding transformer of Fig. 2.11c is a *three-port*.

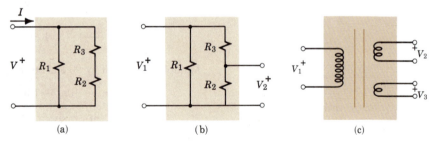

(a)   (b)   (c)

Figure 2.11   One-port, two-port, and multiport networks.

Equivalence

First, we consider how complicated two-terminal circuits can be reduced to simpler circuits that are *equivalent*. By definition:

Two one-ports are equivalent if they present the same v-i characteristics.

Two one-ports consisting of various resistances connected in series and parallel are equivalent if they have the same input resistance or the same input conductance. To calculate the response I to an input V in Fig. 2.11a, what single resistance could replace the three resistances?

Network Reduction

Replacing a complicated network with a simpler equivalent is advantageous in network analysis. Applying Kirchhoff's voltage law to the series circuit of Fig. 2.12a, ("closing the loop" by going through V),

$$V = V_1 + V_2 = IR_1 + IR_2 = I(R_1 + R_2) = IR_{EQ}$$

In general, for n resistances in series the equivalent resistance is

$$R_{EQ} = R_1 + R_2 + \cdots + R_n \tag{2-25}$$

Applying Kirchhoff's current law to the parallel circuit of Fig. 2.12b,

$$I = I_1 + I_2 = VG_1 + VG_2 = V(G_1 + G_2) = VG_{EQ}$$

In general, for n conductances in parallel the equivalent conductance is

$$G_{EQ} = G_1 + G_2 + \cdots + G_n \tag{2-26}$$

For two resistances in parallel (Fig. 2.12c), a convenient relation is

$$R_{EQ} = \frac{1}{G_{EQ}} = \frac{1}{G_1 + G_2} = \frac{1}{(1/R_1) + (1/R_2)} = \frac{R_1 R_2}{R_1 + R_2} \tag{2-27}$$

The corresponding relationship for two conductances in series is

$$G_{EQ} = \frac{1}{(1/G_1) + (1/G_2)} = \frac{G_1 G_2}{G_1 + G_2} \tag{2-28}$$

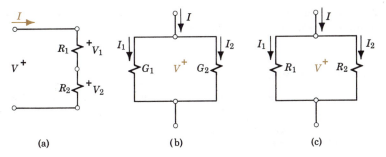

| (a) | (b) | (c) |

Figure 2.12 Resistances in series and parallel.

Practice Problem 2-3

In Fig. 2.11a, the resistances are $R_1 = 75\ \Omega$, $R_2 = 30\ \Omega$, and $R_3 = 20\ \Omega$. Calculate:
(a) The equivalent resistance of the R_2-R_3 branch.
(b) The conductances of the two parallel branches.
(c) The equivalent conductance of the one-port.
(d) The equivalent resistance of the one-port by using the $1/x$ button of your hand calculator. (*Hint*: Extend Eq. 2-27 to fit this circuit.)

Answers: (a) 50 Ω; (b) 0.02 S, 0.01333 S; (c) 0.03333 S; (d) 30 Ω.

By following a similar approach, we can derive expressions for series and parallel combinations of inductances and capacitances.

For n inductances in series,

$$L_{EQ} = L_1 + L_2 + \cdots + L_n \qquad (2\text{-}29)$$

For n capacitances in parallel,

$$C_{EQ} = C_1 + C_2 + \cdots + C_n \qquad (2\text{-}30)$$

Other useful tools in network analysis include the *voltage divider* and the *current divider*. Referring again to Fig. 2.12a, we see that $V_2 = IR_2$ and $V = IR_{EQ} = I(R_1 + R_2)$. Therefore, in the voltage divider,

$$V_2 = \frac{R_2}{R_1 + R_2} V = \frac{R_2}{R_{EQ}} V \qquad (2\text{-}31)$$

In the current divider, Fig. 2.12b,

$$I_2 = \frac{G_2}{G_1 + G_2} I = \frac{G_2}{G_{EQ}} I \qquad (2\text{-}32a)$$

because $I_2 = VG_2$ and $I = VG_{EQ} = V(G_1 + G_2)$. Another useful form of the current-divider equation can be obtained by using $R = 1/G$ as in Fig. 2.12c. Then

$$I_2 = \frac{G_2}{G_1 + G_2} I = \frac{1/R_2}{(1/R_1) + (1/R_2)} I \cdot \frac{R_1 R_2}{R_1 R_2} = \frac{R_1}{R_1 + R_2} I \qquad (2\text{-}32b)$$

When we want to determine a particular voltage or current in a circuit, using these relations is preferable to writing and solving loop or node equations. (See Example 5.)

Thévenin's Theorem

The one-port networks discussed so far have been *passive*; they absorb energy from the source. In contrast, *active* networks include energy sources. A valuable method for representing active networks by simpler equivalents is described in the following statement of *Thévenin's theorem*:

Insofar as a load is concerned, any one-port network of resistance elements and energy sources can be replaced by a series combination of an ideal voltage source V_T and a resistance R_T where V_T is the open-circuit voltage of the one-port and R_T is the ratio of the open-circuit voltage to the short-circuit current.

The box in Fig. 2-14a represents "any one-port network"; a "load" R_L is to be connected across the terminals of the one-port. The series combination of V_T and R_T in Fig. 2.14d is the "Thévenin equivalent" circuit that is to be connected to the same load R_L. Voltages and currents are to be measured at the terminals. We are not concerned with the internal behavior of the two circuits; we are concerned only with the external behavior "insofar as a load is concerned."

If the two networks are to be equivalent for all values of load resistance, they must be equivalent for extreme values such as $R_L = \infty$ and $R_L = 0$. The value $R_L = \infty$ corresponds to the open-circuit condition; with R_L removed in both networks, the

EXAMPLE 5

In the circuit of Fig. 2.13a, determine the current in the 12-Ω resistance.

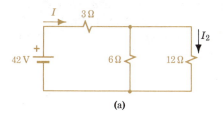

(a)

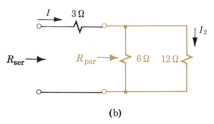

(b)

Figure 2.13 Current-divider application.

For resistances in parallel as in Fig. 2.13b,

$$R_{\text{par}} = \frac{1}{(1/R_1) + (1/R_2)} = \frac{1}{\frac{1}{6} + \frac{1}{12}} = 4\ \Omega$$

For resistances in series,

$$R_{\text{ser}} = R + R_{\text{par}} = 3 + 4 = 7\ \Omega$$

and

$$I = \frac{V}{R_{\text{ser}}}$$

Using the current-divider principle, Eq. 2-32,

$$I_2 = \frac{R_1}{R_1 + R_2} I = \frac{R_1}{R_1 + R_2} \cdot \frac{V}{R_{\text{ser}}}$$

$$= \frac{6}{6 + 12} \cdot \frac{42}{3 + \frac{1}{\frac{1}{6} + \frac{1}{12}}} = \frac{1}{3} \cdot \frac{42}{3 + 4} = 2\ \text{A}$$

By this method I_2 is obtained directly without writing simultaneous equations.

---

open-circuit voltage V_{OC} of the original network is equal to V_T of the equivalent. The value $R_L = 0$ corresponds to the short-circuit condition; with the terminals shorted in both networks, the short-circuit current I_{SC} of the original network is equal to V_T/R_T of the equivalent. Therefore, the equations

$$V_T = V_{\text{OC}} \qquad \text{and} \qquad R_T = \frac{V_T}{I_{\text{SC}}} = \frac{V_{\text{OC}}}{I_{\text{SC}}} \qquad (2\text{-}33)$$

define the components of the Thévenin equivalent network.

In some circuits, the calculation of V_{OC} or I_{SC} is difficult, and we look for an easier way of determining R_T. In Fig. 2.14d, we see that if V_T is shorted out (reducing the

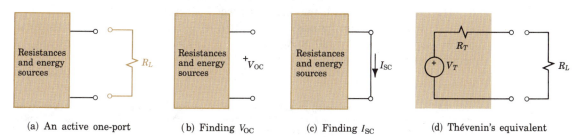

(a) An active one-port   (b) Finding V_{OC} (c) Finding I_{SC} (d) Thévenin's equivalent

Figure 2.14 Finding the Thévenin equivalent of an active one-port network.

effective value of V_T to zero), the resistance measured at the terminals would be R_T. The same thing is true of the general circuit of Fig. 2.14a. Usually it is easier to take advantage of the fact that R_T is the "resistance seen by looking in" at the terminals with all independent[†] energy sources "removed." A source is removed by reducing its effect to zero; voltage sources must be removed by short-circuiting, and current sources must be removed by open-circuiting. By way of explanation, the resistance "seen" by the network in Fig. 2.14d is R_L; the resistance that we see "looking in" at the terminals of the Thévenin equivalent when V_T is short-circuited is R_T. The procedure for applying Thévenin's theorem is outlined in Practice Problem 2-4.

Practice Problem 2-4

Find current I_2 in the circuit of Fig. 2.13 by using Thévenin's theorem.

(a) Disconnect the 12-Ω resistance (R_L) and redraw the remainder of the circuit as an active one-port with terminals ab.
(b) Use the voltage-divider concept to find $V_{OC} = V_{ab}$ of the one-port.
(c) Short terminals ab (shorting out the 6-Ω resistance) and calculate $I_{SC} = I_{ab}$.
(d) Determine R_T from V_{OC}/I_{SC} and also from the concept of "the resistance looking in at terminals ab with all energy sources removed."
(e) Draw the Thévenin equivalent, reconnect the 12-Ω load resistance, and calculate current I_2.

Answers: (b) 28 V; (c) 14 A; (d) 2 Ω; (e) 2 A.

Thévenin's theorem is particularly useful when the load is to take on a series of values or when a general analysis is being performed with literal numbers. We shall use it in analyzing such devices as transistor biasing circuits and feedback amplifiers. No general proof of the theorem is given here, but Example 6 is a demonstration of its validity in a relatively complicated situation. (As a check on the result, solve for I by the loop-current method.)

The significance of the restriction "insofar as a load is concerned" is revealed if we attempt to use the Thévenin equivalent to determine internal behavior in Example 6. The power developed internally in the Thévenin equivalent is $P_T = V_T I$ $= 28 \times 4 = 112$ W. In the original circuit, $I_6 = 3$ A and $I_{12} = 1$ A, and the total power supplied is $P = 30 \times 3 + 24 \times 1 = 114$ W. The Thévenin circuit is "equivalent" only in a restricted sense.

[†]All energy sources considered up to this point have been *independent* of voltages and currents in other parts of the circuit. *Controlled* sources, for example those used in representing transistors (Chapter 13), must not be removed in applying Thévenin's theorem.

EXAMPLE 6

Using Thévenin's theorem, determine the current in the 3-Ω resistance of Fig. 2.15a.

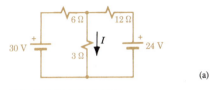

(a)

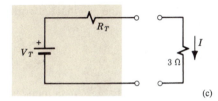

(b)

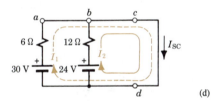

(c)

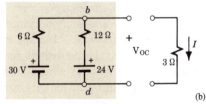

(d)

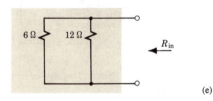

(e)

Figure 2.15 Thévenin's theorem.

First, the resistance of interest is isolated as a load and disconnected from the active network (Fig. 2.15b). Next, the Thévenin equivalent is drawn (Fig. 2.15c). Then we determine the components of the equivalent circuit.

Finding V_{OC}: Under open-circuit conditions (Fig. 2.15b), the node-voltage method yields

$$\Sigma i_b = 0 = \frac{v_b - 30}{6} + \frac{v_b - 24}{12}$$

or

$$2v_b - 60 + v_b - 24 = 0 \quad \text{and} \quad v_b = 28 \text{ V} = V_{OC}$$

Finding I_{SC}: Under short-circuit conditions (Fig. 2.15d),

$$\Sigma V_{dacd} = 0 = 30 - 6I_1 = 0 \therefore I_1 = 5 \text{ A}$$

$$\Sigma V_{dbcd} = 0 = 24 - 12I_2 = 0 \therefore I_2 = 2 \text{ A}$$

and

$$I_1 + I_2 = 5 + 2 = 7 \text{ A} = I_{SC}$$

Finding R_T: Hence, by Eq. 2-33,

$$R_T = \frac{V_{OC}}{I_{SC}} = \frac{28}{7} = 4 \ \Omega$$

Alternatively, the resistance looking into Fig. 2-15e is

$$R_T = R_{in} = 6 \| 12 = \frac{1}{\frac{1}{6} + \frac{1}{12}} = 4 \ \Omega$$

Solving for I: Therefore (Fig. 2-15c),

$$I = \frac{V_T}{R_T + 3} = \frac{28}{4 + 3} = 4 \text{ A}$$

and the current in the 3-Ω resistance is 4 A downward.

Note: If we need a general expression that holds for all values of R_L, we can write

$$I = \frac{V_T}{R_T + R_L} = \frac{28}{4 + R_L}$$

Norton's Theorem

Our experience with networks leads us to expect that there is a corollary to Thévenin's theorem. An alternative proposition is described in the following statement called *Norton's theorem*:

Insofar as a load is concerned, any one-port network of resistance elements and energy sources can be replaced by a parallel combination of an ideal current source I_N and a conductance G_N where I_N is the short-circuit current of the one-port and G_N is the ratio of the short-circuit current to the open-circuit voltage.

In Fig. 2.16, the parallel combination of current source and conductance represents the Norton equivalent of the general active one-port. (We use conductance because of its convenience in parallel circuits.) If the networks of Figs. 2.16a and b are to be equivalent for all values of load conductance, they must be equivalent for extreme values such as $G_L = \infty$ and $G_L = 0$. The value $G_L = \infty$ ($R_L = 0$) corresponds to the short-circuit condition; we see that the short-circuit current I_{SC} of the original must be equal to I_N of the equivalent. The value $G_L = 0$ ($R_L = \infty$) corresponds to the open-circuit condition; we see that the open-circuit voltage V_{OC} of the original network must be equal to $I_N R_N = I_N/G_N$ of the equivalent. Therefore,

$$I_N = I_{SC} \quad \text{and} \quad G_N = \frac{I_N}{V_{OC}} = \frac{I_{SC}}{V_{OC}} \quad (2\text{-}34)$$

define the components of the Norton equivalent network. Note that G_N in siemens is just equal to $1/R_T$ in ohms. Also, G_N can be determined by evaluating the conductance that we see looking in at the terminals with all independent energy sources removed.

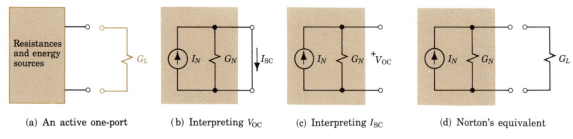

(a) An active one-port   (b) Interpreting V_{OC} (c) Interpreting I_{SC} (d) Norton's equivalent

Figure 2.16   Finding the Norton equivalent of an active one-port network.

Practice Problem 2-5

Disconnect the 12-Ω resistance in Fig. 2.13a, redraw the circuit, evaluate I_N and G_N of the Norton equivalent, and find I_2.

Answers: 14 A, 0.5 S, 2 A.

EXAMPLE 7

Use Norton's theorem to determine the current in the 3-Ω resistance of Example 6.

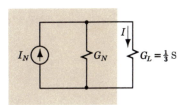

Figure 2.17   Application of Norton's theorem.

The values of $V_{OC} = 28$ V and $I_{SC} = 7$ A are not affected by the choice of the equivalent. By Eqs. 2-34, the parameters in Fig. 2.17 are

$$I_N = I_{SC} = 7 \text{ A}$$

and

$$G_N = \frac{I_{SC}}{V_{OC}} = \frac{7}{28} = 0.25 \text{ S}$$

Alternatively,

$$G_N = G_6 \| G_{12} = \frac{1}{6} + \frac{1}{12} = 0.25 \text{ S} = \frac{1}{R_T}$$

Applying the current-divider principle to the equivalent circuit,

$$I = \frac{G_L}{G_L + G_N} I_N = \frac{0.333}{0.333 + 0.25} \times 7 = 4 \text{ A}$$

Source Transformation

As a by-product of the development of Thévenin's and Norton's theorems, we have obtained another general principle:

Any voltage source V with its associated series resistance R can be replaced by an equivalent current source I with an associated parallel conductance G where I = V/R and G = 1/R. Any parallel combination of I and G can be replaced by an equivalent series combination of V and R.

This principle of source transformation can be very useful in network analysis; for example, it permits us to choose either an ideal voltage source or an ideal current source in representing a real energy source.

Practice Problem 2-6

Find the current I in Fig. 2.18 by repeated use of source transformation. Look for source-resistance combinations that can be replaced by alternative forms until you have a simple series circuit that can be solved by inspection.

Answer: 0.5 A.

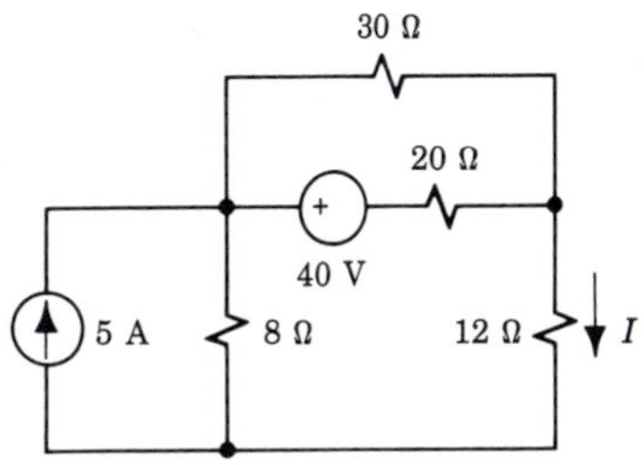

Figure 2.18   Source transformation.

Superposition Theorem

The concept of *linearity* is illustrated in Fig. 2.19; an element of force ΔF produces the same element of deflection Δy at any point in the linear region. In general, where effect y is a function f of cause F, or $y = f(F)$, function f is *linear* if

$$f(F + \Delta F) = f(F) + f(\Delta F) = y + \Delta y \qquad (2\text{-}35)$$

In the nonlinear region, however, an equal element of force ΔF produces a different element of deflection $\Delta y'$.

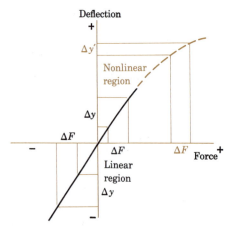

Figure 2.19   The concept of linearity.

In any situation where effect is directly proportional to cause, it is permissible to consider several causes individually and then combine the resulting individual effects to find the total effect. In a high-fidelity (linear) sound system, the pressure waves from a singer are superimposed on the beat from a drum. The output of the speaker can be predicted by adding the response of the system to high notes and the response of the system to low notes. The general principle is called the *superposition theorem* and may be stated as follows:

> *If cause and effect are linearly related, the total effect of several causes acting simultaneously is equal to the sum of the effects of the individual causes acting one at a time.*

In electrical circuits, the causes are excitation voltages and currents and the effects are response voltages and currents. When there are several sources in a complicated circuit, it is convenient to "remove" all except one source and determine the desired response (typically a voltage or current) to that source. The contribution from each individual source is determined in the same way. The response in the original circuit is the sum of the contributions of the individual sources. This approach reduces a complicated problem to a series of relatively simpler problems. An elemen-

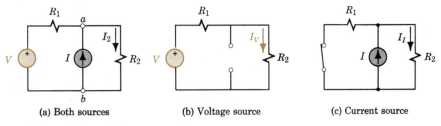

(a) Both sources   (b) Voltage source   (c) Current source

Figure 2.20   Superposition components.

tary example is the circuit of Fig. 2.20a. If the base of the current source is chosen as the reference node, the potential of node a is $I_2 R_2$ and the sum of the currents into a is

$$\frac{V - I_2 R_2}{R_1} + I - I_2 = 0$$

Then

$$V - I_2 R_2 + I R_1 - I_2 R_1 = 0$$

or

$$I_2 = \frac{V + I R_1}{R_1 + R_2} = \frac{V}{R_1 + R_2} + \frac{R_1}{R_1 + R_2} I \qquad (2\text{-}36)$$

Equation 2-36 indicates that the current I_2 consists of two parts, one due to V and the other due to I. If the current source is removed (by open-circuiting) as in Fig. 2.20b, a component of current $I_V = V/(R_1 + R_2)$ flows in R_2. If the voltage source is removed (by short-circuiting) as in Fig. 2.20c, the current divides and a component $I_I = I R_1/(R_1 + R_2)$ flows in R_2. As indicated by the superposition theorem, I_2 is equal to $I_V + I_I$, the sum of the effects produced by the two causes separately.

Practice Problem 2-7

Determine the current in the 6-Ω resistance of Example 6 (Fig. 2.15a) by finding the contribution of each voltage source and applying the principle of superposition. Take advantage of the network reduction theorems, Eqs. 2-25 to 2-32.

Answers: 3.57; -0.57; 3.0 A.

Note that ideal voltage sources are removed by short-circuiting and ideal current sources by open-circuiting. Real energy sources always have internal resistances that must remain in the circuit. (See Example 8 on page 48.)

EXAMPLE 8

A load of 6 Ω is fed by a parallel combination of two batteries. Battery A has an open-circuit voltage of 42 V and an internal resistance of 12 Ω; battery B has an open-circuit voltage of 35 V and an internal resistance of 3 Ω. Determine the current I supplied by battery B, and calculate the power dissipated internally by the battery.

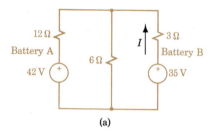

(a)

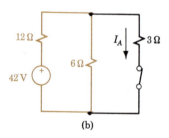

(b)

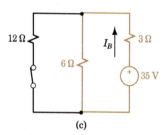

(c)

Figure 2.21 Application of the superposition principle.

If the batteries are represented by linear models, the circuit is as shown in Fig. 2.21a. With the 35-V source removed (Fig. 2.21b), the component of current due to battery A can be written in one step as

$$I_A = \frac{6}{6 + 3} \frac{42}{\left(12 + \dfrac{3 \times 6}{3 + 6}\right)} = \frac{2}{3}\left(\frac{42}{12 + 2}\right) = 2 \text{ A downward}$$

if the principles of current division (Eq. 2-32) and resistance combination (Eqs. 2-25 and 2-27) are applied simultaneously.

If the 42-V source is removed (the internal resistance of the battery remains), the component of current due to battery B (Fig. 2.21c) is

$$I_B = \frac{35}{3 + \dfrac{6 \times 12}{6 + 12}} = \frac{35}{3 + 4} = 5 \text{ A upward}$$

Then, by superposition,

$$I = -I_A + I_B = -2 + 5 = 3 \text{ A upward}$$

To calculate the power dissipated in the 3-Ω resistance, we might note that with the 35-V source removed, the power is

$$P_A = I_A^2 R = 2^2 \times 3 = 12 \text{ W}$$

With the 42-V source removed, the power is

$$P_B = I_B^2 R = 5^2 \times 3 = 75 \text{ W}$$

But, the actual power dissipated is *not* 12 + 75 = 87 W. Why not? What principle would be violated in such a calculation?

The answer is that power is not linearly related to voltage or current and therefore superposition cannot be applied in power calculations. Superposition is applicable only to linear effects such as the current response. Having found the current by superposition, we can calculate the power in the 3-Ω resistance as

$$P = I^2 R = (-I_A + I_B)^2 R = 3^2 \times 3 = 27 \text{ W}$$

NONLINEAR NETWORKS

In the preceding discussions we assumed linearity, but real devices are never strictly linear. In some cases, nonlinearity can be disregarded; we can approximate the v-i characteristic of the device by a straight line and predict the behavior of the real devices within acceptable limits. In other cases, nonlinearity is annoying and special steps must be taken to avoid or eliminate its effect; later we shall learn how to use feedback to minimize the distortion introduced in nonlinear electronic amplifiers. Sometimes nonlinearity is desirable or even essential; the distortion that is annoying in an amplifier is necessary in the harmonic generator for obtaining output signals at frequencies that are multiples of the input signal.

Nonlinear Elements

A dissipative element for which voltage is not proportional to current is a *nonlinear resistor*. An ordinary incandescent lamp has a characteristic similar to that in Fig. 1.19a; the "resistance" of the filament increases with temperature and, therefore, under steady-state conditions, with current.

As we shall see, most electronic devices are inherently nonlinear. Because of its v-i characteristic (Fig. 2.22a), the semiconductor *diode* is useful in discriminating between positive and negative voltages. For positive voltages, the diode acts like a closed switch and current flows readily; for negative voltages, the "switch" is open. If a sinusoidal voltage is applied to such a diode, the resulting current has a large dc component and the diode is a *rectifier*.

The i-v characteristics of one type of transistor are shown in Fig. 2.22b. This device is highly nonlinear, but it can be used as a linear amplifier if, by careful design of the associated circuitry, we maintain operation on the linear portion of the curves.

The *transfer characteristic*, the relation between output current and input voltage, for another type of transistor is shown in Fig. 2.22c. This characteristic is approximately parabolic, but this nonlinear device finds wide application in information processing. Two units of this type form a fast-acting electronic switch that is basic to many digital devices.

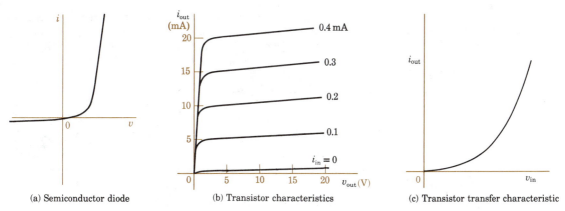

(a) Semiconductor diode   (b) Transistor characteristics   (c) Transistor transfer characteristic

Figure 2.22   Characteristics of nonlinear elements.

Methods of Analysis

The method of analysis employed depends on the nature of the problem, the form of the data, and the computation aids available. The modern digital computer enables us to solve quickly problems that were hopelessly time consuming a few years ago, but here we are concerned only with simple series and parallel circuits.

Analytical Solution. If an analytical expression for the v-i characteristic can be obtained from physical principles or from experimental data, some simple problems can be solved algebraically. A useful model is the power series

$$i = a_0 + a_1 v + a_2 v^2 + a_3 v^3 + \cdots \tag{2-37}$$

The first two terms provide a linear approximation. The first three terms give satisfactory results in many practical nonlinear problems, although more terms may be necessary in some cases.

EXAMPLE 9

A voltage $v = V_m \cos \omega t$ is applied to the semiconductor device of Fig. 2.23. Determine the nature of the resulting current.

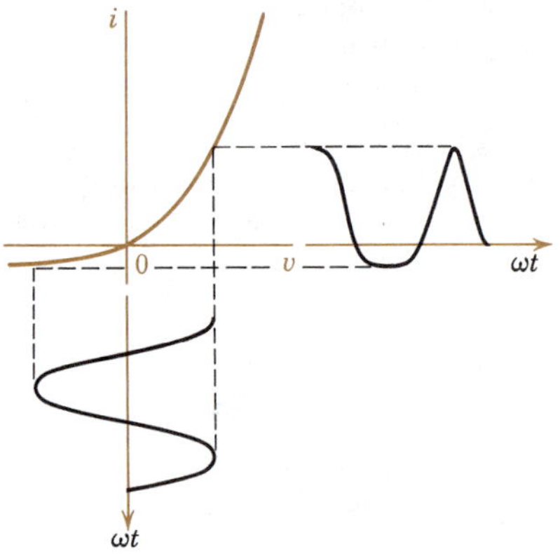

Figure 2.23 Power-series analysis.

Assuming that the v-i characteristic can be represented by the first three terms of the power series of Eq. 2-37, the current is

$$i = a_0 + a_1 v + a_2 v^2$$

But we note that for $v = 0$, $i = 0$; therefore, $a_0 = 0$. For the given voltage,

$$i = a_1 V_m \cos \omega t + a_2 V_m^2 \cos^2 \omega t$$

or

$$i = a_1 V_m \cos \omega t + \frac{a_2 V_m^2}{2}(1 + \cos 2\omega t)$$

$$= \frac{a_2 V_m^2}{2} + a_1 V_m \cos \omega t + \frac{a_2 V_m^2}{2} \cos 2\omega t \tag{2-38}$$

Equation 2-38 indicates that the resulting current consists of a steady or dc component, a "fundamental" component of the same frequency as the applied voltage, and a "second-harmonic" component at the new frequency 2ω.

Depending on the relative magnitudes of a_1 and a_2 and the associated circuitry, the device of Example 9 could be used as a rectifier, an amplifier, or a harmonic generator. In a specific problem, the coefficients in the power series can be determined by choosing a number of points on the v-i characteristic equal to the number of terms to be included, substituting the coordinates of each point in Eq. 2-37, and solving the resulting equations simultaneously.

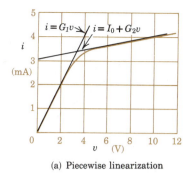
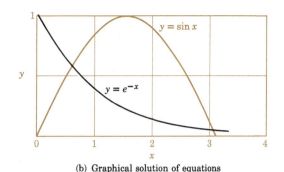

(a) Piecewise linearization   (b) Graphical solution of equations

Figure 2.24   Analysis based on graphical data

Piecewise Linearization. Where approximate results are satisfactory, a convenient approach is to represent the actual characteristic by a series of linear "pieces." The characteristic of the transistor of Fig. 2.24a can be divided into two regions, each approximated by a straight line. Below the "knee" of the curve, the behavior is adequately described by $i = G_1v$; above the knee, a different expression must be used. Usually a trial solution will indicate in which region operation is occurring, and the problem is reduced to one in linear analysis. (This approach is used in developing models for electronic devices in Chapter 5.)

Practice Problem 2-8

(a) Determine the parameters G_1, G_2, and I_0 needed to define the characteristic of Fig. 2.24a by piecewise linearization.
(b) If this device is R_N in Fig. 2.26a where $V_1 = 15$ V, $R_1 = 6$ kΩ, and $R_2 = 12$ kΩ, predict I.
(c) After studying the section on networks with one nonlinear element, use the load-line method to predict I in part (b).

Answers: (a) 1 mS, 0.1 mS, 3 mA; (b) 2 mA; (c) same as (b).

Graphical Solution. In piecewise linearization we use graphical data to obtain analytical expressions that hold over limited regions. Sometimes it is desirable to plot given analytical functions to obtain a solution graphically. For example, the equation $e^{-x} - \sin x = 0$ is difficult to solve analytically. If the functions $y = e^{-x}$ and $y = \sin x$ are plotted as in Fig. 2.24b, the simultaneous solutions of these two equations are represented by points whose coordinates satisfy both equations. Therefore, the intersections of the two curves are the solutions to the original equation and can be read directly from the curves. This method is also useful when the characteristic of some portion of the circuit is available in graphical form or in a table of experimental data.

Networks with One Nonlinear Element

If there is a single nonlinear resistance in an otherwise linear resistive network, and if the v-i characteristic of the nonlinear element is known, a relatively simple method of solution is available. This situation occurs frequently in practice and the method of attack deserves emphasis.

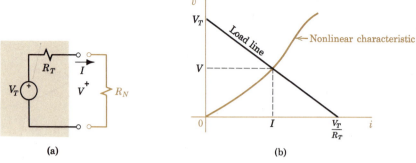

(a)   (b)

Figure 2.25 Load-line analysis of a nonlinear network.

The first step is to replace all except the nonlinear element with the Thévenin equivalent shown in Fig. 2.25a as V_T and R_T. The combination of V_T and R_T is equivalent "insofar as a load is concerned." For any value of load resistance, the terminal voltage is

$$V = V_T - R_T I \qquad (2\text{-}39)$$

This equation is that of a straight line, the *load line*, with intercepts V_T and V_T/R_T and slope $-R_T$. (For the original network, the intercepts are V_{OC} and I_{SC}.) The graph of the nonlinear characteristic is v as a function of i for the load. The simultaneous satisfaction of these two relations, which yields the values of V and I at the terminals, is the intersection of the two curves.

EXAMPLE 10

The nonlinear element whose characteristic is given in Fig. 2.25b is connected in the circuit of Fig. 2.26a. Determine the current I.

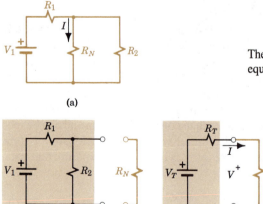

(a)

(b)   (c)

Figure 2.26 Load-line analysis.

First, the nonlinear element is isolated (Fig. 2.26b) and the Thévenin equivalent determined. Considering R_1 and R_2 as a voltage divider,

$$V_T = V_{OC} = \frac{R_2}{R_1 + R_2} V_1$$

The short-circuit current is just $I_{SC} = V_1/R_1$. Therefore, the equivalent resistance is

$$R_T = \frac{V_{OC}}{I_{SC}} = \frac{R_2 V_1}{R_1 + R_2} \frac{R_1}{V_1} = \frac{R_1 R_2}{R_1 + R_2}$$

which is also the resistance seen looking in at the terminals with the voltage source removed.

Next, the load line is drawn with intercepts V_T and V_T/R_T as shown in Fig. 2.25b. The intersection of the load line with the nonlinear characteristic gives the desired current I.

SUMMARY

- The algebraic sum of the currents into a node is zero ($\Sigma i = 0$).
  The algebraic sum of the voltages around a loop is zero ($\Sigma v = 0$).

- The general procedure for formulating equations for circuits is:
    1. Arbitrarily assume a consistent set of currents and voltages.
    2. Write the element equations by applying the element definitions and write the connection equations by applying Kirchhoff's laws.
    3. Combine the element and connection equations to obtain the governing circuit equations in terms of the unknowns.

  The resulting simultaneous equations can be solved by:
    Successive substitution to eliminate all but one unknown, or
    Use of determinants and Cramer's rule.

- Skillful use of loop currents or node voltages may greatly reduce the number of unknowns and simplify the solution.
  Checking is essential because of the many opportunities for mistakes in sign and value.

- Two one-ports are equivalent if they present the same v-i characteristics.
  Two passive one-ports are equivalent if they have the same input resistance.
  For resistances in series, $R_{EQ} = R_1 + R_2 + \cdots + R_n$.
  For conductances in parallel, $G_{EQ} = G_1 + G_2 + \cdots + G_n$.
  For resistances in parallel, $R_{EQ} = 1 \Big/ \left(\dfrac{1}{R_1} + \dfrac{1}{R_2} + \cdots + \dfrac{1}{R_n} \right)$.

- For the two-element voltage divider and current divider,

$$V_2 = \frac{R_2}{R_1 + R_2} V \quad \text{and} \quad I_2 = \frac{G_2}{G_1 + G_2} I = \frac{R_1}{R_1 + R_2} I$$

- Insofar as a load is concerned, any one-port network of resistance elements and energy sources can be replaced by:
  a series combination (Thévenin) of an ideal voltage source V_T and a resistance R_T, or
  a parallel combination (Norton) of an ideal current source I_N and a conductance G_N, where $V_T = V_{OC}$, $R_T = V_{OC}/I_{SC}$, $I_N = I_{SC}$, and $G_N = I_{SC}/V_{OC} = 1/R_T$.

- A series combination of voltage V and resistance R can be replaced by a parallel combination of current $I = V/R$ and conductance $G = 1/R$.

- If cause and effect are linearly related, the principle of superposition applies, and the total effect of several causes acting simultaneously is equal to the sum of the effects of the individual causes acting one at a time.

- Networks involving nonlinear elements can be solved in various ways, depending on the nature of the problem and the form of the data.
  If there is a single nonlinear element in an otherwise linear network, construction of a load line permits a simple graphical solution.

REVIEW QUESTIONS

1. Write out in words a physical interpretation of Eq. 2-18.
2. Outline a three-step procedure for applying the node-voltage method.
3. Outline a three-step procedure for applying the loop-current method.
4. If two passive one-ports are equivalent, is the same power dissipated in both? Prove your answer.
5. What is the difference between "passive" and "active" networks?
6. Under what circumstances is Thévenin's theorem useful?
7. Is the same power dissipated in an active circuit and the Norton equivalent?
8. How are the Norton and Thévenin components related?

9. Outline a four-step procedure for applying Thévenin's theorem.
10. Cite two nonelectrical examples where superposition applies.
11. Under what circumstances is the superposition principle useful?
12. How are voltage and current sources "removed"?
13. Cite nonelectrical examples of linear and nonlinear devices.
14. Give two engineering examples where nonlinearity is desirable.
15. The v-i characteristic of an active device is $v = 100 - 2i - i^2$. Outline two methods for determining the current that flows when a resistance R is connected to the device.
16. What is meant by "piecewise linearization"? A "load line"?

EXERCISES

1. In Fig. 2.27, $i_1 = 2$ A and $i_2 = -3$ A. Predict current i_3.

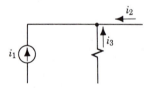

Figure 2.27

2. In Fig. 2.28, $v_2 = 12$ V, $R_1 = 4$ Ω, and $R_2 = 2$ Ω. Calculate current i, voltage v_1 (across R_1), and source voltage v.

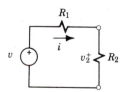

Figure 2.28

3. In Fig. 2.29, determine i_1 and i_2 in terms of the other quantities.

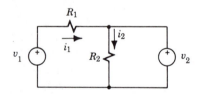

Figure 2.29

4. In Fig. 2.30, given $I_a = 3$ A, $I_b = 2$ A, and $I_c = -8$ A, determine i_1, i_2, and v.

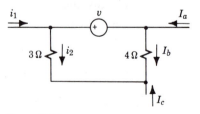

Figure 2.30

5. In Fig. 2.30, given $I_a = -1$ A, $I_c = -4$ A, and $v = +5$ V, determine i_1, i_2, and I_b.

6. In Fig. 2.31, $i_1 = 4$ A, $i_2 = 10 \sin t$ A, and $v_C = 3 \cos t$ V. Find:
(a) i_L and v_L
(b) v_{ac}, v_{ab}, and v_{cd}, all as functions of time.

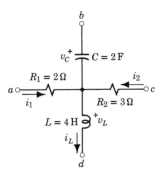

Figure 2.31

7. Solve Exercise 6 for $i_1 = 10\,e^{-2t}$ A, $i_2 = 4$ A, and $v_C = 3\,e^{-2t}$ V.

8. In Fig. 2.32, evaluate:
(a) v_{ba}, v, i_2, and v_{de}
(b) The power dissipated in the 3-Ω resistance.
(c) The power supplied by the current source.

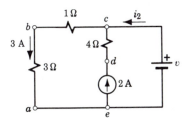

Figure 2.32

9. Given the circuit of Fig. 2.33, calculate the voltage across the 3-Ω resistance by using: (a) element currents and (b) loop currents. Include an adequate check. (*Hint:* Arrange loops so that only one includes the resistance of interest.)

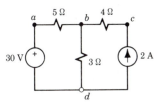

Figure 2.33

10. Given the circuit of Fig. 2.34, calculate the current I using: (a) element currents and (b) loop currents. Include an adequate check. (See *Hint* in Exercise 9.)

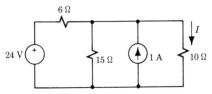

Figure 2.34

11. Use the node-voltage method to calculate: (a) The voltage across the 3-Ω resistance in Fig. 2.33. (b) The current I in Fig. 2.34. (*Hint:* Use one end of the resistance of interest as the reference node.)

12. Given the circuit of Fig. 2.35, calculate the current in the 2-Ω resistance using: (a) element currents, (b) loop currents, and (c) node voltages. Draw a conclusion about the methods. (See *Hints* in Exercises 9 and 11.)

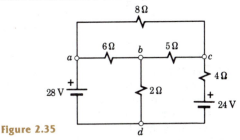

Figure 2.35

13. Given the circuit of Fig. 2.35, calculate the voltage across the 5-Ω resistance using: (a) element currents, (b) loop currents, and (c) node voltages. Draw a conclusion about the methods. (See *Hints* in Exercises 9 and 11.)

14. In Fig. 2.36, V_O is an initial voltage on C. Using only the specified variables, formulate the voltage equation for loop *abcda* with the switch closed.

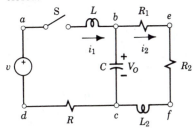

Figure 2.36

15. Repeat Exercise 14 for loop *befcb*
16. Device X requires 4 V at 1.5 mA, whereas device Y operates at 2 V and 1 mA. The two devices are to be operated from a 9-V battery as shown in Fig. 2.37. Design the circuit; i.e., specify the values of R_1 and R_2.

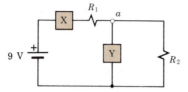

Figure 2.37

17. The heating element in a "hotplate" consists of two resistances that can be connected separately, in series, or in parallel to provide four "heat" settings.
    (a) Draw a wiring diagram showing R_1 as an inside loop and R_2 as an outside loop with terminals for series or parallel connection.
    (b) For 120-V operation and $R_1 = 30\ \Omega$, specify R_2 for a "low" setting of 160 W.
    (c) Calculate the other three power values.
18. Find current I in the ladder network of Fig. 2.38 by network reduction. Reduce series and parallel combinations to a single equivalent resistance, find the battery current, and then use current division to find I.

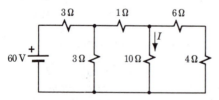

Figure 2.38 Ladder network.

19. Find current I in the ladder network of Fig. 2.38 by the following ingenious method: Assume the current is 1 A and work backward to find the necessary source voltage; determine actual I by simple proportion.
20. Applying the current-divider principle twice to Fig. 2.38, write down a general expression for current I as a function of the battery voltage. Your expression should yield the value of I in one continuous calculator operation.

21. In Fig. 2.39, $R_2 = 2\ \text{k}\Omega$, $R_1 = 10\ \text{k}\Omega$, and $v = 24$ V.
    (a) Predict v_2.
    (b) A practical voltmeter can be represented by a 10-kΩ resistance in series with an ideal (resistanceless) meter calibrated to indicate the voltage across the 10-kΩ meter resistance. Predict the reading of the voltmeter when connected across R_2.
    (c) Repeat part (b) if the effective voltmeter resistance is 100 kΩ, and draw a conclusion about the desirable internal resistance of a voltmeter.

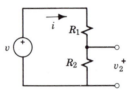

Figure 2.39 Voltage divider.

22. The open-circuit output voltage of an audio oscillator is 5 V. The terminal voltage drops to 4 V when a 2000-Ω resistor is connected across the terminals. Determine the internal resistance of the oscillator.
23. Use source transformation to replace the circuit of Fig. 2.38 by a simpler equivalent that will enable you to find the current in the 1-Ω resistor by a single calculation. Calculate the current.
24. In Fig. 2.34, use Thévenin's theorem to find the current in the 15-Ω resistance.
25. In Fig. 2.38, replace the 10, 6, and 4-Ω resistances by a single equivalent resistance. Then replace the 60-V source and the two 3-Ω resistances by a Thévenin equivalent. Use the voltage-divider principle to predict current I.
26. In Fig. 2.40, determine the current in the 3-Ω resistance without using simultaneous equations. (*Hint:* First replace all except the 3-Ω resistance and the 35-V battery by a Thévenin equivalent.)

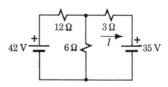

Figure 2.40

27. In Fig. 2.34, use Norton's theorem to find current I.

28. In Fig. 2.40, use Norton's theorem to find the current in the 6-Ω resistance.

29. In Fig. 2.41, with switch S closed, the ammeter AM reads 60 mA. Predict the ammeter reading with switch S open. (*Hint:* Does the AM reading with S closed give you a clue to a Norton equivalent?)

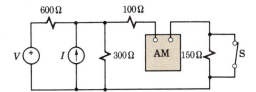

Figure 2.41

30. Repeat Exercise 9 using superposition.

31. Repeat Exercise 12 using superposition.

32. Use superposition to find the current in the 15-Ω resistance in Fig. 2.34.

33. In Fig. 2.34, use superposition to find current I.

34. Find the current in the 12-Ω resistance in Fig. 2.40 by using superposition.

35. Measurements on a human nerve cell indicate an open-circuit voltage of 80 mV, and then a current of 5 nA through a 6-MΩ load. You are to predict the behavior of this cell with respect to an external load.
    (a) What simplifying assumption must be made?
    (b) Devise an appropriate model.
    (c) Predict the current through a 10-MΩ load.

36. A device that can be represented by the mathematical model $v = 6i + 3i^2$ is connected in series with a source that can be represented by 36 V in series with 6 Ω.
    (a) Predict the resulting device current mathematically.
    (b) Sketch the v-i characteristic for $0 < i < 3$ and predict the device current graphically. Compare to the result for part (a).

37. Device A in Fig. 2.42 is connected in series with a 20-V battery and a 5-kΩ resistance. Predict the current that flows in two ways:

(a) Reproduce the i-v characteristic and represent device A by piecewise linearization.
(b) Use the load-line method. Compare results.

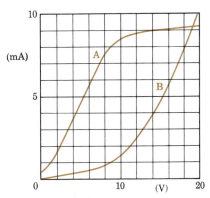

Figure 2.42 Nonlinear device characteristics.

38. For electronic device A in Fig. 2.42,
    (a) Reproduce the v-i characteristic and represent it by piecewise linearization.
    (b) Predict the approximate current in response to an applied voltage $v = 15 + 5 \cos \omega t$ V.

39. Assume that device B in Fig. 2.42 can be represented by the equation $i\,(\text{mA}) = a_1 v + a_2 v^2$ in the region $10 \le v \le 18$ V.
    (a) Evaluate a_1 and a_2 by simultaneous solution of two equations for i at the two values of v.
    (b) Use your mathematical model to determine i for $v = 14$ V and check graphically.

40. Device B in Fig. 2.42 is connected in parallel with a 10-mA current source and a 1.8-kΩ resistance; draw the circuit and predict the voltage and the current in device B.

41. Device A in Fig. 2.42 is connected in the circuit of Fig. 2.43 where $R = 1000\ \Omega$ and $R_1 = 600\ \Omega$.
    (a) If $I = 16$ mA, predict the device current I_D.
    (b) Specify the value of I so that the power dissipated in D is 60 mW.

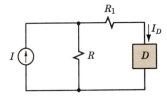

Figure 2.43

PROBLEMS

1. Given the circuit of Fig. 2.44, consider various methods to calculate the voltage across the 5-Ω resistance. Use the method that appears to require the least algebra.

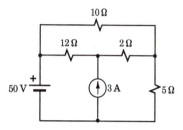

Figure 2.44

2. In Fig. 2.45, find the current through the 20-Ω resistor. Choose your variables so that only one equation is necessary.

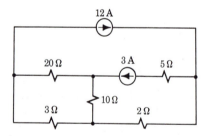

Figure 2.45

3. A signal generator, consisting of a complicated network of nearly linear passive and active elements, supplies a variable load resistance R_L.
   (a) Use Thévenin's theorem to replace the generator by a combination of V_G and R_G.
   (b) Prove the *maximum power transfer* theorem, which states that: "For maximum power transfer from a source to a load, the resistance of the load should be made equal to the Thévenin equivalent resistance of the source."

4. In an amplifier, a transistor can be represented by the two-port model in Fig. 2.46, where $R_1 = 10$ kΩ, $R_2 = 75$ kΩ, $R_3 = 150$ kΩ, and $\beta = 100$. (βI_1 is a "controlled" current source.)
   (a) Simplify the model (shown in color) and voltage source and predict the voltage amplification V_L/V_S as a function of R_L.
   (b) Calculate V_L/V_S for $R_L = 20$ kΩ.

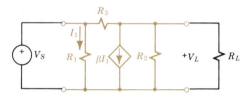

Figure 2.46

5. The behavior of a certain germanium diode is defined by $i = I_S(e^{bv} - 1)$ where $I_S = 10$ μA and $b = 40$ V^{-1}.
   (a) Expand the exponential term in a power series and represent $i(v)$ using the first three terms of Eq. 2-37.
   (b) Predict the current if a voltage $v = +0.05$ V is applied to the diode in series with a 500-Ω resistance.
   (c) Check the accuracy of the method by calculating the current using the diode voltage from part (b) in the given exponential equation.

6. Devices A and B in Fig. 2.42 are connected in series and the combination in series with a 30-V battery and a 5-kΩ resistance.
   (a) Plot the composite v-i characteristic for A and B in series.
   (b) Predict i, v_A, and v_B by the load-line method.
   (c) Predict the power dissipated in devices A and B.

CHAPTER **3**

Signal Processing Circuits

Signal Waveforms

Periodic Waveforms

Electrical Instruments

Ideal Amplifiers

Ideal Diodes

Waveshaping Circuits

A principal application of electronics is information processing. Typically, the information is available in the form of *signals*. An electrical signal may be a voltage or current varying with time in a manner that conveys information; the pattern of ones and zeros in a digital code and the varying currents in a telephone receiver are well-known examples. In processing such information, electronic *devices* are used to detect and amplify weak signals or to convert signals from one form to another—from a high-frequency signal that is easily radiated by an antenna to a low-frequency signal that is audible to humans, for example. The electronic *circuits* that carry signals between devices range in size from the intercontinental cable circuits that span oceans to the microscopic integrated circuits built into a tiny calculator chip of silicon.

In this chapter we look at some elementary signal processing circuits. First we study some common signal waveforms and learn how they can be handled. Next we examine the characteristics of ideal amplifiers and diodes and see how they perform their basic functions. Then we use the circuit principles of Chapters 1 and 2 in learning to analyze and design some electronic circuits that are widely used in communication and control.

SIGNAL WAVEFORMS

In a signal, it is the *change* in current or voltage that carries information. Certain patterns of time variation or *waveforms* are of special significance in electronics because they are encountered so frequently. Some of these are shown in Fig. 3.1.

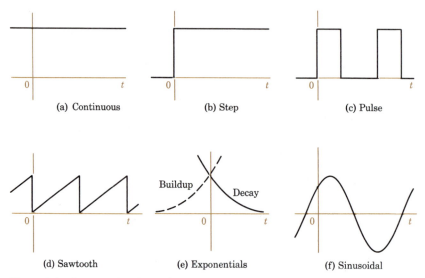

Figure 3.1 Common signal waveforms.

A *direct* or *continuous* voltage is supplied by a storage battery or a direct-current (dc) generator. A *step* of current flows when a switch is thrown, suddenly applying a direct voltage to a resistance; the current in the battery when a pocket radio is turned on is an example. A *pulse* of current flows if the switch is turned **ON** and then **OFF**; the flow of information in a computer consists of very short pulses. A *sawtooth* wave increases linearly with time and then resets; this type of voltage variation causes the electron beam to move repeatedly across the screen of a television picture tube. A decaying *exponential* current flows if energy is stored in the electric field of a capacitor and allowed to leak off through a resistor; in an unstable system, a voltage may build up exponentially. A *sinusoidal* voltage is generated when a coil is rotated at a constant speed in a uniform magnetic field; also, oscillating circuits are frequently characterized by sinusoidal voltages and currents.

Exponentials

Exponential and sinusoidal waveforms are easy to generate and easy to analyze because of their simple derivatives and integrals; we shall pay them special attention because they are encountered frequently and also because they are useful in the analysis of more complicated waves, such as the electrocardiograms of Fig. 3.2.

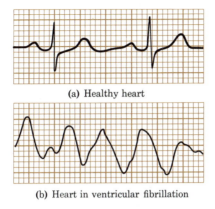

(a) Healthy heart

(b) Heart in ventricular fibrillation

Figure 3.2 Electrocardiograms.

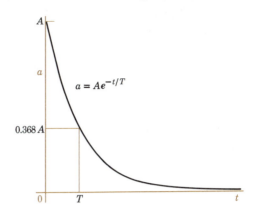

Figure 3.3 The decaying exponential.

The General Exponential. In general, an exponentially decaying quantity (Fig. 3.3) can be expressed as

$$a = A \, e^{-t/T} \qquad\qquad (3\text{-}1)$$

where a = instantaneous value
A = amplitude or maximum value
e = base of natural logarithms = 2.718 . . .
T = time constant in seconds
t = time in seconds

The temperature of a hot body in cool surroundings, the angular velocity of a freely spinning bicycle wheel, and the current of a discharging capacitor can all be approximated by decaying exponential functions of time.

Time Constant. Since the exponential factor in Eq. 3-1 only *approaches* zero as t increases without limit, such functions theoretically last forever. In the same sense, all radioactive disintegrations last forever. To distinguish between different rates of disintegration, physicists use the term "half-life," where the half-life of radium, say, is the time required for any amount of radium to be reduced to one-half of its initial amount. In the case of an exponentially decaying current, it is more convenient to use the value of time that makes the exponent -1. When $t = T = $ the *time constant*, the value of the exponential factor is

$$e^{-t/T} = e^{-1} = \frac{1}{e} = \frac{1}{2.718} = 0.368 \qquad\qquad (3\text{-}2)$$

In other words, after a time equal to the time constant, the exponential factor is reduced to approximately 37% of its initial value. (See Example 1 on p. 62.)

EXAMPLE 1

A capacitor is charged to a voltage of 6 V and then, at time $t = 0$, it is switched across a resistor and observations are made of current as a function of time. The experimental data are plotted in Fig. 3.4a. Draw the circuit, determine the time constant for this circuit, and derive an exponential equation for the current.

A smooth curve is drawn through the experimental points. The amplitude of the exponential is determined by noting that at $t = 0$, $i = I_0 = 6$ mA.

The time constant is determined by noting that when $i = I_0 e^{-1} = 6(0.368) \cong 2.2$ mA, $t = T \cong 2$ ms.

Then $1/T = 1/0.002 = 500$ and the current equation is

$$i = I_0 e^{-t/T} = 6 e^{-500t} \text{ mA} \qquad (3\text{-}3)$$

for time $t > 0$, i.e., *after* switching.

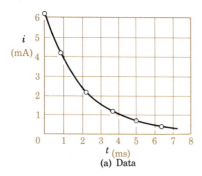

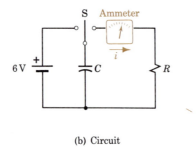

(a) Data   (b) Circuit

Figure 3.4   Observation of time constant.

Normalized Exponentials. All exponential equations can be reduced to a common form by expressing the variables in terms of dimensionless ratios or *normalizing* the variables. When this is done, Eq. 3-1 becomes

$$\frac{a}{A} = e^{-t/T} \qquad (3\text{-}4)$$

Note that after 2 time constants, $t = 2T$ and the function is down to $1/e^2$ times its initial value, or $a/A = (0.368)^2 = 0.135$. After 5 time constants, $t = 5T$ and the normalized function is down to $1/e^5 = 0.0067$. This value is less than the error expected in many electrical measurements, and it may be assumed that after 5 time constants an exponential current or voltage is practically negligible.

When plotted on semilog graph paper, an exponential function becomes a straight line with slope equal to the reciprocal of the time constant. As another interpretation of the time constant, it can be demonstrated (see Problem 1) that, if the exponential decay continues at its *initial rate*, the duration of the function is just equal to T.

Practice Problem 3-1

Calculate the natural log of i (in mA) for the first four points of the curve in Fig. 3.4a. Draw a new vertical scale and plot ln i versus t; verify that the slope of the line is the reciprocal of the time constant.

Answer: Slope $= \Delta(\ln i)/\Delta t = 1.8/3.6 = 0.5/$ms.

Pulse Response of *RC* Circuits

If we connect a battery to a circuit by closing a switch, we apply a "voltage step" to the circuit. If we disconnect the battery, removing the voltage, we complete a voltage "pulse." In digital systems the signals are of this character, and the "pulse response" tells us how a circuit reacts to such an input.

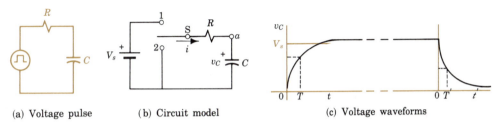

(a) Voltage pulse   (b) Circuit model   (c) Voltage waveforms

Figure 3.5   Pulse response of an *RC* circuit.

In Fig. 3.5b, closing switch S to position 1 connects voltage source V_s, thereby applying a voltage step to the series *RC* circuit, where *C* is initially uncharged. To determine the response, we first apply Kirchhoff's law and write the governing equation

$$\Sigma v = 0 = V_s - iR - \frac{1}{C} \int i \, dt \tag{3-5}$$

Differentiating to eliminate the integral and the constant and then rearranging the terms, we obtain the *homogeneous differential equation*

$$R\frac{di}{dt} + \frac{1}{C}i = 0 \tag{3-6}$$

This equation is homogeneous because it contains only terms relating to the circuit itself; there are no source terms or constants.

The solution of a differential equation consists in finding a function that satisfies the equation. Equation (3-6) says that the combination of a function $i(t)$ and its derivative di/dt must equal zero; the function and its derivative must cancel somehow. This implies that the form of the function and that of its derivative must be the same and suggests an exponential. If we let

$$i = A \, e^{-t/T} \qquad \text{then} \qquad \frac{di}{dt} = -\frac{A}{T}e^{-t/T} \tag{3-7}$$

where *A* is an amplitude, and *T* is a time constant. Substituting these expressions into the homogeneous equation, we obtain

$$-\frac{R}{T}A \, e^{-t/T} + \frac{1}{C}A \, e^{-t/T} = \left(\frac{1}{C} - \frac{R}{T}\right) A \, e^{-t/T} = 0 \tag{3-8}$$

for all values of *t*. In looking for solutions to this equation, we find that $A = 0$ is one possibility, but this is an unimportant case. The other possibility is to let $(1/C) - (R/T) = 0$. This yields $T = RC$ and the desired solution is

$$i = A \, e^{-t/RC} \tag{3-9}$$

To evaluate coefficient A, we note that the energy stored in a capacitance, $\frac{1}{2}Cv_C^2$, cannot change instantaneously. Therefore, the voltage on the capacitance cannot change instantaneously; just after the switch is thrown, v_C must still be zero. At $t = 0^+$, Kirchhoff's voltage law tells us that

$$\Sigma v = 0 = V_s - iR + 0$$

or

$$i(0^+) = \frac{V_s}{R} = A\,e^0 = A$$

Therefore, $A = V_s/R$, and for all values of time after the switch is thrown,

$$i = \frac{V_s}{R}\,e^{-t/RC} \tag{3-10}$$

The voltage on the capacitance is $V_s - Ri$ and, therefore,

$$v_C = V_s - V_s\,e^{-t/RC} = V_s(1 - e^{-t/RC}) \tag{3-11}$$

As shown in Fig. 3.5c, the voltage across the capacitance starts from zero and approaches V_s as a limit. After a few time constants, v_C is essentially equal to V_s.

EXAMPLE 2

Switch S in Fig. 3.6a is closed at time $t = 0$. Predict the voltage v and the current i.

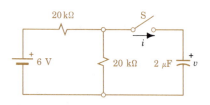

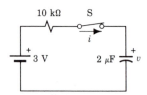

Figure 3.6 *RC* step response.

The expressions for step response assume a single-loop circuit consisting of an equivalent R and an equivalent C. This two-loop circuit can be converted to a single loop by using Thévenin's theorem. As shown in Fig. 3.6b,

$$V_T = 6 \times \frac{20}{20 + 20} = 3\text{ V}$$

$$R = R_T = \frac{20 \times 20}{20 + 20} = 10\text{ k}\Omega$$

and the time constant is

$$T = RC = 10 \times 10^3 \times 2 \times 10^{-6} = 0.02\text{ s}$$

For an initially uncharged capacitor,

$$v = V_s(1 - e^{-t/RC}) = 3(1 - e^{-50t})\text{ V}$$

The capacitance current is

$$i = C\frac{dv}{dt} = C\left(+\frac{1}{RC}\right)V_s\,e^{-t/RC}$$

$$= \left(\frac{V_s}{R}\right)e^{-t/RC}$$

$$= 0.3\,e^{-50t}\text{ mA}$$

If at a later time $t' = 0$, the voltage in Fig. 3.5b is removed by throwing the switch to position 2, the voltage pulse is completed. To determine this part of the response, let node 2 be a reference and write Kirchhoff's current law for currents out of node a. The homogeneous equation is

$$\Sigma i_a = 0 = C \frac{dv_C}{dt'} + \frac{v_C}{R} \tag{3-12}$$

Again an exponential equation is suggested and we let

$$v_C = A' e^{-t'/T'} \qquad \text{and} \qquad \frac{dv_C}{dt'} = -\frac{A'}{T'} e^{-t'/T'}$$

Substituting these expressions into Eq. 3-12 and solving, we find that $T' = RC$ satisfies the equation and

$$v_C = A' e^{-t'/T'} \tag{3-13}$$

Now, however, the capacitance is initially charged and at time $t' = 0^+$,

$$v_C (0^+) = V_s = A' e^0 = A'$$

Therefore, for all values of time after the switch is thrown to position 2,

$$v_C = V_s \, e^{-t'/RC} \tag{3-14}$$

The current in the capacitance is $C \, dv_C/dt$, or

$$i_C = -\frac{V_s}{R} e^{-t'/RC} \tag{3-15}$$

where the minus sign indicates that current is flowing out of the capacitor.

EXAMPLE 3

For an initial voltage $V_0 = 6$ V, the experimental data of Example 1 indicated an initial current $I_0 = 6$ mA and a time constant $T = 2$ ms. Determine the values of R and C.

For $t > 0$, the capacitor is connected across the resistor and $v_R = v_C$. Since v_C cannot change instantaneously, at $t = 0^+$

$$v_C = V_0 = v_R = iR = I_0 R$$

$$\therefore R = \frac{V_0}{I_0} = \frac{6}{0.006} = 1000 \ \Omega$$

For this circuit, time constant $T = RC$

$$\therefore C = \frac{T}{R} = \frac{2 \times 10^{-3}}{1 \times 10^3} = 2 \times 10^{-6} = 2 \ \mu F$$

Practice Problem 3-2

Given a 12-V battery, design a circuit in which the current in one element will jump to 5 mA at time $t = 0^+$ and then decay exponentially with a time constant of 30 ms.

Answer: $R = 2.4$ kΩ; $C = 12.5 \ \mu F$.

PERIODIC WAVEFORMS

A periodic function is one that repeats itself after a period of time. The simple sine waves generated by a laboratory oscillator and the complex triggering signals controlling a television picture are periodic waveforms.

Sinusoids

There are three principal reasons for emphasis on sinusoidal functions of time. First, many natural phenomena are sinusoidal in character; the vibration of a guitar string, the projection of a satellite on the rotating earth, and the current in an oscillating circuit are examples. Second, sinusoids are important in the generation and transmission of electric power and in the communication of intelligence; because the derivatives and integrals of sinusoids are themselves sinusoidal, a sinusoidal source always produces sinusoidal responses in any linear circuit. Third, other periodic waves can be represented by a series of sinusoidal components by means of Fourier analysis; this means that techniques for handling sinusoids are useful in predicting the behavior of circuits involving pulses and sawtooth waves as well. Clearly, an efficient technique for analyzing circuits involving sinusoidal voltages and currents is essential. Since the application of Kirchhoff's laws involves summations of currents and voltages, we need a convenient method for adding quantities that are continually varying functions of time.

The General Sinusoid. In general, a sinusoidally varying quantity (Fig. 3.7) can be expressed as

$$a = A \cos (\omega t + \alpha) \tag{3-16}$$

where a = instantaneous value
A = amplitude or maximum value
ω = frequency in radians per second (omega)
t = time in seconds
α = phase angle in radians (alpha)

Since an angle of 2π rad corresponds to one complete cycle, at ω rad/s,

$$f = \frac{\omega}{2\pi} = frequency \text{ in cycles per second or hertz[†] (Hz)} \tag{3-17}$$

The time for one complete cycle is

$$T = \frac{1}{f} = period \text{ in seconds} \tag{3-18}$$

The same sinusoid can be expressed by the sine-wave function

$$a = A \sin \left(\omega t + \alpha + \frac{\pi}{2} \right) \tag{3-19}$$

since $\sin (x + \pi/2) = \cos x$ for all values of x. In either case, the instantaneous value varies from positive maximum to zero to negative maximum to zero repeatedly; a current that varies sinusoidally is called an *alternating current*.

[†] Heinrich Hertz, a German physicist, first demonstrated the existence of radio waves.

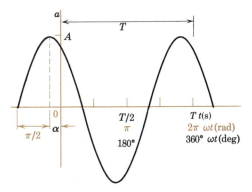

Figure 3.7 The general sinusoid.

EXAMPLE 4

A sinusoidal current with a frequency of 60 Hz reaches a positive maximum of 20 A at $t = 2$ ms (Fig. 3.8). Write the equation of current as a function of time.

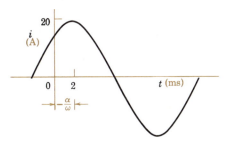

Figure 3.8 A sinusoidal current.

In the form of Eq. 3-16, $A = 20$ A and $\omega = 2\pi f = 2\pi(60) = 377$ rad/s. The positive maximum is reached when $(\omega t + \alpha) = 0$ or any multiple of 2π. Letting $(\omega t + \alpha) = 0$,

$$\alpha = -\omega t = -377 \times 2 \times 10^{-3} = -0.754 \text{ rad}$$

or

$$\alpha = -0.754 \times \frac{360°}{2\pi} = -43.2°$$

Therefore, the current equation is

$$i = 20 \cos (377t - 43.2°) \text{ A}$$

Note that the angle in parentheses is a convenient hybrid; the variable angle is always in radians, but engineers prefer to express the phase angle α in degrees. To avoid confusion, the degree symbol is essential.

Periodic Functions

By definition, $F(t)$ is a *periodic* function of time if

$$F(t + nT) = F(t) \tag{3-20}$$

where T is the period in seconds and n is any integer. The sinusoidal current shown in Fig. 3.9a is periodic with a period of $T = 1/f = 2\pi/\omega$ seconds since

$$i = 10 \cos \omega \left(t + \frac{2\pi}{\omega} \right) = 10 \cos (\omega t + 2\pi) = 10 \cos \omega t$$

Always looking for ways to simplify calculations, we ask: What single value of current can be used to represent this continuously varying current?

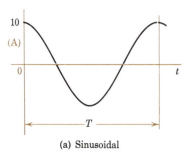 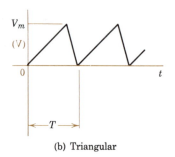

(a) Sinusoidal   (b) Triangular

Figure 3.9 Examples of periodic functions.

The value to be used to represent a varying current depends on the function to be performed by the current. If the current in Fig. 3.9a is used to actuate a relay, the maximum or *peak* value of 10 A is critical. If it is rectified (so that current flow is always in the same direction) and used to deposit silver in an electrolytic plating operation, the *average* value of 6.37 A is the significant quantity. If it is used to develop power in a resistor, the *effective* value of 7.07 A is the significant quantity. The term "peak value" is self-explanatory, but the other terms deserve consideration.

Average Value

The average value of a varying current $i(t)$ over the period T is the steady value of current I_{av} that in period T would transfer the same charge Q. By definition,

$$I_{av}T = Q = \int_t^{t+T} i(t)\, dt = \int_0^T i\, dt$$

or

$$I_{av} = \frac{1}{T}\int_0^T i\, dt \qquad (3\text{-}21)$$

In a similar way, average voltage is defined as

$$V_{av} = \frac{1}{T}\int_0^T v\, dt \qquad (3\text{-}22)$$

In dealing with periodic waves, it is understood that the average is over one complete cycle (or an integral number of cycles) unless a different interval is specified. For example, the average value of any triangular wave (Fig. 3.9b) is equal to half the peak value since the area under the curve ($\int v\, dt$) for one cycle is $\frac{1}{2}V_m T$.

Practice Problem 3-3

The height of the voltage pulses in Fig. 3.1c is 6 V, the width of the pulses is 2 ms, and the period is 5 ms. Calculate the average value of a "train" of such pulses.

Answer: 2.4 V.

For a sinusoidal wave, the average value over a cycle is zero; the charge transferred during the negative half-cycle is just equal and opposite to that transferred during the positive half-cycle. In certain practical problems we are interested in the *half-cycle* average (Fig. 3.10) given by

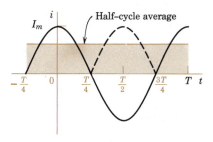

Figure 3.10   The half-cycle average.

$$I_{\text{half-cycle}} = \frac{1}{\frac{1}{2}T} \int_{-T/4}^{+T/4} I_m \cos \frac{2\pi t}{T} \, dt$$

$$= \frac{2I_m}{T} \left(\frac{T}{2\pi} \right) \sin \frac{2\pi t}{T} \Bigg|_{-T/4}^{+T/4}$$

$$= \frac{I_m}{\pi} \left[\sin \left(\frac{\pi}{2} \right) - \sin \left(-\frac{\pi}{2} \right) \right]$$

$$= \frac{2}{\pi} I_m = 0.637 I_m \qquad (3\text{-}23)$$

Effective Value

In many problems we are interested in the energy-transfer capability of an electric current. By definition, the average value of a varying power $p(t)$ is the steady value of power P_{av} that in period T would transfer the same energy W. If

$$P_{\text{av}}T = W = \int_{t}^{t+T} p(t) \, dt = \int_{0}^{T} p \, dt \quad \text{then} \quad P_{\text{av}} = P = \frac{1}{T} \int_{0}^{T} p \, dt \qquad (3\text{-}24)$$

By convention, P always means average power and no subscript is necessary.

If electrical power is converted into heat in a resistance R,

$$P = \frac{1}{T} \int_{0}^{T} p \, dt = \frac{1}{T} \int_{0}^{T} i^2 R \, dt = I_{\text{eff}}^2 R \qquad (3\text{-}25)$$

where I_{eff} is defined as the steady value of current that is equally *effective* in converting power. Solving Eq. 3-25,

$$I_{\text{eff}} = \sqrt{\frac{1}{T} \int_{0}^{T} i^2 \, dt} = I_{\text{rms}} \qquad (3\text{-}26)$$

and I_{eff} is seen to be the "square root of the mean squared value" or the *root-mean-square* current I_{rms}.

The effective or rms value of a sinusoidal current can be found from Eq. 3-26. Where $i = I_m \cos (2\pi/T)t$,

$$I_{\text{rms}}^2 = \frac{1}{T} \int_{0}^{T} I_m^2 \cos^2 \frac{2\pi t}{T} \, dt = \frac{1}{T} \frac{I_m^2}{2} \int_{0}^{T} \left(1 + \cos \frac{4\pi t}{T} \right) dt = \frac{I_m^2}{2}$$

Therefore, the effective value of a sinusoidal current is

$$I_{\text{rms}} = \frac{I_m}{\sqrt{2}} = 0.707 I_m \qquad (3\text{-}27)$$

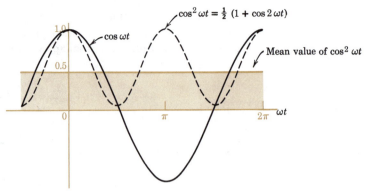

Figure 3.11   Calculating the root-mean-square value of a sinusoid.

In Fig. 3.11, note that the mean value of $\cos^2 \omega t$ is just equal to 0.5 since the average of the $\cos 2\omega t$ term is zero over a full cycle. In words, Eq. 3-27 says: "For sinusoids, the effective value is $1/\sqrt{2}$ times the maximum value." For all other functions Eq. 3-26 must be used.

Similarly, the effective value of a sinusoidal voltage is $1/\sqrt{2} = 0.707$ times the maximum voltage. In the United States, a voltage of 120 V effective at a frequency of 60 Hz is standard for residential power. The instantaneous voltage is $v(t) = 120\sqrt{2} \cos 2\pi60t \cong 170 \cos 377t$ V.

Practice Problem 3-4

Calculate the effective value of the pulse train described in Practice Problem 3-3.

Answer: 3.79 V.

ELECTRICAL INSTRUMENTS

Basically, a measuring instrument converts a physical effect into an observable quantity. The cathode-ray oscilloscope (CRO) converts an applied voltage into a deflection of the spot of impact of an electron beam. The beam has little inertia and will follow rapid voltage variations, and therefore the CRO indicates *instantaneous* values (see Figs. 4.5 and 4.10).

Moving-Coil Instruments

In the common dc ammeter (Fig. 3.12a), a current in the meter coil (suspended in a magnetic field) produces a torque opposed by a spiral spring. Although the torque is directly proportional to the instantaneous current, the high inertia of the movement (coil, support, and needle) prevents rapid acceleration and the observed deflection is proportional to the *average* current.

In a common type of ac ammeter (Fig. 3.12b), the magnetic field is set up by the current itself and the torque produced is proportional to the square of the current.

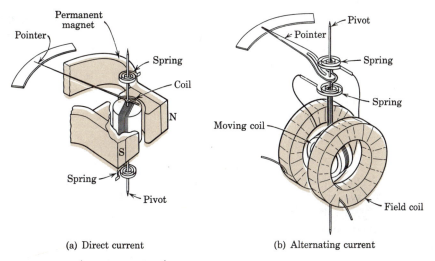

Permanent magnet

Pointer

Spring

Coil

N

S

Spring

Pivot

(a) Direct current

Pivot

Pointer

Spring

Spring

Moving coil

Field coil

(b) Alternating current

Figure 3.12   Ammeter construction.

Because of the inertia of the movement, the observed deflection is proportional to the *average squared* current and the scale is calibrated to indicate *rms* values.

If a periodic voltage is applied to a rectifier in series with a capacitance, the capacitance will tend to charge up until the applied voltage reaches its maximum. When the applied voltage decreases, the rectifier prevents a current reversal and the capacitance voltage remains at the *peak* value, which can be read by an appropriate dc instrument.

DC Meter Applications

The d'Arsonval movement, the mechanism of the dc ammeter, was invented more than one hundred years ago but it is still one of our best current-measuring devices. It is also used to measure voltage and resistance and other physical variables that are directly related to currents.

The Ammeter. Current in a branch is measured by an ammeter connected in series with the branch. One way of characterizing an ammeter is to give the value of current required to develop the torque needed to move the pointer across the scale. If a meter provides full-scale deflection with a current of 1 mA, how could it be used in a 0- to 5-A ammeter? In the connection of Fig. 3.13a, some of the meter current

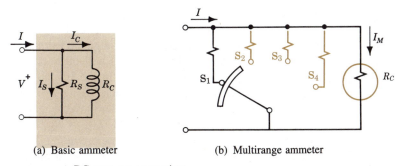

(a) Basic ammeter

(b) Multirange ammeter

Figure 3.13   DC ammeter connections.

is *shunted* through R_S and only a part of the current to be measured flows through the moving coil. If the coil resistance R_C is 20 Ω, then full-scale deflection requires a voltage of

$$V = I_C R_C = 0.001 \times 20 = 0.020 \text{ V}$$

For full-scale deflection with a meter current of 5 A, the current in the *shunt* R_S is $5 - 0.001 = 4.999$ A. The resistance of the shunt should be

$$R_S = \frac{V}{I_S} = \frac{V}{I - I_C} = \frac{0.02}{4.999} \cong 0.004 \text{ Ω} \tag{3-28}$$

A sturdy and precisely adjusted resistance may be mounted inside the meter case or be available for external connection. A circuit for a multirange ammeter is shown in Fig. 3.13b. Four shunts are available to give four ammeter ranges; a make-before-break selector switch selects the appropriate range.

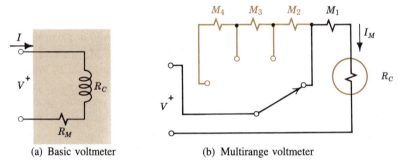

(a) Basic voltmeter   (b) Multirange voltmeter

Figure 3.14   DC voltmeter connections.

The Voltmeter. Voltage across a branch is measured by a voltmeter in parallel with the branch. Could the movement just described be used in a voltmeter? The d'Arsonval mechanism is essentially a current-measuring instrument, but it can be used as a voltmeter by adding a series resistance R_M as in Fig. 3.14a. For a 0- to 300-V voltmeter, full-scale deflection is obtained with 1 mA in the moving coil and a voltage drop of 0.02 V; the remainder of the voltage must appear across the *multiplier* resistance. Therefore,

$$R_M = \frac{V_M}{I_C} = \frac{V - V_C}{I_C} = \frac{300 - 0.02}{0.001} \cong 300,000 \text{ Ω} \tag{3-29}$$

Connecting a voltmeter or an ammeter in a circuit changes the voltages and currents in the circuit. Some current must flow through the voltmeter and some voltage must be lost across the ammeter. These effects will be negligible if the ammeter resistance is small compared to the resistance in the branch and if the voltmeter resistance is large compared to that of the branch. The *sensitivity* of a commercial voltmeter is given in ohms/volt; the 1-mA ammeter movement and the derived 300-V voltmeter have a sensitivity of $R_C/V_C = 20/0.02 = R_M/V_M = 300,000/300 = 1000$ Ω/V. For lab work, d'Arsonval movements providing 20,000 Ω/V are common; electronic instruments provide sensitivities of several megohms/volt.

Practice Problem 3-5

A meter movement with a coil resistance of 25 Ω requires 0.8 mA for full-scale deflection. Specify:

(a) The coil voltage required for full-scale deflection.
(b) The shunt resistance required for a 0- to 1-A ammeter.
(c) The multiplier resistance required for a 0- to 10-V voltmeter.

Answers: (a) 0.02 V; (b) 0.02002 Ω; (c) 12,475 Ω.

A circuit for a multirange voltmeter is shown in Fig. 3.14b. Four multiplier resistances are available to provide four voltmeter ranges; their values can be calculated from the equations

$$\frac{V_1}{I_C} = R_C + M_1 \qquad \frac{V_2}{I_C} = R_C + M_1 + M_2 \qquad (3\text{-}30)$$

and so forth.

The Ohmmeter. A useful application of the d'Arsonval movement is in measuring the dc resistance of passive circuits. A simple ohmmeter is shown in Fig. 3.15. The battery (usually 1.5 V), a fixed resistance R_1, and a variable "Ohms Adjust" resistor R_A are mounted within the ohmmeter case. With the test prods shorted, R_A is varied until the voltmeter reads full scale (usually 1 V). When the test prods are across an unknown resistance R_X, the voltage read is the voltage across R_1. Assuming $R_A + R_C + R_M$ is very large compared to R_1, the deflection as a fraction of full scale will be

$$D = \frac{V_M}{V_{\text{full scale}}} = \frac{R_1}{R_1 + R_X}$$

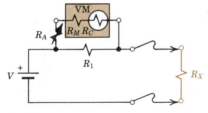

Figure 3.15   A simple ohmmeter.

Solving,

$$R_X = \frac{R_1 - DR_1}{D} = \frac{R_1}{D} - R_1 \qquad (3\text{-}31)$$

Full-scale deflection, $D = 1$, indicates zero resistance. At half scale, $D = 0.5$, $R_X = R_1/0.5 - R_1 = R_1$, and the scale midpoint is marked $R_1 = 1000\ \Omega$, say. The quarter-scale point would be marked $R_X = R_1/0.25 - R_1 = 3R_1 = 3000\ \Omega$, say.

Very high resistance values are crowded into the lower part of the scale. Because the voltmeter reading changes most rapidly with a given percentage change in resistance when $R_X = R_1$, midscale readings are most precise. A selector switch is used to change scales by selecting values of R_1; typical ranges are for midscale readings of 10 Ω, 1 kΩ, and 100 kΩ. Practical ohmmeters use a more complicated circuit but the general principle of resistance measurement is the same.

The Multimeter. Since the same d'Arsonval movement can serve for current, voltage, and resistance measurements, an effective arrangement is to include shunts, multipliers, and battery in the same case along with the necessary selector switch. Such a combination volt-ohm-milliammeter (VOM) is called a *multimeter*. In addition to dc quantities, ac voltages can be measured by using a rectifier circuit.

Rectifier Instruments. Semiconductor diodes can be used to convert alternating current to direct current, which is measured by a d'Arsonval movement. In the full-wave bridge circuit (Fig. 3.16), an alternating current at the terminals produces a unidirectional current through the moving coil and the deflection is proportional to the average current. These instruments are calibrated on sinusoidal waves to read rms values. Since the rms value of the sinusoid is $0.707/0.636 = 1.11$ times the average value, the instrument indicates 1.11 times the average value of the rectified waveform of any applied current. It is accurate for sinusoids only; for any other waveform a correction must be applied. (See Exercise 17.)

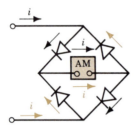

Figure 3.16 A rectifier-type instrument.

IDEAL AMPLIFIERS

Amplification is one of the principal functions of electronic circuits. When voltage or current signals are applied to the input terminals of an amplifier, larger voltage or current signals are available at the output terminals.

Ideal Transformers

As Henry discovered (see p. 13), a changing magnetic field induces a voltage in any coil that it links; the induced voltage is proportional to the number of turns in the coil and the rate of change of magnetic flux. If the two coils of a *transformer* are wound on the same highly permeable iron core, both coils are linked by essentially the same magnetic flux ϕ. Because $v = N\, d\phi/dt$, a sinusoidally varying flux in the core induces sinusoidal voltages in the windings proportional to the number of turns. In the ideal transformer of Fig. 3.17 (closely approximated by a real transformer), the voltage ratio (effective values) is equal to the turn ratio a, or

$$\frac{V_2}{V_1} = \frac{N_2}{N_1} = a \tag{3-32a}$$

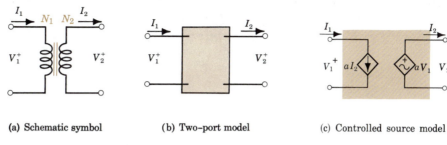

(a) Schematic symbol   (b) Two–port model   (c) Controlled source model

Figure 3.17   An ideal transformer.

The current ratio can be predicted by noting that in an ideal transformer there is no power loss, and the power in must equal the power out. Therefore, a *step-down* in current must accompany a *step-up* in voltage or

$$\frac{I_2}{I_1} = \frac{V_1}{V_2} = \frac{N_1}{N_2} = \frac{1}{a} \tag{3-32b}$$

To simplify the analysis of an electrical device, we may replace the real device by a *model*. The *circuit model* of a device consists of circuit elements arranged so that the terminal behavior of the model is the same as that of the device. In Fig. 3.17c, the voltage relation, Eq. 3-32a, is satisfied by using a *controlled voltage source*; the output voltage is *a* times the input voltage. The current relation, Eq. 3-32b, is satisfied by using a *controlled current source*. These controlled or *dependent* sources, indicated by diamond symbols, show that a variable in one part of a circuit is determined by a variable in another part of the circuit.

Practice Problem 3-6

A transformer with 2000 turns on the "primary" and 400 turns on the "secondary" is connected to a "load" represented by $R_L = 12 \ \Omega$. The primary is connected to 120 V ac. Draw the circuit model for this situation and predict the output voltage and the input current.

Answers: 24 V; 0.4 A.

Real and Ideal Amplifiers

The transformer is a *passive* device; it provides voltage *gain,* but the output power can never exceed the input power. To obtain power gain, we depend on *active* electronic devices such as transistors; by supplying a transistor with easily available dc power from a battery, say, we can amplify the power of an ac signal. Using a few transistors, we can design a receiver that can pick up a weak radio signal and amplify it until it will drive a loudspeaker.

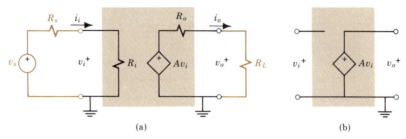

Figure 3.18   Real and ideal amplifiers.

In the circuit model of a real amplifier in Fig. 3.18a, input resistance R_i determines how much input current results from an input voltage v_i. The diamond symbol tells us that Av_i is a controlled voltage source; the voltage available in one part of the circuit is determined by a voltage in another part of the circuit. Here the voltage appearing in the output circuit, consisting of output resistance R_o and load resistance R_L, is controlled by the input voltage v_i. With $R_L = \infty$, output current $i_o = 0$, and voltage Av_i appears at the output terminals; factor A is the open-circuit voltage amplification or gain of the amplifier.

In a practical case, the fraction of the source voltage v_s that appears at the input terminals is $R_i/(R_s + R_i)$. The fraction of the internally developed voltage appearing at the output terminals is $R_L/(R_o + R_L)$. The gain of the real amplifier is

$$A_r = \frac{v_o}{v_s} = \frac{Av_s[R_i/(R_s + R_i)][R_L/(R_L + R_o)]}{v_s} = A \cdot \frac{R_i}{R_s + R_i} \cdot \frac{R_L}{R_L + R_o} \qquad (3\text{-}33)$$

For good performance, R_i should be large and R_o should be small. R_i should be large so that $v_i \simeq v_s$ and is unaffected by R_s; R_o should be small so that $v_o \simeq Av_i$ and is unaffected by R_L. In the *ideal amplifier* of Fig. 3.18b, $R_i = \infty$ and $R_o = 0$; therefore, $v_o = Av_i = Av_s$ or voltage gain $= v_o/v_s = v_o/v_i = A$. In this ideal amplifier A is assumed to be constant, independent of the magnitude or the frequency of the input signal.

EXAMPLE 5

A real amplifier is characterized by $R_i = 1$ MΩ, $R_o = 10$ Ω, and $A = 100$. Determine the voltage gain in a circuit where $R_s = R_L = 1$ kΩ.

By Eq. 3-33, the gain of the real amplifier is

$$A_r = A \frac{R_i}{R_s + R_i} \cdot \frac{R_L}{R_L + R_o}$$

$$= 100 \frac{10^6}{10^3 + 10^6} \cdot \frac{10^3}{10^3 + 10}$$

$$= 100 \frac{1000 \times 10^3}{1001 \times 10^3} \cdot \frac{1000}{1010} \simeq 99$$

With little error, we can assume that this amplifier is ideal.

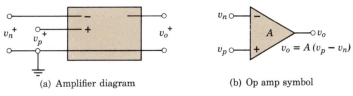

| (a) Amplifier diagram | (b) Op amp symbol |

Figure 3.19   Operational amplifier diagram and symbol.

Operational amplifiers (op amps), first used in analog computers to perform mathematical operations, are now available in high-performance, low-cost, integrated-circuit form. In the *differential* amplifier of Fig. 3.19a, the output voltage is A times $v_p - v_n$, the difference in potential between positive and negative input terminals. In the conventional symbol of Fig. 3.19b, it is understood that all voltages are measured with respect to ground and the ground terminals at input and output are omitted for convenience.

We can buy, for less than a dollar, differential op amps in the form of Fig. 3.19 with very high R_i, very low R_o, and voltage gains A of 10^5 or more. They are widely used in amplification, instrumentation, and waveform generation. In many applications inexpensive op amps can be assumed to be "ideal" circuit elements. The design of an amplifier, for example, is reduced to selecting the commercial unit that is nearly ideal in the parameters that are significant in a given situation.

Basic Inverting Circuit

In the amplifier circuit[†] of Fig. 3.20, the input signal is applied to the negative or inverting terminal and the positive or noninverting terminal is grounded. The input

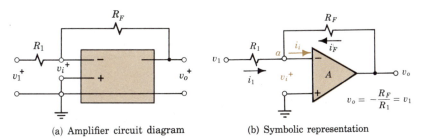

| (a) Amplifier circuit diagram | (b) Symbolic representation |

Figure 3.20   Basic inverting amplifier circuit.

voltage v_1 is applied in series with resistance R_1, and the output voltage v_o is "fed back" through resistance R_F. Because the op amp gain A is very, very large and v_o is practically limited to 10 V or so, $v_i = -v_o/A \cong 0$. Because input resistance R_i is very large, $i_i = v_i/R_i \cong 0$.

For an ideal amplifier, closely approximated by a practical op amp, $R_i = \infty$, $R_o = 0$, $A = \infty$, $v_i = 0$, and $i_i = 0$.

[†] Unfortunately, the term *operational amplifier* is applied to the basic high-gain amplifier as well as to the functional circuit in which it is employed. To lessen confusion, we shall start off by specifying *circuit* when the latter is intended. Later on, we shall use op amp for both, as is done in practice.

Therefore, the sum of the currents into node a in Fig. 3.20b is

$$i_1 + i_F = \frac{v_1}{R_1} + \frac{v_o}{R_F} = 0 \qquad (3\text{-}34)$$

and

$$v_o = -\frac{R_F}{R_1} v_1 \quad \text{or} \quad \frac{v_o}{v_1} = A_F = -\frac{R_F}{R_1} \qquad (3\text{-}35)$$

This relation describes an *inverting amplifier* circuit; the circuit gain with feedback is $A_F = -R_F/R_1$. (As we shall see, feedback is widely used in electronic circuits to improve their performance.)

From this analysis, we can draw the following conclusions:

1. For an ideal amplifier in the inverting amplifier circuit, $v_i = 0$; input current $i_1 = v_1/R_1$ and is independent of R_F. Therefore, the input is isolated from the output. The input resistance of this feedback circuit is just $R_{iF} = v_1/i_1 = R_1$.
2. For an ideal amplifier with infinite gain or a practical op amp with very high gain, the gain of the amplifier circuit A_F is determined by the external resistors R_1 and R_F only.

The first conclusion leads to a simple circuit model for the inverting amplifier circuit; the output voltage $v_o = -R_F(v_1/R_1) = -i_1 R_F$ is represented by a controlled voltage source in Fig. 3.21. The second conclusion indicates an easy way for us to design an amplifier.

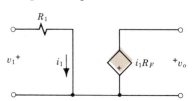

Figure 3.21 Circuit model for an inverting amplifier.

EXAMPLE 6

Design an amplifier with a voltage gain of -100, an input resistance of 1000 Ω, and a flat response over the audiofrequency range.

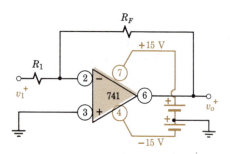

Figure 3.22 Amplifier design.

In this case, "design" means selecting an appropriate op amp, specifying the feedback network, and providing the necessary power supply.

The type 741 (see appendix p. 462) is a high-performance op amp suitable for a variety of analog applications at audio frequencies. It has dual-in-line pin connections as shown in Fig. 3.22 and operates from ± 15-V power supplies or batteries.

The network design proceeds as follows:

$$R_1 = R_{iF} = 1000 \ \Omega$$

$$R_F = -R_1 \times A_F = R_1 \times 100 = 100 \ k\Omega$$

For noncritical applications, this circuit is adequate.

The gain $A_F = -R_F/R_1$ is determined by external resistors and is independent of A as long as A is high.

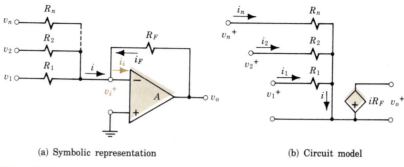

(a) Symbolic representation   (b) Circuit model

Figure 3.23   Summing circuit; $v_o = -[v_1(R_F/R_1) + v_2(R_F/R_2) + \cdots + v_n(R_F/R_n)]$.

Summing Circuit

If the circuit of Fig. 3.20 is modified to permit multiple inputs, the operational amplifier can perform addition. In Fig. 3.23, the input current is supplied by several voltages through separate resistances. In the practical case, $v_i = 0$ and $i_i = 0$ and the circuit model is as shown in Fig. 3.23b. The sum of the input currents is just equal and opposite to the feedback current or

$$i = i_1 + i_2 + \cdots + i_n = -i_F \tag{3-36}$$

Hence

$$\frac{v_1}{R_1} + \frac{v_2}{R_2} + \cdots + \frac{v_n}{R_n} = -\frac{v_o}{R_F}$$

or

$$v_o = -\left(\frac{R_F}{R_1}v_1 + \frac{R_F}{R_2}v_2 + \cdots + \frac{R_F}{R_n}v_n\right) \tag{3-37}$$

As would be expected from looking at the circuit model, the output of the *summing circuit* is the weighted sum of the inputs. Note that a signal can be subtracted by first passing it through an analog *inverter* consisting of an inverting amplifier with $R_F = R_1$.

Basic Noninverting Circuit

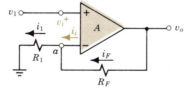

Figure 3.24   The noninverting amplifier circuit.

If the basic amplifier is reconnected as in Fig. 3.24, the behavior is modified. Here the signal is applied to the non-inverting + terminal and a fraction of the output signal is fed back to the − terminal. Since $v_i = 0$ and $i_i = 0$, the governing equations are

$$v_1 = v_a + v_i = v_a \quad \text{and} \quad v_a = \frac{R_1}{R_1 + R_F}v_o \tag{3-38}$$

where R_1 and R_F constitute a voltage divider connected across the output voltage.

The gain of the *noninverting amplifier* circuit is

$$A_F = \frac{v_o}{v_1} = \frac{R_1 + R_F}{R_1} \tag{3-39}$$

Here again the gain is determined by the feedback network elements R_1 and R_F. For the noninverting circuit, however, the gain is positive and equal to or greater than unity.

Practice Problem 3-7

An ideal amplifier is used in the circuit of Fig. 3-24 with $R_1 = 5$ kΩ and $R_F = 500$ kΩ.

(a) For an input voltage $v_1 = 0.01$ V, what is v_a? What is i_i?
(b) Considering the voltage divider R_1-R_F, what must v_o be?
(c) Calculate the voltage gain v_o/v_i by Eq. 3-39.

Answers: (a) 0.01 V, 0; (b) 1.01 V; (c) 101.

EXAMPLE 7

An op amp is used in the circuit of Fig. 3.25, where input signal v_s is a function of time (a sinusoid or a combination of sinusoids). Derive an expression for $i_o(t)$.

Because no current flows into an ideal op amp, there is no drop across R_s, and voltage $v_s(t)$ appears between the $+$ terminal and ground.

Because there is no input voltage across an ideal op amp, $v_a = v_s = i_1 R_1$. But current i_1 must be supplied by the feedback current i_o. Hence

$$i_o(t) = i_1(t) = \frac{1}{R_1} v_s(t)$$

This noninverting circuit is serving as a voltage-to-current converter. The output current is proportional to the input voltage and independent of R_s and R_o.

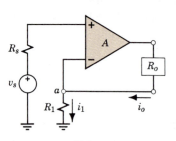

Figure 3.25 Voltage-current converter.

IDEAL DIODES

A useful addition to our repertoire of electronic circuit elements is the *diode* or *rectifier*. A diode is a two-terminal device that acts as a switch; it permits current to flow readily in one direction but it tends to prevent the flow of current in the other

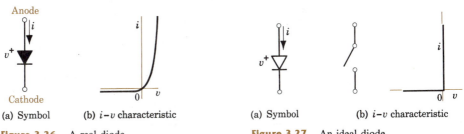

(a) Symbol   (b) i–v characteristic   (a) Symbol   (b) i–v characteristic

Figure 3.26 A real diode.   **Figure 3.27** An ideal diode.

direction. The direction of easy current flow is indicated by the arrowhead in the symbol of Fig. 3.26a. This is a nonlinear circuit element, and it is useful because it is *not* linear. However, the analysis of circuits containing diodes may be complicated because network theorems based on linearity cannot be used. Figure 3.26b shows the i–v characteristic of a semiconductor diode consisting of a junction of dissimilar materials that we will study later. Just as we use ideal models to represent real R, L, and C components, so we can approximate a real diode by the ideal characteristics shown in Fig. 3.27b. When the anode is positive with respect to the cathode, that is, when voltage v is positive, the "switch" is closed and unlimited current i can flow with no voltage drop. In contrast, when the cathode is positive with respect to the anode, the "switch" is open and no current flows even for large negative values of voltage v.

Rectifiers

The nonlinear characteristic of a diode is used to convert alternating current into unidirectional, but pulsating, current in the process called *rectification*. The pulsations are removed in a frequency-selective circuit called a *filter*. Rectifier circuits employ one, two, or four diodes to provide various degrees of rectifying effectiveness. Filter circuits use the energy-storage capabilities of inductors and capacitors to smooth out the pulsations and to provide a steady output current. A combination of ac source, rectifier, and filter is called a *power supply*.

Half-Wave Rectifier. Ideally, a diode should conduct current freely in the forward direction and prevent current flow in the reverse direction. Practical diodes only approach the ideal. Semiconductor diodes, for example, present a small but appreciable voltage drop in the forward direction and permit a finite current to flow in the reverse direction. For most calculations, the reverse current flow is negligibly small and the forward voltage drop can be neglected with little error.

A practical circuit for *half-wave rectification* is shown in Fig. 3.28a. A transformer supplied from 120-V, 60-Hz house current provides the desired operating voltage, which is applied to a series combination of diode and load resistance R_L. For approximate analysis, the actual diode is represented by an ideal diode; the internal

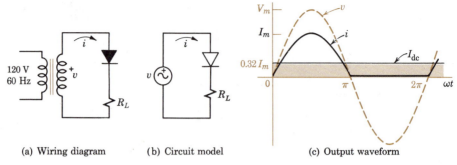

(a) Wiring diagram   (b) Circuit model   (c) Output waveform

Figure 3.28   A half-wave rectifier.

resistance of the transformer is neglected. For $V = V_m \sin \omega t$, the resulting current is

$$
\begin{cases}
i = \dfrac{v}{R_L} = \dfrac{V_m \sin \omega t}{R_L} & \text{for } 0 \leq \omega t \leq \pi \\[2ex]
i = 0 & \text{for } \pi \leq \omega t \leq 2\pi
\end{cases}
\tag{3-40}
$$

as shown in Fig. 3.28c.

The purpose of rectification is to obtain a unidirectional current. The dc component of the load current is the average value or

$$
I_{dc} = \frac{1}{2\pi} \int_0^{2\pi} i\, d(\omega t) = \frac{1}{2\pi} \int_0^{\pi} \frac{V_m \sin \omega t}{R_L} \, d(\omega t) + 0
$$

$$
= \frac{1}{2\pi} \frac{V_m}{R_L} \left[-\cos \omega t \right]_0^{\pi} = \frac{V_m}{\pi R_L} = \frac{I_m}{\pi}
\tag{3-41}
$$

The current through the load resistance consists of half-sinewaves, and the dc component is approximately 30% of the maximum value.

EXAMPLE 8

An ideal diode is connected in the circuit of Fig. 3.29. For $v = 170 \sin \omega t$ V, predict the current through $R = 5$ kΩ.

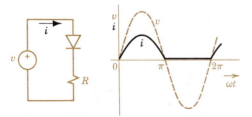

Figure 3.29   Application of diode.

For $0 < \omega t < \pi$, v is positive, the diode switch is closed, and

$$
i(t) = \frac{v}{R} = \frac{170 \sin \omega t}{5000} = 34 \sin \omega t \text{ mA}
$$

For $\pi < \omega t < 2\pi$, v is negative, the diode switch is open, and current $i = 0$. The resulting current is shown by the solid line. The sinusoidal voltage wave has been *rectified*.

The dc component of the rectified current is (by Eq. 3-41)

$$
I_{dc} = \frac{I_m}{\pi} = \frac{34}{\pi} = 10.8 \text{ mA}
$$

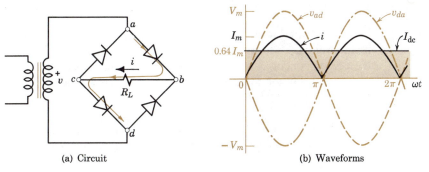

(a) Circuit (b) Waveforms

Figure 3.30 A full-wave bridge rectifier.

Full-Wave Rectifier. The bridge rectifier circuit of Fig. 3.30 provides a greater dc value from the same transformer voltage. When the transformer voltage $v = v_{ad}$ is positive, the current flow is along path $abcd$ as shown and a half-sinewave of current results. When the applied voltage reverses, the voltage $v_{da} = -v_{ad}$ is positive and the current flow is along path $dbca$. The current through the load resistance is always in the same direction, and the dc component is twice as large as in the half-wave rectifier or

$$I_{dc} = \frac{2}{\pi}\frac{V_m}{R_L} = \frac{2I_m}{\pi} \qquad (3\text{-}42)$$

The bridge circuit is disadvantageous because four diodes are required, and two diodes and their power-dissipating voltage drops are always in series with the load. The full-wave rectifier circuit of Fig. 3.31 uses a more expensive transformer to produce the same result with only two diodes and with higher operating efficiency. The second output winding on the transformer provides a voltage v_2 that is 180° out of phase with v_1; such a *center-tapped* winding serves as a *phase inverter*. While v_1 is positive, current i_1 is supplied through diode 1; while v_1 is negative, no current flows through diode 1 but v_2 is positive and, therefore, current i_2 is supplied through diode 2. The current through the load resistance is $i_1 + i_2$ and $I_{dc} = 2I_m/\pi$ as in the bridge circuit.

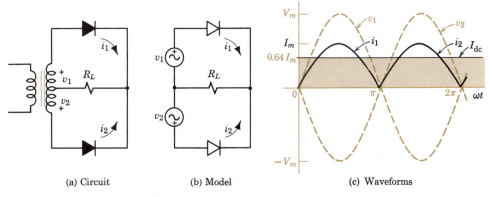

(a) Circuit (b) Model (c) Waveforms

Figure 3.31 A full-wave rectifier with phase inverter.

Filters

The desired result of rectification is direct current, but the output currents of the rectifier circuits described obviously contain large alternating components along with the dc component. Using full-wave instead of half-wave rectification reduces the ac component, but the remaining *ripple voltage* across R_L is still unsatisfactory for most electronic applications.

Capacitor Filter. The ripple voltage can be greatly reduced by a *filter* consisting of a capacitor shunted across the load resistor. The capacitor can be thought of as a tank that stores charge during the period when the diode is conducting and releases charge to the load during the nonconducting period. From Eq. 1-12, we know that the stored charge in Fig. 3.32a is $q = Cv_L$.

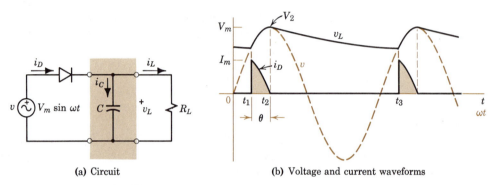

(a) Circuit   (b) Voltage and current waveforms

Figure 3.32   A capacitor filter.

If the diode is nearly ideal and if the steady state has been reached, the operation is as shown in Fig. 3.32b. At time $t = 0$, the source voltage v is zero but the load voltage $v_L = v_C$ is appreciable because the previously charged capacitor is discharging through the load. At $t = t_1$, the increasing supply voltage v slightly exceeds v_L and the diode conducts. The diode current i_D rises abruptly to satisfy the relation $i_C = C\, dv/dt$ and then decreases to zero; the diode switches off when v drops below v_L. During the charging period, $t_1 < t < t_2$,

$$v_L = V_m \sin \omega t \tag{3-43}$$

During the discharging period, $t_2 < t < t_3$, the capacitor voltage decays exponentially so that

$$v_L = V_2\, e^{-(t-t_2)/R_L C} \tag{3-44}$$

with a time constant defined by the RC circuit. At time t_3, the supply voltage again exceeds the load voltage and the cycle repeats.

The load current i_L is directly proportional to load voltage v_L. Because i_L never goes to zero, the average value or dc component is relatively large as compared to the half-wave rectifier alone and the ac component is correspondingly lower. The ripple voltage is greatly reduced by the use of the capacitor.

Capacitor Filter–Approximate Analysis. If the time constant $R_L C$ is large compared to the period T of the supply voltage, the decay in voltage $v_C = v_L$ will be small, and the ripple voltage $V_r = \Delta v_C$ will be small. The magnitude of the ripple can be estimated by assuming that V_r is small, the charging interval $t_2 - t_1$ is small, and v_C is nearly constant. Under this assumption, all the load current is supplied by the capacitor, and the charge transferred to the load is

$$\Delta q = I_{dc} T = C \Delta v_C = C V_r \qquad (3\text{-}45)$$

Solving for the ripple voltage,

$$V_r = \frac{I_{dc} T}{C} = \frac{I_{dc}}{fC} = \frac{V_{dc}}{fR_L C} \qquad (3\text{-}46)$$

This relation holds for a half-wave rectifier; a similar equation can be derived for a full-wave rectifier. By using this approximation, the performance of an existing filter can be predicted or a filter can be designed to meet specifications as in Example 9.

EXAMPLE 9

A load ($R_L = 3330\ \Omega$) is to be supplied with 50 V at 15 mA with a ripple voltage no more than 1% of the dc voltage. Design a rectifier–filter to meet these specifications.

Assuming a 120-V, 60-Hz supply and the half-wave rectifier with capacitor filter of Fig. 3.32a, Eq. 3-46 is applicable. Solving for C,

$$C = \frac{V_{dc}/V_r}{fR_L} = \frac{100}{60 \times 3330} = 500\ \mu F$$

For a peak value of 50 V, the rms value of transformer secondary voltage should be $50/\sqrt{2} = 35.4$ V. The transformer turn ratio should be

$$\frac{N_1}{N_2} = \frac{V_1}{V_2} = \frac{120}{35.4} = 3.39$$

Practice Problem 3-8

Redraw Fig. 3.32b to show the load voltage V_L and the diode current i_D for a full-wave rectifier-capacitor filter supply. For a full-wave rectifier:

(a) How many current pulses are there per second?
(b) Derive the equation for V_r for this case.
(c) What size capacitor is needed to meet the specifications of Example 9?

Answers: (a) $2f$; (b) $V_r = V_{dc}/2fR_L C$; (c) $250\ \mu F$.

WAVESHAPING CIRCUITS

A major virtue of electronic circuits is the ease, speed, and precision with which voltage and current waveforms can be controlled. Some of the basic waveshaping functions are illustrated by a radar pulse-train generator. The word *radar* stands for *radio detection and ranging*. A very short burst of high-intensity radiation is transmitted in a given direction; a return echo indicates the presence, distance, direction, and speed of a reflecting object. The operation of a radar system requires a precisely formed series of timing pulses. Typically, these may be of 5-μs duration with a repetition rate of 500 pulses per second. Starting with a 500-Hz sinusoidal generator, the pulse train could be developed as shown in Fig. 3.33. Can you visualize some relatively simple electronic circuits that would perform the indicated functions?

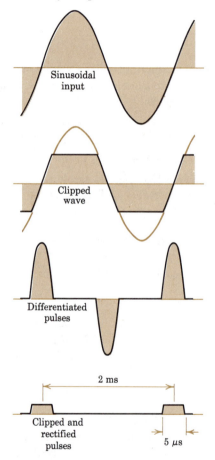

Figure 3.33 Generation of a train of timing pulses.

Clipping

To remove an undesired portion of a signal, we can use a *clipping circuit* consisting of a diode, a resistance, and a voltage source. In Fig. 3.34, the input signal v_1 varies with time as shown; we are interested in output signals v_R and v_D. The sum of the voltages around the loop is zero and the voltage relations are:

$$v_D + v_R = v_1 - V \qquad v_R = v_1 - V - v_D \qquad v_D = v_1 - V - v_R \qquad (3\text{-}47)$$

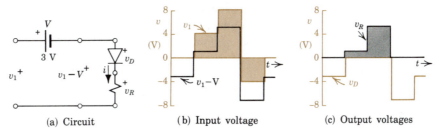

(a) Circuit   (b) Input voltage   (c) Output voltages

Figure 3.34 A clipping circuit.

The behavior of the circuit depends on the state of the diode switch; it is closed when $v_D + v_R = v_1 - V$ (plotted in Fig. 3.34b) is positive. With the diode switch closed, $v_D = 0$ and $v_1 - V$ appears across R. At all other times, v_D is negative and the diode is open; no curent flow in R and $v_R = 0$; this part of the signal has been *clipped*. In other words, the battery shifts the signal down and the diode cuts it off. The wave-shaping function performed by the circuit depends on how we arrange the circuit components.

Practice Problem 3-9

(a) In the circuit of Fig. 3.34a, $v_1 = 6 \sin \omega t$ V. Sketch v_R for one cycle.
(b) Reverse the diode connections and repeat part (a).

Answers: (a) The signal below $v_1 = +3$ V is clipped; (b) $v_R = 0$ for $30° < \omega t < 150°$, $v_R = 6 \sin \omega t - 3$ V elsewhere.

One common type of *clipping* circuit provides an output voltage v_2 equal to (or proportional to) the input voltage v_1 up to a certain value V; above V the wave is clipped off. If both positive and negative peaks are to be clipped, the desired *transfer characteristic* v_2 versus v_1 is as shown in Fig. 3.35a. The diode circuit for this type of clipping is shown in Fig. 3.35b. The *bias* voltages are set so that diode A conducts

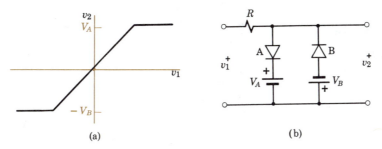

(a)   (b)

Figure 3.35 Clipping characteristic and circuit.

whenever $v_1 > V_A$ and diode B conducts whenever $v_1 < -V_B$. When $-V_B < v_1 < V_A$, neither diode conducts and voltage v_1 appears across the output terminals. When either diode is conducting, the difference between v_1 and v_2 appears as a voltage drop across R. The effect of an asymmetric clipping circuit on a sinewave is illustrated in Example 10.

EXAMPLE 10

A sinewave $v_1 = 20 \sin \omega t$ V is applied to the circuit of Fig. 3.36a. Draw the transfer characteristic and predict the output voltage v_2.

In this circuit the first diode and the 10-V battery provide clipping for voltages greater than $+10$ V. With $V_B = 0$, the second diode provides clipping of all negative voltages or rectification. The transfer characteristic and the resulting output are as shown in Fig. 3.36b.

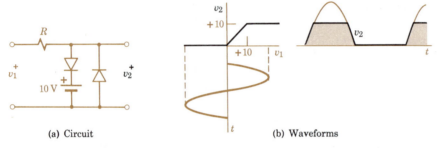

| (a) Circuit | (b) Waveforms |

Figure 3.36   A clipping and rectifying circuit.

Clamping

To provide satisfactory pictures in television receivers, the peak values of certain variable signal voltages must be held or *clamped* at predetermined levels. In passing through ordinary amplifiers, the dc reference level is lost and a *clamper* or *dc restorer* is necessary to return the signal to its original form.

In Fig. 3.32 on page 84, when applied voltage v is positive the diode allows current to flow until capacitor C is charged to the maximum voltage V_m. In the circuit of Fig. 3.37a, if R is small the capacitor tends to charge up to the positive peak value of the input wave, just as in the half-wave rectifier with capacitor. When the polarity

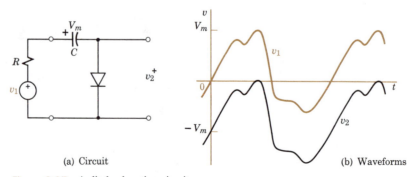

| (a) Circuit | (b) Waveforms |

Figure 3.37   A diode-clamping circuit.

of v_1 reverses, the capacitor voltage remains at V_m because the diode prevents current flow in the opposite direction. Neglecting the small voltage across R, the output voltage across the diode is

$$v_2 = v_1 - V_m \tag{3-48}$$

The signal waveform is unaffected, but a dc value just equal to the peak value of the signal has been introduced. The positive peak is said to be clamped at zero.

Practice Problem 3-10

The signal v_1 in Fig. 3.37 has a positive maximum $V_m = +12$ V and a negative maximum of -8 V.

(a) To what voltage v_C does the capacitor charge up?
(b) After C is charged, what is the value of v_2 when $v_1 = +6$ V?
(c) Reverse the diode connections in Fig. 3.37a and repeat parts (a) and (b).

Answers: (a) 12 V; (b) -6 V; (c) -8 V, -14 V.

If the amplitude of the input signal in Fig. 3.37 changes, the dc voltage across C also changes (after a few cycles[†]) and the output voltage again just touches the axis. If the diode is reversed, the negative peaks are clamped at zero. If a battery is inserted in series with the diode, the reference level of the output may be maintained at voltage V_B.

Clamping and rectifying are related waveshaping functions performed by the same combination of diode and capacitor. In the rectifier the variable component is rejected and the dc value is transmitted; in the clamper the variable component is transmitted and the dc component is restored.

EXAMPLE 11

Design a circuit that will clamp the minimum point of any periodic signal to -5 V.

By Eq. 3-48, the output is to be

$$v_2 = v_1 + (V_{min} - 5)$$

Therefore, the capacitor must charge up to the voltage $v_C = V_{min} - 5$ with the polarity shown in Fig. 3.38.

When the input signal is negative with a magnitude greater than 5 V, current must flow through the diode, which must be connected as shown.

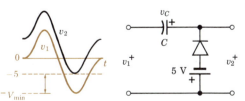

Figure 3.38 Designing a clamping circuit.

[†]In practical clampers, a high resistance across the diode allows the capacitor to change voltage slowly.

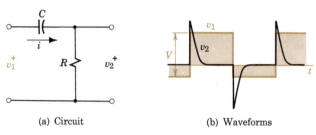

(a) Circuit   (b) Waveforms

Figure 3.39   Differentiating circuit.

Differentiating

Within limits the simple circuit of Fig. 3.39 provides an output that is the derivative of the input. For the special case of a rectangular input voltage wave, the output voltage is proportional to the capacitor charging current in response to a step input voltage. In this case a linear circuit transforms a rectangular wave into a series of short pulses if the time constant RC is small compared to the period of the input wave.

The general operation of this circuit is revealed if we make some simplifying assumptions. Applying Kirchhoff's voltage law to the left-hand loop,

$$v_1 = v_C + v_R \cong v_C \tag{3-49}$$

if v_R is small compared to v_C. Because $i_C = C\, dv_C/dt$,

$$v_2 = v_R = Ri = RC\frac{dv_C}{dt} \cong RC\frac{dv_1}{dt} \tag{3-50}$$

The output is approximately proportional to the derivative of the input.

Integrating

From our previous experience with circuits we expect that if differentiating is possible, integrating is also. In Fig. 3.40, a square wave of voltage has been applied long enough for a cyclic operation to be established. The time constant RC is a little greater than the half-period of the square wave. The capacitor C charges and discharges on alternate half-cycles, and the output voltage is as shown.

If the time constant RC is large compared to the period T of the square waves, only the straight portion of the exponential appears and the output is the *sawtooth* wave in which voltage is directly proportional to time. In general,

$$v_1 = v_R + v_C \cong v_R = iR \tag{3-51}$$

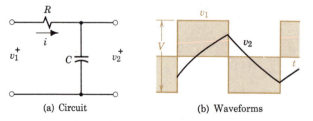

(a) Circuit   (b) Waveforms

Figure 3.40   Integrating circuit.

if v_C is small compared to v_R (i.e., $RC > T$). Because capacitor voltage is proportional to the integral of current $i \cong v_1/R$,

$$v_2 = \frac{1}{C} \int i \, dt \cong \frac{1}{RC} \int v_1 \, dt \qquad (3\text{-}52)$$

and the output is approximately proportional to the integral of the input. If it is necessary, the magnitude of the signal can be restored by linear amplification.

Op Amp Integrator and Differentiator

In the previous discussion of op amps we assumed that the feedback network is purely resistive. In general, however, the network may contain capacitances, inductances, and resistances. Because of the high amplifier gain in combination with feedback, the mathematical operations performed are precise.

In Fig. 3.41a, the feedback element is a capacitor. For the op amp, $v_i \cong 0$, node n is at ground potential, $i_i \cong 0$, and the sum of the currents into node n is

$$\frac{v_1}{R_1} + C \frac{dv_o}{dt} = 0 \qquad \text{or} \qquad dv_o = -\frac{v_1}{R_1 C} \, dt \qquad (3\text{-}53)$$

Integrating each term with respect to time and solving,

$$v_o = -\frac{1}{R_1 C} \int v_1 \, dt + \text{a constant} \qquad (3\text{-}54)$$

and the device is an *integrator*. The analog integrator is very useful in computing, signal processing, and signal generating. (See Example 12 on page 92.)

If the resistance and capacitance are interchanged as in Fig. 3.41b, the sum of the currents is

$$C_1 \frac{dv_1}{dt} + \frac{v_o}{R} = 0 \qquad (3\text{-}55)$$

Solving,

$$v_o = -RC_1 \frac{dv_1}{dt} \qquad (3\text{-}56)$$

and the output voltage is proportional to the derivative of the input. For practical reasons involving instability and susceptibility to noise, the differentiator is not so useful as the integrator.

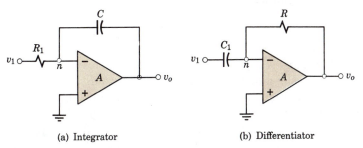

(a) Integrator   (b) Differentiator

Figure 3.41   Op amp integration and differentiation circuits.

EXAMPLE 12

Predict the output voltage of the circuit shown in Fig. 3.42 where the block represents an ideal amplifier with A very large.

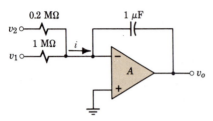

Figure 3.42 Integrator application.

For A very large, v_i is very small and representation as an integrator is accurate. Here $i = i_1 + i_2$ and

$$v_o = -\frac{1}{C} \int \left(\frac{v_1}{R_1} + \frac{v_2}{R_2} \right) dt$$

$$= -\frac{1}{CR_1} \int \left(v_1 + v_2 \frac{R_1}{R_2} \right) dt$$

$$= -\int (v_1 + 5v_2) \, dt$$

The output is the integral of a weighted sum.

SUMMARY

- Exponentials and sinusoids are important waveforms because they occur frequently, they are easy to handle mathematically, and they are used to represent other waves.
  The general decaying exponential is $a = A\, e^{-t/T}$.
  The time constant T is a measure of the rate of decay.
  For $t = T$, $a/A = 1/e = 0.368$; for $t = 5T$, $a/A = 0.0067$ (negligible).

- The response of a series RC circuit to a voltage step is

$$v_C = V_s(1 - e^{-t/RC}) \qquad \text{and} \qquad i_C = \frac{V_s}{R} e^{-t/RC}$$

- In a periodic function of time, $f(t + T) = f(t)$, where T is the period.
  The general sinusoid is $a = A \cos(\omega t + \alpha)$.
  Frequency $\omega = 2\pi f$ rad/s; $f = 1/T$ Hz, where period T is in seconds.
  The average value of a periodic current is $I_{av} = (1/T) \int_0^T i\, dt$.
  For a sinusoid, the half-cycle average is $2I_m/\pi = 0.637I_m$.
  The effective or rms value of a periodic current is $I = \sqrt{(1/T) \int_0^T i^2\, dt}$.
  For a sinusoid, the effective value is $I = I_m/\sqrt{2} = 0.707I_m$.

- In an ideal transformer, $P_{in} = P_{out}$ and

$$\frac{V_2}{V_1} = \frac{N_2}{N_1} = a \qquad \frac{I_2}{I_1} = \frac{V_1}{V_2} = \frac{N_1}{N_2} = \frac{1}{a}$$

- An ideal amplifier is characterized by infinite input resistance, zero output resistance, and constant gain; an ideal op amp has infinite gain.
  Feedback circuits are used with op amps to obtain inverting and noninverting amplifiers, summing circuits, integrators, and differentiators.

- Essentially, a diode discriminates between forward and reverse voltages.
  The ideal diode presents zero resistance in the forward direction and infinite resistance in the reverse direction; it functions as a selective switch.

■ A rectifier converts alternating current into unidirectional current.
In half-wave rectification with a resistive load, $I_{dc} \cong V_m/\pi R_L = I_m/\pi$.
A bridge circuit or phase inverter permits full-wave rectification; $I_{dc} = 2I_m/\pi$.

■ A capacitor filter stores charge on voltage peaks and delivers charge during voltage valleys; with half-wave rectification, the ripple voltage is $V_r \cong V_{dc}/fCR_L$.

■ Waveforms can be shaped easily, rapidly, and precisely.
A diode–resistor–battery circuit can perform clipping.
A diode and peak-charging capacitor can clamp signals to desired levels.
An op amp circuit can perform integration or differentiation precisely.

REVIEW QUESTIONS

1. Write the general equation of an exponentially *increasing* function.
2. Why is the "time constant" useful?
3. Justify the statement: "In a linear circuit consisting of resistances, inductances, and capacitances, if any voltage or current is an exponential, all voltages and currents will be exponentials with the same time constant."
4. Write a statement for sinusoids similar to the foregoing statement for exponentials and justify it.
5. Distinguish between average and effective values.
6. Why are ac ammeters calibrated to read rms values?
7. What is the reading of a dc ammeter carrying a current $i = 10 \cos 377t$ A?
8. What is "amplification"? "Gain"?
9. Why are *high* input resistance and *low* output resistance desirable in an amplifier?
10. What is a differential amplifier? What are its advantages?
11. What is an operational amplifier?

12. Why is the gain of an amplifier circuit independent of the gain of the op amp employed?
13. In circuit analysis of an ideal op amp with feedback, what values are assumed for v_i and i_i? Why?
14. How can an op amp be used to add two analog signals?
15. Why do we wish to replace actual devices with fictitious models?
16. Explain the operation of a full-wave bridge rectifier.
17. Explain the operation of a capacitor filter after a full-wave rectifier.
18. During capacitor discharge, why doesn't some of the current flow through the diode?
19. What is the effect on the ripple factor of the circuit in Fig. 3.32 of doubling R_L? Of halving C? Of doubling V_m?
20. Explain the operation of a diode clipper.
21. Sketch a clamping circuit and explain its operation.
22. Explain in words the operation of a differentiator and an integrator.

EXERCISES

1. Sketch the following signals and write mathematical expressions for $v(t)$:
   (a) A step of 5 V occurring at $t = 2$ s.
   (b) A sawtooth of amplitude 20 V that resets 50 times a second.
   (c) A train of pulses of amplitude 100 V, duration 10 μs, and frequency 40 kHz.
   (d) An exponential voltage with an initial amplitude of 6 V that decays to 1 V in 0.18 s. Predict its value at $t = 0.6$ s.
2. An exponential voltage has a value of 15.5 V at $t = 1$ s and a value of 2.1 V at $t = 5$ s.

   (a) Derive the equation for $v(t)$.
   (b) How many time constants elapse between $t = 5$ s and $t = 9$ s?
   (c) Predict the value at $t = 9$ s.
3. An exponential current has a value of 10.6 mA at $t = 2$ s and value of 5 mA at $t = 5$ s. Write the equation for $i(t)$ and predict the value at 10 s.
4. Throwing a switch in an experimental circuit introduces a transient voltage $v = 120 \, e^{-10t}$ V. If measurements in the circuit are accurate to 0.5%, approximately how long does the switching transient "last"?

5. Throwing a switch in an experimental circuit develops a voltage $v = 6\,e^{-500t}$ V across a parallel combination of $R = 40\,\Omega$, and $C = 20\,\mu F$. Predict the current in each element and the total current.

6. A current $i = 2\,e^{-250t}$ mA flows through a series combination of $R = 3\,k\Omega$ and $L = 4$ H. Determine the voltage across each element and the total voltage.

7. A 1-MΩ resistor is connected across a 500-μF capacitor initially charged to 100 V. Predict how long it will take for the voltage to drop to 14 V.

8. In the circuit of Fig. 3.5b, $V_s = 6$ V, $R = 2\,k\Omega$, and $C = 100\,\mu F$. If C is initially uncharged, predict the current as a function of time after closing the switch in position 1. Calculate the current i and voltage v_C at $t = 0.3$ s.

9. In the circuit of Fig. 3.5b, $V_s = 12$ V, $R = 50\,k\Omega$, $C = 0.1\,\mu F$, and switch S has been in position 1 a "long" time. Predict the voltage on the capacitance as a function of time t after moving the switch to position 2. Calculate v_C and current i at $t = 10$ ms.

10. Sketch, approximately to scale, the following functions:
    (a) $i_1 = 2\cos(377t + \pi/4)$ A.
    (b) $v_1 = 10\cos(300t - \pi/3)$ V.
    (c) $i_2 = 1.5\sin(400t - 60°)$ A.

11. Express the following sinusoidal signals as specific functions of time:
    (a) A current with a frequency of 50 Hz and an amplitude of 2 A and passing through zero with positive slope at $t = 2$ ms.
    (b) A voltage with a period of 40 ms and passing through a positive maximum of 10 V at $t = 10$ ms.
    (c) A current reaching a positive maximum of 2 A at $t = 5$ ms and the next negative maximum at $t = 25$ ms.
    (d) A voltage having a positive maximum of 50 V at $t = 0$ and decreasing to a value of 25 V at $t = 2$ ms.

12. A sinusoidal signal has a value of -5 A at time $t = 0$ and reaches its first maximum (negative) of -10 A at $t = 2$ ms. Write the equation for $i(t)$ and predict the value at $t = 5$ ms. Determine the frequency and period.

13. A current $i = 2\cos(2000t - \pi/4)$ A flows through a series combination of $R = 20\,\Omega$ and $L = 10$ mH.
    (a) Determine the voltage across each element.

(b) Sketch the two voltages, $v = f(\omega t)$, approximately to scale and estimate the amplitude and phase angle of $v_R + v_L$.

14. For each of the periodic voltage and current signals shown in Fig. 3.43:
    (a) Calculate average values of voltage and current.
    (b) Calculate effective values of voltage and current.

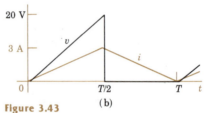

(a)

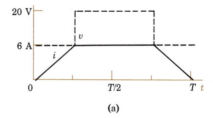

(b)

Figure 3.43

15. Calculate average and effective values of voltage and current for each of the periodic signals shown:
    (a) In Fig. 3.44a.
    (b) In Fig. 3.44b.

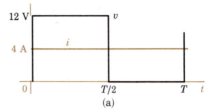

(a)

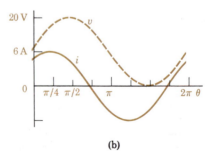

(b)

Figure 3.44

16. Ammeter AM and voltmeter VM, connected as shown in Fig. 3.45, measure instantaneous current and voltage. An ammeter reads upscale or positive when current flows into the meter at the + terminal. Is power flowing *into* or *out of* device D when:
   (a) AM reads positive and VM reads negative?
   (b) AM reads negative and VM reads positive?
   (c) AM reads negative and VM reads negative?

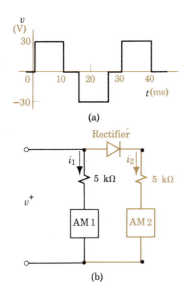

Figure 3.45

17. The signal of Fig. 3.46a is applied to the circuit shown. The rectifier permits current flow only in the direction indicated by the triangle. Predict the readings on the two meters, if they are:
   (a) DC ammeters.
   (b) AC ammeters of the types described in the text.

Figure 3.46

18. Repeat Exercise 17 if the voltage v is as shown in Fig. 3.47.

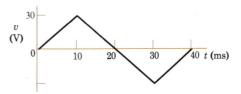

Figure 3.47

19. A d'Arsonval movement provides full-scale deflection with a current of 5 mA or an applied voltage of 30 mV. Specify the shunt and multiplier to provide ranges of 0 to 1 A and 0 to 50 V.

20. Repeat Exercise 19 for ranges of 0 to 5 A and 0 to 300 V.

21. Using a 100-μA d'Arsonval movement with a 10,000-Ω internal resistance:
   (a) Design a multirange dc milliammeter with ranges of 1, 10, and 100 mA.
   (b) Design a multirange voltmeter with ranges of 3, 10, 50, and 150 V.

22. Design an ohmmeter, using the movement of Exercise 21 with a half-scale reading of:
   (a) 50 Ω. (b) 5000 Ω.

23. A transformer with 600 primary turns at 120 V is to provide 120 mA at 500 V at the secondary. Specify the number of turns on the secondary and predict the primary current.

24. A transformer is to "step down" from 120 V to 7.5 V across a load $R_2 = 625\ \Omega$. Draw the circuit model, specify the turn ratio, and predict the primary current.

25. An op amp has a very high gain, a very high input resistance, and a very low output resistance. Design and draw an amplifier circuit (with $R_F = 3\ \text{M}\Omega$) providing:
   (a) $A_F = -150$.
   (b) $A_F = -30$.

26. In Fig. 3.23a, $R_1 = 1\ \text{k}\Omega$, $R_2 = 2\ \text{k}\Omega$, $R_n = 3\ \text{k}\Omega$, and $R_F = 90\ \text{k}\Omega$. If $v_1 = +1$ mV, $v_2 = -5$ mV, and $v_n = +2$ mV, predict v_o.

27. (a) Using an ideal amplifier, design a circuit (let $R_F = 1000\ \Omega$) to provide an output $v_o(t) = -v_1(t) - 10\ v_2(t)$.
   (b) Redesign the circuit to provide an output $v_o(t) = +v_1(t) - 10\ v_2(t)$.

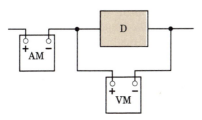

(a)

Rectifier

i_1 i_2

5 kΩ 5 kΩ

v^+

AM 1   AM 2

(b)

28. Given the op amp in Fig. 3.48:
(a) Define v_o in terms of the given inputs.
(b) Describe the operation performed assuming sinusoidal inputs.

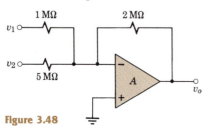

Figure 3.48

29. Given the amplifier circuit of Fig. 3.49:
(a) Define v_o in terms of v_1.
(b) Define the operation performed.

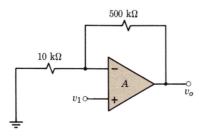

Figure 3.49

30. Repeat Exercise 25 for:
(a) $A_F = +40$.
(b) $A_F = +120$.

31. Explaining your reasoning and stating any simplifying assumptions, predict current I in Fig. 3.50. Reverse diode D_2 and repeat.

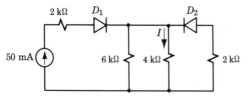

Figure 3.50

32. In Fig. 3.51, $v = 10 \sin \omega t$ V and $V_B = 6$ V. Under what circumstances will current i flow? Sketch v and i as functions of time on the same axes.

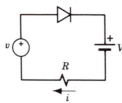

Figure 3.51

33. Four ideal diodes are connected in the full-wave rectifier of Fig. 3.30 where $R_L = 1000\ \Omega$. Specify the effective value of v so that the *average* current through R_L is 20 mA.

34. An ideal diode is used in a half-wave rectifier with power supplied at 120 V (rms) and 60 Hz. For a load $R_L = 500\ \Omega$, predict I_{dc}, V_{dc}, and the power delivered to the load.

35. Repeat Exercise 34 assuming a full-wave bridge rectifier circuit.

36. In Fig. 3.32, $C = 100\ \mu F$ and $R_L = 10\ k\Omega$. For $V_m = 20$ V at 60 Hz, predict:
(a) The dc load current through R_L.
(b) The percent ripple in v_L.
(c) The reading of a dc ammeter in series with R_L if C is disconnected. Compare to part (a).

37. A load requires 10 mA at 30 V dc with no more than 0.5 V ripple.
(a) Draw the circuit of a power supply consisting of a transformer with 120-V 60-Hz input, half-wave rectifier, capacitor filter, and effective R_L; specify C and the turn ratio of the transformer.
(b) Repeat for a full-wave bridge rectifier circuit. Draw a conclusion.

38. In Fig. 3.34a, $v_1 = 7 \sin \omega t$ V. On the same graph, show v_1, $v_1 - V$, v_R, and v_D.

39. For $v_1 = 10 \sin \omega t$ V, $R = 1$ kΩ, and $V = 4$ V in Fig. 3.52a, sketch $v_2(t)$.

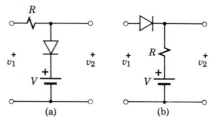

Figure 3.52

40. For $v_1 = 10 \sin \omega t$ V, $R = 1$ kΩ, and $V = 4$ V in Fig. 3.52b, sketch $v_2(t)$.

41. In Fig. 3.53, $v_1 = 10 \sin \omega t$ V.
(a) Sketch $v_1(t)$ and $v_2(t)$ on the same graph.
(b) Draw the transfer characteristic v_2 versus v_1.
(c) What function is performed by this circuit?

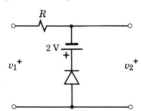

Figure 3.53

42. The periodic voltage v_1 of Fig. 3.54 is applied to the input of Fig. 3.53. Show the input signal v_i and the output signal v_o on the same graph.

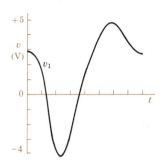

Figure 3.54

43. Repeat Exercise 42 for the circuit of Fig. 3.55.

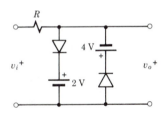

Figure 3.55

44. In Fig. 3.56, the input signal is $v_i = 5 \sin \omega t$ V. What voltage v_C appears across the capacitance? Show v_i and v_o on the same graph.

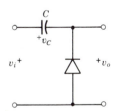

Figure 3.56

45. The periodic voltage v_1 of Fig. 3.54 is applied to the input of Fig. 3.56. Show the input signal v_i and the output signal v_o on the same graph.

46. Repeat Exercise 44 for the circuit of Fig. 3.57 where $V_B = 2$ V.

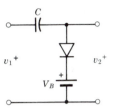

Figure 3.57

47. Repeat Exercise 45 for the circuit of Fig. 3.57 where $V_B = 2$ V.

48. Design a circuit that will clamp the minimum point of a periodic signal to -5 V.

49. In Fig. 3.58, when the voltage $v_1 = V_m(0.5 + \sin \omega t)$ V, the high-resistance dc voltmeter VM reads 70 V.
   (a) What function is performed by each of the three sections of the circuit?
   (b) How is v_2 related to v_1? Show v_1, v_2, and v_3 on the same graph.
   (c) Determine V_m and define the function of this instrument.

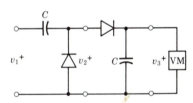

Figure 3.58

50. In Fig. 3.41a, $R_1 = 1$ MΩ and $C = 0.1$ μF. How is v_o related to v_1? For $v_1 = 10 \sin 20t$ V, what is v_o?

51. In Fig. 3.41a, $R_1 = 50$ kΩ and $C = 0.05$ μF (initially uncharged). At $t = 0$, v_1 is connected to $V_B = -2$ V; how long will it take for v_o to reach 10 V?

52. Devise a sweep circuit (see Fig. 4.10a) using an op amp to provide an output voltage proportional to time. Show a reset switch to ensure that $v_o = 0$ at $t = 0$.

PROBLEMS

1. Demonstrate analytically that the tangent to any exponential function at time t intersects the time axis at $t + T$ where T is the time constant.
2. The voltage across a 10-μF capacitor is observed on an oscilloscope. In 0.5 s after a 10-V source is removed, the voltage has decayed to 1.35 V. Derive and label a circuit model for this capacitor.
3. A voltage consists of a dc component of magnitude V_0 and a sinusoidal component of effective value V_1; show that the effective value of the combination is $(V_0^2 + V_1^2)^{1/2}$.
4. The circuit of Fig. 3.59 is a practical means for obtaining a high dc voltage without using a transformer. Stating any simplifying assumptions, predict voltages V_1 and V_L.

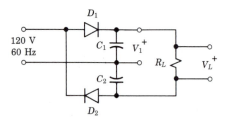

Figure 3.59

5. A diode is connected in series with a 30-V rms source to charge a 12-V battery with an internal resistance of 0.1 Ω. Specify the series resistance necessary to limit the peak current to 2 A. Estimate the time required to recharge a 10-A·hr battery.
6. A diode and battery are used in the "voltage regulator" circuit of Fig. 3.60, where $R_S = 1000\ \Omega$ and $R_L = 2000\ \Omega$. If V_1 increases from 16 to 24 V (a 50% increase), calculate the corresponding variation in load voltage V_L. Is the "regulator" doing its job?

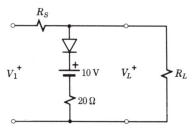

Figure 3.60

7. The circuit of Fig. 3.36a is used in an electronic voltmeter. With C initially uncharged, v_1 is connected to an unknown voltage V_x for 100 μs. (How much charge Q_x is stored in C?) Then v_1 is connected to a reference voltage $V_r = 2.000$ V. The time required to reduce v_o to zero is shown on a clock as 186 μs. What is V_x?
8. In Fig. 3.61, input voltages A and B are restricted to either 0 or +5 V. Tabulate the four possible combinations of A and B and the corresponding values of output voltage V_o. Define in words the output in terms of the inputs. Why is this called an **OR** circuit? Where is it useful?

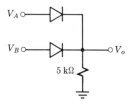

Figure 3.61

4

Cathode-Ray Tubes

Electron Motion

Cathode-Ray Tubes

Oscilloscopes

The discovery of cathode rays a century ago marked the beginning of the electronic era, and the invention of the multielectrode vacuum tube brought electronics into our daily living. As we learned to build efficient devices for generating and controlling streams of electrons, these electron tubes were applied in communication, entertainment, industrial control, and instrumentation. In many of these applications, the tube has been replaced by the semiconductor devices emphasized in the remainder of this book. However, sophisticated versions of the basic cathode-ray tube are widely used in oscilloscopes, data display devices, television cameras and receivers, and radar scanners.

In this chapter we study some of the physical principles that underlie the operation of electronic devices. First we examine the behavior of electrons in a vacuum and derive the equations for motion in electric and magnetic fields. Then we consider electron emission, acceleration, and deflection in a cathode-ray tube. Finally, we see how cathode-ray oscilloscopes can be used for precise observations over a wide range of conditions.

ELECTRON MOTION

The model of the electron as a negatively charged particle of finite mass but negligible size is satisfactory for many purposes. Based on many careful measurements, the accepted values for charge and mass of the electron are

$$e = 1.602 \times 10^{-19} \text{ C} \cong 1.6 \times 10^{-19} \text{ C}$$

$$m = 9.109 \times 10^{-31} \text{ kg} \cong 9.1 \times 10^{-31} \text{ kg}$$

In contrast, the hydrogen ion, which carries a positive charge of the same magnitude, has a mass approximately 1836 times as great. If the mass, charge, and initial velocity are known, the motion of individual electrons and ions in electric and magnetic fields can be predicted using Newton's laws of mechanics.

Motion in a Uniform Electric Field

A uniform electric field of strength $\mathcal{E}$ is established between the parallel conducting plates of Fig. 4.1a by applying a potential difference or voltage. By definition (Eq. 1-5)

$$\mathcal{E} = -\frac{dv}{dl} = -\frac{V_b - V_a}{L} \text{ volts/meter} \tag{4-1}$$

if the spacing is small compared to the dimensions of the plates.

By definition (Eq. 1-4), the electric field strength is the force per unit positive charge. Therefore, the force in newtons on a charge q in coulombs is

$$\mathbf{f} = q\mathcal{E} \tag{4-2}$$

In Fig. 4.1b, $V_b > V_a$, $V_b - V_a$ is a positive quantity, and the electric field is negative (directed in the $-x$ direction). Considering the energy dw gained by a charge q moving a distance dl against the force of the electric field (Eq. 1-3), the voltage of point b with respect to point a is

$$V_{ba} = \frac{1}{q} \int_a^b dw = \frac{1}{q} \int_a^b f \, dl = \frac{1}{q} \int_a^b q(-\mathcal{E}) \, dl = -\int_a^b \mathcal{E} \, dl \tag{4-3}$$

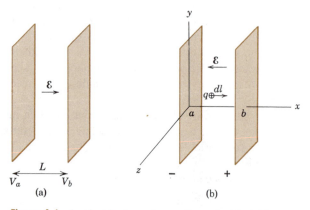

Figure 4.1 An electric charge in a uniform electric field.

In words, the voltage between any two points in an electric field is the line integral of the electric field strength. Equation 4-3 is the corollary of Eq. 4-1.

In general, an electron of charge $-e$ in an electric field $\mathcal{E}$ experiences a force

$$f = (-e)(\mathcal{E}) = -e\,\mathcal{E} = ma$$

and an acceleration

$$a = \frac{f}{m} = -\frac{e\mathcal{E}}{m} \qquad (4\text{-}4)$$

We see that an electron in a uniform electric field moves with a constant acceleration. We expect the resulting motion to be similar to that of a freely falling mass in the earth's gravitational field. Where u is velocity and x is displacement, the equations of motion in a field $\mathcal{E}_x$ are

$$u_x = \int_0^t a_x\, dt = a_x t + U_O = -\frac{e\,\mathcal{E}_x}{m}t + U_O \qquad (4\text{-}5)$$

$$x = \int_0^t u_x\, dt = \frac{a_x t^2}{2} + U_O t + X_O = -\frac{e\,\mathcal{E}_x}{2m}t^2 + U_O t + X_O \qquad (4\text{-}6)$$

Application of these equations is illustrated in Example 1.

EXAMPLE 1

A voltage V_D is applied to an electron deflector consisting of two horizontal plates of length L separated a distance d, as in Fig. 4.2. An electron with initial velocity U_O in the positive x direction is introduced at the origin. Determine the path of the electron and the vertical displacement at the time it leaves the region between the plates.

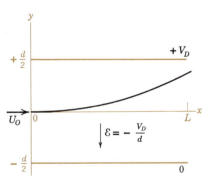

Figure 4.2 Calculation of electron deflection in a uniform electric field.

Assuming no electric field in the x direction and a uniform electric field $\mathcal{E}_y = -V_D/d$, the accelerations are

$$a_x = 0 \qquad \text{and} \qquad a_y = -\frac{e\,\mathcal{E}_y}{m} = \frac{eV_D}{md}$$

There is no acceleration in the x direction and the electron moves with constant velocity to the right. There is a constant upward acceleration and the electron gains a vertical component of velocity. The path is determined (Eq. 4-6) by

$$x = U_O t \qquad \text{and} \qquad y = -\frac{e\,\mathcal{E}_y}{2m}t^2 = \frac{eV_D}{2md}t^2$$

Eliminating t,

$$y = \frac{eV_D}{2mdU_O^2}x^2 \qquad (4\text{-}7)$$

or the electron follows a parabolic path.

At the edge of the field, $x = L$ and the vertical displacement is

$$y_L = \frac{eV_D}{2mdU_O^2}L^2$$

If V_D exceeds a certain value, displacement y exceeds $d/2$ and the electron strikes the upper plate.

Energy Gained by an Accelerated Electron

When an electron is accelerated by an electric field it gains kinetic energy at the expense of potential energy, just as does a freely falling mass. Since voltage is energy per unit charge, the potential energy "lost" by an electron in "falling" from point a to point b is, in joules,

$$PE = W = q(V_a - V_b) = -e(V_a - V_b) = eV_{ba} \qquad (4\text{-}8)$$

where V_{ba} is the potential of b with respect to a.

The kinetic energy gained, evidenced by an increase in velocity, is just equal to the potential energy lost, or

$$KE = \tfrac{1}{2}mu_b^2 - \tfrac{1}{2}mu_a^2 = PE = eV_{ba} \qquad (4\text{-}9)$$

This important equation indicates that the kinetic energy gained by an electron in an electric field is determined only by the voltage difference between the initial and final points; it is independent of the path followed and the electric field configuration. (We assume that the field does not change with *time*.)

Frequently we are interested in the behavior resulting from a change in the energy of a single electron. Expressed in joules, these energies are very small; a more convenient unit is suggested by Eq. 4-8. An *electron volt* is the potential energy lost by 1 electron falling through a potential difference of 1 volt. By Eq. 4-8,

$$1 \text{ eV} = (1.6 \times 10^{-19} \text{ C})(1 \text{ V}) = 1.6 \times 10^{-19} \text{ J} \qquad (4\text{-}10)$$

For example, the energy required to remove an electron from a hydrogen atom is about 13.6 eV. The energy imparted to an electron in a linear accelerator may be as high as 24 BeV (24 billion electron volts).

For the special case of an electron starting from rest $(u_a = 0)$ and accelerated through a voltage V, Eq. 4-9 can be solved for $u_b = u$ to yield

$$u = \sqrt{2(e/m)V} = 5.93 \times 10^5 \sqrt{V} \text{ m/s} \qquad (4\text{-}11)$$

In deriving Eq. 4-11 we assume that mass m is a constant; this is true only if the velocity is small compared to the velocity of light, $c \cong 3 \times 10^8$ m/s.

EXAMPLE 2

Find the velocity reached by an electron accelerated through a voltage of 3600 V.

Assuming that the resulting velocity u is small compared to the velocity of light c, Eq. 4-11 applies and

$$u = 5.93 \times 10^5 \sqrt{3600} = 3.56 \times 10^7 \text{ m/s}$$

In this case,

$$\frac{u}{c} = \frac{3.56 \times 10^7}{3 \times 10^8} \cong 0.12$$

At the velocity of Example 2 the increase in mass is appreciable, and the actual velocity reached is about 0.5% lower than that predicted. For voltages above 4 or 5 kV, a more precise expression should be used. (See Problem 1.)

Motion in a Uniform Magnetic Field

One way of defining the strength of a magnetic field (Eq. 1-6) is in terms of the force exerted on a unit charge moving with unit velocity normal to the field. In general, the force in newtons is

$$\mathbf{f} = q\mathbf{u} \times \mathbf{B} \tag{4-12}$$

where q is charge in coulombs and $\mathbf{B}$ is magnetic flux density in teslas (or webers/meter^2). The vector cross product is defined by the right-hand screw rule illustrated in Fig. 4.3. Rotation from the direction of $\mathbf{u}$ to the direction of $\mathbf{B}$ advances

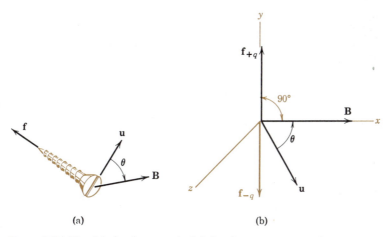

(a)   (b)

Figure 4.3 The right-hand screw rule defining the vector cross product.

the screw in the direction of $\mathbf{f}$. The magnitude of the force is $quB \sin \theta$ and the direction is always normal to the plane of $\mathbf{u}$ and $\mathbf{B}$.

Equation 4-12 is consistent with three observable facts:

1. A charged particle at rest in a magnetic field experiences no force ($u = 0$).
2. A charged particle moving parallel with the magnetic flux experiences no force ($\theta = 0$).
3. A charged particle moving with a component of velocity normal to the magnetic flux experiences a force that is normal to u and therefore the magnitude of velocity (or speed) is unchanged.

From the third statement we conclude that no work is done by a magnetic field on a charged particle and its kinetic energy is unchanged.

Figure 4.4 shows an electron entering a finite region of uniform flux density. For an electron ($q = -e$) moving in the plane of the paper ($\theta = 90°$), the force is in the direction shown with a magnitude

$$f = eU_oB \tag{4-13}$$

Applying the right-hand rule, we see that the initial force is downward and, therefore, the acceleration is downward and the path is deflected as shown. A particle

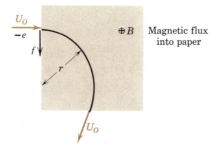

Figure 4.4   Electron motion in a uniform magnetic field.

moving with constant speed and constant normal acceleration follows a circular path. The centrifugal force due to circular motion must be just equal to the centripetal force due to the magnetic field, or

$$m\frac{U_o^2}{r} = eU_oB$$

and the radius of the circular path is

$$r = \frac{mU_o}{eB} \tag{4-14}$$

The dependence of radius on the mass of the charged particle is the principle underlying the *mass spectrograph* for analyzing unknown substances.

Motion in Combined ℰ and B Fields

In the general case, both electric and magnetic fields are present and exert forces on a moving charge. The total force is

$$\mathbf{f} = q(\,\mathcal{E} + \mathbf{u} \times \mathbf{B}) \tag{4-15}$$

The special cases previously described can be derived from this general relation. If both ℰ and B are present, the resulting motion depends on the relative orientation of ℰ and B, and also on the initial velocities. An interesting case is that in which an electron starts from rest in a region where ℰ and B are mutually perpendicular. The electron is accelerated in the direction of $-$ℰ, but as soon as it is in motion there is a reaction with B and the path starts to curve. Curving around, the electron is soon traveling in the $+$ℰ direction and experiences a decelerating force that brings it to rest. Once the electron is at rest, the cycle starts over; the resulting path is called a *cycloid* and resembles the path of a point on a wheel as it rolls along a line.

If the fields are not uniform, mathematical analysis is usually quite difficult. A case of practical importance is the "electron lens" used in electron microscopes and for electric-field focusing of the electron beam in the cathode-ray tube.

CATHODE-RAY TUBES

The television picture tube and the precision electron display tube used in oscilloscopes are modern versions of the evacuated tubes used by Crookes and Thomson to study cathode rays. The electrons constituting a "cathode ray" have little mass or inertia, and therefore they can follow rapid variations; their ratio of charge to mass is high, so they are easily deflected and controlled. The energy of high-velocity electrons is readily converted into visible light; therefore, their motion is easily observed. For these reasons, the C-R tube or CRT is a unique information processing device; from our standpoint, it is also an ingenious application of the principles of electron motion and electron emission.

Electron Emission

The Bohr model of the atom is satisfactory for describing how free electrons can be obtained in space. As you may recall, starting with the hydrogen atom consisting of a single proton and a single orbital electron, models of more complex atoms are built up by adding protons and neutrons to the nucleus and electrons in orbital groups or shells. In a systematic way, shells are filled and new shells started. The chemical properties of an element are determined by the *valence* electrons in the outer shell. Good electrical conductors like copper and silver have one highly mobile electron in the outer shell.

Only certain orbits are allowed, and atoms are stable only when the orbital electrons have certain discrete energy levels. Transfer of an electron from an orbit corresponding to energy W_1 to an orbit corresponding to a lower energy W_2 results in the radiation of a *quantum* of electromagnetic energy of frequency f given by

$$W_1 - W_2 = hf \qquad (4\text{-}16)$$

where h is Planck's constant $= 6.626 \times 10^{-34}$ J·s.

The energy possessed by an orbital electron consists of the kinetic energy of motion in the orbit and the potential energy of position with respect to the positive ion representing all the rest of the neutral atom. If other atoms are close (as in a solid), the energy of an electron is affected by the charge distribution of the neighboring atoms. In a crystalline solid, there is an orderly arrangement of atoms and the permissible electron energies are grouped into *energy bands*. Between the permissible bands there may be ranges of energy called *forbidden bands*.

For an electron to exist in space it must possess the energy corresponding to motion from its normal orbit out to an infinite distance; the energy required to move an electron against the attractive force of the net positive charge left behind is the *surface barrier energy* W_B. Within a metal at absolute zero termperature, electrons possess energies varying from zero to a maximum value W_M. The minimum amount of work that must be done on an electron before it is able to escape from the surface of a metal is the *work function* W_W where

$$W_W = W_B - W_M \qquad (4\text{-}17)$$

For copper, $W_W = 4.1$ eV, while for cesium $W_W = 1.8$ eV.

The energy required for electron emission may be obtained in various ways. The beta rays given off spontaneously by *radioactive* materials (along with alpha and gamma rays) are emitted electrons. In *photoelectric* emission, the energy of a quantum of electromagnetic energy is absorbed by an electron. In *high-field* emission, the potential energy of an intense electric field causes emission. In *secondary* emission, a fast-moving electron transfers its kinetic energy to one or more electrons in a solid surface. All these processes have possible applications, but the most widely used process is *thermionic* emission in which thermal energy is added by heating a solid conductor.

The temperature of an object is a measure of the kinetic energy stored in the motion of the constituent molecules, atoms, and electrons. The energies of the individual constituents vary widely, but an average energy corresponding to temperature T can be expressed as kT, where $k = 8.62 \times 10^{-5}$ eV/K is the Boltzmann constant. At a temperature above absolute zero, the distribution of electron energy in a metal is modified and some electrons possess energies appreciably above W_M. Statistical analysis shows that the probability of an electron receiving sufficient energy to be emitted is proportional to $e^{-W_W/kT}$. At high temperatures, many electrons possess energies greater than W_B and emission current densities of the order of 1 A/cm^2 are practical. Commercial cathodes make use of special materials that combine low work function with high melting point.

CRT Components

The essential components of a CRT are shown in Fig. 4.5; an *electron gun* produces a focused beam of electrons, a *deflection system* determines the direction of the beam, and a *fluorescent screen* converts the energy of the beam into visible light.

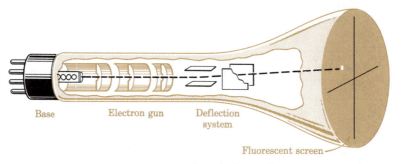

Base   Electron gun   Deflection system

Fluorescent screen

Figure 4.5 The essential components of a cathode-ray tube.

Electron Gun. Electrons are emitted from the hot cathode (Fig. 4.6a) and pass through a small hole in the cylindrical control electrode; a negative voltage (with respect to the cathode) on this electrode tends to repel the electrons and, therefore, the voltage applied controls the intensity of the beam. Electrons passing the control electrode experience an accelerating force due to the electric field established by the positive voltages V_F and V_A on the focusing and accelerating anodes. The space between these anodes constitutes an electron lens (Fig. 4.6b); the electric flux lines and equipotential lines resulting from the voltage difference $V_A - V_F$ provide a precise focusing effect. A diverging electron is accelerated forward by the field, and at the same time it receives an inward component of velocity that brings it back to the axis of the beam at the screen.

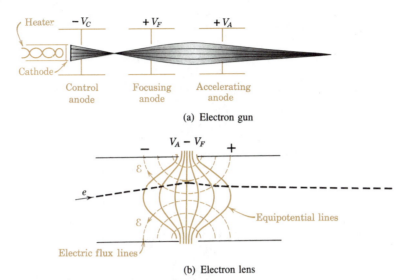

(a) Electron gun

(b) Electron lens

Figure 4.6 An elementary electron gun with electric-field focusing.

Deflection System. In a television picture tube, the beam is moved across the screen 15,750 times per second, creating a picture consisting of 525 horizontal lines of varying intensity. The deflection of the beam may be achieved by a magnetic field or an electric field. Figure 4.7 shows the beam produced in the electron gun entering the vertical deflection plates of an electric-field system.

We can calculate the beam deflection at the screen by using our knowledge of electron motion. For the coordinate system of Fig. 4.7, Eq. 4-11 gives an axial velocity

$$U_z = \sqrt{2eV_A/m} \tag{4-18}$$

where V_A is the total accelerating potential. Within the deflecting field, the parabolic path (Eq. 4-7) is defined by

$$y = \frac{eV_D}{2mdU_z^2}z^2 = kz^2 \tag{4-19}$$

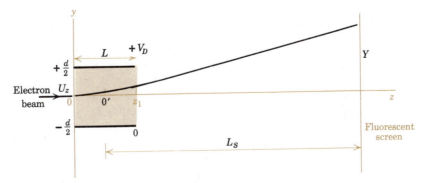

Figure 4.7 An electric-field vertical deflection system.

The slope of the beam emerging from the deflecting field at $z = z_1 = L$ is

$$\frac{dy}{dz} = 2kz = 2kL \qquad (4\text{-}20)$$

The equation of the straight-line path followed by the beam to the screen is

$$y - y_1 = \frac{dy}{dz}(z - z_1) = 2kL(z - z_1)$$

where $z_1 = L$ and $y_1 = kz_1^2 = kL^2$. Substituting these values and solving,

$$y = 2kL(z - L) + kL^2 = 2kL\left(z - \frac{L}{2}\right) \qquad (4\text{-}21)$$

Since for $y = 0$, $z = L/2$, Eq. 4-21 leads to the conclusion that the electron beam appears to follow a straight-line path from a virtual source at $0'$. At the screen, $z = L_S + L/2$ and (by Eqs. 4-19 and 4-18) the deflection is

$$Y = 2kLL_S = \frac{eV_DLL_S}{mdU_z^2} = \frac{eV_DLL_S}{md} \cdot \frac{m}{2eV_A} = \frac{LL_S}{2dV_A}V_D \qquad (4\text{-}22)$$

or the vertical deflection at the screen is directly proportional to V_D, the voltage applied to the vertical deflecting plates. A second set of plates provides horizontal deflection.

EXAMPLE 3

Determine the *deflection sensitivity* in centimeters of deflection per volt of signal for a CRT in which $L = 2$ cm, $L_S = 30$ cm, $d = 0.5$ cm, and the total accelerating voltage is 2 kV.

From Eq. 4-22, the deflection sensitivity is

$$\frac{Y}{V_D} = \frac{LL_S}{2dV_A} = \frac{0.02 \times 0.3}{2 \times 0.005 \times 2000}$$

$$= 0.0003 \text{ mV} = 0.03 \text{ cm/V}$$

To obtain a reasonable deflection, say 3 cm, a voltage of 100 V would be necessary. In a practical CRO, amplifiers are provided to obtain reasonable deflections with input signals of less than 0.1 V.

Fluorescent Screen. Part of the kinetic energy of the electron beam is converted into luminous energy at the screen. Absorption of kinetic energy results in an immediate *fluorescence* and a subsequent *phosphorescence*. The choice of screen material depends on the application. For laboratory oscilloscopes, a medium persistence phosphor with output concentrated in the green region is desirable; the eye is sensitive to green and the persistence provides a steady image of a repeated pattern. For color television tubes, short persistence phosphors that emit radiation at various wavelengths are available. For radar screens, a very long persistence is desirable.

OSCILLOSCOPES

The CRT provides a controlled spot of light whose x and y deflections are directly proportional to the voltages on the horizontal and vertical deflecting plates. The cathode-ray oscilloscope (CRO), consisting of the tube and appropriate auxiliary apparatus, enables us to "see" the complex waveforms that are critical in the performance of electronic circuits.

The block diagram of Fig. 4.8 indicates the essential components of a CRO. A signal applied to the "Vert" terminal causes a proportional vertical deflection of the spot; the calibration of the *vertical amplifier* can be checked against an internal calibrating signal. The *attenuator* precisely divides large input voltages. An external *intensity control* varies the accelerating potentials in the electron gun, and a *focus control* determines the potentials on the focusing electrodes.

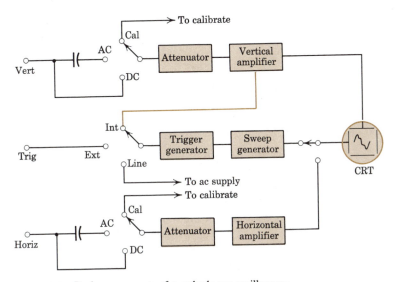

Figure 4.8   Basic components of a cathode-ray oscilloscope.

The *sweep generator* causes a horizontal deflection of the spot proportional to time; it is *triggered* to start at the left of the screen at a particular instant on the internal vertical signal ("Int"), an external signal ("Ext"), or the ac supply ("Line"). Instead of the sweep generator, the *horizontal amplifier* can be used to cause a deflection proportional to a signal at the "Horiz" terminal.

The particular point on the triggering waveform that initiates the sawtooth sweep is determined by setting the *slope* and *level* controls. In Fig. 4.9, the input to the vertical amplifier provides the triggering waveform. The level control determines the instantaneous voltage level at which a trigger pulse is produced by the trigger generator. With the slope switch in the "+" position, triggering occurs only on a positive slope portion of the triggering waveform. By proper adjustment of these two controls, it is possible to initiate the sweep consistently at almost any point in the triggering waveform. For example, if slope is set to "−" and level is set to "0," the sweep will begin when the triggering signal goes down through zero.

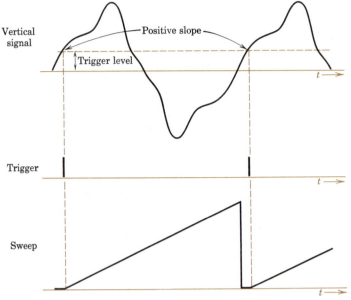

Figure 4.9 Trigger operation.

CRO Applications

In its practical form, with controls conveniently arranged for precise measurements over a wide range of test conditions, the CRO is the most versatile laboratory instrument.

Voltage Measurement. An illuminated scale dividing the screen into 1-cm divisions permits use of the CRO as a voltmeter. With the vertical amplifier sensitivity set at 0.1 V/cm, say, a displacement of 2.5 cm indicates a voltage of 0.25 V. Currents can be determined by measuring the voltage across a known resistance.

Time Measurement. A calibrated sweep generator permits time measurement. With the sweep generator set at 5 ms/cm, two events separated on the screen by 2 cm are separated in time by 10 ms or 0.01 s. By measuring the period of a wave, the frequency can be determined by calculation.

Waveform Display. A special property of the CRO is its ability to display high-frequency or short-duration waveforms. If voltages varying with time are applied to vertical (y) and horizontal (x) input terminals, a pattern is traced out on the screen; if the voltages are periodic and one period is an exact multiple of the other, a stationary pattern can be obtained. A sawtooth wave from the sweep generator (Fig. 4.10a) applied to the x-deflection plates provides an x-axis deflection directly proportional to time. If a signal voltage wave is applied to the vertical-deflection plates, the projection of the beam on the y-axis is directly proportional to the amplitude of this signal. If both voltages are applied simultaneously, the pattern displayed on the screen is the signal as a function of time (Fig. 4.10c). A *blanking circuit* turns off the electron beam at the end of the sweep so the return trace is not visible. Nonrepetitive voltages are made

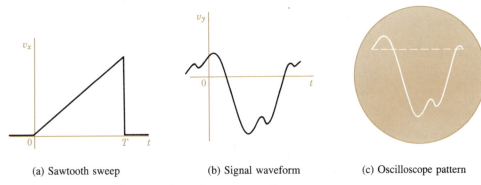

(a) Sawtooth sweep   (b) Signal waveform   (c) Oscilloscope pattern

Figure 4.10 Display of a repeated waveform on an oscilloscope.

more visible by using a long persistence fluorescent material, or a high-speed camera can be used for a permanent record. New CROs provide digital waveform storage.

X-Y Plotting. The relation between two periodic variables can be displayed by applying a voltage proportional to x to the horizontal amplifier and one proportional to y to the vertical amplifier. The characteristics of diodes or transistors are quickly displayed in this way. The hysteresis loop of a magnetic material can be displayed by connecting the induced voltage (proportional to B) to the vertical amplifier and an iR drop (proportional to H) to the horizontal amplifier. Since the two amplifiers usually have a common internal ground, some care is necessary in arranging the circuits.

Phase-Difference Measurement. If two sinusoids of the same frequency are connected to the X and Y terminals, the phase difference is revealed by the resulting pattern (Fig. 4.11). For applied voltages $v_x = V_x \cos \omega t$ and $v_y = V_y \cos (\omega t + \theta)$, it can be shown that the phase difference is

$$\theta = \sin^{-1}\frac{A}{B} \qquad (4\text{-}23)$$

where A is the y deflection when the x deflection is zero and B is the maximum y deflection. The parameters of the circuit are useful in determining whether the angle is leading or lagging.

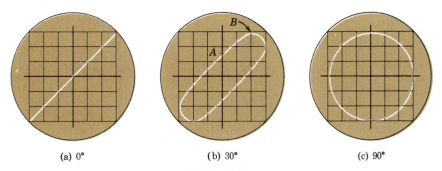

(a) 0°   (b) 30°   (c) 90°

Figure 4.11 Phase difference as revealed by CRO patterns.

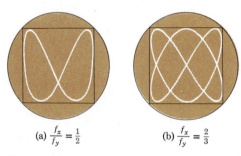

(a) $\dfrac{f_x}{f_y} = \dfrac{1}{2}$ (b) $\dfrac{f_x}{f_y} = \dfrac{2}{3}$

Figure 4.12   Lissajous figures for comparing the frequencies of two sinusoids.

Frequency Comparison. When the frequency of the sinusoid applied to one input is an exact multiple of the frequency of the other input, a stationary pattern is obtained. For a 1:1 ratio, the so-called Lissajous patterns are similar to those in Fig. 4.11. For the 1:2 and 2:3 ratios, the Lissajous figures might be as shown in Fig. 4.12. For a stationary pattern, the ratio of the frequencies is exactly equal to the ratio of the numbers of tangencies to the enclosing rectangle. Patterns can be predicted by plotting x and y deflections from the two signals at corresponding instants of time.

Square-Wave Testing. The unique convenience of a CRO is illustrated in the measurement technique called *square-wave testing*. Any periodic wave can be represented by a Fourier series of sinusoids. As indicated in Fig. 4.13b, the sum of the first three odd harmonics (with appropriate amplitude and phase) begins to approximate a square wave; additional higher harmonics would increase the slope of the leading edge and smooth off the top. If a square-wave input to a device under test results in an output resembling Fig. 4.13c, it can be shown that low-frequency components have been attenuated and shifted forward in phase. If the output resembles Fig. 4.13d, it can be shown that high-frequency components have been attenuated and shifted backward in phase. The frequency response of an amplifier, for example, can be quickly determined by varying the frequency of the square-wave input until these distortions appear; the useful range of the amplifier is bounded by the frequencies at which low-frequency and high-frequency defects appear. As another example, two devices giving similar responses to a square-wave input can be expected to respond similarly to other waveforms.

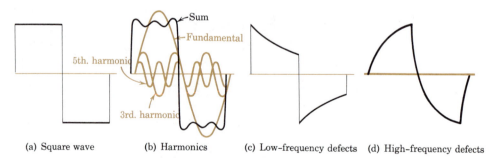

(a) Square wave   (b) Harmonics   (c) Low–frequency defects   (d) High–frequency defects

Figure 4.13   Square-wave testing with a CRO.

CRO Features

Every year sees new advances in CRO design as the instrument manufacturers strive to meet new needs of the laboratory and the field by taking advantage of newly developed devices and techniques. Among the features offered by modern CROs are:

Differential Inputs. Each amplifier channel has two terminals in addition to the ground terminal. By amplifying only the difference between two signals, any common signal such as hum is rejected. (See p. 439.)

Optional Probes. To improve the input characteristics of a CRO or to expand its functions, the input stage can be built into a small unit (connected to the CRO by a shielded cable) that can be placed at the point of measurement. *Voltage probes* insert impedances in series with the CRO input to increase the effective input impedance. *Current probes* use transformer action or the Hall effect to convert a current into a proportional voltage for measurement or display.

Dual Channels. The waveforms of two different signals can be displayed simultaneously by connecting the signals alternately to the vertical deflection system. The more expensive *dual beam* feature requires separate electron guns and deflection systems but permits greater display flexibility.

Delayed and Expanded Sweep. An auxiliary delayed sweep with a faster sweep speed displays a magnified version of a selected small portion of a waveform.

Storage. If the input signal is a single, nonrepetitive event, the image can be *stored* by using a long persistence phosphor. By placing a storage mesh directly behind the phosphor and controlling the rate at which the charge pattern leaks off the mesh, a *variable persistence* is obtained.

Sampling. To display signals at frequencies beyond the limits of the CRO components, very short *sample* readings are taken on successive recurrences of the waveform. Each amplitude sample is taken at a slightly later instant on the waveform and the resulting dots appear as a continuous display. With this sampling technique, 18-GHz signals can be displayed.

Digital Readout. The addition of a built-in *microprocessor* provides direct digital readout of time interval, frequency, or voltage in addition to the conventional CRO display. The operator sets two markers to indicate a horizontal or vertical displacement on the waveform; the microprocessor, a computer-on-a-chip, interrogates the function switches and the scale switches, calculates the desired variable, and converts it to digital form for display by light-emitting diodes.

Programmability. The oscilloscope is preeminent in displaying information; the digital computer is preeminent in processing information. The new digitizing oscilloscope is an open-ended instrument that accepts either preprogrammed or user-programmable instructions and into which the user enters his or her choice of parameters, functions, frequency ranges, data reduction operations, and display characteristics by means of a single keyboard just as we enter numbers in a hand-held calculator. It can store, display, measure, and analyze complex signal waveforms.

SUMMARY

- Individual charged particles in electric and magnetic fields obey Newton's laws. For electrons (of primary interest here),

$$\mathbf{f} = -e(\boldsymbol{\mathcal{E}} + \mathbf{u} \times \mathbf{B})$$

- In any electric field, KE gained = PE lost = $W_{ab} = eV_{ab}$. For an electron starting from rest (and for $V < 4$ kV),

$$u = \sqrt{2(e/m)V} = 5.93 \times 10^5 \sqrt{V} \text{ m/s}$$

In a uniform electric field where $\mathcal{E}_x = \mathcal{E}$, $f_x = -e\mathcal{E}$, and

$$a_x = -\frac{e\mathcal{E}}{m} \qquad u_x = -\frac{e\mathcal{E}}{m}t + U_O \qquad x = -\frac{e\mathcal{E}}{2m}t^2 + U_Ot + X_O$$

- In any magnetic field, $\mathbf{f}$ is normal to $\mathbf{u}$ and no work is done. In a uniform magnetic field, the path is circular with $r = mU_O/eB$.

- Electron emission from a solid requires the addition of energy equal to the work function; this energy can be obtained in various ways. The probability of an electron possessing energy eV_T varies as $e^{-eV_T/kT}$.

- A cathode-ray tube consists of an electron gun producing a focused beam, a magnetic or electric deflection system, and a fluorescent screen for visual display. For deflecting voltage V_D, the deflection is

$$Y = \frac{LL_s}{2dV_A}V_D$$

- A cathode-ray oscilloscope includes display tube, intensity and focus controls, amplifiers and attenuators, sweep generator, and triggering circuit. A cathode-ray oscilloscope measures voltage and time, displays waveforms, and compares phase and frequency.

REVIEW QUESTIONS

1. What quantities are analogous in the equations of motion of a mass in a gravitational field and motion of an electron in an electric field?

2. Define the following terms: electron-volt, electron gun, vector cross product, work function, and virtual cathode.

3. Justify the statement that "no work is done on a charged particle by a steady magnetic field."

4. Describe the motion of an electron starting from rest in a region of parallel electric and magnetic fields.

5. Sketch the pattern expected on a CRT screen when the deflection voltages are $v_x = V \sin \omega t$ and $v_y = V \cos 2\omega t$.

6. What is the effect on a TV picture of varying the voltage on the control anode (Fig. 4.6)? On the accelerating anode?

7. Sketch a magnetic deflection system for a CRT.

8. Explain qualitatively the process of thermionic emission.

9. How can a CRO be used to measure voltage? Frequency? Phase?

10. Explain the operation of CRO sweep, trigger, and blanking circuits.

EXERCISES

1. An electron is accelerated from rest by a potential of 200 V applied across a 5-cm distance under vacuum. Calculate the final velocity and the time required for transit. Repeat for a hydrogen ion. Express the energy gained by each particle in electron-volts.

2. In Fig. 4.14, an electron is introduced at point P in an evacuated space. It is accelerated from rest toward anode 1 at voltage V_1 and passes through a small hole at point b. In terms of the given quantities and the properties of an electron:
   (a) What is the acceleration at point a?
   (b) What is the velocity at point b?
   (c) What voltage V_2 would just bring the electron to rest at point c?

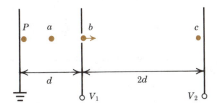

Figure 4.14

3. In a region where $\varepsilon = 200$ V/m, an electron is accelerated from rest. Predict the velocity, kinetic energy gained, displacement, and potential energy lost after 2 ns. Compare the PE lost $(e\,\Delta V)$ and the KE gained $(\frac{1}{2}mu^2)$.

4. An electron is to be accelerated from rest to a velocity $u = 12 \times 10^6$ m/s after traveling a distance of 5 cm.
   (a) Specify the electric field required and estimate the time required.
   (b) Repeat part (a) for a hydrogen ion.
   (c) Express the energy gained by each particle in joules.

5. For the electron of Exercise 3, derive expressions for velocity as a function of time and as a function of distance.

6. An electron with an energy of 200 eV is projected at an angle of 45° into the region between two parallel plates carrying a voltage V and separated a distance $d = 5$ cm. (See Fig. 4.15.)
   (a) If $V = -200$ V, determine where the electron will strike.
   (b) Determine the voltage V at which the electron will just graze the upper plate.

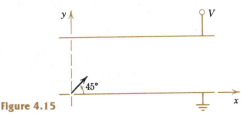

Figure 4.15

7. If the polarity of the upper plate in Fig. 4.15 is reversed so that $V = +200$ V, determine where the electron will strike.

8. Two electrons, e_1 traveling at velocity u and e_2 traveling at velocity $2u$, enter an intense magnetic field directed out of the paper (Fig. 4.16). Sketch the paths of the electrons.

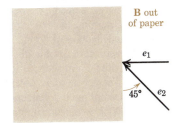

Figure 4.16

9. An electron with a velocity of 30 km/s is injected at right angles to a uniform magnetic field where $B = 0.02$ T into the paper.
   (a) Sketch the path of the electron.
   (b) Predict the radius of the path and the time required to traverse a semicircle.

10. In a *mass spectrograph*, isotopes of charge e and unknown mass are accelerated through voltage V and injected into a transverse magnetic field B. Derive an expression for the mass m in terms of the radius r of the circle described by the particle.

11. In a cyclotron (see any physics book), hydrogen ions are accelerated by an electric field, curved around by a magnetic field, and accelerated again. If B is limited to 2 T, what cyclotron diameter is required for a velocity half that of light?

12. An electron moves in an electric field $\varepsilon = 100$ V/m with a velocity of 10^6 m/s normal to the earth's magnetic field ($B = 5 \times 10^{-5}$ T). Compare the electric and magnetic forces with that due to the earth's gravitational field. Is it justifiable to neglect gravitational forces in practical problems?

13. The work function of copper is 4.1 eV and the melting point is 1356 K. Thorium has a work function of 3.5 eV and melts at 2120 K. Compare the factor of $e^{-W_W/kT}$ for these two metals at 90% of their respective melting points and decide which is more likely to be used as a cathode.

14. The Richardson-Dushman equation indicates that cathode emission current is proportional to $T^2 e^{-W_W/kT}$. For a special cathode coating with a work function of 1 eV, the emission current density at 1100 K is 200 mA/cm². Estimate the emission current densities at 800 and 1200 K.

15. In the CRT deflection system of Fig. 4.7, the accelerating voltage is 1000 V, the deflecting plates are 2.5 cm long and 1 cm apart, and the distance to the screen is 40 cm.
    (a) Determine the velocity of the electrons striking the screen.
    (b) Determine the deflecting voltage required for a deflection of 4 cm.

16. The vertical deflecting plates of a CRT are 2 cm long and 0.5 cm apart; length L_s in Fig. 4.7 is 50 cm. The acceleration potential is 5 kV.
    (a) Determine the deflection sensitivity.
    (b) What is the maximum allowable deflection voltage?
    (c) For a brighter picture, the acceleration potential is increased to 10 kV. What is the new deflection sensitivity?

17. A CRO with a 10 × 10-cm display has vertical amplifier settings of 0.1, 0.2, and 0.5 V/cm and horizontal sweep settings of 1, 2, and 5 ms/cm. Assuming the sweep starts at $t = 0$, select appropriate settings and sketch the pattern observed when a voltage $v = 2 \sin 60\pi t$ V is applied to the vertical input.

18. Repeat Exercise 17 for an applied voltage $v = 0.5 \cos 1000t$ V.

19. On the CRO of Exercise 17, determine the frequency of a square-wave signal that occupies:
    (a) 2.5 cm per cycle at a sweep setting of 5 ms/cm.
    (b) 4.0 cm for 10 cycles at a sweep setting of 1 ms/cm.

20. On a CRO vertical and horizontal amplifiers are set at 1 V/cm. Sketch the pattern observed when the voltages applied to the vertical and horizontal inputs are:
    (a) $v_v = 5 \cos 400t$ and
        $v_h = 5 \cos 100t$ V.
    (b) $v_v = 5 \cos 200t$ and
        $v_h = 5 \cos (700t - \pi/2)$ V.

21. Design a circuit for displaying the v-i characteristic of a diode on a CRO. Assume that the horizontal and vertical amplifiers have a common internal ground.

22. The CRO of Exercise 21 has an 8 × 8-cm screen. Anticipating an I-V characteristic similar to that shown in Fig. 5.8, specify the significant CRO settings.

PROBLEMS

1. As Einstein pointed out, the actual mass m of a moving particle is $m = m_O/\sqrt{1 - u^2/c^2}$ where m_O is the rest mass, u is velocity, and c is the velocity of light. Since energy and mass are equivalent, the potential energy lost in falling through a voltage V must correspond to an increase in mass $m - m_O$.
    (a) Equate the potential energy lost to the equivalent energy gained and calculate the voltage required to accelerate an electron to 99% of the speed of light.
    (b) Calculate the electron velocity for $V = 24$ kV, a typical value in a television picture tube, and determine the percentage of error in Eq. 4-11.

2. The magnetic deflecting *yoke* of a TV picture tube provides a field $B = 0.002$ T over an axial length $l = 2$ cm. The *gun* provides electrons with a velocity $U_Z = 10^8$ m/s. Determine:
    (a) The radius of curvature of the electron path in the magnetic field.
    (b) The approximate angle at which electrons leave the deflecting field.
    (c) The distance from yoke to screen for a 3-cm *positive* deflection.
    (d) A general expression for deflection Y in terms of acceleration potential V_a and compare with Eq. 4-22. (Assume small angular deflections where $Y/L = \tan \alpha \cong \alpha$.)

3. How long is an electron in the deflecting region of the CRT of Exercise 16? If the deflecting voltage should not change more than 10% while deflection is taking place, approximately what frequency limit is placed on this deflection system?

4. The useful range of an amplifier is bounded by frequencies f_1 and f_2 at which low-frequency and high-frequency defects occur.

   (a) In a certain amplifier, signals at $3f_1$ are amplified linearly, but at f_1 sinusoidal components are reduced to 70% of their relative value and shifted forward 45° in phase. For a square-wave input of frequency f_1, draw the fundamental and third harmonic components in the output and compare their sum to Fig. 4.13c.

   (b) In the same amplifier, signals at $f_2/3$ are amplified linearly, but at f_2 sinusoidal components are reduced to 70% of their relative value and shifted backward 45° in phase. For a square-wave input at frequency $f_2/3$, repeat part (a) and compare to Fig. 4.13d.

5

Semiconductor Diodes

Conduction in Solids

Doped Semiconductors

Junction Diodes

Special-Purpose Diodes

The cathode-ray tube is a good example of the virtues and the disadvantages of vacuum tubes. By using carefully constructed electric and magnetic fields, the flow of mobile electrons can be precisely controlled, and the result is a versatile information processing device. However, high power is necessary for thermionic emission and large dimensions are required for practical operation of vacuum tubes; these are serious disadvantages. In contrast, semiconductor electronic devices require no cathode power and their dimensions are extremely small. The emphasis in this book will be on semiconductor diodes and transistors in discrete and integrated forms.

First we consider conduction in solids where electron motion is influenced by the fixed ions of a conductor or the doping atoms added to a semiconductor. Using our knowledge of conduction mechanisms, we then examine the operation of semiconductor diodes and develop a quantitative understanding of diode junction behavior. Finally we derive circuit models for real diodes and look at some applications where the ideal diode representation of Chapter 3 is inadequate. We shall use the knowledge of conduction and junction phenomena that we gain here in analyzing the operation of transistors in Chapter 6 and the operation of more sophisticated devices in Chapters 7 and 8.

CONDUCTION IN SOLIDS

Conduction occurs in a vacuum if free electrons are available to carry charge under the action of an applied field. In an ionized gas, positively charged ions as well as electrons contribute to the conduction process. In a liquid, the charge carriers are positive and negative ions moving under the influence of an applied field. Solids vary widely in the type and number of charge carriers available and in the ease with which the carriers move under the action of applied fields. In electronic engineering we are interested in *insulators*, which have practically no available charge carriers; *conductors*, which have large numbers of mobile charge carriers; and *semiconductors*, which have conductivities intermediate between those of insulators and conductors.

Metallic Conductors

In a typical metal such as copper or silver, the atoms are arranged in a systematic array to form a *crystal*. The atoms are in such close proximity that the outer, loosely bound electrons are attracted to numerous neighboring nuclei and, therefore, are not closely associated with any one nucleus (Fig. 5.1). These *conduction electrons* are visualized as being free to wander through the crystal structure or *lattice*. At absolute zero temperature, the conduction electrons encounter no opposition to motion and the resistance is zero.

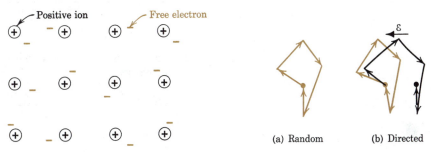

Figure 5.1 A simplified two-dimensional representation of a metallic crystal.

(a) Random   (b) Directed

Figure 5.2 The motion of an electron in a metallic crystal.

At ordinary temperatures, the electron-deficient atoms called *ions* possess kinetic energy in the form of vibration about their neutral positions in the lattice; this vibrational energy is measured by temperature. There is a continual interchange of energy between the vibrating ions and the free electrons in the form of elastic and inelastic collisions. The resulting electron motion is random, there is no net motion, and the net current is zero (Fig. 5.2a).

If a uniform electric field of intensity $\mathcal{E}$ (V/m) is applied, the electrons are accelerated; superimposed on the rapid random motion, there is a small component of velocity in the direction of $-\mathcal{E}$. At each inelastic collision with an ion, the electron loses its kinetic energy; it then accelerates again, gains a component of velocity in the $-\mathcal{E}$ direction, and loses its energy at the next inelastic collision (Fig. 5.2b). The time between collisions is determined by the random velocity and the length of the mean free path. On the average, the electrons gain a directed *drift velocity u* that is directly proportional to $\mathcal{E}$ (since the acceleration is constant over an average increment of time), and

$$u = \mu(-\mathcal{E})\qquad\qquad(5\text{-}1)$$

where μ (mu) is the *mobility* in meters per second/volts per meter or $m^2/V \cdot s$.

$R = \frac{l\rho}{A}$

The resulting flow of electrons carrying charge $-e$ at drift velocity u constitutes a current. If there are n free electrons per cubic meter, the current density $J\,(\mathrm{A/m^2})$ is

$$J = n(-e)u = ne\mu\mathcal{E} = \sigma\mathcal{E} \qquad (5\text{-}2)$$

where

$$\sigma = ne\mu$$

is the *conductivity* of the material in siemens per meter. The reciprocal of σ (sigma) is the *resistivity* ρ (rho) in ohm-meters.

$\rho = \frac{RA}{l} = \text{ohm-units}$

$\sigma = \frac{l}{RA}$

EXAMPLE 1

$\sigma = \frac{1}{\rho} = ne\mu$

$u = \frac{1}{\rho n e}$

The resistivity of copper is 1.73×10^{-8} $\Omega \cdot \mathrm{m}$ at 20 °C. Use Avogadro's law to find the number of atoms of copper in a cubic meter. Then find the average drift velocity in a copper conductor with a cross-sectional area of $10^{-6}\,\mathrm{m^2}$ carrying a current of 4 A.

For copper with an atomic weight of 63.6 and a density of 8.9 $\mathrm{g/cm^3}$, by Avogadro's law the number of atoms per $\mathrm{meter^3}$ is

$$n_A = \frac{6.022 \times 10^{23}\ \mathrm{atoms/g\text{-}atom} \times 8.9 \times 10^6\ \mathrm{g/m^3}}{63.6\ \mathrm{g/g\text{-}atom}}$$

$$= 8.43 \times 10^{28}/\mathrm{m^3}$$

Assuming that there is one free electron per atom, $n = n_A$. By Eq. 5-2, the velocity is

$$u = \frac{J}{ne} = \frac{I/A}{ne} = \frac{4 \times 10^6}{8.43 \times 10^{28} \times 1.6 \times 10^{-19}}$$

$$\cong 3 \times 10^{-4}\ \mathrm{m/s}$$

The average drift velocity in a good conductor is very slow compared to the random thermal electron velocities of the order of 10^5 m/s at room temperatures. As temperature increases, random thermal motion increases, the time between energy-robbing collisions decreases, mobility decreases, and therefore conductivity decreases. It is characteristic of metallic conductors that resistance increases with temperature.

Note that Eq. 5-2 is another formulation of Ohm's law; it says that the current density is directly proportional to the voltage gradient. For a conductor of area $A\,(\mathrm{m^2})$ and length l (m),

$$I = JA = \sigma A\mathcal{E} = \sigma A \frac{V}{l} = \sigma \frac{A}{l} V \qquad (5\text{-}3)$$

where $\sigma A/l$ is the conductance G in siemens. Alternatively,

$$R = \frac{V}{I} = \frac{1}{G} = \frac{l}{\sigma A} = \rho \frac{l}{A} \qquad (5\text{-}4)$$

where R is the resistance in ohms.

Practice Problem 5-1

In a certain conductor with a cross-sectional area of 10^{-7} m², there are 10^{23} electrons/m³ with mobility equal to 0.4 m²/V · s. Predict the conductivity and the resistance of 10 cm of this conductor. $R = \dfrac{\ell}{\sigma A} = 1.56 \times 10^{-12}\,\Omega$

Answers: 6400 S/m; 156 Ω.

Semiconductors

The two semiconductors of greatest importance in electronics are silicon and germanium. These elements are located in the fourth column of the periodic table and have four valence electrons. The crystal structure of silicon or germanium follows a tetrahedral pattern with each atom sharing one valence electron with each of four neighboring atoms. The *covalent bonds* are shown in the two-dimensional representation of Fig. 5.3. At temperatures close to absolute zero, the electrons in the outer shell are tightly bound, there are no free carriers, and silicon is an insulator.

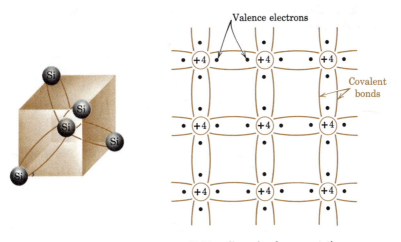

(a) Tetrahedral pattern   (b) Two–dimensional representation

Figure 5.3 Arrangement of atoms in a silicon crystal.

The energy required to break a covalent bond is about 1.1 eV for silicon and about 0.7 eV for germanium. At room temperature (300 K), a few electrons have this amount of thermal energy and are excited into the conduction band (see p. 105), becoming free electrons. When a covalent bond is broken (Fig. 5.4), a vacancy or *hole* is left. The region in which this vacancy exists has a net positive charge; the region in which the freed electron exists has a net negative charge. In such semiconductors *both electrons and holes contribute to electrical conduction*. If a valence electron from another covalent bond fills the hole (without ever gaining sufficient energy to become

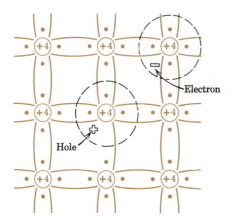

Figure 5.4 Silicon crystal with one covalent bond broken.

"free"), the vacancy appears in a new place and the effect is as if a positive charge (of magnitude e) has moved to the new location.[†]

Consideration of hole behavior as the motion of a positively charged particle of definite mass and mobility is consistent with the quantum mechanical properties of valence electrons. Attempting to describe this behavior in terms of classical physics leads to results that are contrary to the observed facts. For our purposes we shall consider that:

> *Conduction in semiconductors is due to two separate and independent particles carrying opposite charges and drifting in opposite directions under the influence of an applied electric field.*

With two charge-carrying particles, the expression for current density is

$$J = (n\mu_n + p\mu_p)e\mathcal{E} = \sigma\mathcal{E} \qquad (5\text{-}5)$$

where n and p are the concentrations of electrons and holes (number/m^3) and μ_n and μ_p are the corresponding mobilities. Conductivity depends on the number of charge carriers and their mobility; for a semiconductor, the conductivity is

$$\sigma = (n\mu_n + p\mu_p)e \qquad (5\text{-}6)$$

Some of the properties of silicon and germanium are shown in Table 5-1.

Because holes and electrons are created simultaneously, in a pure semiconductor the number of holes is just equal to the number of conduction electrons, or

$$n = p = n_i \qquad (5\text{-}7)$$

where n_i is the *intrinsic* concentration. New electron-hole pairs are generated continuously and, under equilibrium conditions, they are removed by recombination at the same rate. At ordinary operating temperatures, only a very small fraction of the valence electrons are in the conduction state at any instant. (See Example 2.)

[†] There is an analogy in our economic system. In payment for a sweater, you may give a retailer $30 in cash or accept a bill from him for $30. You feel the same with $30 *subtracted* from your wallet or with $30 *added* to your debts. If you send the bill to your father for payment, the "vacancy appears in a new place" and the effect is as if your father had sent you $30. The sum of cash to you and bills to your father constitutes the net financial "current" from your father to you.

Table 5-1 Properties of Silicon and Germanium at 300 K[†]

	Silicon	Germanium
Energy gap (eV)	1.1	0.67
Electron mobility μ_n (m²/V·s)	0.135	0.39
Hole mobility μ_p (m²/V·s)	0.048	0.19
Intrinsic carrier density n_i (/m³)	1.5×10^{16}	2.4×10^{19}
Intrinsic resistivity ρ_i (Ω·m)	2300	0.46
Density (g/m³)	2.33×10^6	5.32×10^6

[†] Quoted by Esther M. Conwell in "Properties of Silicon and Germanium II," *Proc. IRE*, June 1958, p. 1281.

EXAMPLE 2

Estimate the relative concentration of silicon atoms and electron-hole pairs at room temperature, and predict the intrinsic resistivity. (80 °F ≅ 27 °C = 300 K.)

By Avogadro's law, the concentration of atoms is

$$n_A = \frac{6.022 \times 10^{23} \text{ atoms/g-atom} \times 2.33 \times 10^6 \text{ g/m}^3}{28.09 \text{ g/g-atom}}$$

$$\cong 5 \times 10^{28} \text{ atoms/m}^3$$

From Table 5-1, the intrinsic concentration at 300 K is 1.5×10^{16} electron-hole pairs per cubic meter; therefore, there are

$$\frac{n_A}{n_i} \cong \frac{5 \times 10^{28}}{1.5 \times 10^{16}}$$

$$\cong 3.3 \times 10^{12} \text{ silicon atoms per electron-hole pair}$$

Since in the pure semiconductor $n = p = n_i$, by Eq. 5-6 the intrinsic conductivity is

$$\sigma_i = (n\mu_n + p\mu_p)e = (\mu_n + \mu_p)n_i e$$

$$= (0.135 + 0.048) \times 1.5 \times 10^{16} \times 1.6 \times 10^{-19}$$

$$= 4.4 \times 10^{-4} \text{ S/m}$$

and the intrinsic resistivity is

$$\rho_i = \frac{1}{\sigma_i} = \frac{1}{4.4 \times 10^{-4}} \cong 2280 \text{ Ω·m}$$

The result of Example 2 tells us that intrinsic silicon is a poor conductor as compared to copper ($\sigma = 5.8 \times 10^7$ S/m); semiconductors are valuable, not for their conductivity, but for two unusual properties. First, the concentration of free carriers, and consequently the conductivity, increases exponentially with temperature (approximately 5% per degree Celsius at ordinary temperatures). The *thermistor*, a temperature-sensitive device that utilizes this property, is used in instrumentation and control, and it also may be used to compensate for the *decrease* in conductivity with temperature in the metallic parts of a circuit. Second, the conductivity of a semiconductor can be increased greatly, and to a precisely controlled extent, by adding small amounts of impurities in the process called *doping*. Since there are two types of mobile charge carriers, of opposite sign, extraordinary distributions of charge carriers can be created. The semiconductor diode and the transistor use this property.

DOPED SEMICONDUCTORS

We can demonstrate the startling effect of doping by adding to silicon or germanium an impurity element from the adjacent third or fifth column of the periodic table. (See Table 5-2.) Assume that a small amount of a *pentavalent* element (antimony, phosphorus, or arsenic) is added to otherwise pure silicon. Such a doping element has

Table 5–2 Semiconductor Elements in the Periodic Table
(with atomic number and atomic weight)

III	IV	V
$+3$	$+4$	$+5$
5 B	6 C	7 N
BORON	CARBON	NITROGEN
10.82	12.01	14.008
13 Al	14 Si	15 P
ALUMINUM	SILICON	PHOSPHORUS
26.97	28.09	31.02
31 Ga	32 Ge	33 As
GALLIUM	GERMANIUM	ARSENIC
69.72	72.60	74.91
49 In	50 Sn	51 Sb
INDIUM	TIN	ANTIMONY
114.8	118.7	121.8

five valence electrons and the effective ionic charge is $+5e$. When a pentavalent atom replaces a silicon atom in the crystal lattice (Fig. 5.5), only four of the valence electrons are used to complete the covalent bonds; the remaining electron, with the addition of a small amount of thermal energy, becomes a free electron available for conduction. We call the resulting material an *n-type* semiconductor because of the

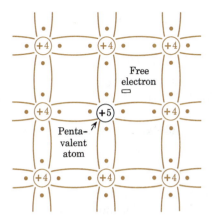

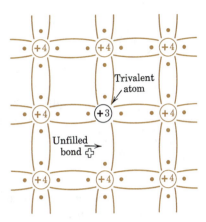

Figure 5.5 Effect of *n*-type doping.

Figure 5.6 Effect of *p*-type doping.

presence of negative charge carriers in an electrically neutral crystal. The net positive charges left behind are immobile and cannot contribute to current.

If we add a small amount of a *trivalent* element (aluminum, boron, gallium, or indium) to otherwise pure silicon, we get a *p-type* semiconductor. When a trivalent atom replaces a silicon atom in the crystal lattice (Fig. 5.6), only three valence electrons are available to complete covalent bonds. If the remaining unfilled covalent bond is filled by a valence electron from a neighboring atom, a mobile hole is created and there is the possibility of current conduction by the motion of positive charges. Used in this way, a trivalent atom is called an *acceptor* atom because it accepts an electron. (The net negative charge created is immobile.) In the same sense, a pentavalent atom is called a *donor* atom.

By adding donor or acceptor atoms in very small amounts, we can greatly increase the conductivity of a semiconductor. To predict the magnitude of the effect, we need a qualitative understanding of the process of carrier generation and recombination.

The *thermal generation rate g*, in electron-hole pairs per second per cubic meter, depends on the material and the temperature. The energy needed in the "generation" of a conduction electron is expressed in electron volts as eV_g. The energy corresponding to temperature T can be expressed as kT. Statistical analysis shows that the probability of a valence electron becoming free for conduction is proportional to $e^{-eV_g/kT}$. If the energy gap eV_g is low or the temperature T is high, the thermal generation rate will be high.

In a semiconductor, the mobile electrons and holes tend to recombine and disappear. The *rate of recombination R*, in electron-hole pairs/s $\cdot$ m^3, depends on the number of reacting carriers present. If there are few free electrons and holes, the rate of recombination is low; if there are many it is high.[†] If, as in *n*-type material, there

[†] As a crude analogy, consider a dance in which couples dance together for a while, then separate and wander among the other couples until they find new partners. The rate of recombination depends on the number of unattached partners available.

are few holes but many, many electrons, the rate of recombination will be high. In general,

$$R = rnp \tag{5-8}$$

where r is a proportionality constant for the material.

Under equilibrium conditions, the rate of generation is just equal to the rate of recombination. The new electron-hole pairs are created by thermal generation and, even in a doped crystal, practically all the atoms are pure silicon (or germanium). Therefore,

$$g = g_i = R_i = rn_ip_i = rn_i^2 \tag{5-9}$$

The concentration of mobile holes and electrons must satisfy the general recombination rule that $R = rnp$. Equating recombination rate to generation rate tells us that, in a doped semiconductor,

$$np = n_i^2 \tag{5-10}$$

In words, the product of electron and hole concentrations is a constant; if one is increased (by doping), the other must decrease.

If we add impurity atoms to intrinsic semiconductor, practically all the donor or acceptor atoms are ionized at ordinary temperatures. In typical n-type material, donor atoms in concentration $N_d \gg n_i$ provide a mobile electron concentration

$$n_n \cong N_d \tag{5-11}$$

By Eq. 5-10, the concentration of holes is greatly reduced and

$$p_n = \frac{n_i^2}{n_n} \cong \frac{n_i^2}{N_d} \tag{5-12}$$

If, instead, we create p-type material by adding an acceptor impurity, the carrier concentrations will be

$$p_p \cong N_a \quad \text{and} \quad n_p \cong \frac{n_i^2}{N_a} \tag{5-13}$$

To summarize:

In a typical doped semiconductor, the concentration of mobile charge carriers (n_n or p_p) is approximately equal to the concentration of doping atoms (N_d or N_a); the conductivity is determined by the doping concentration alone because the concentration of carriers of the opposite polarity (p_n or n_p) is greatly reduced.

We can use these relations to predict the effect of doping as in Example 3.

Practice Problem 5-2

Predict the new conductivity of the silicon crystal in Example 3 if the doping impurity is arsenic instead of indium.

Answer: 108 S/m.

EXAMPLE 3

Predict the effect on the electrical properties of a silicon crystal at room temperature if every 10-millionth silicon atom is replaced by an atom of indium.

Since there are 5×10^{28} silicon atoms/m³, the necessary concentration of acceptor atoms is

$$N_a = 5 \times 10^{28} \times 10^{-7} = 5 \times 10^{21} \cong p_p$$

The intrinsic concentration is $p_i = n_i = 1.5 \times 10^{16}/m^3$. Therefore, there are $(5 \times 10^{21}) \div (1.5 \times 10^{16}) = 3.3 \times 10^5$ as many holes. The new electron concentration is only

$$n_p = \frac{n_i^2}{p_p} \cong \frac{n_i^2}{N_a} = \frac{2.25 \times 10^{32}}{5 \times 10^{21}} = 4.5 \times 10^{10}/m^3$$

Therefore, the electron concentration has been reduced by the factor $(1.5 \times 10^{16}) \div (4.5 \times 10^{10}) = 3.3 \times 10^5$ and Eq. 5-13a is justified.

The new conductivity is

$$\sigma_p = (n_p \mu_n + p_p \mu_p)e \cong N_a(\mu_p)e$$

$$= 5 \times 10^{21} \times 0.048 \times 1.6 \times 10^{-19} = 38.4 \text{ S/m}$$

as compared to 0.00044 S/m in intrinsic silicon. In other words, the resistivity of the doped silicon is

$$\rho_p = 1/\sigma_p = 1/38.4 = 0.026 \ \Omega \cdot m$$

as compared to the intrinsic value $\rho_i = 2300 \ \Omega \cdot m$ in Table 5-1.

We conclude from Example 3 that a small concentration of impurity greatly modifies the electrical properties of a semiconductor. Even more startling effects occur in regions where the impurity concentration is highly nonuniform, as it is at a junction of n-type and p-type materials.

Diffusion

If the doping concentration is nonuniform, the concentration of charged particles is also nonuniform, and it is possible to have charge motion by the mechanism called *diffusion*. In an analogous situation, if a bottle of perfume is opened at one end of an apparently still room, in a short time some of the perfume will have reached the other end. For mass transfer by diffusion, two factors must be present. There must be random motion of the particles and there must be a concentration gradient. The thermal energy of apparently still air provides the first, and opening a bottle of concentrated perfume provides the second. Note that no force is required for this type of transport; the net transfer is due to a statistical redistribution.

If a nonuniform concentration of randomly moving electrons or holes exists, diffusion will occur. If, for example, the concentration of electrons on one side of an imaginary plane is greater than that on the other, and if the electrons are in random motion, after a finite time there will be a net motion of electrons from the more dense side. This net motion of charge that constitutes a *diffusion current* is proportional to the concentration gradient dn/dx. The diffusion current density due to electrons is given by

$$J_n = eD_n \frac{dn}{dx} \qquad (5\text{-}14)$$

where D_n = the diffusion constant for electrons (m^2/s). If dn/dx is positive, the resulting motion of negatively charged electrons in the $-x$ direction constitutes a positive current in the $+x$ direction. The corresponding relation for holes is

$$J_p = -eD_p \frac{dp}{dx} \qquad (5\text{-}15)$$

Note that in each case the charged particles move away from the region of the greatest concentration, but the motion is not due to any force of repulsion. Like mobility, diffusion is a statistical phenomenon and it is not surprising to find that they are related by the Einstein equation

$$\frac{\mu_n}{D_n} = \frac{\mu_p}{D_p} = \frac{e}{kT} \qquad (5\text{-}16)$$

or that the factor e/kT, which appeared in thermionic emission and thermal generation, shows up again here.

JUNCTION DIODES

Semiconductors, pure or doped, p-type or n-type, are bilateral; current flows in either direction with equal facility. If, however, a p-type region exists in close proximity to an n-type region, there is a carrier density gradient that is unilateral; current flows easily in one direction only. The resulting device, a *semiconductor diode*, exhibits a very useful control property.

pn Junction Formation

The fabrication of semiconductor devices is a developing technological art. For satisfactory results, the entire device must have a common crystalline structure. The pure crystals necessary may be "grown" by touching a "seed" crystal to the surface of molten silicon and then slowly raising it (Fig. 5.7a). The silicon crystallizes onto the seed and grows in the same crystalline orientation as the seed. If the melt contains a donor element such as arsenic, an n-type semiconductor is obtained. If at a certain point in the growing process an excess of acceptor element is added to the melt, the subsequent crystal will be p-type. The metallurgical boundary between the two materials is called a *grown junction*.

If a dot of trivalent indium is placed on an n-type silicon wafer (sliced from a grown crystal) and heated to the proper temperature, a small drop of molten mixture will form. Upon cooling, the alloy solidifies into p-type silicon that follows the crystalline pattern of the host crystal (Fig. 5.7b). The resulting device is an *alloy-junction* diode. By careful control of the alloying procedure, it is possible to obtain

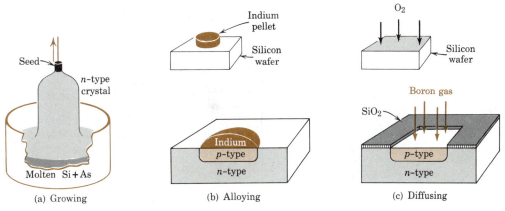

(a) Growing   (b) Alloying   (c) Diffusing

Figure 5.7 Methods of forming *pn* junctions.

diodes in which the transition from *p*-type to *n*-type material occurs in a distance corresponding to a few atomic layers.

Precisely controlled geometries and improved device properties are possible with the *planar* technology illustrated in Fig. 5.7c. First a thin layer of an inert oxide (such as glassy silicon dioxide) is formed on the surface of the *n*-type semiconductor wafer by heating in the presence of oxygen. Then the oxide layer is removed from a selected area by etching through a mask. When the heated semiconductor is exposed to boron gas, for example, acceptor atoms penetrate the selected area of the crystalline structure by solid-state diffusion and a precisely defined *pn* junction is formed. (The unetched oxide layer is impervious.) In another technique, the junction is formed by deposition of the doped material from a vapor. The atoms formed by this *epitaxial growth* (grown "upon, in the same form") follow the orientation of the underlying *substrate* crystal, but they may possess entirely different electrical characteristics.

pn Junction Behavior

The external behavior of a typical *pn* junction diode is shown in Fig. 5.8. To use a diode effectively, we must understand (at least qualitatively) the internal mechanism that results in these remarkable properties. An understanding of diode behavior is directly transferable to the analysis of more complicated devices, such as the transistor, so a detailed study at this point is justified. We assume that the diode consists of a single crystal with an abrupt change from *p*- to *n*-type at the junction in the *y-z* plane; only variations with respect to *x* are significant.

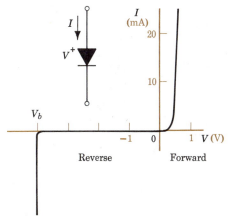

Figure 5.8 Junction diode symbol and *I-V* characteristic for germanium.

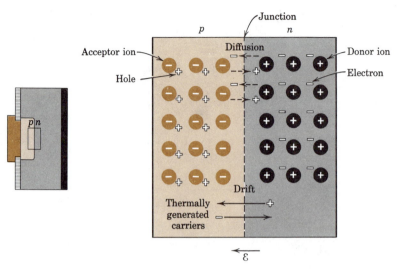

Figure 5.9 Schematic representation of charges and conduction mechanisms in a section of a *pn* diode.

The distribution of charges near the junction in such a diode is represented in Fig. 5.9. In the enlarged view of the small *pn* rectangle, the circles represent immobile impurity ions with their net charges; the separate $+$ and $-$ signs represent mobile holes and electrons. We imagine that a junction is created instantaneously; because of the concentration gradient at the junction, holes diffuse from the *p*-type material to the right across the junction and recombine with some of the many free electrons in the *n*-type material. Similarly, electrons diffuse from the *n*-type material to the left across the junction and recombine. The diffusion of holes leaves *uncovered* bound negative charges on the left of the junction, and the diffusion of electrons leaves *uncovered* bound positive charges on the right.

To avoid repeating every statement, we call holes in the *p*-region and electrons in the *n*-region *majority carriers*. Thus the result of the diffusion of majority carriers is the uncovering of bound charge in the *depletion* or *transition region* and the creation of an internal electric field $\mathcal{E}$. At normal temperatures, electron-hole pairs are being created continually in both regions. Electrons in the *p*-region and holes in the *n*-region are called *minority carriers*. Under the action of the electric field, minority carriers drift across the junction. With two types of carriers (reduced from four by the use of the terms "majority" and "minority") and two different conduction mechanisms (drift and diffusion), we expect the explanation of the behavior displayed in Fig. 5.8 to be complicated.

Open Circuit. The simpler representation of the diode in Fig. 5.10 shows only the uncovered charge in the transition region; the remainder of the open-circuited diode is electrically neutral. The separation of charge at the junction creates an electric field $\mathcal{E}$ in the $-x$ direction. We use $v = -\int \mathcal{E}\, dx$ to obtain the potential distribution across the depletion region by integrating the field. The potential barrier or "potential hill" of height V_O acts to oppose diffusion of majority carriers (holes to the right and electrons to the left) and to encourage drift of minority carriers across the junction.

Under open-circuit conditions the net current is zero; the tendency of majority

$d\mathcal{E} = \dfrac{k \, \int \rho \, dv}{r^2}$

$\int$ of charge density

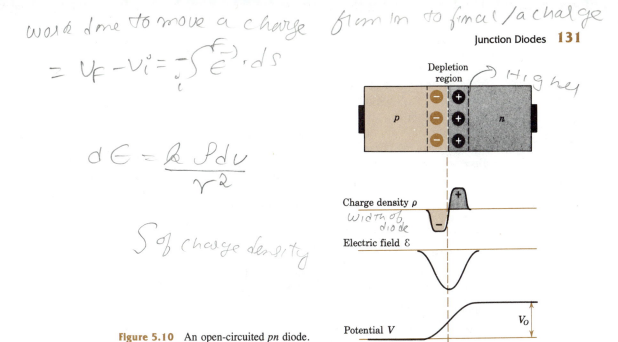

Figure 5.10   An open-circuited *pn* diode.

carriers to diffuse as a result of the density gradient is just balanced by the tendency of minority carriers to drift across the junction as a result of the electric field. The height of the potential barrier and the magnitude of V_O are determined by the necessity for equilibrium; for no net current, the diffusion component must be just equal to the drift component. The equilibrium value V_O is also called the *contact potential* for the junction. This potential, typically a few tenths of a volt, cannot be used to cause external current flow. Connecting an external conductor across the terminals creates two new contact potentials that just cancel the potential across the junction.[†]

Forward Bias. The diffusion component of current is very sensitive to the barrier height; only those majority carriers having kinetic energies in excess of eV_O diffuse across the junction. If equilibrium is upset by a decrease in the junction potential to $V_O - V$ (Fig. 5.11), the probability of majority carriers possessing sufficient energy to

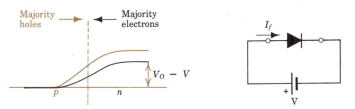

Figure 5.11   A forward-biased *pn* diode.

pass the potential barrier is greatly increased. The result is a net flow of carriers. Holes from p-type material cross the junction and are *injected* into n-type material where they combine with electrons; electrons move in the reverse direction.

[†] You may find it helpful to consider an analogy. In a vertical column of gas, molecules drift downward under the action of gravity, but they diffuse upward because of their random thermal energy. A pressure difference exists between top and bottom, but there is no net flow. The pressure difference cannot be used to cause any external flow.

In other words, a reduction in the potential barrier encourages carriers to flow from regions where they are in the majority to regions where they are in the minority. If the junction potential is reduced by the application of an external *biasing* voltage V, as shown in Fig. 5.11, a net current I flows. Under forward bias, the current is sustained by the continuous generation of majority carriers and recombination of the injected carriers. The external battery supplies electrons at the n terminal and removes electrons at the p terminal.

Practice Problem 5-3

At room temperature, $e/kT \cong 40$ V^{-1}. If the contact potential $V_O = 0.7$ V, estimate the factor by which the probability of diffusion of a charge carrier is increased by a forward-biased voltage $V = 0.2$ V.

Answer: $e^{-20}/e^{-28} \cong 3000$.

Reverse Bias. If the junction potential is raised to $V_O + V$ by the application of a reverse bias (Fig. 5.12), the probability of majority carriers possessing sufficient energy to pass the potential barrier is greatly decreased. The net diffusion of majority carriers across the junction, is reduced to practically zero by a reverse bias of a tenth of a volt or so.

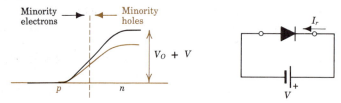

Figure 5.12 A reverse-biased *pn* diode.

A very small reverse current I_r does flow. The minority carriers thermally generated in the material adjacent to the transition region drift in the direction of the electric field. Because all minority carriers appearing at the edge of the transition region are forced across by the field (holes "roll down" a potential hill; electrons "roll up"), this reverse current depends on only the rate of thermal generation and is independent of the barrier height. (This current reaches its maximum value I_s or "saturates" at a low value of reverse bias.)

At ordinary temperatures the reverse current is very small, measured in μA (germanium) or nA (silicon). Ideally the reverse current in a diode should be zero; a low value of I_s is one of the principal advantages of silicon over germanium. Practically I_s is an important parameter that is sensitive to temperature. It is directly proportional to the thermal generation rate $g = rn_i^2$ which varies exponentially with temperature. As a rough approximation near room temperature, for silicon or germanium diodes, I_s doubles with each increase of 10° C.

General Behavior

To calculate the current flow in a *pn* junction diode, we can assume that the diode current consists of an injection component I_i sensitive to junction potential and a reverse saturation component I_s independent of junction potential. As we expect from our previous experience with thermionic emission and thermal generation (p. 125), the injection process is related to the statistical probability of electrons and holes having sufficient energy. Under a forward bias voltage V, this probability increases; the critical factor becomes

$$e^{-e(V_0-V)/kT} = e^{-eV_0/kT} \cdot e^{eV/kT} = A_1 e^{eV/kT} \qquad (5\text{-}17)$$

At a given temperature, A_1 is just a proportionality constant; the injection current I_i is of the form $A_1 A_2 e^{eV/kT} = A e^{eV/kT}$, where A is another constant. The reverse current is negative and nearly constant at value I_s. The total current is

$$I = I_i - I_s = A e^{eV/kT} - I_s \qquad (5\text{-}18)$$

Under open-circuit conditions where $V = 0$,

$$I = 0 = A e^0 - I_s = A - I_s \quad \therefore A = I_s$$

In general, then, the junction diode current is

$$I = I_s(e^{eV/kT} - 1) \qquad (5\text{-}19)$$

At room temperature (say, $T = 20°\,C = 293\,K$), $e/kT \cong 40\,V^{-1}$ and a convenient form of the equation is

$$I = I_s(e^{40V} - 1) \qquad (5\text{-}20)$$

EXAMPLE 4

The current of a germanium diode at room temperature is 100 μA at a voltage of -1 V. Predict the magnitude of the current for voltages of -0.2 V and $+0.2$ V at room temperature.

The current at -1 V can be assumed to be the reverse saturation current or $I_s = 100\ \mu A$.

At $V = -0.2$ V, $e^{40V} = e^{-8} \cong 0$; therefore,

$$I = I_s (0 - 1) \cong -I_s = -100\ \mu A.$$

At $V = +0.2$ V, $e^{40V} = e^8 \cong 3000$; therefore,

$$I = 100 \times 10^{-6}(3000 - 1) \cong +300\ mA.$$

Repeat the prediction for operation at 20 °C above room temperature.

At 20 °C above room temperature, I_s will double twice and be approximately four times as great. The new exponent eV/kT will be

$$8\,\frac{293}{293 + 20} = 7.49$$

The two values of current will be

$$I \cong -I_s = -400\ \mu A \quad \text{and} \quad I = I_s e^{7.49} = +715\ mA$$

respectively.

We can draw some general conclusions from Example 4. Since $e^{40(-0.1)} = e^{-4} = 0.02 \ll 1$ and $e^{40(0.1)} = e^{+4} \gg 1$, at room temperature,

For a reverse bias of more than 0.1 V or so, $I \cong -I_s$
For a forward bias of more than 0.1 V or so, $I \cong I_s e^{40V}$

Equation 5-19 holds quite well for germanium diodes, but it holds only approximately for silicon diodes. At normal currents, silicon diodes follow the relation

$$I = I_s(e^{eV/\eta kT} - 1) \qquad (5\text{-}21)$$

where factor η (eta) is approximately 2. For silicon diodes at high currents (and for germanium diodes in general), $\eta \cong 1$.

Practice Problem 5-4

The voltage at which significant diode current flows is called the *threshold voltage*. Assuming that the criterion is 1% of rated current, use Eq. 5-21 to estimate the threshold voltage at room temperature for two diodes rated at 1 A: a germanium diode with $I_s = 3 \ \mu A$ and a silicon diode with $I_s = 60$ nA.

Answers: 0.2 V; 0.6 V.

Breakdown Diodes

Along with the change in junction potential due to forward and reverse biasing, there is a significant change in the width of the transition or depletion region. For a given junction potential, a definite number of bound charges must be uncovered; therefore, the width of the depletion region depends on the density of doping. Under forward bias, the junction potential is reduced and the depletion region narrows; under reverse bias, the depletion region widens. (This behavior is utilized in the field-effect transistor described in the next chapter.)

If the reverse bias is increased sufficiently, a sudden increase in reverse current is observed (see Fig. 5.8). This behavior is due to the *Zener* effect or the *avalanche* effect. In Zener breakdown, the electric field in the junction becomes high enough to pull electrons directly out of covalent bonds. The electron-hole pairs thus created then contribute a greatly increased reverse current. The avalanche effect occurs at voltages higher than Zener breakdown voltages. At these high voltages, carriers gain sufficient energy between collisions to knock electrons from covalent bonds; these too can reach ionizing energies in a cumulative process.

In either type of breakdown, the reverse current is large and nearly independent of voltage. If the power dissipated is within the capability of the diode, there is no damage; reduction of the reverse bias voltage below V_b reduces the current to I_s. By controlling doping densities, it is possible to manufacture Zener diodes[†] with breakdown voltages from a few volts to several hundred volts. These are valuable because of their ability to maintain a nearly constant voltage under conditions of widely varying current.

[†]In commercial practice, the term "Zener diodes" includes devices employing both effects.

Circuit Models

The electrical model of a device is frequently called a *circuit model* because it is composed of ideal circuit elements. We derive the models from the characteristic curves of the real devices (Fig. 5.13) and, within limits, we can use them to predict the behavior of real devices in practical applications.

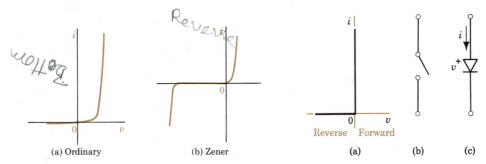

(a) Ordinary	(b) Zener

Figure 5.13 Real diode characteristics.

(a)	(b)	(c)

Figure 5.14 Ideal diode models.

The important diode characteristic is the discrimination between forward and reverse voltages. The ideal diode presents no resistance to current flow in the forward direction and an infinite resistance to current flow in the reverse direction. A selective switch, which is closed for forward voltages and open for reverse voltages, is equivalent to the ideal diode because it has the same current-voltage characteristic. Such a switch is represented in circuit diagrams by the symbol shown in Fig. 5.14c; the triangle points in the direction of forward current flow.

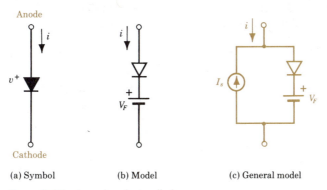

(a) Symbol	(b) Model	(c) General model

Figure 5.15 A semiconductor diode.

In a semiconductor diode, only a few tenths of a volt (0.3 V for germanium, 0.7 V for silicon) of forward bias is required for appreciable current flow. The combination of an ideal diode and a voltage source shown in Fig. 5.15b is usually satisfactory for predicting diode performance. If the reverse saturation current in a semiconductor diode is appreciable, this fact must be incorporated into the model. One possibility is to use a current source with a magnitude I_s directed as in Fig. 5.15c.[†]

[†]For example, the manufacturers' ratings for the IN4004 subminiature silicon rectifier indicate conservative values as follows: 1 V drop at 1 A in the forward direction and only 10 μA reverse current under rated conditions.

EXAMPLE 5

A common application of silicon diodes is shown in Fig. 5.16. Predict I_2 as a function of V_1.

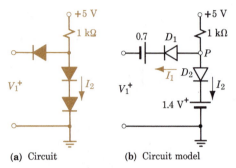

(a) Circuit   **(b)** Circuit model

Figure 5.16 Diode switching circuit.

As a first step in the analysis, replace the diodes by appropriate circuit models. The two diodes in series are represented by one ideal diode and a source of $2 \times 0.7 = 1.4$ V.

V_P cannot exceed 1.4 V, because at that voltage D_2 is forward biased and current I_2 flows readily. (With diode switch 2 "closed," P is held at 1.4 V.)

Also, V_P cannot exceed $V_1 + 0.7$, because at that voltage D_1 is forward biased and current I_1 flows readily.

The critical value is

$$V_P = 1.4 = V_1 + 0.7$$

or

$$V_1 = 1.4 - 0.7 = 0.7 \text{ V}$$

For $0 < V_1 < 0.7$ V, D_1 is forward biased and $I_2 = 0$.
For $V_1 > 0.7$ V, D_2 is forward biased and

$$I_2 = V/R = (5 - 1.4)/1000 = 3.6 \text{ mA}$$

Conclusion: A very small change in V_1 can switch I_2 from 0 to 3.6 mA.

The characteristic curve for a Zener diode can be linearized as in Fig. 5.17. When forward biased (v positive), current flows freely; the forward resistance is small and can be neglected. For reverse bias voltages greater than the breakdown potential, the resistance is estimated to be

$$R_Z = \frac{\Delta v}{\Delta i} = \frac{12 - 10}{0.01 - 0} = 200 \ \Omega$$

The circuit model includes a voltage source to indicate that reverse current does not flow until the negative voltage across the diode exceeds 10 V.

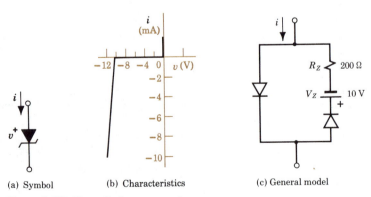

(a) Symbol   **(b)** Characteristics   **(c)** General model

Figure 5.17 Zener-diode representation.

Zener-Diode Voltage Regulator

Filters are designed to minimize the rapid variations in load voltage due to cyclical variations in rectifier output voltage (Fig. 3.32). There are other reasons for load voltage variation. If the amplitude of the supply voltage (V_m) fluctuates, as it does in practice, the dc voltage will fluctuate. If the load current changes due to a change in the demands of the connected equipment, the dc voltage will change due to the IR drop in the transformer, rectifier, and inductor (if any). A filter cannot prevent these types of variation and, if the load voltage is critical, a *voltage regulator* must be employed. Zener diodes are suitable for voltage regulation.

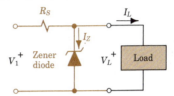

Figure 5.18   A Zener-diode voltage regulator circuit.

A simple Zener-diode voltage regulator (Fig. 5.18) consists of a series voltage dropping resistance R_S and a Zener diode in parallel with the load. Voltage V_1 is the dc output of a rectifier filter. The function of the regulator is to keep V_L nearly constant with changes in supply voltage V_1 or load current I_L. The operation is based on the fact that, in the Zener breakdown region, small changes in diode voltage are accompanied by large changes in diode current (see Fig. 5.17). The large currents flowing through R_S produce voltages that compensate for changes in V_1 or I_L.

EXAMPLE 6

A load draws a current varying from 10 to 100 mA at a nominal voltage of 100 V that can vary a maximum of 1%.

Specify V_1 for a regulator consisting of $R_S = 200 \ \Omega$ and a Zener diode represented by $V_Z = 100$ V and $R_Z = 20 \ \Omega$ (Fig. 5.19).

For $I_L = 50$ mA, determine the increase in dc input voltage V_1 corresponding to the permissible 1% variation in output voltage V_L.

To place the diode well into the Zener breakdown region, it is good design practice to use a minimum current $I_Z(\text{min})$ equal to 10% of $I_L(\text{max})$. Therefore, let $I_Z(\text{min}) = 0.1 \ I_L(\text{max}) = 0.1 \times 100 = 10$ mA here. Then

$$V_L = V_Z + I_Z R_Z = 100 + 0.01(20) = 100.2 \text{ V}$$

and

$$V_1 = V_L + (I_L + I_Z)R_S$$
$$= 100.2 + (0.05 + 0.01)200 = 112.2 \text{ V}$$

If V_L increases by 1% or 1 V, $V_L' = 101.2$ V,

$$I_Z' = \frac{V_L' - V_Z}{R_Z} = \frac{101.2 - 100}{20} = 0.06 \text{ A}$$

With $I_L = 50$ mA,

$$V_1' = V_L' + (I_L + I_Z')R_S$$
$$= 101.2 + (0.05 + 0.06)200 = 123.2 \text{ V}$$

Conclusion: A change in V_1 of 11 V produces a change in V_L of only 1 V.

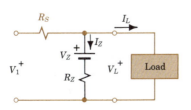

Figure 5.19   Performance of a voltage regulator.

SPECIAL-PURPOSE DIODES

The *pn* junction diode described by Eq. 5-19 is widely used in rectifying and waveshaping functions as discussed in Chapter 3. Other diodes,[†] similar in basic structure but presenting quite different external characteristics, can be explained in terms of our knowledge of junctions.

Tunnel Diodes

If the doping concentrations are increased, thereby increasing the concentration of uncovered charges, the width of the depletion region and the associated potential barrier are decreased. If the doping is greatly increased so that the depletion region is less than 10 nm wide, a new conduction mechanism is possible and the device characteristics are radically changed.

As explained by Leo Esaki in his 1958 announcement, for very thin potential barriers quantum mechanics theory indicates that there is a finite probability that an electron may *tunnel* through the barrier without ever possessing enough energy to climb over it. The *V-I* characteristics of an *Esaki diode* are shown in Fig. 5.20. The solid colored line indicates the tunneling effect. At voltages well below the threshold for normal forward current, electrons can tunnel from the *n* region to the *p* region if there are available holes of appropriate energy level into which the electrons can move without ever possessing sufficient energy to become free.[‡]

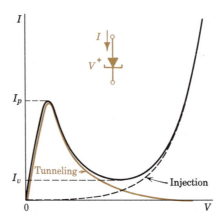

Figure 5.20 Tunnel diode characteristics.

The tunneling current increases with voltage until the effect of forward bias begins to alter the energy relationship and reduce the available holes. After the peak current I_p is reached, the tunneling current *decreases* with increased voltage and the normal injection current begins to dominate.

[†]Millman and Halkias, *Integrated Electronics*, McGraw-Hill Book Co., New York, 1972, has an excellent discussion in Chapter 3.

[‡]Millman and Halkias, op. cit.

The peak current I_p and the valley current I_v are stable operating points. Since tunneling is a wave phenomenon, electron transfer occurs at the speed of light and switching between I_p and I_v is fast enough for computer applications. Furthermore, between I_p and I_v there is a region in which the resistance $r = dV/dI$ is *negative*. By offsetting losses in L and C components, such a negative resistance permits oscillation and the tunnel diode is used as a very high-frequency oscillator.

Metal-Semiconductor Diodes

The leads to a semiconductor diode should make *ohmic contact* to provide for the easy flow of current without creating additional potential barriers. Aluminum in contact with silicon acts as a p-type impurity, and the flow of hole current is easily accomplished by recombination with electrons supplied by the external circuit. Aluminum in contact with n-type material, however, would create a *rectifying contact* instead of the desired ohmic contact. In practice, therefore, contacts to n-type material are made of special alloys or a layer of heavily doped n^+ material is used to provide a low-potential transition between the semiconductor and the metal.

The rectifying contact created between metal and semiconductor differs from a *pn* junction in two important respects. First, in such a *Schottky diode* the forward voltage drop is only about half that across a *pn* junction at the same current. Second, since in an Al-n diode there are only majority carriers, switching is very fast because there is no wait for the recombination of injected minority carriers. These characteristics make Schottky diodes useful in integrated circuits.

Photodiodes

In a *photodiode*, radiant energy is used to create electron-hole pairs in the region near the junction. Under reverse-biased conditions, the light-injected minority carriers cross the junction and augment the normal reverse saturation current due to thermally generated carriers. The resulting current (a fraction of a milliampere) is approximately proportional to the total illumination, and the ratio of "illuminated" current to "dark" current is very high. These characteristics are desirable in a phototransducer; also, photodiodes are used in the conversion of solar energy to electricity.

Light-Emitting Diodes

The creation of electron-hole pairs is a reversible process; energy is released when an electron recombines with a hole. In silicon and germanium, recombination usually occurs at defects in the crystal that can *trap* a moving electron or hole, absorb its energy, and hold it until a recombination partner comes along. Only occasionally in silicon or germanium but frequently in a *III-V* compound such as gallium arsenide, an electron drops directly into a hole and a photon of energy is generated. Gallium-arsenide junctions providing optimum conditions for the generation of radiation in the visible range are called *light-emitting diodes* (LED). Under special conditions, the light emitted is coherent and the device is a *junction laser*.

SUMMARY

- In a metal, electrons of concentration n, charge e, and mobility μ constitute a current density $J = ne\mu\mathcal{E} = \sigma\mathcal{E}$, defining conductivity σ.

  In a semiconductor, thermally generated electrons and holes constitute a drift current density $J = (n\mu_n + p\mu_p)e\mathcal{E} = \sigma\mathcal{E}$.

- Low-density doping with a pentavalent (trivalent) element produces an n-type (p-type) semiconductor of greatly increased conductivity that is highly temperature dependent.

  For equilibrium, generation rate equals recombination rate and $np = n_i^2$.

  In n-type, $n_n \cong N_d$ and $p_n \cong n_i^2/N_d$; in p-type, $p_p \cong N_a$ and $n_p \cong n_i^2/N_a$.

- The diffusion current density due to nonuniform concentrations of randomly moving electrons and holes is

$$J = J_n + J_p = eD_n\frac{dn}{dx} - eD_p\frac{dp}{dx}$$

- A pn junction diode is a single crystal with an abrupt change from p- to n-type material at the junction.

  Diffusion of majority carriers across the junction uncovers bound charge, creates a potential hill, and encourages an opposing drift of minority carriers.

  Forward biasing reduces the potential hill and encourages diffusion.

  Reverse biasing increases the potential hill and discourages diffusion.

  In general,

$$I = I_s(e^{eV/kT} - 1)$$

  Normal reverse current is very small and nearly constant.

  At breakdown, reverse current is large and nearly independent of voltage.

- Essentially, a diode discriminates between forward and reverse voltages.

  Actual diodes differ significantly from ideal diodes.

  Circuit models of the required precision can be derived from the characteristic curves.

- A voltage regulator minimizes changes in dc load voltage.

  In Zener-diode regulators, small voltage changes cause large current changes that produce voltage drops to compensate for variations in V_{in} or I_{out}.

- Heavily doped Esaki tunnel diodes provide unique characteristics.

  Schottky metal-semiconductor diodes have special applications.

  Photodiodes use radiant energy to create electron-hole pairs.

  Light-emitting diodes use the energy released in direct recombination.

REVIEW QUESTIONS

1. How does resistance vary with temperature for a conductor? For a semiconductor? Why?
2. How does the speed of transmission of a "dot" along a telegraph wire compare with the drift velocity?
3. Define the following terms: mobility, covalent bond, electron-hole generation, recombination, doping, and intrinsic.
4. Describe the formation of a hole and its role in conduction.
5. How do you expect the term n_i^2 to vary with temperature?
6. In a semiconductor, what is the effect of doping with donor atoms on electron density? On hole density?
7. Is a semiconductor negatively charged when doped with donor atoms?
8. Is the approximation of Eq. 5-12 justified if $N_d = 100 n_i$?
9. Why does the factor eV/kT appear in thermionic emission, electron-hole generation, and pn junction current flow?
10. Cite two nonelectrical examples of diffusion.
11. What are the two charge carriers and the two conduction mechanisms present in semiconductors?
12. Describe the forward, reverse, and breakdown characteristics of a junction diode.
13. Define the following terms: junction, uncovered charge, majority carriers, transition region, injection, and potential hill.
14. Describe with sketches the operation of an np diode.
15. How is contact potential utilized in a thermocouple?
16. Sketch curves near the origin to distinguish between the v-i characteristics of ideal and semiconductor diodes.
17. Why does reverse saturation current vary with temperature?
18. Describe the behavior of a Zener diode under forward and reverse bias.
19. Why do we wish to replace actual devices with fictitious models? Why *linear* models?
20. Cite an example from aeronautical, chemical, civil, industrial, and mechanical engineering of a process or a device that is customarily analyzed in terms of a mathematical or physical model.
21. Explain what is meant by a "linearized characteristic curve."
22. Given a *real* semiconductor diode in a half-wave rectifier circuit, sketch the voltage waveform across a load resistor. Repeat for an *ideal* diode.
23. Explain the operation of a voltage regulator using the diode of Fig. 5.8.

EXERCISES

1. If electrons drift at a speed of 0.3 mm/s in ordinary residential wiring (No. 12 copper wire, diameter = 0.081 in.), what current does this represent?
2. Estimate the mobility of electrons in copper at 20 °C and compare to that in silicon.
3. For a copper conductor 1 mm in diameter and 10 cm in length, calculate the total number of charge carriers and the resistance. Repeat for an intrinsic silicon conductor of the same dimensions.
4. Estimate the relative concentration of germanium atoms and electron-hole pairs at room temperature and predict the intrinsic resistivity.
5. For $V = 1$ V, sketch a graph of $e^{-eV/kT}$ for $200 < T < 400$ K and write a brief statement describing the relation between thermal generation rate and temperature.
6. A silicon crystal is doped with 10^{20} boron atoms/m³.
   (a) Justify *quantitatively* the statement that: "The metallurgical properties of the doped crystal are not significantly altered but the electrical properties are greatly changed."
   (b) Determine the factor by which the conductivity has been changed by doping.
7. It is desired to increase the conductivity of silicon by a factor of 1000 by doping with arsenic. Specify the doping density and identify the important charge carriers.
8. Consider three specimens: a pure metal, a pure semiconductor, and a moderately doped semiconductor, at room temperature. Predict the effect on the conductivity of each specimen by a 10% increase in absolute temperature (K). Write an appropriate expression for conductivity for

each; predict the new conductivity by selecting the most appropriate phrase from the following:
(a) About the same.
(b) Increased by about 10%.
(c) Increased by more than 10%.
(d) Decreased by about 10%.
(e) Decreased by more than 10%.
Explain, briefly, your reasoning.

9. A certain block of pure silicon has a resistance of 200 Ω. It is desired to reduce this resistance to 0.2 Ω by doping with phosphorus.
   (a) What are the charge carriers of importance in the doped semiconductor?
   (b) What fraction of the silicon atoms must be replaced by phosphorus atoms?

10. A chip of intrinsic silicon is 1 mm $\times$ 2 mm in area and 0.1 mm thick. What voltage is required for a current of 2 mA between opposite faces?

11. The block of Exercise 10 is doped by adding 1 atom of phosphorus per 10^7 atoms of silicon.
    (a) How many atoms per hole-electron pair are there in intrinsic silicon?
    (b) Is the doped material n-type or p-type? Why?
    (c) In the doped material, what is the density of majority and minority carriers? Can one be neglected?
    (d) What voltage is required for a current of 2 mA between faces?
    (e) By what factor has doping increased the conductivity?

12. Repeat Exercise 11, assuming that 1 in every 10^8 atoms of silicon is replaced by an atom of boron.

13. The resistivity of a p-type silicon specimen is specified to be 0.12 $\Omega \cdot$m.
    (a) Estimate electron and hole concentrations.
    (b) Repeat for an n-type specimen.

14. A 5000-Ω resistor is to be fabricated (in an "integrated circuit") as a narrow strip of p-type silicon 4 μm thick. If the strip is 25 μm wide and 600 μm long, what concentration of acceptor atoms is required?

15. (a) Calculate the diffusion constants D_n and D_p for silicon at room temperature.
    (b) At a junction created by heating an n-type silicon crystal in the presence of boron gas (see Fig. 5.7c), the density of electrons drops from 5×10^{20} to $4 \times 10^{20}/m^3$ in a distance of 0.1 μm. Estimate the diffusion current density across the junction.

16. In a germanium pn junction diode, the density of holes drops from 10^{21} to $0.9 \times 10^{21}/m^3$ in 2 μm. Estimate the diffusion current density due to holes across the junction at room temperature.

17. A diode consists of a left-hand portion doped with 10^{20} antimony atoms/m^3 and a right-hand portion doped with 4×10^{20} boron atoms/m^3.
    (a) Sketch the diode, indicating the junction and identifying the minority and majority carriers in each portion.
    (b) For electrical neutrality, how do the widths of the depletion layers in n and p regions compare?
    (c) Sketch a graph of charge density across the unbiased diode.

18. A semiconductor diode consists of a left-hand portion doped with arsenic and a right-hand portion doped with indium.
    (a) Sketch the diode, indicating the junction. Below the sketch, draw graphs of charge density, electric field, and potential distribution across the unbiased diode.
    (b) Draw labeled arrows indicating directions of drift and diffusion of electrons, holes, and net currents.
    (c) For forward bias, indicate the necessary polarity.

19. In a reverse-biased junction, when all the available charge carriers are being drawn across the junction, increasing the reverse junction voltage does not increase the reverse current.
    (a) At what voltage will the reverse current in a germanium pn junction at room temperature reach 98% of its saturation value?
    (b) If the reverse saturation current is 20 μA, calculate the current for a forward-biased voltage of the magnitude determined in part (a).

20. For a germanium diode carrying 10 mA, the required forward bias is about 0.2 V. Estimate the reverse saturation current and the bias voltages required for currents of 1 and 100 mA, respectively. Comment on the range of voltages required for a 100 to 1 change in current.

21. At a reverse voltage of 0.7 V, the reverse current in a germanium diode is 0.01 μA. Predict the current at voltages of -1.4, 0, and $+0.35$ V. Estimate the voltage at a forward current of 18 mA.

22. A real semiconductor diode in the circuit of Fig. 3.29 has a reverse saturation current of 20 μA.

Predict the current i at room temperature under the following conditions:

(a) $v = +0.15$ V, $R = 0$.
(b) $v = -5.0$ V, $R = 1000$ Ω.
(c) $v = +5.0$ V, $R = 1000$ Ω.

23. (a) Represent the real diode of Fig. 5.21 by a circuit model consisting of an ideal diode and a voltage source.
(b) Use your model to estimate the current if the diode is placed in series with a 6-V battery and a 3-Ω resistance.
(c) Obtain a graphical solution and compare with the result of part (b).

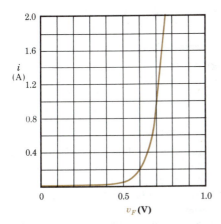

Figure 5.21 Silicon diode characteristics.

24. Derive approximate circuit models for the diodes characterized in Fig. 5.22.
25. Derive circuit models for the devices characterized in Fig. 5.23.

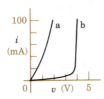

Figure 5.22

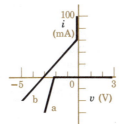

Figure 5.23

26. The germanium diode whose characteristics are shown in Fig. 5.8 is connected in the three different circuits of Fig. 5.24. In each case, select the appropriate model and predict the current i.

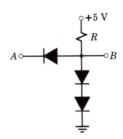

Figure 5.24

27. The switching circuit of Fig. 5.25 contains three silicon diodes. Predict V_B for the following values of V_A: +2, +1, 0, and −1 V.

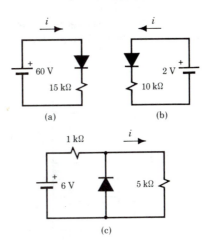

Figure 5.25

28. The switching circuit of Fig. 5.26 contains two silicon diodes.
(a) For switch S closed, estimate I_A and I_B for $V_A = -1$, −0.1, +1, and +2 V.
(b) For switch S open, estimate I_A and I_B for $V_A = +1$ V.

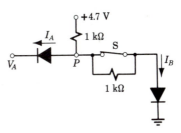

Figure 5.26

29. Devise a circuit model to represent, approximately, the Zener diode of Fig. 5.27.

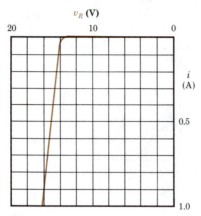

Figure 5.27 Zener region characteristics.

30. The Zener diode whose reverse characteristics are given in Fig. 5.27 is connected in the circuit of Fig. 5.28 where $R_L = 10 \ \Omega$.
(a) Replace the diode by a model and redraw the circuit.
(b) For $V_L = 14$ V and $V_1 = 26$ V, specify R_S.
(c) Under what circumstances could superposition concepts be used in solving this nonlinear circuit?
(d) If V_1 drops to 24 V, what is the new V_L?

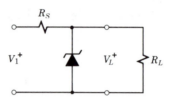

Figure 5.28

PROBLEMS

1. Stating any necessary assumptions, show that the spacing between atoms in copper and germanium crystals is on the order of 1 Å (0.1 nm).
2. Sketch in perspective a rectangular block of semiconductor of unknown type in a magnetic field B out of the paper. When a current I is introduced from right to left, it is observed that the upper surface exhibits a positive potential with respect to the lower (the Hall effect).
   (a) Is the specimen a p-type or n-type semiconductor?
   (b) How does this experiment justify the concept of the hole as a mobile positive charge?
3. An experimental melt contains 100 g of pure germanium, 5×10^{-5} g of indium, and 3×10^{-5} g of antimony. Determine the conductivity of a sample of this germanium alloy.
4. A silicon sample contains boron at a concentration of 2×10^{20} atoms/m³.

   (a) Estimate the hole and electron concentrations at room temperature. Is this p-type or n-type material?
   (b) Repeat part (a) for a temperature of 300 °C where $n_i = 3 \times 10^{21}/m^3$.
5. For a nonlinear device, the incremental resistance is defined as $r = dv/di$. For an ideal semiconductor diode, derive a general expression for r in terms of I_s and an approximate expression in terms of I the forward current. Evaluate r for a reverse current of 5 μA and for a forward current at 5 mA.
6. The Zener diode of Fig. 5.17b is used in the voltage regulator circuit of Fig. 5.28, where $R_S = 10,000 \ \Omega$ and $R_L = 20,000 \ \Omega$. Derive an expression for V_L in terms of V_1. If V_1 increases from 16 to 24 V (a 50% increase), calculate the corresponding variation in load voltage V_L. Is the "regulator" doing its job?

6

Transistors and Integrated Circuits

Field-Effect Transistors

Bipolar Junction Transistors

Integrated Circuits

When DeForest added a third electrode to a vacuum diode and created the first electronic amplifier, he started a revolution in the field of communications. Amplifiers made it possible to generate high-frequency signals, send them around the world, and restore them to usable levels at the destination. Also, amplifiers provided the flexibility and sensitivity needed in electronic control and instrumentation. Forty years later, the search for a simpler device capable of performing the same functions, but without the disadvantages of the thermionic tube, led to the discovery of the *transistor*.

The transistor is available in several basic forms for a wide variety of applications ranging from implantable heart pacers to powerful computer-aided systems for designing complete automobiles. It may be just one of a million similar devices in a memory chip, or it may serve as a discrete switch controlling many amperes.

In this chapter we develop the principles necessary for the analysis and design of circuits containing transistors. We start with the simplest form of "unipolar" transistor and derive the i-v characteristics that determine its behavior. Then we study the more complicated "bipolar" transistor and derive its characteristics. Finally, we see how combinations of transistors, diodes, resistors, and capacitors can be fabricated into "integrated circuits."

FIELD-EFFECT TRANSISTORS

The control of current in a vacuum tube by varying an electric field has its semiconductor counterpart in the *field-effect transistor* (FET). Shockley proposed the device in 1952, but a decade passed before the fabrication techniques necessary for dependable operation became available. Now such transistors are widely used in a variety of forms.

JFET

To understand the operation of the junction-gate field-effect transistor or JFET (Fig. 6.1), we must keep in mind the following concepts:

A thin conducting channel from *source* S to *drain* D is created by selective doping.

The channel is insulated by setting up, along its borders, reverse-biased *pn* junctions with their accompanying depletion layers.

For a given junction potential, a definite number of bound charges must be uncovered (see Fig. 5.9); the width of the depletion layer increases as the junction voltage is made more negative.

The conductance of the channel, *n*-type here, is determined by its dimensions, primarily by its width.

The width of the channel can be varied by changing the width of the carrier-free depletion layers along the reverse-biased junctions.

The width of the depletion layers and therefore the conductance of the channel can be controlled externally by the voltage applied to the *gate* G.

The current that flows in the doped channel consists of mobile carriers, electrons here. (The arrow in the symbol always points toward *n*-type material; *p*-channel

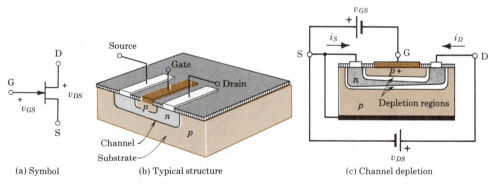

(a) Symbol   (b) Typical structure   (c) Channel depletion

Figure 6.1 An *n*-channel, depletion-mode JFET.

devices are also possible.) With $v_{DS} > 0$, the drain end is positive with respect to the source, and electrons flow from the source to the drain; in our convention, this is equivalent to a flow of positive charges from D to S and the drain current i_D is positive.

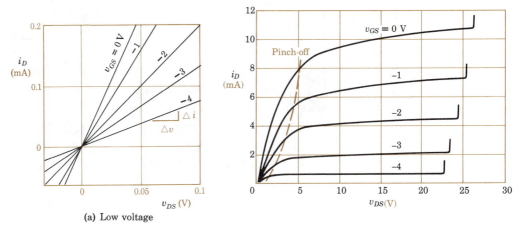

(a) Low voltage

Figure 6.2   Depletion-mode FET characteristics.

With $v_{GS} = 0$ (the gate as well as the p-type substrate connected to the source) and a very small voltage applied to the drain, the current that flows is directly proportional to voltage v_{DS}. The $v_{GS} = 0$ line in Fig. 6.2a indicates a resistance of $\Delta v / \Delta i = 0.05/0.0002 = 250\ \Omega$. If the gate-source voltage is changed to -2 V, the depletion layer is widened, the channel width is decreased, and the indicated resistance is $\Delta v / \Delta i = 0.05/0.0001 = 500\ \Omega$. We see that for a given drain-source voltage, the channel current can be controlled by an external gate voltage. For current flow from drain to source, v_{DS} must be positive; to reverse-bias the pn junction, v_{GS} must be negative. Figure 6.2a shows the behavior of a JFET at low values of v_{DS}.

For larger voltages, the behavior is complicated by the fact that the depletion layer is asymmetric. (See Fig. 6.1c.) The source is positive with respect to the gate and the drain is positive with respect to the source. Therefore, near the drain end the channel is most positive with respect to the gate, the reverse bias is greatest, and the width of the depletion layer is greatest. With increasing v_{DS}, the reverse bias increases until the two depletion regions almost meet, tending to "pinch off" the conducting channel. In Fig. 6.2b, the *pinch-off voltage* V_p for $v_{GS} = 0$ is about 5 V. Above pinch-off, an increase in v_{DS} decreases the channel width, offsetting the increase in current density expected from a higher drain-source voltage, and the i_D curve flattens out.

Because the gate-channel voltage determines the width of the depletion layer, with a negative voltage applied to the gate, pinch-off occurs at a lower drain-source voltage and the drain current is limited to a lower value. Note that for $v_{GS} = 0$ in Fig. 6.2b, a value of $v_{DS} \cong 5$ V establishes a channel-gate voltage of 5 V and pinch-off occurs; if $v_{GS} = -4$ V, pinch-off is evident at $v_{DS} \cong 1$ V where the channel-gate voltage is equal to V_p. Above pinch-off, the current curves are relatively flat until the gate-drain voltage reaches the point where avalanche *breakdown* occurs. The portion of the i-v characteristics where i_D is nearly independent of v_{DS} is called the *constant-current* or *saturation* region. For many applications, the normal operating range extends from pinch-off to breakdown. Note that the i-v characteristics for this two-port device are represented by a *family* of curves.

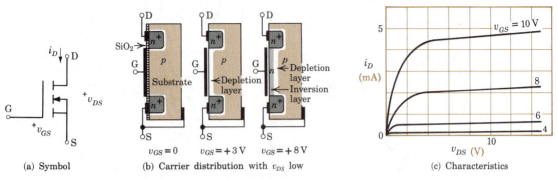

(a) Symbol   (b) Carrier distribution with v_{DS} low   (c) Characteristics

Figure 6.3   An n-channel enhancement-mode MOSFET.

MOSFET

In the metal-oxide-semiconductor FET or MOSFET, a thin layer of SiO_2 insulates the gate contact from the channel. The *n-channel enhancement-mode* transistor in Fig. 6.3 offers the highest performance. No channel is built into this device; here a conducting channel is *induced* by an electric field established between the gate and the p-type substrate. With no gate voltage, little current flows through the two back-to-back pn junctions. With a small positive gate voltage, holes in the adjacent p material are repelled and a depletion layer formed. At a slightly more positive voltage, an *inversion layer* of mobile electrons is formed at the surface of the p-type region, which becomes n-type. The inversion results from the fact (Eq. 5-10) that in a doped semiconductor, $np = n_i^2$. If the density of holes is driven down, the density of mobile electrons goes up. When the gate voltage exceeds a threshold level v_T (about 4 V in Fig. 6.3c), the conductivity of the region has been "enhanced," the transistor has been "turned on," and current can flow readily along a continuous n-type path from drain to source.

The drain current is not proportional to v_{DS}, however. As the potential at the drain end of the channel becomes more positive, the effective gate-to-channel voltage and the accompanying electric field are reduced. The carrier density in the inversion layer is reduced and the current levels off.

Similar p-channel devices, in which holes are the mobile charges, are widely used. However, the greater mobility of electrons permits smaller n-type channels and therefore lower capacitances for the same resistance. N-channel transistors offer faster switching in digital systems and higher frequency response in amplifiers.

In another form of MOSFET, a narrow lightly doped conducting layer is built into the channel region between the high conductivity n^+ contacts (Fig. 6.4b). At $v_{GS} = 0$, an appreciable drain current flows. We can produce either depletion *or* enhancement of the conducting channel by applying the appropriate gate voltage. In the n-channel device shown, a negative gate voltage causes a shrinking of the channel; a positive gate voltage causes a widening. The characteristic curves are similar to those for the JFET with the added flexibility of permitting positive or negative control voltages. For the device shown in Fig. 6.4c, the pinch-off voltage V_p is about -4 V. Such DE MOSFETs are available with either n- or p-type channels.[†]

[†] The FET symbols are related to their distinctive characteristics. The gate terminal on a JFET represents a junction; the gap on MOSFETs identifies an insulated gate. The channel bar is continuous (normally conducting) for JFET and DE MOSFET, broken (normally "open") for the E MOSFET.

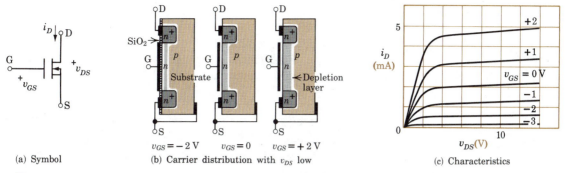

(a) Symbol

(b) Carrier distribution with v_{DS} low

$v_{GS} = -2$ V $v_{GS} = 0$ $v_{GS} = +2$ V

(c) Characteristics

Figure 6.4 An n-channel depletion- or enhancement-mode (DE) MOSFET.

Transfer Characteristics

The i-v characteristics of field-effect transistors show that output current is controlled by input voltage, and the FET can be used as a voltage-controlled *switch*. If the output current is sent through a resistance, the voltage developed may be much larger than the input voltage and the FET can be used as an *amplifier*. Because the characteristics of individual devices are never precisely known, approximate methods of analysis are acceptable. Within the saturation region, that is, between pinch-off or turn-on and breakdown, drain current i_D is nearly independent of drain-source voltage v_{DS}, and the *transfer characteristics*, relating output current to input voltage, are as shown in Fig. 6.5.

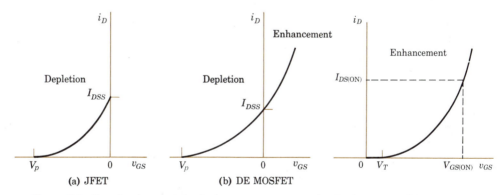

(a) JFET

(b) DE MOSFET

Figure 6.5 Transfer characteristics in the constant-current region for three types of FET.

From theoretical analysis and practical measurement, it can be shown that the transfer characteristics are approximately parabolic in all three cases. For the JFET, the drain current in the constant-current region is

$$i_{DS} = I_{DSS}(1 - v_{GS}/V_p)^2 \tag{6-1}$$

where i_{DS} = the drain current in the constant-current region,
I_{DSS} = the value of i_{DS} with gate shorted to source, and
V_p = the pinch-off voltage.

Practice Problem 6-1

With gate-source voltage equal to zero, the drain current in the saturation region of a certain JFET is 8 mA. If the pinch-off voltage is -4 V, sketch the transfer characteristic and estimate the drain current at $V_{GS} = -2$ V.

Answer: 2 mA.

The depletion-*or*-enhancement MOSFET is also described by Eq. 6-1, and both positive and negative values of v_{GS} are permitted. For the enhancement-only MOSFET, the transfer characteristic is

$$i_{DS} = K(v_{GS} - V_T)^2 \tag{6-2}$$

where K is a device parameter and V_T is the turn-on or threshold voltage.

These simple relations are useful in predicting the dc behavior of FETs. For the JFET or DE MOSFET, the manufacturer usually specifies typical values of I_{DSS} and the *gate-source cutoff voltage* $V_{GS(OFF)}$, which is approximately equal to V_p since the same pinch-off effect is created between gate and channel. For the enhancement-only MOSFET, the manufacturer specifies V_T and a particular value of $I_{DS(ON)}$ corresponding to a specified value of $V_{GS(ON)}$.

EXAMPLE 1

For the n-channel enhancement-only MOSFET of Fig. 6.6, the manufacturer specifies $V_T = 4$ V and $I_{DS} = 7.2$ mA at $V_{GS} = 10$ V. For $V_{DD} = 24$ V and $R_G = 100$ MΩ, specify R_D for operation at $V_{DS} = 8$ V.

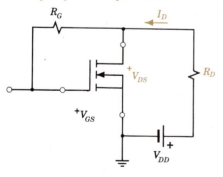

Figure 6.6 MOSFET analysis.

The arrowhead pointing toward the channel identifies an n-channel device.

Substituting the given data in Eq. 6-2,

$$I_{DS} = 0.0072 = K(V_{GS} - V_T)^2 = K(10 - 4)^2$$

$$\therefore K = 0.0072/(6)^2 = 0.0002 \text{ A/V}^2$$

Because $i_G = 0$, there is no drop across R_G; therefore, $V_{GS} = V_{DS} = 8$ V, and

$$I_D = 0.0002(8 - 4)^2 = 3.2 \text{ mA}$$

Around the drain-source loop,

$$\Sigma V = 0 = V_{DD} - I_D R_D - V_{DS}$$

or

$$R_D = \frac{V_{DD} - V_{DS}}{I_D} = \frac{24 - 8}{0.0032} = 5 \text{ k}\Omega$$

Because the gate-source junction of the JFET is reverse biased, the input signal current is very small; in other words, the input resistance is very high and little input power is required. In the MOSFET or *insulated-gate* FET, the input resistance may be as high as 10^{15} Ω.[†]

[†] The nearly infinite gate resistance creates a practical handling problem. If a small static charge (10^{-10} C) builds up on the gate-substrate capacitance (10^{-12} F), the voltage produced ($V = Q/C = 100$ V) may rupture the gate. External gate leads from an *IC* are usually protected by a Zener diode.

FETs are particularly useful in digital systems where thousands of units, some acting as resistors or capacitors, can be fabricated on a single silicon chip at very low cost. The primary advantages of the FET are its high element density and low power requirement in comparison to the more complex "bipolar" transistor.

EXAMPLE 2

The MOSFET of Fig. 6.4 is to be used as an amplifier. Assuming $I_{DSS} = 2$ mA and $V_{GS(OFF)} \cong -4$ V, predict the voltage gain with a load resistance $R_L = 4$ kΩ.

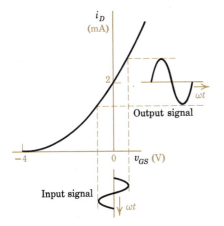

The transfer characteristic (Eq. 6-1) becomes

$$i_D = 0.002(1 + v_{GS}/4)^2$$

For a signal $v_{GS}(t) = V_m \sin \omega t$ (Fig. 6.7), the output voltage across $R_L = 4$ kΩ will be

$$v_L = i_D R_L = 8(1 + 0.25 V_m \sin \omega t)^2$$

$$= 8 + 4V_m \sin \omega t + 0.5 V_m^2 \sin^2 \omega t$$

The output consists of an 8-V dc component, a $4V_m$ signal component at the input frequency, and a $0.5V_m^2$ "distortion" component resulting from the nonlinear (parabolic) transfer characteristic.

For small values of V_m, this is an amplifier with a voltage gain of 4.

For large values of V_m, this is a "square-law device" with special virtues described in Chapter 13.

Figure 6.7   A MOSFET as an amplifier.

BIPOLAR JUNCTION TRANSISTORS

Another widely employed electronic device is the *bipolar junction transistor* or BJT.[†] In contrast to the "unipolar" FET, in the "bipolar" device both majority and minority carriers play significant roles. In comparison to the FET, the BJT permits much greater gain and provides better high-frequency performance. Essentially, this transistor consists of two *pn* junctions in close proximity. Like the diode, it is formed of a single crystal, with the doping impurities distributed so as to create two abrupt changes in carrier density.

Fabrication

The structure of an *npn* alloy-diffused transistor is shown in Fig. 6.8a. An *n*-type semiconductor chip less than 1 mm square serves as the *collector* and provides mechanical strength for mounting. A *p*-type *base* region is created by diffusion and

[†] Semiconductor experiments at Bell Telephone Laboratories led to new theoretical concepts, and William Shockley proposed an idea for a semiconductor amplifier that would critically test the theory. The actual device had far less amplification than predicted, and John Bardeen suggested a revision of the theory. In December 1947, Bardeen and Walter Brattain discovered a new phenomenon and created a novel device—the *point-contact transistor*. The next year Shockley invented the *junction transistor*.

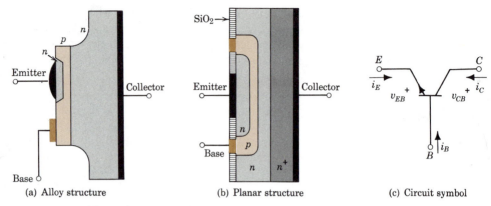

Figure 6.8 Types of *npn* bipolar junction transistors.

a connection is provided by a metallic contact. An *n*-type *emitter* region is then alloyed to the base region. The result is a pair of *pn* junctions separated by a base region that is thinner than this paper.

In the planar structure (Fig. 6.8b), a lightly doped film (n) is grown epitaxially upon a heavily doped (n^+) substrate. After oxidation of the surface, a window is opened by etching and an impurity (p) is allowed to diffuse into the crystal to form a junction. After reoxidation, a smaller window is opened to permit diffusion of the emitter region (n). Finally, contacts are deposited and leads attached.

In the conventional symbol (Fig. 6.8c), the emitter lead is identified by the arrow that points in the direction of positive charge flow in normal operation. (In a *pnp* transistor, which works just as well, the arrow in the symbol is reversed; it always points toward *n*-type material.) Although an *npn* transistor will work with either *n* region serving as the emitter, it should be connected as labeled because doping densities and geometries are intentionally asymmetric.

Operation

Figure 6.9a represents a thin horizontal slice through an *npn* transistor. The operation can be explained qualitatively in terms of the potential distributions across the junctions (Fig. 6.9b). The emitter junction is forward biased; the effect of bias voltage V_{EB} is to reduce the potential barrier at the emitter junction and to facilitate the injection of electrons (in this example) into the base where they are minority carriers. The collector junction is reverse biased; the effect of bias voltage V_{CB} is to increase the potential barrier at the collector junction. The base is so thin that almost all the electrons injected into the base from the emitter diffuse across the base and are swept up the potential hill into the collector where they

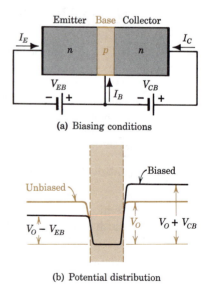

Figure 6.9 Operation of an *npn* transistor.

recombine with holes "supplied" by the external battery. (Actually, electrons are "removed" by the external battery, leaving a supply of holes available for re-combination.)

The net result is the transfer from the emitter circuit to the collector circuit of a current that is nearly independent of the collector-base voltage. As we shall see, this transfer permits the insertion of a large load resistance in the collector circuit to obtain voltage amplification. Alternatively, variation of the base current (the small difference between emitter and collector currents) can be used to control the relatively larger collector current to achieve current amplification or to perform the switching operation used in digital signal processing.

DC Behavior

The dc behavior of a BJT can be predicted on the basis of the motion of charge carriers across the junctions and into the base. With the emitter junction forward biased and the collector junction reverse biased, in so-called *normal* operation, the carrier motion of an *npn* transistor is as shown in the symbolic diagram of Fig. 6.10.

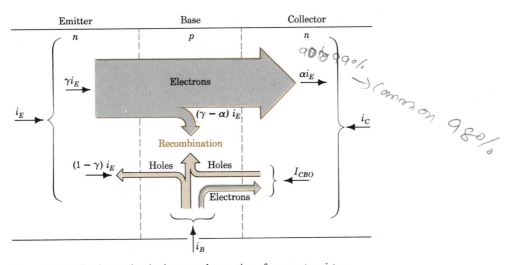

Figure 6.10   Carrier motion in the normal operation of an *npn* transistor.

The emitter current i_E (negative in an *npn* transistor) consists of electrons injected across the *np* junction and holes injected from the base. To optimize transistor per-formance, the base is doped relatively lightly, emitter efficiency γ (gamma) is almost unity, and most of the current consists of electrons injected from the emitter. Some of the injected electrons recombine with holes in the *p*-type base, but the base is made very narrow so that most of the electrons (minority carriers in this *p* region) diffuse across the base and are swept across the collector junction (up the potential hill). The factor α (alpha) varies from 0.90 to 0.999; a typical value is 0.98.

The electron current αi_E constitutes the major part of the collector current. In addition, there is a reverse current across the collector junction due to thermally generated minority carriers, just as in a diode (see Eq. 5-19). In a transistor, this *collector cutoff current* is given by

$$-I_{CBO}(e^{eV/kT} - 1) \cong I_{CBO} \tag{6-3}$$

when the reverse biasing exceeds a few tenths of a volt. The total collector current is then

$$i_C = -\alpha i_E + I_{CBO} \tag{6-4}$$

where α, the *forward current-transfer ratio,* increases slightly with increased collector biasing voltage v_{CB}. The collector current is normally positive in an *npn* transistor.

The base current (Fig. 6.10) consists of holes diffusing to the emitter, positive charges to supply the recombination that occurs in the base, and the reverse saturation current to the collector. By Kirchhoff's law,

$$i_B = -i_E - i_C \tag{6-5}$$

and we see that i_B is the small difference between two nearly equal currents.

Common-Base Characteristics

The transistor connection of Fig. 6.11 is called the *common-base* (*CB*) configuration because the base is common to input and output ports. The *i-v* characteristics

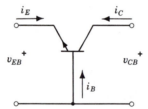

Figure 6.11 A transistor in the common-base configuration.

of a BJT in this configuration can be derived from our knowledge of diode characteristics and transistor operation. Because the emitter-base section is essentially a forward-biased diode, the input characteristics in Fig. 6.12b are similar to those of the first quadrant of Fig. 6.12a; the effect of collector-base voltage v_{CB} is small. With v_{CB} positive and the emitter open-circuited, $i_E = 0$ and the base-collector section is essentially a reverse-biased junction. (Negative values of v_{CB} would forward-bias the *CB* junction and negative i_C would flow.) For $i_E = 0$, $i_C \cong I_{CBO}$ (exaggerated in Fig. 6.12c) and the collector characteristic is similar to the third quadrant of Fig. 6.12a. For $i_E = -5$ mA, say, the collector current is increased by an amount $-\alpha i_E \cong +5$ mA (see Eq. 6-4) and the curve is as shown. The slope of the curves in Fig. 6.12c is due to an effective increase in α as v_{CB} increases. Because the factor α is always less than 1, the common-base configuration is not good for practical current amplification.

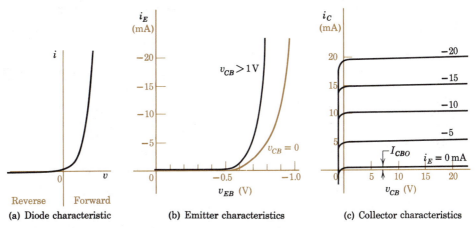

(a) Diode characteristic   (b) Emitter characteristics   (c) Collector characteristics

Figure 6.12 Common-base characteristics of an *npn* transistor.

Practice Problem 6-2

The *npn* transistor of Fig. 6.11 has the emitter characteristics of Fig. 6.12b. If the emitter current is -10 mA and the collector-base voltage is 15 V, estimate the emitter-base voltage. If 0.96 and the collector cutoff current $= 0.01$ mA, estimate the collector current.

Answers: -0.75 V; 9.61 mA.

Common-Emitter Characteristics

If the *npn* transistor is reconnected in the *common-emitter* (*CE*) configuration of Fig. 6.13, current amplification is possible. The input current is now base current i_B, and emitter current $i_E = -(i_C + i_B)$; therefore, collector current is

$$i_C = -\alpha i_E + I_{CBO} = +\alpha(i_C + i_B) + I_{CBO}$$

Solving,

$$i_C = \frac{\alpha}{1-\alpha}i_B + \frac{I_{CBO}}{1-\alpha} \tag{6-6}$$

To simplify Eq. 6-6, we define the *current-transfer ratio* (*CE*) as

$$\beta = \frac{\alpha}{1-\alpha} \tag{6-7}$$

and note that the *collector cutoff current* (*CE*) is

$$\frac{I_{CBO}}{1-\alpha} = (1+\beta)I_{CBO} = I_{CEO} \tag{6-8}$$

The simplified equation for output (collector) current in terms of input (base) current and the current-transfer ratio (beta) is

$$i_C = \beta i_B + I_{CEO} \tag{6-9}$$

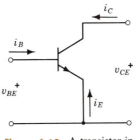

Figure 6.13 A transistor in the common-emitter configuration.

EXAMPLE 3

A silicon *npn* transistor with $\alpha = 0.99$ and $I_{CBO} = 10^{-11}$ A is connected as shown in Fig. 6.14. Predict i_C, i_E, and v_{CE}. (*Note:* It is convenient and customary in drawing electronic circuits to omit the battery, which is assumed to be connected between the $+10$ V terminal and ground.)

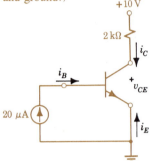

Figure 6.14   Transistor operation.

For this transistor,

$$\beta = \frac{\alpha}{1 - \alpha} = \frac{0.99}{1 - 0.99} = \frac{0.99}{0.01} = 99$$

and the collector cutoff current is

$$I_{CEO} = (1 + \beta)I_{CBO} = (1 + 99)10^{-11} = 10^{-9} \text{ A}$$

The collector current is

$$i_C = \beta i_B + I_{CEO} = 99 \times 2 \times 10^{-5} + 10^{-9} \cong 1.98 \text{ mA}$$

As expected for a silicon transistor, I_{CEO} is a very small part of i_C.

The emitter current is

$$i_E = -(i_B + i_C) = -(0.02 + 1.98)10^{-3} = -2 \text{ mA}$$

The collector-emitter voltage is

$$v_{CE} = 10 - i_C R_C \cong 10 - 2(\text{mA}) \times 2(\text{k}\Omega) = 6 \text{ V}$$

Since $v_{CB} = v_{CE} - v_{BE} \cong 6 - 0.7 = +5.3$ V, the *np* collector-base junction is reverse biased as required.

Typical common-emitter characteristics are shown in Fig. 6.15. The input current i_B is small and, for a collector-emitter voltage of more than a volt or so, depends on only the emitter-base junction voltage. For a silicon BJT, about 0.7 V of forward bias provides adequate base current.

The collector characteristics are in accordance with Eq. 6-9; for $i_B = 0$, the collector current is small and nearly constant at a value I_{CEO} (exaggerated in the drawing). For each increment of current i_B, the collector current is increased an amount βi_B. For $\alpha = 0.98$, $\beta = \alpha/(1 - \alpha) = 0.98/(1 - 0.98) = 49$, and a small increase in i_B corresponds to a large increase in i_C. A small increase in α produces a much greater change in β, and the effect of v_{CE} on i_C is more pronounced here than in the *CB* configuration (Fig. 6.12c).

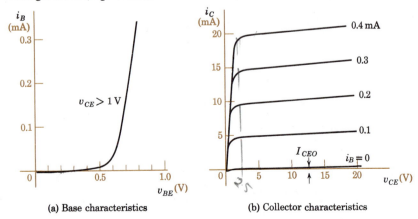

(a) Base characteristics   (b) Collector characteristics

Figure 6.15   Common-emitter characteristics of an *npn* transistor.

Practice Problem 6-3

The transistor of Practice Problem 6-2 is connected in the circuit of Fig. 6.14 with $i_B = 0.15$ mA instead of the 20 μA shown. Calculate β and I_{CEO} and predict current i_C and voltage v_{CE}.

Answers: 24; 0.25 mA; 3.85 mA; 2.3 V.

Current Amplification

The collector characteristics of a bipolar junction transistor indicate the possibility of current amplification. Graphical analysis of an *npn* BJT amplifier is shown in Fig. 6.16. The input signal i_i (assumed sinusoidal) is applied in parallel with the dc base-biasing current I_B. The reverse bias on the collector is maintained by V_{CC}; note that most of this voltage appears across the collector junction because the voltage

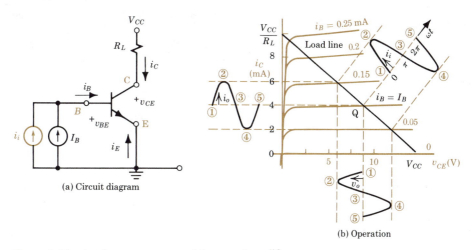

(a) Circuit diagram

(b) Operation

Figure 6.16 An elementary *npn* transistor current amplifier.

across the forward-biased emitter junction is quite small. (See Fig. 6.15a.) We anticipate that signal current i_i will cause a variation in total base current $i_B = I_B + i_i$ that will produce a variation in collector current i_C; the varying component of i_C constitutes an amplified output current.

By Kirchhoff's voltage law around the output "loop,"

$$v_{CE} = V_{CC} - i_C R_L \tag{6-10}$$

This is the equation of the load line shown in Fig. 6.16b. For no signal input, $i_i = 0$ and $i_B = I_B$, the *quiescent* value. The quiescent point Q lies at the intersection of the load line and the characteristic curve for $i_B = I_B$. The intersections of the load line and the characteristic curves of the nonlinear transistor represent graphical solutions of Eq. 6-10 and define the instantaneous values of i_C and v_{CE} corresponding to instantaneous values of the input signal current i_i.

If the signal current is sinusoidal, the base current takes successive values corresponding to points 1, 2, 3, 4, and 5 on the waveform. The corresponding values of collector current and voltage are obtained graphically from the load line. The signal output i_o is the sinusoidal component of i_C, the total current in R_L. The current amplification, or current gain, is

$$A_I = \frac{i_o}{i_i} = \frac{I_{om}}{I_{im}} = \frac{I_o}{I_i} \tag{6-11}$$

depending on whether the current ratio is expressed in terms of instantaneous, maximum, or rms values.

EXAMPLE 4

For the elementary amplifier of Fig. 6.16, the collector-battery voltage $V_{CC} = 15$ V. The quiescent point Q is at $i_B = I_B = 0.1$ mA and $v_{CE} = V_{CE} = 9$ V. Specify R_L and predict the current gain and the output voltage for an input current $i_i = 0.05 \sin \omega t$ mA.

By Eq. 6-10, the load resistance is

$$R_L = \frac{V_{CC} - V_{CE}}{I_C} = \frac{15 - 9}{0.004} = 1500 \ \Omega$$

The load line intersects the collector current axis at

$$i_C = V_{CC}/R_L = 15/1500 = 0.01 = 10 \text{ mA}$$

At point 2, $\omega t = \pi/2$, the base current is $i_B = 0.15$ mA, the signal current is $i_B - I_B = 0.15 - 0.1 = 0.05$ mA, and the corresponding collector current is $i_C = 6$ mA. A signal current swing of 0.05 mA causes a collector current swing of $6 - 4 = 2$ mA. The current gain is

$$A_I = \frac{I_{om}}{I_{im}} = \frac{2 \times 10^{-3}}{5 \times 10^{-5}} = 40$$

The sinusoidal component of voltage across R_L is

$$v_o = R_L i_o = 1500 \times 0.002 \sin \omega t = 3 \sin \omega t \text{ V}$$

Switching

In contrast to the continuous signals amplified in the circuit of Fig. 6.16, computer information is generated, processed, and stored as discrete signals; each electronic element is either **ON** or **OFF**. Because a small base current can control a much larger collector current, the transistor is attractive as a possible control device. The fact that base current is an exponential function of base-emitter voltage permits a BJT to operate as a sensitive switching element in Example 5. In Chapter 7 we explore the application of transistors and diodes to logic gates and memory elements.

① iB use curve, ② E is ground so CE
③ output is C because CE
Integrated Circuits **159**

EXAMPLE 5 ✓

The transistor of Fig. 6.15 ($\beta \cong 50$ and $I_{CEO} \cong 1$ nA) is used in the *switching circuit* of Fig. 6.17. Predict the output voltage for input voltages of 0.2, 0.7, and 0.8 V.

CE, input B

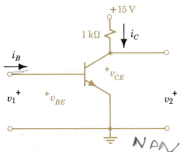

Figure 6.17   A transistor switch.

NPN

For $v_1 = v_{BE} = 0.2$ V, base current $i_B \cong 0$ and $i_C = I_{CEO} \cong 1$ nA. The voltage drop across the 1-kΩ resistor is only 1 μV and voltage $v_2 \cong 15$ V. $V_2 = 15 - i_C(1k\Omega) \cong 15 v, \ To 66$

For $v_1 = v_{BE} = 0.7$ V, $i_B \cong 0.12$ mA, the transistor is in the normal operating range, and $i_C = \beta i_B = 50 \times 0.12 = 6$ mA. Therefore, voltage $v_2 = v_{CE} = 15 - 1000(0.006) = 9$ V.

For $v_1 = v_{BE} = 0.8$ V, $i_B \cong 0.4$ mA. Assuming $\beta = 50$, collector current would be $i_C = \beta i_B = 50 \times 0.4 = 20$ mA. But this is impossible because only 15 V is available to drive collector current through the 1-kΩ resistor.

Figure 6.15b indicates that the $\beta = i_C/i_B$ concept is not valid for small values of v_{CE}; as i_B increases, v_{CE} approaches as a limit a value of about 1 V.

Conclusion: A change in input from 0.2 to 0.8 V switches the output from 15 to 1 V. We say: for $v_{be} < 0.2$ V, the switch is OPEN; for $v_{be} > 0.8$ V, the switch is CLOSED.

INTEGRATED CIRCUITS

In 1958 J. S. Kilby developed an "integrated circuit," a single monolithic chip of semiconductor in which active and passive circuit elements were fabricated by successive diffusions and depositions. Shortly thereafter Robert Noyce fabricated a complete circuit including the interconnections on a single chip. The ability to control precisely the dimensions and doping concentration of diffused or deposited regions combined with advances in photolithographic techniques have made possible the mass production of sophisticated devices of very small size but exceedingly high reliability.

With the exception of inductance, all the ordinary electronic circuit elements can be fabricated in semiconductor form. A resistor (Fig. 6.18a) is obtained in the form of a thin filament of specified conductivity, terminated in metallic contacts and isolated by a reverse-biased junction. Resistances from a few ohms to 40,000 Ω are possible. A low-loss capacitor is formed by depositing a thin insulating layer of silicon dioxide on a conducting region and then providing a metalized layer to form the second plate of the capacitor (Fig. 6.18b). Alternatively, a reverse-biased diode may

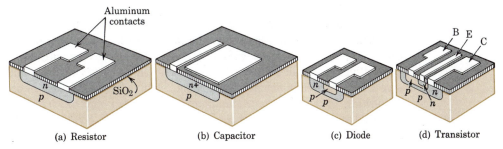

(a) Resistor   (b) Capacitor   (c) Diode   (d) Transistor

Figure 6.18   Basic components of microcircuits.

be used to provide capacitance; capacitances up to 50 pF are possible. Typically, capacitors require the largest areas, resistors next, and active devices such as diodes, BJTs, and FETs the smallest. Interconnections are provided by a layer of aluminum that is deposited, masked, and etched to leave the desired pattern.

Design Implications

The availability of sophisticated devices of high reliability in small packages at low cost has changed the character of electronic design. Instead of developing optimum circuits, buying components, and fabricating such units as amplifiers and digital registers, the design engineer selects mass-produced, high-performance functional units and incorporates them into specific products. Originally conceived as a means of reducing the size of complex equipment, integrated circuits now are equally noted for their low cost and high reliability.[†]

Since processing cost is proportional to area, circuits are designed to use more *active devices* (transistors and diodes) and fewer resistors and capacitors. (This is in sharp contrast to vacuum tube technology, where the active devices are the most expensive items by far.) The small circuit dimensions decrease the likelihood of pickup of unwanted noise; therefore signal levels can be lower. This in turn means lower voltages and lower power requirements. On the other hand, component values are hard to control precisely, and they may be voltage or temperature dependent. Also, unintentional capacitances may provide undesired coupling between circuit components.

Fabrication Processes

The fabrication of monolithic integrated circuits follows the planar technology used in transistor manufacture. (See Fig. 6.8b.) A thin slice or *wafer* of *p*- type silicon 2 to 6 in. (50 to 150 mm) in diameter and about 20 mils (0.02 in. or 0.5 mm) thick provides the *substrate*. The active and passive components are built within a thin *n*-type *epitaxial layer* on top (Fig. 6.19a). First the surface is oxidized by heating it in the presence of oxygen. Then the cooled wafer is coated with a photosensitive material called *photoresist* and covered with a mask of the desired geometry. Upon exposure, the light-sensitive material hardens and the unexposed material is dissolved away. An etching solution, in which the photoresist is insoluble, is then used to remove the SiO_2, leaving a pattern of windows through which the first diffusion takes place.

One method of insulating components from each other is to provide *diode isolation* by surrounding each element with a reverse-biased *pn* junction. The first diffusion provides *p*-type regions that extend through to the substrate (Fig. 6.19c). The individual circuit elements are then built within these isolated islands by successive steps involving oxidation, masking, etching, and diffusion of *p*- and *n*-type impurities. In the final step, aluminum is evaporated onto the wafer and selectively removed to provide interconnections and relatively large *bonding pads* to which leads are welded.

[†] See References at the end of this chapter.

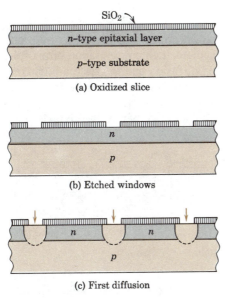

(a) Oxidized slice

(b) Etched windows

(c) First diffusion

Figure 6.19 Insulation of IC elements by diode isolation.

Integrated circuits are cheap because thousands of complex units can be fabricated simultaneously. After the circuit is designed functionally and tested experimentally, it is designed dimensionally by using computer-aided design (CAD) techniques. Then an electron-beam pattern generator places on the wafer hundreds of identical images corresponding to one fabrication step. Each processed wafer (Fig. 6.20) is *diced* to yield several hundred individual chips that are tested and encapsulated. If each of 500 chips contains 10,000 components, and if 20 wafers are processed in a batch, a total of $500 \times 10,000 \times 20 = 10$ million components are fabricated at once. Even if the yield of perfect chips is only 50%, the cost per component is very small.

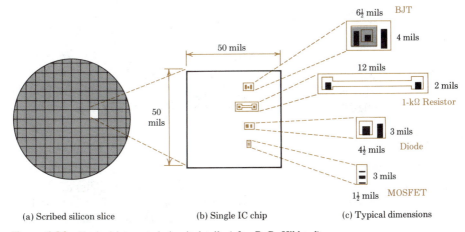

(a) Scribed silicon slice   (b) Single IC chip   (c) Typical dimensions

Figure 6.20 Typical integrated-circuit details (after R.G. Hibberd).

Component Formation

Bipolar transistors in IC form are obtained by successive diffusions of boron (p-type) and phosphorus (n-type). A high concentration of phosphorus is necessary to overcome the previous boron diffusion and to create a high conductivity n^+-type emitter (Fig. 6.21a). Another n^+ area is created to provide a low-resistance contact with the collector. (To minimize the series resistance of the thin collector region, another n^+ layer is formed in the substrate before the epitaxial n layer is grown.)

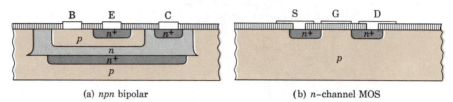

(a) npn bipolar (b) n–channel MOS

Figure 6.21 Structure of IC transistors.

The MOS transistor in IC form (Fig. 6.21b) is similar to its discrete counterpart. MOS transistors can be smaller—and therefore cheaper—than bipolar transistors because they are self-isolating. The source, drain, and channel incorporate insulating pn junctions, and the gate is isolated by oxide. The fabrication process is simple in concept, but MOS device properties are sensitive to surface conditions, which must be very carefully controlled.

Complementary devices, one n-channel and one p-channel, can be combined in a single unit. These CMOS units require only a single power supply, use very little power, and offer special advantages in digital circuits.

Diodes may be formed in various ways. For example, the emitter-base junction of a transistor (with collector tied to the base) provides a low-voltage diode with fast response. If the emitter is omitted, the remainder of a transistor provides a collector-base diode with higher reverse-voltage rating but slower response.

Resistors are usually formed during transistor base diffusion and the conductivity is dependent upon the transistor requirements. The desired value of resistance is obtained by specifying the length and width of the conducting path. Low values of resistance can be obtained by using the higher conductivity n^+-type emitter diffusion. Also, the channel between source and drain of an MOS transistor can be used as a resistor whose value is governed by gate voltage.

Capacitors may be fabricated by the MOS technique (Fig. 6.18b), or the inherent property of a reverse-biased pn junction can be used to provide a capacitor whose value is determined by bias voltage. Junction capacitors can be formed at the same time as the collector junctions of transistors, but the capacitance per unit area is quite low.

Applications

The design of integrated circuits consists in planning an optimum geometrical layout of standard components and connections. Because the fabrication of *microelectronic* devices requires sophisticated techniques and precision equipment, large volume production is necessary. Large volume means standardization of a relatively few products.

Integrated circuits are used in great quantities in digital computers because of their small size, low power consumption, and high reliability. The complexity of digital ICs ranges from simple logic gates and memory units (Chapter 7) to large arrays capable of complete data processing. The small size of MOS elements has led to *large-scale integration* (LSI), in which thousands of elements are created on a single chip. The Intel 80386 32-bit microprocessor is an example of *very large-scale integration* (VLSI). This powerful MOS device incorporates 275,000 transistors on a single chip that can process engineering data faster than a typical minicomputer.

SUMMARY

- In a field-effect transistor, the gate voltage controls the channel conductance.
  In a JFET, the width of the depletion layers controls the conductance.
  In an E MOSFET, the carrier density in the channel induced in the substrate can be enhanced by the gate voltage.
  In a DE MOSFET, the conductivity of a thin channel under the gate can be depleted or enhanced by the gate voltage.
  In the constant-current region, the transfer characteristics are given by:

  For JFET or DE MOSFET, $\qquad i_{DS} = I_{DSS}\left(1 - \dfrac{v_{GS}}{V_p}\right)^2$

  For E MOSFET, $\qquad i_{DS} = K(v_{GS} - V_T)^2$

- A bipolar junction transistor consists of two *pn* junctions in close proximity; normally, the emitter junction is forward biased, the collector reverse biased.
  In common-base operation,

  $$i_C = -\alpha i_E + I_{CBO} \qquad \text{where } \alpha \cong 1$$

  In common-emitter operation, a small base current controls the relatively larger collector current to achieve current amplification.

  $$i_C = \beta i_B + I_{CEO} \qquad \text{where } \beta = \frac{\alpha}{1 - \alpha}$$

- Graphical analysis of nonlinear transistors is based on load-line construction.
  Variations in input quantities about a properly selected quiescent point result in larger output variations that represent amplification.

- A small change in input voltage can switch the output voltage from near zero to a high value as required in digital operations.

- Integrated circuits have changed the character of electronic design.
  Transistors, diodes, resistors, and capacitors are fabricated simultaneously by successive oxidation, masking, etching, and diffusion steps.
  By putting hundreds of thousands of interconnected transistors on a single chip, engineers create tiny information processors with giant capability.

REVIEW QUESTIONS

1. Explain the function of the source, gate, and drain in the JFET.
2. Sketch the cross section of a JFET at pinch off and explain the effect on i_D.
3. Why is drain current nearly constant above pinch-off?
4. Explain the operation of an enhancement-type MOSFET.
5. How are electrons generated in a p-type region under the gate of an E MOSFET?
6. How is channel conductivity controlled in a DE MOSFET?
7. Sketch, from memory, the transfer characteristics of a DE MOSFET.
8. Explain with sketches the operation of a pnp transistor.
9. Why must the base be narrow for BJT action?
10. Describe how amplification and switching are achieved by a BJT.

11. Explain common-base transistor characteristics in terms of diode characteristics.
12. Draw from memory the input and output i vs v characteristics of an npn transistor. Repeat for a pnp transistor.
13. Explain the role of the load line in graphical analysis of amplifiers.
14. List two ways in which the availability of ICs has changed electronic design.
15. Explain how system reliability is increased by using ICs.
16. How many diffusion steps are required in typical IC fabrication?
17. How does diode isolation provide element insulation?
18. Without reference to the text, outline the process of making an IC, including resistors, capacitors, and transistors.

EXERCISES

1. The JFET of Fig. 6.2 is connected as shown in Fig. 6.22. Complete the diagram by indicating the proper polarities for V_{GG} and V_{DD}. For $V_{GG} = 3$ V and $V_{DD} = 20$ V, predict the drain current.

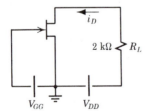

Figure 6.22 V_{GG} ≟ V_{DD}

2. The JFET of Fig. 6.2 is connected as shown in Fig. 6.22. Indicate the proper polarities for V_{GG} and V_{DD}. If V_{DD} is 20 V and R_L is changed to 2.5 kΩ, specify V_{GG} for a drain current of 5 mA.
3. The MOSFET of Fig. 6.3 is to be used as a switch in the circuit of Fig. 6.23 where $V_{DD} = 10$ V and $R_L = 2.3$ kΩ. Draw the load line defining the relation between the nonlinear device and the linear supply. For V_i switched from 2 to 10 V, calculate the corresponding values of I_D and V_o.

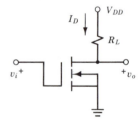

Figure 6.23

4. Assume that your roommate has a good knowledge of physics, including semiconductors, and mathematics, but has never studied electronics. Using labeled sketches and graphs, explain to him or her the operation of a p-channel enhancement-mode MOSFET.
5. A JFET constructed like the one in Fig. 6.1b has the following typical properties: $I_{DSS} = 4$ mA and $V_{GS(OFF)} = -5$ V.
   (a) Sketch a set of i_D vs v_{DS} curves.
   (b) Draw transfer characteristic i_D vs v_{GS}.
   (c) For $V_{GS} = -2$ V in Fig. 6.24, specify R_S for operation in the normal region.
   (d) For the value of R_S in part (c) and $R_D = 3$ kΩ, select a reasonable value of v_{DS} and specify V_{DD}.

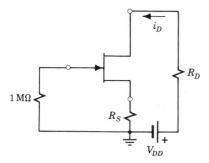

Figure 6.24

6. The FET of Fig. 6.2 is to be used as a voltage amplifier in the circuit of Fig. 6.22. An average gate voltage $V_{GS} = -2$ V is maintained by the battery. The input signal is inserted in series with battery V_{GG} and the output voltage is taken across $R_L = 3$ kΩ. Plot the transfer characteristic for the middle of the normal operating region, $V_{DS} \cong 13$ V. Estimate the output (drain) current variation for an input (gate-source) signal of $0.2 \sin \omega t$ V and the voltage amplification possibilities of this amplifier.

7. The MOSFET of Fig. 6.3 is to be used as a passive load resistor by connecting the gate terminal to the drain.
   (a) Draw a graph of i_D versus v_{DS} for this connection.
   (b) Estimate the resistance offered to a dc current of 2 mA.
   (c) Use piecewise linearization to represent this device by a combination of voltage source and resistance.

8. For an n-channel enhancement-mode MOSFET, threshold voltage $V_T = 1$ V and $I_{DS} = 8$ mA for $V_{GS} = 5$ V.
   (a) Draw the transfer characteristic for the normal operating region.
   (b) For $V_{GS} = +3$ V, $V_{DD} = 16$ V, and $R_D = 3$ kΩ, draw the circuit and predict the drain current and the drain-source voltage.

9. The MOSFET of Fig. 6.3 is to be used as an elementary amplifier.
   (a) Draw an appropriate circuit showing bias voltages V_{GG} and V_{DD} and load resistor R_L.
   (b) For $V_{DD} = 12$ V, $V_{GG} = 8$ V, and $R_L = 2$ kΩ, determine quiescent drain current I_D.
   (c) Estimate the input signal voltage for an output $v_o = 1 \sin \omega t$ V.

10. The n-channel MOSFET shown in Fig. 6.25 can operate in the depletion or enhancement mode. In the normal operation (constant-current) region with gate shorted to source, the drain current is 8 mA, and the pinch-off voltage is -4 V. The MOSFET is to operate at $I_D = 2$ mA and $V_{DS} = 7$ V (the quiescent point).
   (a) Sketch the i_D vs v_{GS} transfer characteristic and specify the quiescent value of v_{GS}.
   (b) Complete the circuit design by specifying R_1 and R_2. (*Note:* Neglect any voltage drop across resistance R_G, which carries no current.)

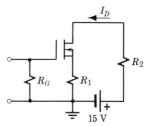

Figure 6.25

11. For a certain DE MOSFET, $I_{DSS} = 2$ mA and $V_p = -4$ V.
   (a) Calculate V_{GS} for operation at $I_{DS} = 1$ mA in the constant-current region.
   (b) For the circuit shown in Fig. 6.24, derive an expression for i_D as a $f(v_{DS})$.
   (c) For operation at $I_D = 1$ mA and $V_{DS} = 8$ V with $R_D = 5$ kΩ, specify R_S and V_{DD}.

12. For a pnp BJT:
   (a) Sketch the charge distribution when unbiased.
   (b) Sketch the potential distribution when unbiased.
   (c) Sketch the potential distribution when properly biased.

13. For high-gain amplification, high values of α and β are desirable. Considering the effect of doping on factor γ (Fig. 6.10), explain why it is desirable to dope the base relatively lightly in comparison to the emitter and collector regions.

14. Assume that your roommate has a good knowledge of physics, including semiconductors, and mathematics, but has never studied electronics. Using labeled sketches and graphs, explain to him or her the "normal" operation of a pnp junction transistor.

15. A BJT is connected as in Fig. 6.26.
  (a) Label the polarities of v_{EB} and v_{CB} to provide for *normal* operation.
  (b) For switch S OPEN, sketch a graph of I_C versus $\pm V_{CB}$.
  (c) For switch S CLOSED and $I_E = -2$ mA, repeat part (b).

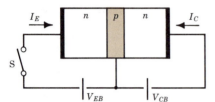

Figure 6.26

16. A BJT has the following parameters: $\alpha = 0.97$ and $I_{CBO} = 10 \ \mu A$. For $0 < i_C < 10$ mA:
  (a) Sketch the common-base collector characteristics.
  (b) Sketch the common-emitter collector characteristics.

17. For a silicon BJT, the base current is 0.3 mA at a base-emitter voltage of 0.7 V. Assuming that the base-emitter junction follows the behavior of a theoretical diode (Eq. 5-20), predict the range of base currents for base-emitter voltages from 0.65 to 0.75 V. Comment on the assumption that $V_{BE} \cong 0.7$ V in many situations.

18. A silicon *pnp* BJT is connected in the circuit shown in Fig. 6.27.
  (a) Complete the drawing by putting the arrowhead on the emitter symbol and indicating the proper battery polarities for a forward-biased emitter-base junction and a reverse-biased collector-base junction.
  (b) Assuming $|V_{BE}| = 0.7$ V, calculate I_B for $V_{BB} = 4$ V and $R_B = 33$ kΩ.
  (c) If $V_{CC} = 15$ V, $R_L = 1.5$ kΩ, and $\beta = 60$, calculate I_C, I_E, and V_{CE}.

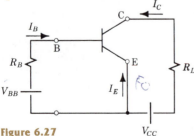

Figure 6.27

19. A silicon *npn* transistor is connected in the circuit shown in Fig. 6.27.
  (a) Complete the drawing by putting the arrowhead on the emitter symbol and indicating the proper battery polarities for a forward-biased emitter-base junction and a reverse-biased collector-base junction.
  (b) Assuming $|V_{BE}| = 0.7$ V, specify R_B for $V_{BB} = 4$ V and $I_B = 0.2$ mA.
  (c) If $V_{CE} = 8.5$ V, $R_L = 500$ Ω, and $\beta = 75$, calculate I_C, I_E, and V_{CC}.

20. A silicon BJT is connected as shown in Fig. 6.28, where $R_C = 3.5$ kΩ. Stating any assumptions (see Exercise 17), predict I_C and specify R_B to establish V_{CE} at 5 V.

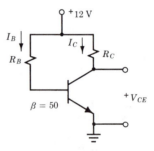

Figure 6.28

21. In the elementary amplifier of Fig. 6.16, $V_{CC} = 10$ V and $R_L = 1$ kΩ. The quiescent collector current I_C is to be 3 mA and the signal component is to be $i_c = 3 \sin \omega t$ mA.
  (a) Reproduce the collector characteristics and draw the load line.
  (b) Estimate I_B and i_i, the quiescent and signal components of the base current.
  (c) Estimate the signal component of the voltage across R_L.

22. The silicon BJT of Fig. 6.15 is used in the circuit of Fig. 6.16 with $I_B = 0.1$ mA, $R_L = 2$ kΩ, and $V_{CC} = 20$ V.
  (a) Estimate β from the graph, and predict I_C at the quiescent point.
  (b) Reproduce the collector characteristics and draw the "load line." Determine I_C graphically.
  (c) Find i_c (graphically) for $i_i = 0, +50$, and $-50 \ \mu A$.
  (d) If i_i is a sinusoid varying between $+50$ and $-50 \ \mu A$, write an expression for $i_C(t)$.
  (e) Estimate the current gain of this amplifier.

23. A subcircuit used in ICs is shown in Fig. 6.29. (Assume $\beta = 30$ for the transistor.)

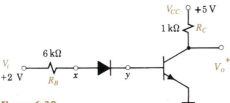

Figure 6.29

(a) Sketch the i-v characteristic of a silicon diode and estimate the voltage at point x with respect to point y if the diode is conducting.

(b) If the base-emitter junction of the transistor is forward biased, estimate the voltage of point y and the voltage at point x with respect to ground.

(c) Estimate the base and collector currents of the transistor and the output voltage V_o under these conditions.

24. The transistor of Fig. 6.15 is used in the circuit of Fig. 6.29 with V_{CC} increased to 12 V; the characteristics of the diode are similar to Fig. 6.15a.

(a) For $i_C = 7$ mA, use the load line to estimate V_o and i_B, and specify the necessary V_i.

(b) For $V_i = 3.4$ V, estimate i_B and predict i_C and $V_o = V_{CE}$.

(c) For $V_i = 1$ V, predict i_C and V_o.

25. A certain IC device would cost $120,000 for the complete design and $2 each to manufacture. Predict:

(a) The unit cost for a custom run of 100.

(b) The unit cost for a standard run of 100,000.

(c) The number of units at which design cost is 10% of the manufacturing cost.

26. It costs $60 to process a 100-mm wafer, of which only the central 80 mm is usable. If the final yield is 50%, estimate the fabrication cost of a single 1.5-mm × 1.5-mm IC chip.

27. Sketch the masks required to fabricate an IC *pnp* transistor.

PROBLEMS

1. Two MOSFETs (Fig. 6.3) are to be used as a switch in a digital computer. In the circuit of Fig. 7.2c, determine the i_D vs V_{DS} characteristic for the upper (load) transistor. For $V_{DD} = 10$ V, plot $V_{DD} - V_{DS(load)}$ on a reproduction of Fig. 6.3c. Estimate the output voltages for inputs of 2 and 10 V. Describe the switching operation.

2. In a conventionally arranged amplifier, the transistor terminals are labeled X, Y, Z. With no signal applied, it is observed that terminal X is 20 V positive with respect to terminal Y and 0.5 V positive with respect to terminal Z. Explain your reasoning:

(a) Which terminal is the emitter?

(b) Which terminal is the collector?

(c) Which terminal is the base?

(d) Whether X is p- or n-type material?

3. The variation in α with v_{CB} in a transistor (Fig. 6.12c) can be explained in terms of the effect of v_{CB} on the *effective width* of the base.

(a) As the reverse bias increases, what is the effect on the width of the transition (or depletion) region adjacent to the base-collector junction?

(b) How does a change in transition width change the effective base width?

(c) How does base width affect α?

(d) Why does an increase in v_{CB} produce the effect shown in Fig. 6.12b?

4. A transistor amplifier using a 2N3114 is to supply a 10-V (rms) output signal across a 10-kΩ load resistance R_L. "Design" an elementary amplifier by drawing an appropriate circuit and specifying the components. Estimate the input signal current required.

5. A nearly constant current source $I \cong 2$ mA is needed in an IC, and the circuit of Fig. 6.30 is proposed with $R = 1$ kΩ. Choose R_y so that the voltage across R_y is about 4 times V_{BE} so that small changes in V_{BE} do not affect the operation. Stating any assumptions, specify R_x. Comment on the "constancy" of this circuit if β changes from 50 to 200, say.

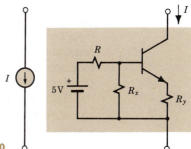

Figure 6.30

CHAPTER 7

Logic Elements

Switching Logic

Electronic Switches

Transistor Logic Gates

Memory Elements

Digital Integrated Circuits

The music on a phonograph record is preserved as a groove that guides a needle whose motion changes in amplitude and frequency as the record turns; these *continuous* or *analog* signals can take any value in a wide range of values. By contrast, the music on a laser disk is preserved as a pattern of flat areas and holes that either reflect light or do not; these *discrete* or *digital* signals are processed by circuits made up of electronic switches that are either on or off. A major virtue of electronic circuits is the ease and speed with which digital signals can be processed, and the use of such signals in control, computation, and communication is the most rapidly developing aspect of electronic engineering.

To process information in digital form we need special circuits, and for the efficient design of digital circuits we use a special numbering system and a special algebra. The circuits must provide for storing instructions and data, receiving new data, performing calculations, making decisions, and communicating the results. For example, an automatic airline reservation system must receive and store information from the airline regarding the number of seats available on flights all over the world, respond to inquiries from travel agents across the country, subtract the number of seats requested from the number available or add the number of seats canceled, handle 50 or so requests per minute, and keep no one waiting more than a minute.

Information processing is an important component of all branches of engineering and science. Aeronautical and chemical engineers may be designing automatic control systems. Civil and industrial engineers may be concerned with data on traffic flow or product flow. Mechanical engineers may be designing "smart" tools or products.

Chemists and medical doctors may be interested in automated laboratory analysis. Biologists and physicists may need remote control of precisely timed events. Everyone engaged in experimental work or in management can benefit from the new data processing techniques.

In this chapter we begin our study of digital systems by learning the basic functions performed by decision-making elements and seeing how we can use semiconductor devices in practical circuits. Then we look at basic memory elements and see how we can construct them from bipolar and MOS transistors. Finally we examine more sophisticated memory elements with desirable operating features. In the next chapter we shall learn to design the data processing circuits and devices that are used in digital computers.

SWITCHING LOGIC

Digital computers, automatic process controls, and instrumentation systems have the ability to take action in response to input stimuli and in accordance with instructions. In performing such functions, an information processing system follows a certain *logic*; the elementary logic operations are described as **AND, OR**, and **NOT**. Tiny electronic circuits consuming very little power can perform such operations dependably, rapidly, and efficiently.

Gates

A *gate* is a device that controls the flow of information, usually in the form of pulses. First we consider gates employing magnetically operated switches called *relays*. If the switches are normally open, they close when input signals in the form of currents are applied to the relay coils.

In Fig. 7.1a, the lamp is turned on if switch A *and* switch B are closed; it is called an **AND** circuit or **AND** gate. In Fig. 7.1b, the lamp is turned on if switch A *or* switch B is closed *or* if both are closed; it is called an **OR** circuit. In Fig. 7.1c, the switch is connected *across* the lamp. For an input at C, the switch is closed, shorting out the lamp, and there is *not* an output; the input has been *inverted* by the **NOT** gate.

In general, there may be several inputs and the output may be fed to several other logic elements. In a typical digital computer there are millions of such logic elements. Although the first practical digital computer (the Harvard Mark I, 1944) employed relays, the success of the modern computer is due to our taking advantage of the switching capability of semiconductor devices.

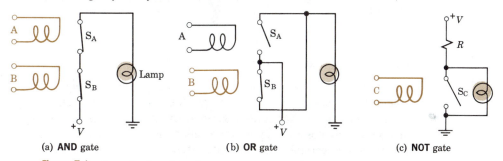

(a) **AND** gate   (b) **OR** gate   (c) **NOT** gate

Figure 7.1   Gate circuits using relays.

Electronic Switching

In the gates of Fig. 7.1, "data" are represented by **ON** and **OFF** switch positions or by the corresponding presence or absence of voltages. How can diodes and transistors be used as switches? If the input to the circuit in Fig. 7.2a is zero (i.e., if the input

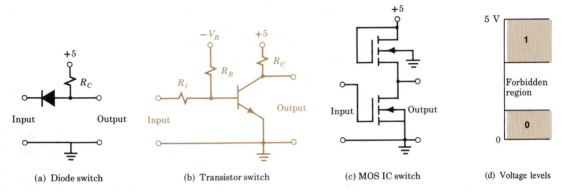

(a) Diode switch   (b) Transistor switch   (c) MOS IC switch   (d) Voltage levels

Figure 7.2   Elementary semiconductor switching circuits.

terminals are shorted), the diode is forward biased, current flows easily, and the output is a few tenths of a volt or approximately zero. If the input is 5 V or more, the diode is not forward biased, no diode current flows, and the output is $+5$ V. The **OFF–ON** positions of this diode switch correspond to output voltage levels near zero and $+5$ V.

In the elementary transistor switch (Fig. 7.2b) with a zero input, the base-emitter junction is reverse biased by $-V_B$ across the voltage divider, the collector current is very small, and the output is approximately $+5$ V. An input of $+5$ V forward biases the base-emitter junction, a large collector current flows, and the output is approximately zero. This circuit is analogous to Fig. 7.1c; the input signal is inverted.

The MOS transistor can operate as a binary switch controlled by changes in gate voltage. Because an MOS transistor can also function as a resistor, the arrangement of Fig. 7.2c is convenient in integrated circuits. With the gate of the upper or "load" transistor connected to the drain terminal, enhancement is high and the resistance of the load is determined by the channel dimensions. With no input signal, the lower or "driver" transistor is **OFF**, there is no IR drop across the load, and the output is approximately $+5$ V. A positive input signal turns the driver **ON** and the output falls to nearly zero.

Basic Logic Operations

To make our work with logic circuits easier, we use the conventional symbols and nomenclature developed by circuit designers and computer architects. Each basic logic operation is indicated by a *symbol*, and its function is defined by a *truth table* that shows all possible input combinations and the corresponding outputs. Logic operations and variables are in boldface type. Electronic circuits are drawn as single-line diagrams with the ground terminal omitted; all voltages are with respect to ground so that "no input" means a shorted input terminal. Two distinct voltage levels, separated by a forbidden region (Fig. 7.2d), electronically represent the *binary* numbers **1** and **0** corresponding to the **TRUE** and **FALSE** of logic.

AND Gate. The symbol for an **AND** gate is shown in Fig. 7.3a, where **A·B** is read "**A AND B**." As indicated in the truth table, an output appears only when there are inputs at **A AND B**. If the inputs in Fig. 7.3b are in the form of positive voltage pulses

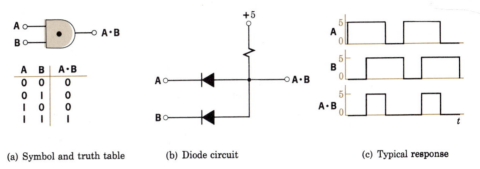

A	B	A·B
0	0	0
0	1	0
1	0	0
1	1	1

(a) Symbol and truth table   (b) Diode circuit   (c) Typical response

Figure 7.3   A two-input **AND** gate.

(with respect to ground), inputs at A and B turn off both diodes, no current flows through the resistance, and there is a positive output (**1**). In general, there may be several input terminals. If any one of the inputs is zero (**0**), current flows through that forward-biased diode and the output is nearly zero (**0**). For two inputs varying with time, a typical response is shown in Fig. 7.3c.

OR Gate. The symbol for an **OR** gate is shown in Fig. 7.4a where **A + B** is read "**A OR B**." As indicated in the truth table, the output is **1** if input **A OR** input **B** is **1**. For no input (zero voltage) in Fig. 7.4b, no current flows, and the output is zero (**0**). An input of +5 V (**1**) at either terminal A or B or both (or at *any* terminal in the general case) forward biases the corresponding diode, current flows through the resistance, and the output voltage rises to nearly 5 V (**1**). For two inputs varying with time, the response is as shown.

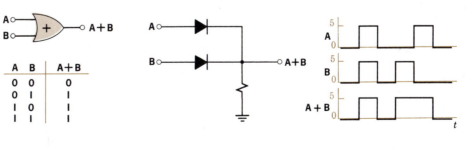

A	B	A+B
0	0	0
0	1	1
1	0	1
1	1	1

(a) Symbol and truth table   (b) Diode circuit   (c) Typical response

Figure 7.4   A two-input **OR** gate.

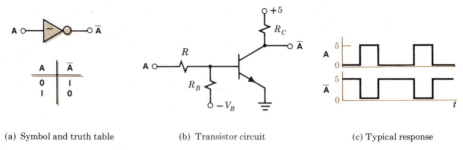

(a) Symbol and truth table   (b) Transistor circuit   (c) Typical response

Figure 7.5   A transistor **NOT** gate.

NOT Gate. The inversion inherent in a transistor circuit corresponds to a logic **NOT** represented by the symbol in Fig. 7.5a where $\overline{A}$ is read "**NOT A**." As indicated in the truth table, the **NOT** element is an *inverter*; the output is the *complement* of the single input. With no input (**0**), the transistor switch is held open by the negative bias voltage and the output is +5 V (**1**). A positive input voltage (**1**) forward biases the base-emitter junction, collector current flows, and the output voltage drops to a few tenths of a volt (**0**). For a changing input, the output response is as shown.

NOR Gate. In the diode **OR** gate of Fig. 7.4b, an input of +5 V at **A OR B** produces a voltage across R and a positive output voltage. But this output is less than the input (by the diode voltage drop), and after a few cascaded operations the signal would decrease below a dependable level. A transistor supplied with +5 V can be used to restore the level as in Fig. 7.6b; however, the inherent inversion results in a

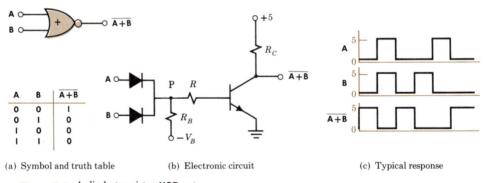

(a) Symbol and truth table   (b) Electronic circuit   (c) Typical response

Figure 7.6   A diode-transistor **NOR** gate.

NOT OR or **NOR** operation. The small circle on the **NOR** element symbol and the bar in the $\overline{A + B}$ output indicate the inversion process.

In the **NOR** circuit with no input, the transistor switch is held **OFF** by the negative bias voltage V_B and the output is +5 V (**1**). An input of +5 V at terminal A or B raises the base potential, forward biases the base-emitter junction, turns the transistor switch **ON**, and drops the output to nearly zero (**0**). As we shall see (p. 176), in addition to restoring the signal level, the transistor provides a relatively low output resistance so that this **NOR** element can supply inputs to many other gates. Another advantage is that all the basic logic operations can be achieved by using only **NOR** gates.

Practice Problem 7-1

(a) Form the truth table for a **NOR** gate with the two input terminals tied together so that **A** = **B**. What logic function is performed by this gate?

(b) A logic circuit consists of a **NOR** gate with **NOT** gates inserted in each input line. Draw the circuit, form the truth table, and state the logic function performed.

Answers: (a) **NOT**; (b) **AND**.

NAND Gate. Diodes and a transistor can be combined to perform an inverted **AND** function. Such a **NAND** gate has all the advantages of a **NOR** gate and is very easy to fabricate, particularly in IC form. In a complex logic system, it is convenient to use just one type of gate, even when simpler types would be satisfactory, so that gate characteristics are the same throughout the system.

The **NAND** gate function is defined by the truth table in Fig. 7.7. The small circle on the **NAND** element symbol and the bar on the $\overline{A \cdot B}$ output indicate the inversion

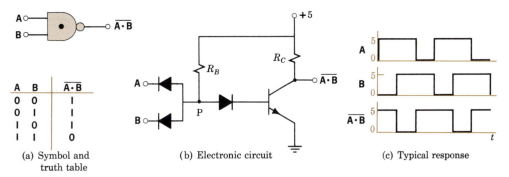

A	B	$\overline{A \cdot B}$
0	0	1
0	1	1
1	0	1
1	1	0

(a) Symbol and truth table

(b) Electronic circuit

(c) Typical response

Figure 7.7   A diode-transistor **NAND** gate.

process. With positive inputs at A and B (**1 AND 1**), the input diodes are reverse biased, and no input current flows; the positive base current supplied through R_B causes heavy collector current and the output is approximately zero (**0**). If either A or B has a zero (**0**) input, at least one input diode conducts to ground, the voltage at point P is too low to supply base current to the transistor, hence no collector current flows, and the output is +5 V (**1**). For better separation of voltage levels (see Fig. 7.2d), a second diode may be placed in series with the base of the transistor.

In Example 1 on p. 174, we see that three **NAND** gates can be used to replace an **OR** gate. The combination of **NAND** gates is equivalent to an **OR** gate in that it performs the same logic operation. In digital nomenclature, the function **f** is defined by

$$f = A + B = \overline{\overline{A} \cdot \overline{B}}$$

We arrived at this relation by considering the desired and available truth tables. A "digital algebra" for the direct manipulation of such expressions is presented in Chapter 8. First, however, let us see how practical semiconductor logic elements function.

EXAMPLE 1

Use **NAND** gates to form a two-input **OR** gate.

The desired function is defined by the truth table of Fig. 7.8a. Comparing this with Fig. 7.7a, we see that if each input were inverted (replaced by its complement), the **NAND** gate would produce the desired result as indicated in the truth table.

To provide simple inversion, we tie both terminals of a **NAND** gate together (the first and last rows of Fig. 7.7a). The desired logic circuit is shown in Fig. 7.8b.

A	B	f	$\overline{A}$	$\overline{B}$	$\overline{\overline{A} \cdot \overline{B}}$
0	0	0	1	1	0
0	1	1	1	0	1
1	0	1	0	1	1
1	1	1	0	0	1

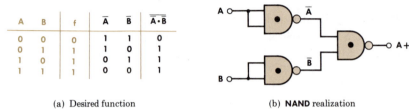

(a) Desired function   (b) **NAND** realization

Figure 7.8   Using **NAND** gates to form an **OR** gate.

ELECTRONIC SWITCHES

Binary digital elements must respond to two-valued input signals and produce two-valued output signals. In practice the "values" are actually ranges of values separated by a *forbidden region*. The two discrete *states* may be provided electrically by switches, diodes, or transistors. Magnetic cores in which the two states are represented by opposite directions of magnetization were used extensively in early computers. Purely fluid switches are sometimes used in hydraulic control applications. In applications requiring high speed and flexibility, however, electronic switches predominate.

Classification of Electronic Logic

Electronic logic circuits are classified in terms of the components employed. Basic operations can be performed by diode logic (DL), resistor-transistor logic (RTL), or diode-transistor logic (DTL). Currently popular are transistor-transistor logic (TTL), metal-oxide-semiconductor (MOS) and complementary MOS (CMOS), and emitter-coupled logic (ECL). In specifying a logic system, the designer selects the type of logic whose characteristics match the requirements.

Logic types vary in *signal degradation, fan-in, fan-out,* and *speed.* A major disadvantage of diode logic (Figs. 7.3 and 7.4) is that the forward voltage drop is appreciable, and the output signal is "degraded" in that the forbidden region is narrowed. The use of transistors minimizes degradation. The number of inputs that can be accepted is called the *fan-in* and is low (3 or 4) in DL and high (8 or 10) in TTL. The number of inputs that can be supplied by a logic element is called the *fan-out.* Fan-out depends on the output current capability (and the input current requirement) and varies from 4 in DL to 10 or more in TTL.

The speed of a logic operation depends on the time required to change the voltage levels, which is determined by the effective time constant of the element. In high-speed diodes, the charge storage is so low that response is limited primarily by wiring and load capacitances. In transistors in the **ON** state, base current is high and the charge stored in the base region is high; before the collector bias can reverse, this charge must be removed. Typically, 5 to 10 ns are required to process a signal. (In ECL, the charge stored is minimized and ECL gates can operate at rates up to 200 MHz.)

In practical design, the important factors are cost, speed, immunity to noise, power consumption, and reliability. In IC manufacture, a gate containing many highly uniform active components may cost no more than a discrete transistor and is no less reliable. As a result, sophisticated logic elements of greatly improved characteristics are now available at reasonable cost.

Transistor Switches

The diode is an automatic "gate," **CLOSED** to reverse voltages and **OPEN** to forward voltages. Figures 7.3 and 7.4 show how we could use diodes to perform basic logic operations. The highly nonlinear characteristic of the diode isolates the several inputs of a logic gate. For example, the output of an **OR** gate (Fig. 7.4b) always follows the most positive of the input signals; diodes with less positive inputs are reverse biased. (The diode model of Fig. 5.15b with $V_F = 0.7$ V is suitable for the analysis of logic circuits using silicon diodes.) After several successive **OR** gates, the output signal will be significantly degraded. Eventually it is necessary to restore the signal, and a transistor inverter can be used for this purpose.

As shown in Fig. 6.16, a transistor amplifier is operated in the *linear* or *normal* region with the emitter-base junction forward biased and the collector-base junction reverse biased. If both junctions are reverse biased, however, practically no collector current flows and the transistor is said to be operating in the *cutoff region*. In the basic switching circuit of Fig. 7.9a, if the input voltage is zero, there is no base current and operation is at point 1 in Fig. 7.9b. The collector current is practically zero ($I_C \cong I_{CEO}$) and the *switch* whose contacts are the collector and emitter terminals is **OPEN**. The cutoff current is exaggerated in Fig. 7.9b; a typical value of collector current of less than 1 μA with an applied voltage of 5 V corresponds to a dc *cutoff resistance* of more than 5 MΩ.

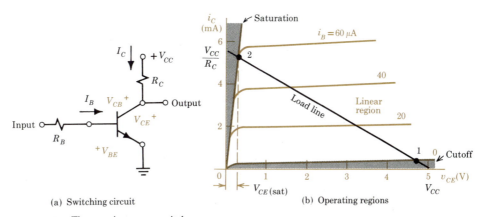

(a) Switching circuit

(b) Operating regions

Figure 7.9 The transistor as a switch.

A positive voltage pulse applied to the input terminal forward biases the emitter-base junction, causes an appreciable base current, and moves operation to point 2. An increase in base current above 60 μA produces no further effect on collector current and the transistor is said to be operating in the *saturation region*. (This limit on collector current is represented in the dc transistor model of Fig. 7.11b by $I_C \leq \beta I_B$.) The voltage drop across the "switch" is called *collector saturation* voltage $V_{CE(sat)}$ and is typically a few tenths of a volt. Note that if V_{CE} is less than V_{BE}, the collector-base junction is also forward biased. As indicated in Fig. 7.9b, a collector current of around 6 mA at a saturation voltage of 0.3 V corresponds to a *saturation resistance* of around 50 Ω; more typical values are 10–20 Ω. When the transistor switch is **CLOSED**, if $R_C \gg R_{sat}$ the collector current is determined primarily by the load resistance and $I_C \cong V_{CC}/R_C$.

Practice Problem 7-2

A transistor with $\beta = 100$ and $V_{CE(sat)} = 0.2$ V is connected in the circuit of Fig. 7.9a with $V_{CC} = 5$ V and $R_C = 1$ kΩ.
(a) Estimate the output voltage and current I_C when the transistor switch is closed.
(b) Estimate the minimum base current to close the switch.
(c) If $V_{BE} = 0.8$ V, specify the value of R_B if the switch is to close when the input voltage exceeds 3 V.

Answers: (a) 0.2 V, 4.8 mA; (b) 48 μA; (c) 45.8 kΩ.

TRANSISTOR LOGIC GATES

The transistor switch of Fig. 7.9 is basically an RTL inverter or **NOT** gate. When the input is near zero (logic **0**), the output is near +5 (logic **1**); a positive input (**1**) causes current flow and the output drops to $V_{CE(sat)}$(**0**). The *transfer characteristic*, that is, the relationship between input and output voltages, is shown in Fig. 7.10. The

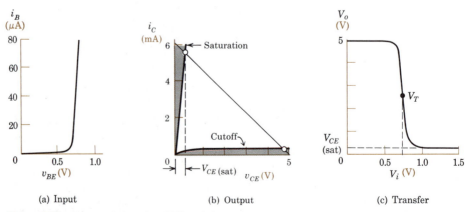

(a) Input   (b) Output   (c) Transfer

Figure 7.10 Characteristics of an RTL switch.

threshold voltage V_T is approximately 0.7 V.[†] Such switches can be connected in series or parallel to provide **NAND** or **NOR** gates.

Diode-Transistor Logic

A more sophisticated, and faster, device is the DTL **NAND** gate of Fig. 7.11. It is faster because the charging currents accompanying a change in transistor state flow through the low forward resistances of diodes instead of through the higher series resistances necessary in RTL operation. A fast-acting Schottky diode may be added in parallel with the collector-base junction to limit the forward bias and reduce the turn-off delay. The greater sophistication is inexpensively realized in IC form.

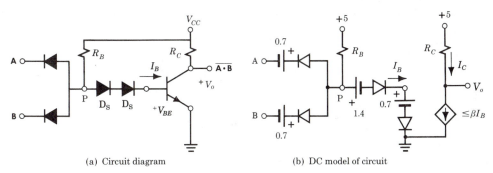

(a) Circuit diagram   (b) DC model of circuit

Figure 7.11   Operation of a DTL **NAND** gate.

The dc model of a silicon **NAND** gate is shown in Fig. 7.11b. Diodes A and B are represented by ideal diodes in series with 0.7-V sources. Series diodes D_S are replaced by one ideal diode and a 1.4-V source. The transistor input characteristics (Fig. 7.10a) resemble those of a junction diode, and the input circuit is modeled just as the other diodes. In general, the collector circuit is modeled as a controlled current source (see p. 75) where $I_C \leq \beta I_B$ if I_{CEO} is neglected. The inequality sign is introduced here because the collector current cannot exceed the saturation value. Implicit in the transistor model is the fact that V_{CE} is always equal to or greater than $V_{CE(\text{sat})}$. The performance of the gate can be predicted on the basis of this model. (See Example 2 on p. 178.)

If this gate follows other DTL gates in the **1** state, inputs A and B are approximately 5 V (**1**), and diodes A and B are reverse biased. Series diodes D_S are forward biased by V_{CC} and the high base current holds the transistor in saturation, so the output is about 0.3 V (**0**). The potential at point P is about $2 \times 0.7 + 0.7 = 2.1$ V. If now input A drops, at $V_P = V_T = 2.1 - 0.7 = 1.4$ V, diode A is forward biased and begins to conduct. The potential at point P drops, the transistor is cut off, and V_o rises to $+5$ V (**1**). Note that a change in input from 5 to 1.4 or 3.6 V is required to switch from the high state, and a change from 0.3 to 1.4 or 1.1 V is required to switch from the low state. The difference between the operating input voltage and the threshold

[†] In practice, the circuit designer may hold V_{BE} below 0.4 V, say, to ensure cutoff and provide $V_{BE} = 0.8$ V to ensure saturation.

EXAMPLE 2

For the DTL **NAND** gate of Fig. 7.11, $V_{CC} = 5$ V and $R_B = R_C = 5$ kΩ. For the transistor, $\beta = 30$ and $V_{CE(\text{sat})} = 0.3$ V. If input A is 5 V and input B is 0.3 V, determine the no-load output voltage V_o. *Note:* In dealing with nonlinear "threshold" devices, circuit equations involve inequalities. The practical approach is to assume a state (**ON, OFF**) and check to see if that state is consistent with the data.

For input $A = 5$ V and $B = 2$ V, determine the base current, the collector current, and the no-load output voltage.

The lowest input governs the state of a **NAND** gate. For $B = 0.3$ V, V_P cannot exceed $0.3 + 0.7 = 1.0$ V.

If current did flow through the series diodes,

$$V_{BE} = V_P - 2(0.7) = 1.0 - 1.4 = -0.4 \text{ V}$$

But a forward bias of $V_{BE} \geq +0.7$ is required; therefore, no base current flows, the transistor is cut off, and

$$V_o = V_{CC} - I_C R_C = 5 - 0 = 5 \text{ V}$$

For $B = 2$ V, V_p cannot exceed $2 + 0.7 = 2.7$ V. The critical value for turn on is $V_P = 0.7 + 1.4 = 2.1$ V. The lowest value governs ∴ base current flows and

$$I_B \cong (V_{CC} - 2.1)/R_B = (5 - 2.1)/5\text{k} = 0.58 \text{ mA}$$

But the largest possible no-load collector current is

$$I_C \cong V_{CC}/R_C = 5/5\text{k} = 1 \text{ mA}$$

which requires a base current of only

$$I_B = I_C/\beta = 1/30 \cong 0.033 \text{ mA}$$

Therefore, the transistor is in saturation and

$$V_o = V_{CE(\text{sat})} = 0.3 \text{ V}$$

voltage is called the *noise margin* because unwanted voltages lower than this value will not cause false switching.

Typically, the output of the **NAND** gate of Example 2 is connected to the inputs of other DTL gates. When V_o is high, the next input diodes are reverse biased and no "load" current is drawn. When V_o is low, however, each input diode carries current equal to $(V_{CC} - V_P)/R_B = (5 - 0.7 - 0.3)/5000 = 0.8$ mA in this example. For this **NAND** gate, the base current I_B must put the transistor into saturation while supplying additional collector current to "drive" the load represented by following gates. The maximum fan-out for DTL gates is about 8. (See Problem 2.)

Practice Problem 7-3

In one version of DTL **NAND** gate (Fig. 7.11a), the threshold voltage for diodes and transistor is 0.8 V.
(a) If the output is logic **0**, what is the voltage at point P?
(b) For $V_A = 4$ V, what is the critical value of V_B to change the output to logic **1**?

Answers: (a) 2.4 V; (b) <1.6 V.

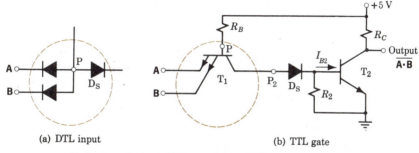

(a) DTL input   (b) TTL gate

Figure 7.12   Derivation of a basic TTL gate from a DTL gate.

Transistor-Transistor Logic

The emitter-base section of a transistor is essentially a *pn* junction diode. (See Fig. 6.12b.) Therefore, the input diodes of the DTL gate of Fig. 7.11a can be replaced by a multiemitter transistor. As shown in Fig. 7.12b, the collector-base junction of T_1 provides an offset voltage equal to that of one series diode. The resulting *transistor-transistor logic* gate requires less silicon area and is faster in operation than a similar DTL gate.

TTL and DTL gates are similar in operation. In Fig. 7.12b, if we assume that the output is **LOW**, then T_2 is **ON** and the emitter-base junction of T_2 and diode D_S must be forward biased. Therefore, voltage V_{P2} must be approximately $0.7 + 0.7 = 1.4$ V, and the base-collector junction of T_1 must be forward biased to supply adequate base current to T_2. This can be true only if V_P is approximately $1.4 + 0.7 = 2.1$ V, and this condition requires that both inputs A and B be **HIGH**, that is, greater than $2.1 - 0.7 = 1.4$ V.

If we assume that either A or B is **LOW** (0.3 V corresponding to $V_{CE(\text{sat})}$ of a preceding stage), then a base-emitter junction of T_1 is forward biased and V_P cannot exceed $0.3 + 0.7 = 1.0$ V. Under this condition, D_S cannot be forward biased, I_{B2} cannot be adequate, and T_2 must be cut off, allowing the output to go **HIGH**. The logic relation is that the output is **LOW** for both inputs **HIGH** and **HIGH** for either input **LOW**. In other words, the output is **NOT (A AND B)** or $\overline{\mathbf{A \cdot B}}$ and this is a **NAND** gate.

A more sophisticated, and very popular, version of the TTL **NAND** gate is shown in Fig. 7.13. Here the output stage consisting of R_4, T_4, D, and T_3 is called a *totem pole*. The operation can be explained as follows. If we assume that the output is **LOW**, then T_3 is **ON**, the emitter-base junctions of T_3 and T_2 must be forward biased, and the collector-base junction of T_1 must be forward biased to supply base current to T_2 and, in turn, to T_3. This can be true only if V_P is approximately $0.7 + 0.7 + 0.7 = 2.1$ V, and this requires that inputs **A**, **B**, and **C** be **HIGH**.

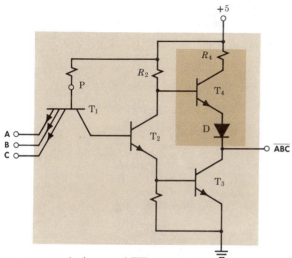

Figure 7.13   An integrated TTL **NAND** gate.

Note that if T_2 is solidly **ON** (saturated) to supply adequate base current to T_3, then $V_{C2} = V_{B4}$ cannot exceed $V_{CE(sat)} + V_{BE3} = 0.3 + 0.7 = 1$ V. But for T_4 to be **ON**, the voltage at the base of T_4 would have to be $V_{C3} + V_D + V_{BE4} = 0.3 + 0.7 + 0.7 = 1.7$ V. We conclude that with T_2 **ON**, T_4 must be **OFF**. This means that little current flows through R_4, and power loss is small when the output goes **LOW**. (Note that, as shown in Fig. 7.9b, high I_B holds T_2 in saturation and V_{CE} low even if I_C is low.)

If one or more inputs goes **LOW** (0.3 V, say), one base-emitter junction of T_1 is forward biased, and the voltage at point P cannot exceed $0.3 + 0.7 = 1.0$ V. Since $V_P = 2.1$ V is required to turn T_2 and T_3 **ON**, we know that T_2 and T_3 are **OFF**. With T_2 and T_3 essentially open circuits, V_{B4} rises and causes T_4 to conduct. Neglecting the small voltage drop due to I_{B4}, we find that the output voltage is equal to $V_{CC} - I_{B4} R_2 - V_{BE4} - V_D \cong 5 - 0 - 0.7 - 0.7 = 3.6$ V. The logic relation is that the output is **HIGH** if any input is **LOW**—the characteristic of a **NAND** gate.

For fast switching and high fan-out, the output resistance of a logic gate should be low. Because all capacitances associated with the inputs to the next logic level must be charged or discharged as the voltage level changes, the output circuit must be capable of *supplying* large currents in an upward transition or *sinking* large currents in a downward transition. The output resistance of a saturated transistor (T_3) is inherently low (about 10 Ω typically), providing the desired low output resistance when the output goes **LOW**. When the output is **HIGH**, T_4 is active and provides the low output resistance characteristic. In IC form, TTL gates are small, reliable, and cheap; because of their excellent characteristics they are used in a great variety of IC devices.

EXAMPLE 3

For the device shown in Fig. 7.14, $V_{CE(sat)} = 0.2$ V and the threshold voltage $V_{BE} = 0.8$ V. Identify the logic element, complete the truth table of voltages, and predict whether T_5 is **ON** or **OFF**.

V_A	V_B	V_C	V_x	V_y	V_z	T_5
2.0	2.4	3.4	2.4	1.0	0.2	**ON**
3.4	3.4	1.0	1.8	5	3.4	**OFF**

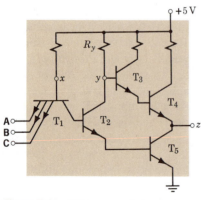

Figure 7.14 TTL gate analysis.

This is another form of three-input TTL **NAND** gate. Multi-emitter transistor T_1 acts as an **AND** gate; V_x is **HIGH** only if $A \cdot B \cdot C = 1$. In other words, the lowest input governs.

Our approach is to assume T_5 is **ON** and determine what voltages are required. For T_5 **ON**, T_2 must be **ON**; T_5, T_2, and the base-collector junction of T_1 must be forward biased, and V_x must be at least $0.8 + 0.8 + 0.8 = 2.4$ V. Therefore, all input voltages must exceed

$$V_x - V_{BE1} = 2.4 - 0.8 = 1.6 \text{ V}$$

In the first case, the lowest input is $V_A = 2.0$ V; therefore V_x is high, and we conclude that T_5 is **ON** and

$$V_y = V_{BE5} + V_{CE2(sat)} = 0.8 + 0.2 = 1.0 \text{ V}$$

$$V_z = V_{CE5(sat)} = 0.2 \text{ V}$$

In the second case, $V_C = 1.0 < 1.6$ V; therefore, T_5 is **OFF** and

$$V_x = V_{ABC(min)} + V_{BE1} = 1.0 + 0.8 = 1.8 \text{ V}$$

$$V_y = V_{CC} - I_{B3}R_y \cong V_{CC} = 5 \text{ V}$$

$$V_z = V_y - V_{BE3} - V_{BE4} = 5 - 0.8 - 0.8 = 3.4 \text{ V}$$

With $V_z = V_{out}$ **HIGH**, T_4 supplies the small reverse currents of the connected gates.

Practice Problem 7-4

A simple form of TTL gate is shown in Fig. 7.16a. The threshold voltages are 0.7 V, $V_{CE(sat)} = 0.3$ V, and $V_{CC} = 5$ V. For $V_A = 4.5$ V, predict V_{out} for $V_B = 1$, 2, and 3 V.
Answers: 5; 0.3; 0.3 V.

Three-State Logic

In a complex digital control or computation system, logic devices "talk" to each other over a "party line" called a "system bus." For example, one of several input devices may communicate with one of several storage devices in a data processing unit over a data bus. This requires that devices *not* selected be disconnected from the bus. In *three-stage logic*, in addition to **0** and **1**, there is a "high-impedance state" in which the output terminal is essentially disconnected from the internal circuitry.

In the typical TTL gate of Fig. 7.13, multiemitter transistor T_1 acts as an **AND** gate; base current flows to T_2 if **ABC = 1**. Now let us assume the output is connected to a bus and consider **C** as a control input. If **C** is held **LOW**, inputs **A** and **B** have no effect; the inputs have been "disconnected" and T_3 turned **OFF**, providing a high-impedance path from the output to ground. If, simultaneously, the base of T_4 is grounded through a transistor switch, T_4 is **OFF**, and there is a high-impedance path (very low current) from the output to +5 V. The output terminal has been isolated and, in effect, the unit has been disconnected from the bus. In a three-state gate, shown symbolically in Fig. 7.15a, the circuit is arranged so that holding the **CONTROL** terminal **LOW** effectively grounds one input emitter and the base of T_4.

Buffers

The performance of an electronic component may be adversely affected if it is heavily loaded, that is, if it is required to supply considerable power to another component. For example, for good performance, an oscillator generating a precise frequency must be isolated from the device that it controls, or a register with limited fan-out must be separated from a bus supplying data to many devices. A *buffer* is an intermediate unit placed between the source of information and the load requiring power or current. In digital systems a buffer may be used to receive data at low current levels and make it available at higher current levels.

The symbol for a three-state inverting buffer is shown in Fig. 7.15b. With several sources connected to a common bus, the **OUTPUT ENABLE** of all but one would be

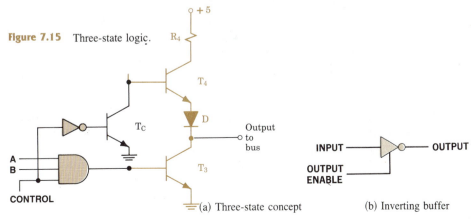

Figure 7.15   Three-state logic.

(a) Three-state concept   (b) Inverting buffer

held **LOW**, effectively disconnecting them. With **OUTPUT ENABLE** held **HIGH** on one unit, that **INPUT** is complemented, and the **OUTPUT** is capable of driving the bus supplying data to one or more loads. Typical TTL buffers have a fan-out of 30.

Open-Collector TTL Gates

Another approach to the problem of interconnecting several devices is to use *open-collector* gates. If two standard TTL gates (Fig. 7.13) have their outputs electrically connected and one output is **LOW** while the other is **HIGH**, there is a low-impedance path from V_{CC} through T_4 of the **HIGH** gate through T_3 of the **LOW** gate; the resulting high current will usually destroy T_3. However, in the **NAND** gate of Fig. 7.16 the collector is left "open." (In practice, it is tied to V_{CC} by *pull-up resistor R* to bring the output to nearly $+5$ V when T_3 cuts off.) Now the outputs of several **NAND** gates can be wired together, and the output of the combination will be **LOW** if any one of the individual outputs is **LOW**. In the "wired logic" shown symbolically in Fig. 7.16, the output is described as $f = \overline{AB} + \overline{CD} + \overline{EF}$ (wire-**NOR**) or as $f = \overline{AB} \cdot \overline{CD} \cdot \overline{EF}$ (wire-**AND**); the electrical connection has performed a logic function at no cost.

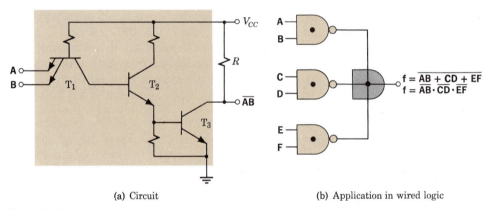

(a) Circuit	(b) Application in wired logic

Figure 7.16   Open-collector TTL gate.

MOS Logic Gates

The n-channel enhancement-mode MOS transistor has several virtues that make it useful in digital circuits; it requires only a small area, it is normally **OFF**, and it has a high input resistance. Furthermore, the basic gates can be fabricated from interconnected MOS structures only and, therefore, they are easy to produce in IC form. MOS logic is attractive in applications where low power consumption is important and extremely high switching speed is not required.

The characteristics in Fig. 7.17b indicate the possibility of using a MOSFET as a switch or inverter. An input voltage $V_i = v_{GS} \cong 0.5$ V (**LOW**) is insufficient for turn-on and the output is $V_o = v_{DS} \cong 5$ V (**HIGH**). A **HIGH** input of $V_i = 5$ V drives the MOSFET into saturation, and $V_o \cong 0.5$ V (**LOW**). The output of one inverter can drive the input of another in logical operations.

In practice, R_L is replaced by an n-channel depletion-mode MOSFET that serves as a passive *load* for the enhancement-mode *driver*. With gate tied to source, the load

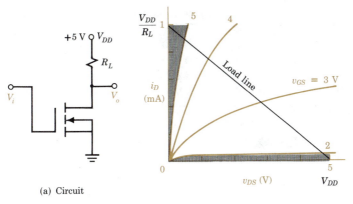

(a) Circuit

Figure 7.17   The enhancement MOSFET as a switch.

element provides an appropriate resistance in a very small chip area. These simple NMOS elements are used in complex LSI circuits.

CMOS Gates

A popular family of IC logic gates employs complementary symmetry, combining p- and n-channel enhancement-mode MOSFETs in the same chip. The basic building block of the CMOS gates is the inverter of Fig. 7.18a. When input voltage $V_i = v_{GS}$ is **LOW**, the n-channel device T_n is **OFF**. For the p-channel device T_p, $v_{GS} = V_i - V_{DD} \cong -V_{DD}$ and, therefore, T_p is **ON**. This state corresponds to Fig. 7.18b with switch S_p closed and switch S_n open. Hence the output is $V_o = V_{DD}$ or **HIGH**.

As the input voltage increases, T_p turns **OFF** and T_n turns **ON**; the output goes **LOW**. By design, the threshold voltage is usually about $V_{DD}/2$. The high resistance presented by T_p **OFF** limits the current drain on the power supply to a very small value. (See Example 4 on p. 185.)

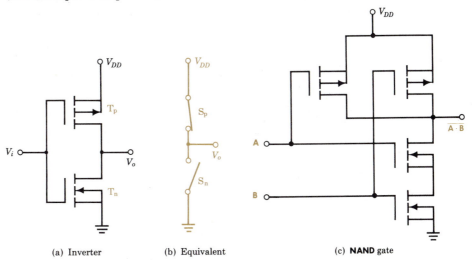

(a) Inverter   (b) Equivalent   (c) **NAND** gate

Figure 7.18   Complementary-symmetry (CMOS) logic gates.

A two-input CMOS logic gate is shown in Fig. 7.18c. If either **A** or **B** is **LOW**, one of the series T_n devices is **OFF** and one of the parallel T_p devices is **ON**, the output is **HIGH**. If both **A** and **B** are **HIGH**, both T_n devices are **ON** and both T_p devices are **OFF**; the output is **LOW**. These are the logic relations of a **NAND** gate. More inputs and other functions can be created by using the same elements.

As shown in Example 4, CMOS circuits consume significant power only during transitions from one state to another; the very low quiescent power consumption is the key factor in many applications, such as battery-powered watches and calculators. Most of the functions available in TTL form are available in CMOS as well. CMOS units can drive low-power TTL gates, but buffers are required to drive other TTL types.

MEMORY ELEMENTS

The outputs of the basic logic gates are determined by the present inputs; in response to the various inputs, these *combinational* circuits make "decisions." Along with these *decision* components we need *memory* components to store instructions and results; the outputs of such *sequential* circuits are affected by past inputs as well as present.

What characteristics are essential in a memory unit? A binary storage device must have two distinct states, and it must remain in one state until instructed to change. It must change rapidly from one state to the other, and the state value (**0** or **1**) must be clearly evident. The *bistable multivibrator* or *flip-flop*, a simple device that meets these requirements inexpensively and reliably, is used in all types of digital data processing systems.

A Logic Gate Memory Unit

First let us analyze a basic memory unit consisting of familiar logic gates. In the **NOR** gate flip-flop of Fig. 7.19, the output of each **NOR** gate is fed back into the input of the other gate. The operation is summarized in Table 7-1 where we assume to start that Q_0, the *present* state of the output **Q**, is **0** and inputs to the *set* terminal S and the *reset* terminal R are both **0**.

To **SET** the flip-flop, a **1** is applied to S only. For $Q_0 = 0$, $\overline{Q} = \overline{Q_0 + S} = 0$, and $Q = \overline{R + \overline{Q}} = 1$, the present state of the output is inconsistent with the input, the system is *unstable*, and **Q** must *flip*. After **Q** changes, Q_0 (the present state **Q**) changes to **1**, and $\overline{Q}$ becomes $\overline{1 + 1} = 0$; hence $Q = \overline{0 + 0} = 1$, a *stable* state. (In this analysis,

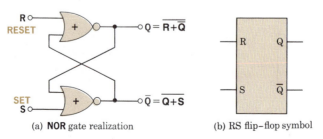

(a) **NOR** gate realization   (b) RS flip–flop symbol

Figure 7.19 The flip-flop, a basic memory unit.

EXAMPLE 4

A CMOS inverter consists of the *n*-channel MOSFET of Fig. 7.17b and its *p*-channel complement. Determine the transfer characteristic and the drain current curve.

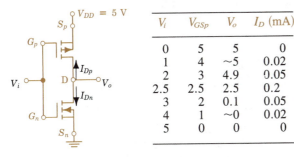

V_i	V_{GSp}	V_o	I_D (mA)
0	5	5	0
1	4	~5	0.02
2	3	4.9	0.05
2.5	2.5	2.5	0.2
3	2	0.1	0.05
4	1	~0	0.02
5	0	0	0

(a) Circuit and tabulated values

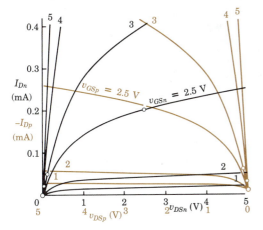

(b) *I-V* characteristics

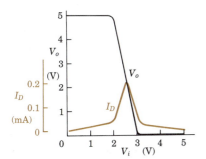

(c) Transition characteristics

Figure 7.20 CMOS inverter operation.

The *p*-channel element has identical characteristics but the polarities of gate and drain voltages and drain current are opposite. As shown in Fig. 7.20a, the two drains are connected together so that $I_D = I_{Dn} = -I_{Dp}$. The other governing relations are

$$V_{GSn} = V_i \qquad V_{GSp} = V_i - V_{DD} = V_{GSn} - 5$$

$$V_{DSn} = V_o \qquad V_{DSp} = V_o - V_{DD} = V_{DSn} - 5$$

The significant part of the *n*-channel characteristics is shown in black in Fig. 7.20b. The *p*-channel characteristics are shown in color, plotted to satisfy the governing circuit relations. Because $V_{DSp} = V_{DSn} - 5$, the *p*-channel curves are mirror images of the *n*-channel curves.

For each value of input voltage $V_i = V_{GSn}$, there is one value of $V_o = V_{DSn}$ at which $I_{Dn} = -I_{Dp}$. Graphically, this is the intersection of the corresponding I_D curves. For example, for $V_i = 2$ V $= V_{GSn}$, $V_{GSp} = 2 - 5 = -3$ V and the intersection is at $V_{DS} = V_o = 4.9$ V and $I_D = 0.05$ mA.

The results, plotted in Fig. 7.20c, show the principal advantages of CMOS inverters: The current is very low and flows only during transition from one state to the other; power consumption, very important in VLSI, is extremely low. The transition is very sharp, which provides immunity against "noise voltages" superimposed on high and low signal voltages.

keep in mind that *if either input to a* **NOR** *gate is* **1**, *the output is* **0**.) Removing the input from S causes no change. We conclude that **Q** = **1**, and $\overline{\mathbf{Q}}$ = **0** is the stable state after being **SET**. Applying another input to S produces no change.

To **RESET** the flip-flop, a **1** is applied to R only. This results in an unstable system and **Q** must *flop* to **0**. (You should perform the detailed analysis.) A change in $\mathbf{Q}_0$ to **0** produces a stable output **Q** = **0**. Removing the input to R or applying another input to R produces no change. We conclude that **Q** = **0** and $\overline{\mathbf{Q}}$ = **1** is the stable state after being **RESET**.

Table 7-1 Analysis of a Memory Unit

Action	$\mathbf{Q}_0$	S	R	$\overline{\mathbf{Q}}$	Q	Conclusion
Assume	0	0	0	1	0	This is a stable state.
Apply **1** to S	0	1	0	0	1	Unstable state; **Q** changes.
($\mathbf{Q}_0$ becomes **1**)	1	1	0	0	1	Stable.
Remove **1** from S	1	0	0	0	1	The stable state after **SET**.
Apply **1** to S again	1	1	0	0	1	No change in **Q**.
Remove **1** from S	1	0	0	0	1	The stable state after **SET**.
Apply **1** to R	1	0	1	0	0	Unstable state; **Q** changes.
($\mathbf{Q}_0$ becomes **0**)	0	0	1	1	0	Stable.
Remove **1** from R	0	0	0	1	0	The stable state after **RESET**.
Apply **1** to S and R	0	1	1	0	0	Unacceptable; $\mathbf{Q} \neq \overline{\mathbf{Q}}$.

Note that only a momentary input is required to produce a complete transition; this means that very short pulses can be used for triggering. Attempting to **SET** and **RESET** simultaneously would create an ambiguous state with both **Q** and $\overline{\mathbf{Q}}$ = **0**. This state ambiguity would be unacceptable in a bistable unit and actual circuits are designed to avoid this condition. (See Fig. 7.23b.)

A Transistor Flip-Flop

The operation of an electronic flip-flop is based on the switching properties of a transistor. With no input to the switch in Fig. 7.21a, the voltage divider R_A-R_B reverse biases the base-emitter junction and the transistor is in cutoff or the switch is **OPEN**; because $I_C R_C = 0$, a positive voltage appears at the output terminal. If a positive signal is applied to the input, V_B rises, the base-emitter junction is forward biased, the switch is **CLOSED**, and the output voltage drops to zero (nearly).

An input signal applied to point P, making the $+$ terminal of R_A more positive, could also forward bias the base-emitter junction and drop the output voltage to zero. In other words, a **1** input at either of two terminals produces a **0** at the output. This switch is a form of **NOR** gate; a flip-flop can be created from two such switches.

To follow the operation of the flip-flop (Fig. 7.21b), assume that T_1 is conducting (**CLOSED**) and T_2 is cut off (**OPEN**). With T_1 conducting, the potential of point P_1 is nearly zero and, in combination with the negative voltage applied to R_{B2}, this ensures

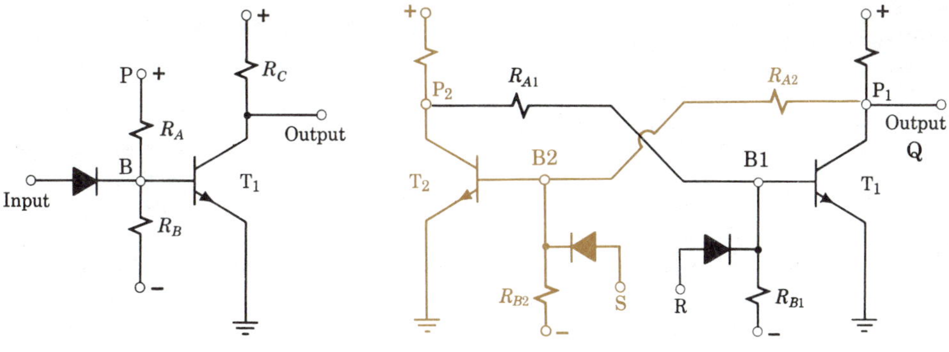

(a) Transistor switch

(b) Elementary flip–flop

Figure 7.21 A transistor RS flip-flop.

that T_2 is cut off. With T_2 cut off, the potential of point P_2 is large and positive, and this supplies the bias current through R_{A1} that ensures that T_1 is conducting and V_{P1} is low. In logic terms, **Q = 0**; this is a stable state that we may designate as the **0** state of this binary memory element.

A positive pulse applied to **RESET** terminal R that would raise V_{B1} has no effect since T_1 is already conducting. However, a positive pulse at **SET** terminal S causes T_2 to begin conducting, the potential of P_2 drops, the forward bias on T_1 is reduced, the potential of P_1 rises, the forward bias on T_2 increases, the potential of P_2 drops further, T_2 goes into saturation, and T_1 is cut off. The output voltage is high, **Q = 1**, and this indicates another stable state that we may designate as the **1** state. If a flip-flop in the **1** state receives a positive pulse at R, transition proceeds in the opposite direction (since the device is symmetric) and the device is **RESET** to the **0** state. In a well-designed flip-flop, these changes in state take place in a few nanoseconds, and we see that this simple device satisfies all the requirements of a binary storage element.

Practice Problem 7-5

For the transistors of Fig. 7.21, $V_{CE(sat)} = 0.3$ V and $V_{CC} = 5$ V. Draw up a table showing the voltages at P_1 and P_2 for the following conditions: (a) T_1 initially conducting, (b) a negative pulse applied to S, (c) a positive pulse applied to S, (d) a negative pulse applied to R, (e) a positive pulse applied to R.

Answers: (a) 0.3, 5; (b) 0.3, 5; (c) 5, 0.3; (d) 5, 0.3; (e) 0.3, 5.

Timing Waveforms

In the RS flip-flop of Fig. 7.20, a **1** input at the S input will **SET** the output **Q** to **1**. To **RESET** the flip-flop, a **1** is applied to input R. The duration of the input signal (as long as it exceeds a certain minimum time) and the time at which an input signal is applied are not significant. Such a flip-flop responds to *asynchronous* inputs.

A more sophisticated flip-flop incorporating two **AND** gates is shown in Fig. 7.22. Here an input is effective only when *enabled* by a **1** input at terminal E. In a digital

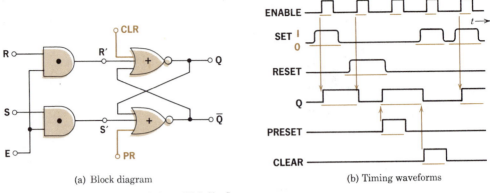

(a) Block diagram

(b) Timing waveforms

Figure 7.22   A more sophisticated RS flip-flop.

system composed of many elements, it is usually necessary for the outputs of all elements to be synchronized. The synchronizing signal may come from a *clock*, and the enabling terminal is frequently designated **CLOCK** (**CK**). In a clocked system, transitions cannot run wild through a circuit; instead, changes occur in an orderly, one-step-at-a-time fashion. In addition to the synchronous inputs **R** and **S**, there may be asynchronous inputs to *clear* or *preset* the flip-flop.

The operation of a *clocked* RS flip-flop is illustrated by the typical waveforms of Fig. 7.22b. Initially, output **Q** = **0**. If a **1** appears at **SET**, when **ENABLE** goes to **1** the flip-flop is set with **Q** = **1**. At the next clock pulse, the presence of a **1** at **RESET** forces the output to **0**. At any time, a **1** at **PRESET** forces the output to **1**; a **1** input at the **CLEAR** terminal overrides other inputs and forces **Q** to **0**.

The Data Latch

The functional symbol for a simple RS flip-flop (without **PRESET** and **CLEAR**) is shown in Fig. 7.23a. One way to avoid the ambiguous state where **R** = **1** and **S** = **1** simultaneously is the circuit modification shown in Fig. 7.23b. By connecting an inverter between the R and S terminals and using only one input signal, the ambiguity is avoided and the number of terminals is reduced (an advantage in IC packages). When the **ENABLE** line is **HIGH**, the output **Q** follows the input **D**. In other words, when this *flip-flop* is enabled, the input data is transferred to the output line. After the **ENABLE** line goes **LOW**, no change in **Q** is possible, and the output is "latched" at the previous data value. This *data latch* is widely used as an element in digital systems; for example, a set of eight such latches could "remember" the eight digits representing a number or an instruction.

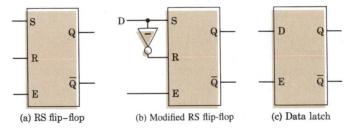

(a) RS flip–flop

(b) Modified RS flip-flop

(c) Data latch

Figure 7.23   Deriving a data latch from an RS flip-flop.

EXAMPLE 5

The enable and data inputs to a data latch are shown below. Predict the waveform of the output.

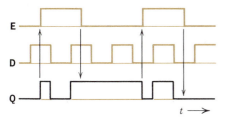

Figure 7.24   Typical data latch waveforms.

In the data latch, the output **Q** follows input **D** whenever enabled (**E = 1**). When **E** goes to **0**, the output remains *latched* in the previous condition.

In Fig. 7.24, when **E** first goes **HIGH**, **D = 1**; therefore, **Q** follows **D** and becomes **1**. As long as **E** is **HIGH**, **Q** follows any changes in **D**. When **E** goes **LOW**, **Q = D = 1** and remains so. The output waveform is as shown.

The D Flip-Flop

In digital systems it is sometimes desirable to delay the transfer of data from input to output. For example, we may wish to maintain a present state at **Q** while we read in a new state that will be transferred to the output at the appropriate time. The D (for *delay*) flip-flop[†] shown in Fig. 7.25 is a refinement of the data latch incorporating a

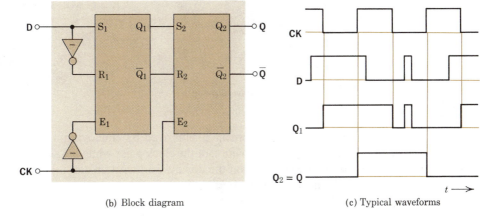

(a) Symbol   (b) Block diagram   (c) Typical waveforms

Figure 7.25   The D flip-flop.

second RS flip-flop. Here the data latch is enabled when the clock signal goes **LOW**, but the following RS flip-flop is enabled when **CLOCK** goes **HIGH**. In other words, Q_1 follows **D** whenever **CK** is **LOW**, but any change in the output of the combination $Q = Q_2$ is delayed until the next upward transition of **CK**. This is an *edge-triggered* flip-flop; Q_1 follows **D** while **CK** is **LOW**, then, on the leading edge of the clock pulse,

---

[†] "Flip-flop" is a general term applied to the basic two-transistor element and to sophisticated, multigate IC devices.

the value of **D** is transferred to output **Q**. On the logic symbol, the small triangle indicates an edge-triggered device.

Because the output can change only at the instant that the clock goes **HIGH**, the output can be synchronized with the outputs of other elements. Furthermore, a sudden spurious change in **D** similar to that shown in Fig. 7.25c will not affect the output. For proper operation of a practical device, the data input must be stable for a few nanoseconds before the device is clocked (the *set-up time*), and it must remain stable for a few nanoseconds after the clocking is initiated (the *hold time*).

The JK Flip-Flop

A widely used memory element is the JK flip-flop shown in Fig. 7.26. In its most common IC form, the output changes state on downward transitions of the clock pulse. The small circle on the symbol identifies this as a *trailing-edge-triggered* flip-flop. The

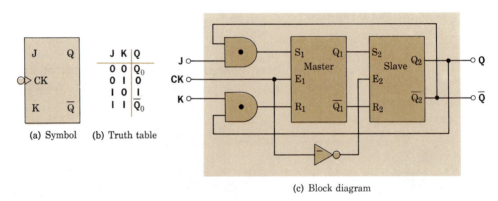

(a) Symbol   (b) Truth table

(c) Block diagram

Figure 7.26   The JK master-slave flip-flop.

operation of this logic element is improved by employing a *master* flip-flop that is enabled on the upward transition of the clock pulse while the *slave* flip-flop is inactive. Then the slave is enabled on the downward transition and follows its master; that is, it takes on the state of the immobilized master.

In addition to avoiding the ambiguity referred to previously, this versatile device provides three different modes of response. Because of the feedback connections from output to input, the output of a JK flip-flop depends on the states of the inputs and the outputs at the instant the clock goes **LOW**. As indicated in the truth table, with **0** inputs at J and K, the clock has no effect, and the flip-flop remains in its present state Q_0. With unequal inputs, the unit behaves like an RS flip-flop. For **J = 1** and **K = 0**, the clock **SETS** the flip-flop to **Q = 1**; for **K = 1** and **J = 0**, the clock **RESETS** the flip-flop to **Q = 0**. (In other words, with **J ≠ K**, **Q = J** or **Q** follows **J**.) With **1** inputs at both J and K, the flip-flop toggles; that is, the output changes each time the clock goes **LOW**. (The operation of the JK flip-flop is revealed by a complete truth table constructed for the eight possible combinations of **J**, **K**, and Q_0. (See Exercise 34.)

Practice Problem 7-6

Two JK flip-flops that respond to downward transitions are connected in tandem (Fig. 7.27). For a 2-kHz square-wave input, determine the output.

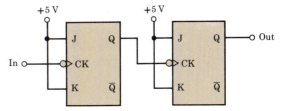

Figure 7.27 JK flip-flops in tandem.

Answer: 500-Hz square wave.

The sophistication of the JK response makes it useful in a variety of digital computer applications such as counters, arithmetic units, and registers. For greater flexibility, some versions include **PRESET** and **CLEAR** capabilities. In the unit shown in Fig. 7.28, the **PR** and **CLR** terminals are normally held **HIGH**. The small circles (inversion or "active **LOW**") indicate that if **PR** goes **LOW**, **Q** is forced to **1**; whereas if **CLR** goes **LOW**, **Q** is forced to **0**.

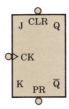

Figure 7.28 A JK flip-flop.

EXAMPLE 6

Connect a JK flip-flop to function as a data latch; that is, when it is **ENABLED**, the **DATA** is to be transferred to **Q** when the **CLOCK** goes **LOW**.

When **ENABLED**, **Q** should follow **J = D**, which requires **K ≠ J**.

When **DISABLED**, **Q** should remain "latched" in its present state, which requires **K = J = 0**. The truth table and the necessary connections are shown in Fig. 7.29.

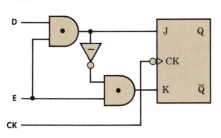

Figure 7.29 A JK flip-flop connected as a data latch.

E	J	K
0	0	0
1	D	J

The T Flip-Flop

With the **J** and **K** inputs tied together and brought out to a single input terminal, the JK unit becomes a T or *toggle* flip-flop (Fig. 7.30). For **T = 0** (**J = K = 0**), the clock pulse has no effect on output **Q**. For **T = 1** (**J = K = 1**), the flip-flop toggles each time **CK** goes to **LOW**. The waveforms show that for **T** held **HIGH**, the output is a square wave of half the frequency of the clock; the device is a frequency divider. If the CK input responds to a sequence of events, the T flip-flop "divides by two."

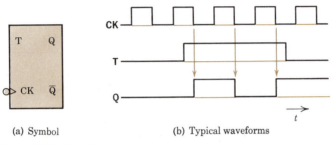

(a) Symbol   (b) Typical waveforms

Figure 7.30   A T flip-flop.

As we shall see in the next chapter, sets of flip-flops can be used to represent *binary numbers* in which each digit corresponds to the value of **Q** (**0** or **1**) of a flip-flop. A *register* is a set of flip-flops in which binary data can be stored. The same flip-flops can be reconnected to serve as a *counter* in which the number stored is the number of events being counted.

DIGITAL INTEGRATED CIRCUITS

The logic gates and memory elements described here are available in IC form with significant advantages in small size, low power consumption, and low cost. Integrated circuits containing fewer than a dozen gates represent *small-scale integration* (SSI), while those with more than a hundred elements represent *large-scale integration* (LSI). In between are *medium-scale integration* (MSI) circuits. Digital ICs in common use range from 14-pin dual 4-input **NAND** gates (SSI) and 24-pin seven-segment lamp drivers (MSI) to 40-pin microprocessors incorporating more than 100,000 transistors (VLSI).

The active elements may be BJTs—the "bipolar" family of which TTL is the most common example—or MOSFETs (the MOS family) using either *n*-channel or *p*-channel devices in enhancement or depletion modes.

TTL. TTL circuits are superior to DTL units in most respects. The logic voltage swing is large (0.2 to 3.3 V), and they are relatively immune to noise. The fan-out (typically 10) is high, power consumption (10 mW/gate) is low, and the speed of operation (9 ns) is very high. These excellent characteristics have made TTL logic very popular, and TTL devices are available in great variety. Newer versions include the S series (high-speed Schottky), the LS series (low-power Schottky), and the ALS series (advanced LS). A Schottky diode (see p. 139) connected between base and collector prevents transistor saturation and eliminates storage-time delay. (See Fig. A6 and the Lancaster reference at the end of this chapter.)

I²L. The necessity for isolating islands in fabricating chips (see p. 160) limits gate density in TTL technology. In the new *integrated injection logic* (I²L), bipolar junction transistors are "merged" to provide interconnections without isolation regions. The high speed of bipolar elements is retained, and the increased density makes I²L technology attractive for LSI applications.

ECL. In emitter-coupled logic, the circuit is arranged so that the transistors never saturate, and the delay associated with removing charge from a saturated transistor is avoided. ECL logic is characterized by very high speed (typically 2-ns propagation delay) and large fan-out (16) but high power consumption (25 mW/gate) and low voltage swing (0.8 V).

MOS. The MOS inverter of Fig. 7.2c is extremely simple in appearance and in fabrication. An MOS gate requires much less "real estate" than a corresponding TTL unit and, therefore, the gate density on a silicon chip can be much higher; MOS is widely used in LSI. The high input resistance means low input currents and power consumption is low (typically 1 mW/gate). Furthermore, as fabrication techniques improve, each generation of MOS devices offers improved performance.

CMOS. In *complementary MOS* devices (CMOS), n- and p-channel MOSFETs are paired so that, during switching, an **ON** transistor is always available for rapid charging of the load capacitance, whereas the **OFF** transistor limits dc current consumption. The circuit complexity is increased, but power is consumed only during switching. The compromise reduces gate density but results in higher speed and very low power consumption (0.01 mW/gate)—characteristics that are desirable in digital watches, for example.

SUMMARY

- Digital logic circuits employ discrete instead of continuous signals.
  The basic logic operations are: **AND, OR, NOT, NAND,** and **NOR.**

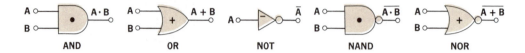

AND	OR	NOT	NAND	NOR

  All operations can be achieved using only **NAND** gates or only **NOR** gates.
  The validity of any logic statement can be demonstrated with a truth table.

- Logic circuits are classified in terms of the components (R, D, T) employed.
  Design factors include: cost, noise immunity, power consumption, signal degradation,
  fan-in, fan-out, speed, and reliability.

- Diodes and transistors are used as switches in electronic logic gates.
  Transistors are switched from cutoff to saturation by changes in i_B or v_{GS}.
  Integrated circuits permit high-performance gates at reasonable cost.

- Memory components store data processed by decision components.
  When instructed, a flip-flop must change rapidly from one distinct stable state to the
  other and clearly evidence the new state.
  The transistor flip-flop is an inexpensive, widely used binary storage device.

- The basic RS flip-flop is **SET** by **S = 1** and **RESET** by **R = 1.**
  When a data latch is enabled, output **Q** follows input **D.**
  In a D flip-flop, transfer of data from input to output is delayed.
  Operation of the versatile JK flip-flop is determined by inputs at J and K.
  The T flip-flop is toggled from one state to the other by successive triggers.
  Basic applications of flip-flops include counters and registers.

REFERENCES

1. A.P. Malvino and Donald P. Leach, *Digital Principles and Applications*, McGraw-Hill Book Co., New York, 2nd ed., 1975.

   An excellent introduction to digital systems with detailed discussions and many illustrative examples. Includes basic concepts and their practical applications in computers.

2. Victor Grinich and Horace Jackson, *Introduction to Integrated Circuits*, McGraw-Hill Book Co., New York, 1975.

   A good undergraduate treatment of IC design of digital and analog circuits.

3. John B. Peatman, *The Design of Digital Systems*, McGraw-Hill Book Co., New York, 1972.

   A well-written introduction to modern digital devices and systems and their engineering applications.

4. Don Lancaster, *TTL Cookbook*, Howard Sams and Co., Indianapolis, 1974.

   A practical guide to understanding and using TTL circuits. Includes construction hints, design techniques, and operating characteristics of popular gates, flip-flops, counters, registers, and memories.

   (Also see the Millman and Holt references at the end of Chapter 6 and the Horowitz and Hill reference at the end of Chapter 11.)

REVIEW QUESTIONS

1. Distinguish between analog and digital signals. Between gates and switches.
2. Why did the successful computer follow the invention of the transistor?
3. Explain how diodes and transistors function as controlled switches.
4. What is a "truth table"? How is it used?
5. Draw symbols for and distinguish between **AND**, **OR**, **NOR**, and **NAND** operations.
6. What is "**NOT** the complement of variable **A** inverted"?
7. Draw a **NAND** gate. Why are they used so widely?
8. How are electronic logic circuits classified? What is DTL? TTL? ECL? MOS? CMOS?
9. List the factors of importance in the practical design of logic circuits.
10. Show with a sketch how cutoff and saturation resistances are calculated.
11. What are the advantages of TTL gates?
12. What is "three-state" logic? How is it achieved? What is a "buffer"?
13. Draw a TTL **NOR** gate and explain its operation.
14. What is the principal advantage of ECL gates?
15. Explain the operation of an MOS gate. A CMOS gate.
16. Distinguish between decision and memory components; give an example of each.
17. What is a flip-flop? What is its function in a computer?
18. Explain the operation of a flip-flop consisting of two **NOR** gates.
19. Explain the operation of a transistor RS flip-flop.
20. Distinguish between a data latch and a D flip-flop.
21. Distinguish between RS, T, and JK flip-flops.
22. Explain how T flip-flops "count." How many are needed to count to ten?

EXERCISES

1. Devise a two-input **NOR** gate using relays (Fig. 7.1) and construct the truth table.
2. An electric light is to be controlled by three switches. The light is **ON** if both switches A and B are closed, or if switch C is closed by itself, or if all three switches are closed.
   (a) Draw up a truth table for this function.
   (b) Describe the operation of the light using the states of the switches A, B, and C and the words "OR" and "AND."
3. The "resistor logic" circuit of Fig. 7.31 has binary inputs of 0 and 4 V.
   (a) For $R = 1$ kΩ, construct a truth table showing actual voltages. If a detector with an adjustable threshold were available, what logic operations could be performed with this circuit?

(b) At what threshold should the detector be set for **AND** operations?
4. Replace the resistors R in Fig. 7.31 by diodes to form a three-input diode **OR** gate. Construct the truth table of voltages.
5. How many rows are needed in the truth table of a 4-input logic circuit? An n-input logic circuit?
6. Assuming 5-V inputs and a diode forward voltage drop of 0.7 V, construct a truth table showing actual voltages for the diode circuits of (a) Fig. 7.3 and (b) Fig. 7.4.
7. The three resistances in Fig. 7.5 are 1 kΩ each and the lower end of R_B is at -1 V. Calculate a few values and plot v_{BE} and v_{out} for v_{in} between 0 and 5 V.
8. Design a two-input **AND** gate using **NOR** gates only; construct the truth table as a check.
9. A portion of a table defining the performance of the basic gates is given:

Gate	Inputs	Output
AND	All inputs I	1
	Any one input 0	0

Complete the table for **NAND**, **OR**, **NOR**, and **NOT** gates.

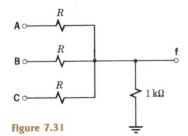

Figure 7.31

10. The multiple-input logic gates commonly available are:

Inputs	2	3	4	5	8	> 8
NAND	X	X	X		X	X
NOR	X	X		X		
AND	X	X	X			
OR	X					

Only 2-input **OR** gates are available; however, 2-input **OR**s can be combined to form an n input **OR** circuit.

(a) Design a 9-input **OR** gate using 2-input **OR** gates only.

(b) The number of logic "levels" in part (a) introduces an undesirable delay. Design a 9-input **OR** gate using commonly available gates arranged so that signals pass through no more than two gates. (*Hint:* Use 3-input gates.)

11. The transistor of Fig. 6.15 is used in the switching circuit of Fig. 7.9a with $V_{CC} = 15$ V, $R_B = 5$ kΩ, and $R_C = 1$ kΩ.

(a) Reproduce the i_C versus v_{CE} characteristics for this transistor, neglecting I_{CEO}, and draw the load line.

(b) For $v_i = 0.5$ V, *estimate* i_B, i_C, and v_o.

(c) Repeat part (b) for $v_i = 1.7$ V and 2.8 V.

(d) If the input (v_i) is "switched" from 0.5 V to 2.8 V, what are the corresponding output voltages?

12. When the transistor of Fig. 7.6b is conducting, $V_{BE} = 0.8$ V and $V_{CE} = 0.3$ V; $R = R_B = R_C = 2$ kΩ and $I_{CEO} = 10$ μA. Voltage $-V_B$ holds the transistor off if A and B are low; if either diode is conducting, the forward voltage drop on the diodes is 0.7 V. After several logic operations, the inputs at A are 1 and 2 V.

(a) If input A is $+3$ V and diode A is conducting, what is V_P?

(b) Assuming 0 input at B, estimate the output voltages for A inputs of 1 V and 2 V.

13. The transistor and diodes of Exercise 12 are used in Fig. 7.7b with $R_C = 2$ kΩ and $R_B = 20$ kΩ. Assuming a 5-V input at B, estimate the output for inputs of 1 and 2 V at A.

14. For a 2N3114 transistor switch (Fig. A8) operating at $V_{CC} = 5$ V and $R_C = 100$ Ω, estimate the typical **ON** and **OFF** resistances at room temperature. Repeat for the worst possible conditions at 150° C. (*Note:* $h_{FE} = \beta$.)

15. For the **NAND** gate of Fig. 7.7b, $R_B = 2$ kΩ, $R_C = 4.7$ kΩ, $\beta = 50$, and the current drawn at the output is negligible.

(a) Sketch the i_B versus v_{BE} characteristic for the transistor and estimate the critical values of v_{BE} for the **ON** and **OFF** states.

(b) Sketch the i_C versus v_{CE} characteristics and estimate the critical values of i_C and output voltage V_o.

(c) Replace the active devices by dc circuit models and redraw the circuit. Assuming V_B is high, estimate the critical values of V_A for V_o **HIGH** and **LOW**, and sketch the transfer characteristic V_o versus V_A.

16. In the silicon **NAND** gate of Fig. 7.11, $R_B = 2$ kΩ, $R_C = 5$ kΩ, and $\beta \cong 50$. The "load" across which V_o appears consists of the **A** inputs of four following DTL **NAND** gates.

(a) For $V_A = 0.2$ V and $V_B = 0.2$ V, predict V_o.

(b) For $V_A = 3$ V and $V_B = 4$ V, predict V_o.

(c) For $V_A = 4$ V and $V_B = 0.3$ V ($= V_{CE}$ of a preceding **ON** transistor), estimate the current flowing out at diode B.

(d) Estimate the base and collector currents for conditions (a) and (b).

17. In Fig. 7.12b, R_2 is "large" and $R_B = 5.8$ kΩ.

(a) For $V_A = 1$ V and $V_B = 4$ V, estimate V_P, V_{P2}, I_{B2}, and V_{out}.

(b) Repeat for $V_A = V_B = 3.5$ V.

18. In an elementary form of TTL, the combination of R_4, T_4, and D in Fig. 7.13 is replaced by a resistance R_C.

(a) Draw the simplified circuit diagram with $R_1 = R_2 = R_C = 2$ kΩ.

(b) What assmptions can you make about threshold and saturation voltages?

(c) For T_3 **ON**, estimate V_o, V_{B3}, V_{C2}, V_{B2}, and V_P.

(d) For T_3 **ON**, estimate I_{R3}, I_{R2}, and I_{R1}.

(e) For T_3 **OFF**, estimate V_o, V_{B3}, and V_{C2}.

(f) For T_3 **OFF**, what can you say about the input voltages?

19. Assume T_3 in Fig. 7.13 is **ON** and estimate the voltage at the base of each transistor. Prove that T_4 is **OFF**.

20. For the circuit in Fig. 7.32:

(a) Draw up a truth table for **F** in terms of **A** and **B**.

(b) Identify the logic classification and the function performed.

(c) For $V_A = 0.3$ and $V_B = 1.5$ V, predict V_x, V_y, and V_F.

(d) For $V_A = 2.3$ and $V_B = 1.5$ V, predict V_x, V_y, and V_F.

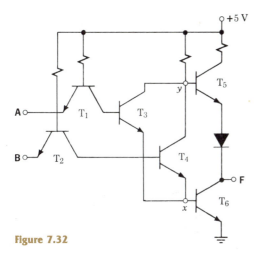

Figure 7.32

21. In the IC logic gate of Fig. 7.33, the threshold value of V_{BE} is 0.8 V and $V_{CE(sat)} = 0.2$ V. Predict voltages at x, y, and F and the condition (**ON**, **OFF**) of T_1 for:
(a) $V_A = V_B = 4$ V.
(b) $V_A = 4$ V, $V_B = 0.2$ V.
(c) $V_A = V_B = 0.2$ V.

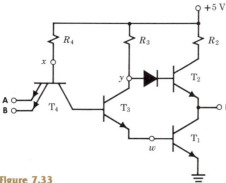

Figure 7.33

22. Classify the element in Fig. 7.33 in terms of components employed and logic function performed. Assuming $V_{BE} = 0.7$ V and $V_{CE(sat)} = 0.3$ V, complete the following table:

Condition			Approximate voltages					
T_3	T_2	T_1	F	w	y	x	A	B
—	—	ON	—	—	—	—		
—	—	—	—	—	—	—	3.0	0.2

23. (a) In Fig 7.13, $V_C = 0$ and $V_A = V_B = 5$ V; what is V_{out}? If V_C is slowly increased to 5 V, what happens to V_{out}? At what value of V_C?
(b) With $V_C = 0$, $V_A = V_B$ is changed from 5 to 0 V. What happens to V_{out}? How is this behavior used in "three-state" logic?

24. In the circuit of Fig. 7.16a, T_3 serves as a "sink" for currents flowing from or to the inputs of subsequent logic gates "driven" by this gate. When the output $(\overline{AB})$ is **LOW**, the current *into* the output terminal is 2 mA; when the output is **HIGH**, the current *out of* terminal $\overline{AB}$ is 0.5 mA.
(a) Assuming $R = 1$ kΩ, $V_{BE} = 0.7$ V, and $V_{CE(sat)} = 0.3$ V, complete the following table ($B_1 =$ base of T_1):

Condition		Approximate voltages					
T_2	T_3	B_1	B_2	B_3	$\overline{AB}$	A	B
—	ON	—	—	—	—		
—	—	—	—	—	—	0.3	3.3

(b) Form a truth table and identify the logic function.

25. Figure 7.34 incorporates two open-collector **NOT** gates and a three-state buffer (**ENABLED** when **LOW** in this case). Draw a truth table showing the states of **g** and **f** for inputs **A**, **B**, and **C**.

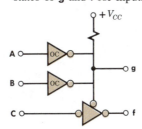

Figure 7.34

26. For the MOSFET of Fig. 7.17, $V_i = V_{GS}$ and $V_o = V_{DS}$. Draw the transfer characteristic V_o vs. V_i. What voltage ranges would be reasonable to define logic **0** and **1**?

27. Design, that is, draw the circuit diagram for, a 2-input **NAND** gate using MOS devices only.

28. Repeat Exercise 27 for a 2-input **NOR** gate.

29. A CMOS digital device is fabricated as shown in Fig. 7.35. Draw a clearly labeled circuit diagram using standard symbols. If $|V_T|$ is 2.3 V for each unit, determine the outputs for inputs of 1 and 9 V. What logic function is performed?

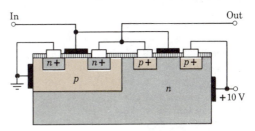

Figure 7.35

30. Design a basic memory unit consisting of two **NAND** gates and describe its operation in a table similar to Table 7-1.
31. (a) Design a clocked RS flip-flop using four **NAND** gates.
    (b) Modify the circuit to allow **PRESET** and **CLEAR** operations.
32. Draw the circuit diagram for a flip-flop using MOS devices only.
33. (a) Draw the block diagram and symbol for a D flip-flop that is "trailing-edge triggered"; that is, the value of **D** is transferred to the output **Q** when **CK** goes **LOW**.
    (b) Reproduce the waveforms for **CK** and **D** (Fig. 7.36) and show the resulting waveforms of Q_1 and of $Q_2 = Q$.

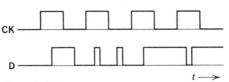

Figure 7.36

34. Reproduce the circuit of Fig. 7.26c. Draw up the truth table showing S_1, R_1, and **Q** (the *next* output state) for all values of **J**, **K**, and Q_0 (the *present* output state). Compare your result to the truth table of Fig. 7.26b.
35. A 500-Hz square wave is applied to the **CK** input of a T flip-flop with **T** held high.
    (a) Draw 6 cycles of the square wave along with the corresponding flip-flop output.
    (b) Draw a circuit showing two T flip-flops in

tandem and the waveform of the output of T2.
    (c) What operation is performed by the combination of flip-flops?
36. Analyze the memory element in Fig. 7.37. Draw up a truth table assuming the input is a series of pulses and **Q** is initially **0**. What function is performed by the **AND** gates?

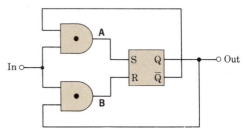

Figure 7.37

37. A JK flip-flop that responds to downward transitions is connected as in Fig. 7.38. For a 2-kHz square-wave input, determine the output.

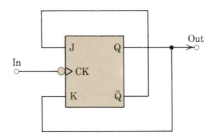

Figure 7.38

PROBLEMS

1. Devise a complete dc circuit model for a BJT in the switching mode, including the effect of $V_{CE(\text{sat})}$.
2. A four-input DTL **NAND** gate similar to Fig. 7.11a "drives" eight similar **NAND** gates (a fan-out of 8). Assume $V_{CC} = 5$ V, $R_B = R_C = 5$ k$\Omega \pm 10\%$, $\beta = 30 \pm 40\%$, $V_{CE(\text{sat})} = 0.3$ V, and $V_{\text{junct}} = 0.7 \pm 0.1$ V. Draw the circuit, define the "worst possible" operating condition, and predict the **LOW** output voltage under this condition.
3. In the DTL **NAND** gate of Fig. 7.11, $R_B = 2$ kΩ, $R_C = 4$ kΩ, and $\beta = 20$. For a "fan-out of 10," that is, the output V_o connected to the inputs of 10 similar gates, predict base and collector currents and output voltage V_o.
4. The data control circuit of Fig. 7.34 is used with devices A and B, which may be too "busy" to consider new data that is available at pin C. Identify the logic elements and form a truth table showing the states of outputs **g** and **f** in terms of the inputs. Explain how this circuit could function in data processing.

5. Following Fig. 6.21a, draw a cross section of the structure of a three-emitter transistor.

6. Two sensors are mounted on a half-white rotating disk as in Fig. 7.39. Sensor output is 5 V for white and 0 V for dark. Specify the digital element or elements to put in the black box so that the LED is **ON** for clockwise rotation. (*Hint:* Look at the waveforms.)

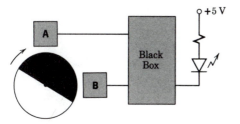

Figure 7.39

7. Four T flip-flops are connected in cascade to form a scale-of-sixteen counter. Draw a schematic diagram indicating the paths taken by pulses. Using feedback links to reintroduce counts into earlier stages, convert the unit into a decimal scalar, that is, a counter with one output pulse for every ten input pulses.

Digital Devices

To process digital information we use special electronic components that respond to binary signals. In Chapter 7 we saw how diodes and transistors can be used as switches in gates and flip-flops. To design efficient digital circuits, we also need a special numbering system and a special algebra. Our next step is to examine the binary number system and learn to apply logic theorems to binary relations. With this background we can analyze given logic circuits or synthesize new ones to realize desired logic functions.

Computers consist of large numbers of logic gates and memory elements organized to process data at high speed. Binary data or instructions are stored temporarily in *registers*. Binary *counters* are used in calculations and to keep track of computer operations. Instructions and data are stored at specified locations in *memory* and can be retrieved at will. With the understanding of registers, counters, memories, and computer organization gained here, we shall be prepared to study, in Chapter 9, the versatile electronic device called a *microprocessor*.

BINARY NUMBERS

In the decimal system a quantity is represented by the *value* and the *position* of a digit. The number 503.14 means

or

$$500 \quad + 0 \quad + 3 \quad + \tfrac{1}{10} \quad + \tfrac{4}{100}$$

$$5 \times 10^2 + 0 \times 10^1 + 3 \times 10^0 + 1 \times 10^{-1} + 4 \times 10^{-2}$$

In other words, 10 is the *base* and each position to the left or right of the decimal point corresponds to a *power* of 10. Perhaps it is unfortunate that we do not have 12 fingers, because in certain ways 12 would be a better base. In fact, such a base-12 or *duodecimal* system was used by the Babylonians, and we still use 12 in subdividing the foot, the year, and the clock face.

In representing data by an **ON-OFF** switch position, there are only two possibilities and the corresponding numbers are **1** and **0**. In such a *binary* system, the base is 2 and the total number of fingers on both hands is written **1010** because

$$1 \times 2^3 + 0 \times 2^2 + 1 \times 2^1 + 0 \times 2^0 = 8 + 0 + 2 + 0 = 10$$

In electronic logic circuits the numbers **1** and **0** usually correspond to two easily distinguished voltage levels specified by the circuit designer. For example, in TTL, **0** corresponds to a voltage near zero and **1** to a voltage near +5 V.

Number Conversion

Binary-to-Decimal Conversion. In a binary number, each position to the right or left of the "binary point" corresponds to a power of 2, and each power of 2 has a decimal equivalent.

*To convert a binary number to its decimal equivalent, add the decimal equivalents of each position occupied by a **1**.*

For example,

$$\mathbf{110001} = 2^5 + 2^4 + 0 + 0 + 0 + 2^0 = 32 + 16 + 1 = 49$$

$$\mathbf{101.01} = 2^2 + 0 + 2^0 + 0 + 2^{-2} = 4 + 1 + \tfrac{1}{4} = 5.25$$

Decimal-to-Binary Conversion. A decimal number can be converted to its binary equivalent by the inverse process, that is, by expressing the decimal number as a sum of powers of 2. An automatic, and more popular, method is the *double-dabble* process in which integers and decimals are handled separately.

To convert a decimal integer to its binary equivalent, progressively divide the decimal number by 2, noting the remainders; the remainders taken in reverse order form the binary equivalent.

To convert a decimal fraction to its binary equivalent, progressively multiply the fraction by 2, removing and noting the carries; the carries taken in forward order form the binary equivalent.

EXAMPLE 1

Convert decimal 28.375 to its binary equivalent.

Using the double-dabble method on the integer (a shorthand notation is shown at the left),

$$
\begin{array}{c}
2 \mid 28 \\
\hline \quad 14 \qquad 0 \qquad 28 \div 2 = 14 \text{ with a remainder of } 0 \\
\quad 7 \qquad 0 \qquad 14 \div 2 = 7 \text{ with a remainder of } 0 \\
\quad 3 \qquad 1 \qquad 7 \div 2 = 3 \text{ with a remainder of } 1 \\
\quad 1 \qquad 1 \qquad 3 \div 2 = 1 \text{ with a remainder of } 1 \\
\quad 0 \qquad 1 \qquad 1 \div 2 = 0 \text{ with a remainder of } 1
\end{array}
$$

The binary equivalent is **11100**.

Then converting the fraction,

$$0.375 \times 2 = 0.75 \text{ with a carry of } 0$$

$$0.75 \times 2 = 1.50 \text{ with a carry of } 1$$

$$0.50 \times 2 = 1.00 \text{ with a carry of } 1$$

The binary equivalent is **.011**.

28.375 is equivalent to binary **11100.011**.

Binary Arithmetic

Since the binary system uses the same concept of value and position of the digits as the decimal system, we expect the associated arithmetic to be similar but easier. (The binary multiplication table is very short.) In *addition*, we add column by column, carrying where necessary into higher position columns. In *subtraction*, we subtract column by column, borrowing where necessary from higher position columns. In subtracting a larger number from a smaller, we can subtract the smaller from the larger and change the sign just as we do with decimals.

EXAMPLE 2

Convert the numbers in color to the other form and perform the indicated operations.

$$
\begin{array}{cc}
14 & 1110 \\
+11 & +\,1011 \\
\hline
25 & 11001
\end{array}
\qquad
\begin{array}{cc}
13 & 1101 \\
-10 & -1010 \\
\hline
3 & 0011
\end{array}
\qquad
\begin{array}{cc}
1010 & 10 \\
-1101 & -13 \\
\hline
-0011 & -3
\end{array}
$$

In *multiplication*, we obtain partial products using the binary multiplication table ($0 \times 0 = 0$, $0 \times 1 = 0$, $1 \times 0 = 0$, $1 \times 1 = 1$) and then add the partial products.[†] In *division*, we perform repeated subtractions just as in long division of decimals.

[†] Digital multiplying circuits are basically adding circuits.

EXAMPLE 3

Convert the numbers in color to the other form and perform all the indicated operations.

After decimal-to-binary conversion, the operations are

(a)
$$
\begin{array}{r}
14.5 \\
\times 1.25 \\
\hline
725 \\
290 \\
145 \\
\hline
18.125
\end{array}
$$

(b)
$$
\begin{array}{r}
101.1 \\
11.1 \overline{)10011.01} \\
\underline{111} \\
1010 \\
\underline{111} \\
111 \\
\underline{111} \\
0
\end{array}
$$

(a)
$$
\begin{array}{r}
1110.1 \\
\times\ 1.01 \\
\hline
11101 \\
00000 \\
11101 \\
\hline
10010.001
\end{array}
$$

(b)
$$
\begin{array}{r}
5.5 \\
3.5 \overline{)19.25} \\
\underline{175} \\
175 \\
\underline{175} \\
0
\end{array}
$$

Bits, Bytes, and Words

A single binary digit is called a "bit." All information in a digital system is represented by a sequence of bits. An 8-bit sequence is called a "byte"; a 4-bit sequence is a "nibble"; a 16-bit sequence is a "word." The number of bits in the data sequences processed by a given computer is a key characteristic. An *8-bit microprocessor* can receive, process, store, and transmit data or instructions in the form of bytes. Eight bits can be arranged in $2^8 = 256$ different combinations.

Practice Problem 8-1

(a) Write in binary the following decimals: 6, 9, 24.
(b) Write in decimal the following binaries: **00010, 01011, 10100**.
(c) Convert 0.625 to binary and **0.0011** to decimal.

Answers: (a) **00110, 01001, 11000**; (b) 2, 11, 20; (c) **0.101**, 0.1875.

Other Notations

The number of years in a century can be written 100D or 100_{10}. In binary notation, this would be written $0 + 2^6 + 2^5 + 0 + 0 + 2^2 + 0 + 0 =$ **01100100B** or **01100100**$_2$; the suffix B or subscript 2 is used whenever necessary to avoid confusion.

Although 8-bit numbers are easy for computers, they are difficult for humans to deal with. In *octal* notation, a single decimal number from 0 to 7 is used to represent each group of three bits. As an octal number, **01100100B** would be written as **01 100 100**→**144Q** = 144_8. Three-digit octal numbers are easier to remember and easier to check than their 8-bit binary equivalents.

In the alternative notation most commonly used in microprocessor work, each group of four bits is represented by a single *hexadecimal* number. In "hex," **01100100B** would be written as **0110 0100**→**64H** = 64_{16}. Because four bits can take on sixteen different values, we supplement the ten decimal digits 0 . . . 9 with the

Table 8-1   Number Systems

Decimal	Binary	Hex	Octal
0	0000	0	00
1	0001	1	01
2	0010	2	02
3	0011	3	03
4	0100	4	04
5	0101	5	05
6	0110	6	06
7	0111	7	07
8	1000	8	10
9	1001	9	11
10	1010	A	12
11	1011	B	13
12	1100	C	14
13	1101	D	15
14	1110	E	16
15	1111	F	17

letters, A, B, C, D, E, and F. For example, $11000011_2 \rightarrow 1100\ 0011 \rightarrow C3_{16}$ and $255_{10} = 11111111_2 \rightarrow 1111\ 1111 \rightarrow FF_{16}$. (See Table 8-1. In this book, binary, octal, and hex numbers are in SANS SERIF type.)

Signed Magnitudes

In binary notation, an n-bit data word can represent the first 2^n nonnegative integers. To allow for both positive and negative numbers, the most significant bit (MSB) can be designated as the *sign bit* (1 for negative numbers). The lower order bits then represent the *magnitude* of the number in "straight" binary notation. Although it is used, this arrangement has two disadvantages: the number zero has two different representations, and two different arithmetic circuits are required to process positive and negative numbers.

Two's Complement Notation

A better notation for computers, one that is easily implemented in hardware, is based on the fact that adding the complement of a number is equivalent to subtracting the number. For example, in evaluating $9 - 3$, to subtract 3 from 9, we can "add the 10's complement" of 3 (i.e., $10 - 3 = 7$) to obtain $9 + 7 = 16 \rightarrow 6$ after discarding the final carry. In the decimal system, the 10's complement of a multidigit number is easily found by taking the 9's complement of each digit (by inspection) and then adding 1. In general, to subtract a two-digit number B from A we use the relationship

$$A - B = A + [100 - B] - 100 = A + [(99 - B) + 1] - 100 \qquad (8\text{-}1)$$

where $(99 - B)$ is the 9's complement. (See Example 4.)

In the binary system, arithmetic is simplified if negative numbers are in *signed 2's complement* notation. In this notation, the MSB is the sign bit: 0 for plus, 1 for minus. To form the 2's complement of any number, positive or negative:

Form the 1's complement by changing 1s to 0s and 0s to 1s.
Add 1.

If the result of an arithmetic operation has a **1** sign bit, it is a negative number in 2's complement notation; to obtain the true magnitude, subtract **1** and form the 1's complement.

EXAMPLE 4

(a) Obtain the 10's complement of 15 and 24.

Form the 9's complement of each digit, then add 1: $15 \rightarrow 84 + 1 = 85$ $24 \rightarrow 75 + 1 = 76$.

(b) Represent -15 and -24 in 8-bit signed 2's complement notation.

Form the 1's complement of each digit, then add 1.

$$-15_{10} \rightarrow -1111 \rightarrow -00001111 \rightarrow 11110000 + 1 \rightarrow 11110001 \rightarrow 1\ 1110001$$

$$-24_{10} \rightarrow -11000 \rightarrow -00011000 \rightarrow 11100111 + 1 \rightarrow 11101000 \rightarrow 1\ 1101000$$

(c) Perform $24 - 15$ and $15 - 24$ directly and by complement notation.

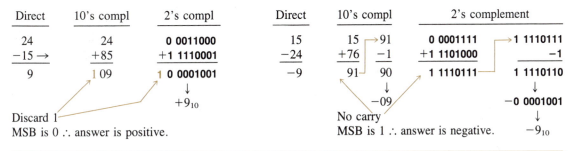

Discard 1
MSB is 0 ∴ answer is positive.

No carry
MSB is 1 ∴ answer is negative.

Binary-Coded Decimals (BCD)

For the convenience of humans, computer input/output devices may accept/provide decimals on the human side and binaries on the computer side. The conversion is simplified by *coding* each decimal digit, that is, replacing it by the 4-bit binary representation of the digit. For example, in the 8421 code $6_{10} \rightarrow$ **0110**, $3_{10} \rightarrow$ **0011**, and $363_{10} \rightarrow$ **0011 0110 0011**.

When a computer is to handle letters as well as numbers, an *alphanumeric code* is used. In the American Standard Code for Information Interchange (ASCII), seven bits are used to represent all the characters and punctuation marks on a teletypewriter keyboard plus some controls signals. (Note that $2^7 = 128$ combinations of 7 bits.) An eighth bit, the MSB, is a *parity bit* used in error detection. In the *even parity* convention, the MSB is set so that the number of **1**s in each ASCII character is even; the presence of an odd number of **1**s indicates an error.

Practice Problem 8-2

(a) Perform $38 - 24$ and $24 - 38$ by using signed 2's complement notation.
(b) Express decimal 541 as a binary-coded digit in the 8421 code.

Answer: (b) **0101 0100 0001**.

BOOLEAN ALGEBRA

We see that binary arithmetic and decimal arithmetic are similar in many respects. In working with logic relations in digital form, we need a set of rules for symbolic manipulation that will enable us to simplify complex expressions and solve for unknowns; in other words, we need a "digital algebra." Nearly 100 years before the first digital computer, George Boole, an English mathematician (1815–1864), formulated a basic set of rules governing the true-false statements of logic. Eighty-five years later (1938), Claude Shannon (at that time a graduate student at MIT) pointed out the usefulness of *Boolean algebra* in solving telephone switching problems and established the analysis of such problems on a firm mathematical basis. From our standpoint, Boolean algebra is valuable in manipulating binary variables in **OR**, **AND**, or **NOT** relations and in the analysis and design of all types of digital systems.

Boole's Theorems

The basic postulates are displayed in Tables 8-2 and 8-3. At first glance, some of the relations in the tables are startling. Properly read and interpreted, however, their validity is obvious. For example, "**1 OR 1**" (**1 + 1**) is just equivalent to **1**, and "**0 AND 1**" (**0 · 1**) is effectively **0**. In other words, for an **OR** gate with inputs of **1** at both terminals, the output is **1**, and for an **AND** gate with inputs of **0** and **1**, the output is **0**.

Table 8-2   Boolean Postulates in 0 and 1

OR	AND	NOT
$0 + 0 = 0$	$0 \cdot 0 = 0$	$\bar{0} = 1$
$0 + 1 = 1$	$0 \cdot 1 = 0$	$\bar{1} = 0$
$1 + 0 = 1$	$1 \cdot 0 = 0$	
$1 + 1 = 1$	$1 \cdot 1 = 1$	

Table 8-3   Boolean Theorems in One Variable

OR	AND	NOT
$A + 0 = A$	$A \cdot 0 = 0$	$\bar{\bar{A}} = A$
$A + 1 = 1$	$A \cdot 1 = A$	
$A + A = A$	$A \cdot A = A$	
$A + \bar{A} = 1$	$A \cdot \bar{A} = 0$	

In general, the inputs and outputs of logic circuits are *variables*; that is, the signal may be present or absent (the statement is true or false) corresponding to the binary numbers **1** and **0**. The validity of the theorems in Table 8-3 may be reasoned out in terms of the corresponding logic circuit or demonstrated in a truth table showing all possible combinations of the input variables.

EXAMPLE 5

Demonstrate the theorems

$$A + \bar{A} = 1 \quad \text{and} \quad A \cdot 1 = A$$

by constructing truth tables.

Using the postulates, the truth tables are

A	$\bar{A}$	$A + \bar{A}$
0	1	1
1	0	1

A	$A \cdot 1$
0	0
1	1

Table 8-4   Boolean Theorems in More Than One Variable

Commutation rules:	Association rules:	DeMorgan's theorems:
A + B = B + A	**A + (B + C) = (A + B) + C**	$\overline{A + B} = \overline{A} \cdot \overline{B}$
A · B = B · A	**A · (B · C) = (A · B) · C**	$\overline{A \cdot B} = \overline{A} + \overline{B}$
Absorption rules:	Distribution rules:	
A + (A · B) = A	**A · (B + C) = (A · B) + (A · C)**	
A · (A + B) = A	**A + (B · C) = (A + B) · (A + C)**	

Some of the more useful theorems in more than one variable are displayed in Table 8-4. The *commutation* rules indicate that the order of the variables in performing **OR** and **AND** operations is unimportant, just as in ordinary algebra. The *association* rules indicate that the order of the **OR** and **AND** operations is unimportant, just as in ordinary algebra. The second *distribution* rule indicates that variables or combinations of variables may be distributed in multiplication (not permitted in ordinary algebra) as well as in addition. The *absorption* rules are new and permit the elimination of redundant terms.

EXAMPLE 6

Derive the absorption rule

$$A + (A \cdot B) = A$$

using other basic theorems.

A	A · B	A + (A · B)
0	0	0
1	B	1

Factoring by the second distribution rule,

$$A + (A \cdot B) = (A + A) \cdot (A + B) = A \cdot (A + B)$$

we see that the two absorption forms are equivalent. Substituting **A · 1** for **A · A** (Table 8-3),

$$A \cdot (A + B) = A \cdot 1 + A \cdot B = A \cdot (1 + B) = A \cdot 1 = A$$

The truth table is shown at the left.

DeMorgan's Theorems

Augustus DeMorgan, a contemporary of George Boole, contributed two interesting and very useful theorems. These concepts are easily interpreted in terms of logic circuits. The first says that a **NOR** gate ($\overline{A + B}$) is equivalent to an **AND** gate with **NOT** circuits in the inputs ($\overline{A} \cdot \overline{B}$). The second says that a **NAND** gate ($\overline{A \cdot B}$) is equivalent to an **OR** gate with **NOT** circuits in the inputs ($\overline{A} + \overline{B}$). As generalized by Shannon, DeMorgan's theorems say:

To obtain the inverse of any Boolean function, invert all variables and replace all ORs by ANDs and all ANDs by ORs.

Application of this rule is illustrated in Example 7 on page 208.

Practice Problem 8-3

Use truth tables to prove Boole's theorem **A + (B · C) = (A + B) · (A + C)** and DeMorgan's theorem $\overline{A + B} = \overline{A} \cdot \overline{B}$.

EXAMPLE 7

Use DeMorgan's theorems to derive a combination of **NAND** gates equivalent to a two-input **OR** gate.

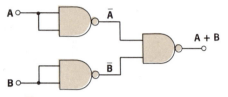

Figure 8.1   **OR** gate from **NAND** gates.

The desired function is $f = A + B$. Applying the general form of DeMorgan's theorems,

$$f = A + B = \overline{\overline{A} \cdot \overline{B}}$$

suggesting a **NAND** gate with **NOT** inputs.

Because $\overline{A \cdot A} = \overline{A}$, a **NAND** gate with the inputs tied together performs the **NOT** operation. The logic circuit is shown in Fig. 8.1.

LOGIC CIRCUIT ANALYSIS

By comparing Example 7 with Example 1 in Chapter 7, we see that the theorems of Boolean algebra permit us to manipulate logic statements or functions directly, without setting up the truth tables. Furthermore, the use of Boolean algebra can lead to simpler logic statements that are easier to implement. This result is important when it is necessary to design a circuit to perform a specified logic function using the available gates—only **NAND** gates, for example. DeMorgan's theorems are particularly helpful in finding **NAND** operations that are equivalent to other operations.

Up to this point we have been careful to retain the specific **AND** sign in $A \cdot B$ to focus attention on the *logic* interpretation. Henceforth, we shall use the simpler equivalent forms **AB** and **A(B)** whenever convenient. Also note that henceforth we shall follow convention and rely on the distinctive shapes of the logic symbols to identify their functions.

The *analysis* of a logic circuit consists in writing a logic statement expressing the overall operation performed in the circuit. This can be done in a straightforward manner, by starting at the input and tracing through the circuit, noting the function realized at each output. The resulting expression can be simplified or written in an alternative form using Boolean algebra. A truth table can then be constructed. (See Example 8.)

LOGIC CIRCUIT SYNTHESIS

One of the fascinating aspects of digital electronics is the construction of circuits that can perform simple mental processes at superhuman speeds. A typical digital computer can perform thousands of additions of 10-place numbers per second. The logic designer starts with a logic statement or truth table, converts the logic function into a convenient form, and then realizes the desired function by means of standard or special logic elements.

EXAMPLE 8

Analyze the logic circuit of Fig. 8.2.

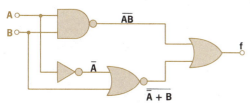

Figure 8.2 Logic circuit analysis.

Construct the truth table to demonstrate that this circuit could be replaced by a single **NAND** gate.

The suboutputs are noted on the diagram. The overall function can be simplified as follows:

$$f = \overline{\overline{AB} + \overline{A} + B}$$

$$= (A + \overline{B}) + A\overline{B} \qquad \text{(DeMorgan's rule)}$$

$$= A + \overline{B}(1 + A) \qquad \text{(Distribution)}$$

$$= A + \overline{B} \qquad (1 + A = 1)$$

$$= \overline{\overline{A}B} \qquad \text{(DeMorgan's rule)}$$

A	B	$\overline{AB}$	$\overline{A} + B$	f
0	0	1	0	1
0	1	1	0	1
1	0	1	1	1
1	1	0	0	0

The Half-Adder

As an illustration, consider the process of addition. In adding two binary digits, the possible sums are as shown in Fig. 8.3a. Note that when **A = 1** and **B = 1**, the *sum* in the first column is **0** and there is a *carry* of **1** to the next higher column. As indicated in the truth table, the half-adder must perform as follows: "**S** is **1** if **A** is **0 AND B** is **1, OR** if **A** is **1 AND B** is **0**; **C** is **1** if **A AND B** are **1**." In logic nomenclature, this becomes

$$S = \overline{A}B + A\overline{B} \qquad \text{and} \qquad C = AB \tag{8-2}$$

(A *full-adder* is capable of accepting the carry from the adjacent column.)

To *synthesize* a half-adder circuit, start with the outputs and work backward. Equation 8-2 indicates that the sum **S** is the output of an **OR** gate; the inputs are obtained from **AND** gates; inversion of **A** and **B** is necessary. Equation 8-2 also indicates that the carry **C** is simply the output of an **AND** gate. The corresponding logic circuit is shown in Fig. 8.3c.

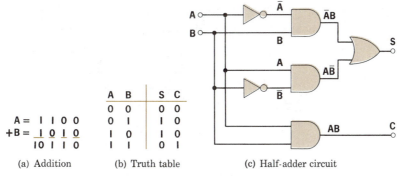

$$A = 1\ 1\ 0\ 0$$
$$+B = \underline{1\ 0\ 1\ 0}$$
$$10\ 1\ 1\ 0$$

A	B	S	C
0	0	0	0
0	1	1	0
1	0	1	0
1	1	0	1

(a) Addition   (b) Truth table   (c) Half-adder circuit

Figure 8.3 Addition of two binary numbers.

There may be several different Boolean expressions for any given logic statement, and some will lead to better circuit realizations than others. Algebraic manipulation of Eq. 8-2a yields

$$\overline{AB} + A\overline{B} = \overline{(A + \overline{B})(\overline{A} + B)} \qquad \text{(DeMorgan's theorems)}$$

$$= \overline{A\overline{A} + AB + \overline{A}\,\overline{B} + B\overline{B}} \qquad \text{(Multiplication)}$$

$$= \overline{\overline{A}\,\overline{B} + AB} \qquad (A\overline{A} = 0 \text{ and } B\overline{B} = 0)$$

$$= (A + B)(\overline{A} + \overline{B}) \qquad \text{(DeMorgan's theorems)}$$

$$= (A + B)\overline{A}\,\overline{B} \qquad \text{(DeMorgan's theorems)}$$

Looking again at the truth table (Fig. 8.3b), we see that another interpretation is: "**S** is **1** if (**A OR B**) is **1 AND** (**A AND B**) is **NOT 1**." Therefore binary addition can be expressed as

$$S = (A + B)\overline{AB} \qquad \text{and} \qquad C = AB \qquad (8\text{-}3)$$

The synthesis of this circuit, working backward from the output, is shown in Fig. 8.4. This circuit is better than that of Fig. 8.3c in that fewer logic elements are used and

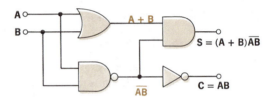

Figure 8.4   Another half-adder circuit.

the longest path from input to output passes through fewer *levels*. The ability to optimize logic statements is an essential skill for the logic designer.

The Exclusive–OR Gate

The function $(A + B)\overline{AB}$ in Equation 8-3 is called the **Exclusive–OR** operation. As indicated by the truth table, it can be expressed as: "**A OR B** but **NOT (A AND B)**." The alternative form (Eq. 8-2a) $\overline{AB} + A\overline{B}$ is called an "inequality comparator" because it provides an output of **1** if **A** and **B** are not equal.

EXAMPLE 9

Show that the inverse of the inequality comparator is an "equality comparator" and synthesize a suitable circuit.

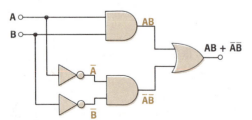

Figure 8.5   An equality comparator.

Algebraic manipulation of the inverse function by using DeMorgan's theorem and Boole's theorem that $A \cdot \overline{A} = 0$ yields

$$\overline{\overline{AB} + A\overline{B}} = (A + \overline{B})(\overline{A} + B) = AB + \overline{A}\,\overline{B} \qquad (8\text{-}4)$$

This is an equality comparator in that the output is **1** if **A** and **B** are equal. This function is highly useful in digital computer operation. Straightforward synthesis results in the circuit of Fig. 8.5.

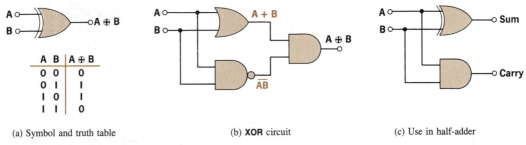

(a) Symbol and truth table   (b) **XOR** circuit   (c) Use in half-adder

Figure 8.6   An **Exclusive–OR** circuit and its application.

The **Exclusive–OR** gate is used so frequently that it is represented by the special symbol $\oplus$ defined by

$$A\ \textbf{XOR}\ B = A \oplus B = (A + B)\overline{AB} \tag{8-5}$$

One realization of the **Exclusive–OR** gate is shown in Fig. 8.6b. Assuming that this gate is available as a logic element, the half-adder takes on the simple form of Fig. 8.6c. As a further simplification, we can treat the half-adder as a discrete logic element and represent it by a rectangular block labeled **HA**.

The Full-Adder

In adding two binary digits or *bits*, the half-adder performs the most elementary part of what may be highly sophisticated computation. To perform a complete addition, we need a *full-adder* capable of handling the carry input as well. The addition process is illustrated in Fig. 8.7a, where **C** is the carry from the preceding column. Each carry of **1** must be added to the two digits in the next column, so we need a logic circuit capable of combining three inputs. This operation can be realized using two half-adders and an **OR** gate. The values shown in Fig. 8.7b correspond to the third column of the addition. The operation of the full-adder can be seen more clearly by construction of a truth table. (See Review Question 12.)

To perform the addition of "4-bit words," that is, numbers consisting of four binary digits, we need a half-adder for the first column and a full-adder for each additional column. In the *parallel binary adder*, the input to the half-adder consists of digits A_1 and B_1 from the first column. The input to the next unit, a full-adder,

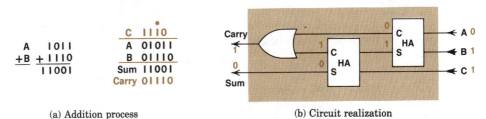

(a) Addition process   (b) Circuit realization

Figure 8.7   Design of a full-adder.

consists of carry C_1 from the first unit and digits A_2 and B_2 from the second column. The sum of the two numbers is represented by $C_4 S_4 S_3 S_2 S_1$. In subtraction, **NOT** gates provide the complement of the subtrahend (see Example 4, p. 205) and the process is reduced to addition.

A Design Procedure

A common problem in logic circuit design is to create a combination of gates to realize a desired function. The basic approach is to proceed from a statement of the function to a truth table and then to a Boolean expression of the function. The experienced logic designer then manipulates the Boolean expression into the simplest form. The realization of the final expression in terms of **AND**, **OR**, and **NOT** gates is straightforward.

For increased reliability on a spacecraft, triple sensing systems are used; no action is taken unless at least two of the three systems call for action. The truth table of the required *vote taker* is shown in Fig. 8.8a. Because the function is **YES (1)** only

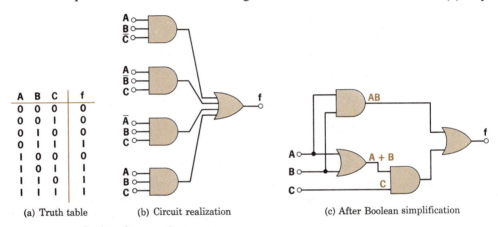

A	B	C	f
0	0	0	0
0	0	1	0
0	1	0	0
0	1	1	1
1	0	0	0
1	0	1	1
1	1	0	1
1	1	1	1

(a) Truth table   (b) Circuit realization   (c) After Boolean simplification

Figure 8.8   Design of a vote taker.

when a majority of the inputs are **YES**, the Boolean expression must contain a product term for each row of the truth table in which the function is **1**. Because the function is **YES** for any one *or* more of these rows, the Boolean expression is

$$f = AB\overline{C} + A\overline{B}C + \overline{A}BC + ABC \tag{8-6}$$

Assuming that the complement of each variable is available, as is true in most computers, the straightforward realization is a combination of four **AND** gates feeding an **OR** gate. (See Fig. 8.8b.)

If the complements were not available, eight logic elements would be required and simplification of the circuit would be desirable. First, the Boolean function of Equation 8-6 can be expanded into

$$f = AB\overline{C} + A\overline{B}C + \overline{A}BC + ABC + ABC$$

because **ABC + ABC = ABC**. Factoring by the distribution rule yields

$$f = AB(\overline{C} + C) + C(AB + A\overline{B} + \overline{A}B)$$

Because $\overline{C} + C = 1$ and $AB + A\overline{B} + \overline{A}B = A + B$, the function becomes

$$f = AB + C(A + B)$$

This function requires only four logic elements (Fig. 8.8c).

Practice Problem 8-4

Given the logic function $f = \overline{A}B + A\overline{B}$, apply DeMorgan's theorem to convert function f into a "**NAND**ed product of **NAND**s," and design a circuit consisting of **NAND** gates only, assuming complements are available.

Answer: $f = \overline{\overline{\overline{A}B} \cdot \overline{A\overline{B}}}$.

MINIMIZATION BY MAPPING

In creating a logic circuit to realize a desired function, the designer seeks the optimum form. The criterion may be maximum speed (fewest logic levels) or minimum cost (fewest gate leads because the number of leads determines the cost of manufacture and the cost of assembly) or minimum design time (if only a few circuits are required). We have seen how Boolean algebra can be used to derive simpler logic expressions. If the truth table is available or if the logic function is expressed as a "sum of products," the designer can go directly to the minimal expression by the mapping technique suggested by Maurice Karnaugh.

Karnaugh Maps

The *map* of the general logic function of three variables is shown in Fig. 8.9a. Each square in the map corresponds to one of the eight possible combinations of the

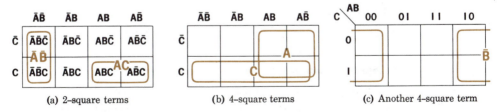

Figure 8.9   Mapping a three-variable logic function.

three variables. The order of the columns is such that *combinations in adjacent squares differ only in the value of one variable.* Therefore, 2-square groups are independent of one variable; that is, $f_1 = \overline{A}\overline{B}\,\overline{C} + \overline{A}\overline{B}C = \overline{A}\,\overline{B}$ and $f_2 = ABC + A\overline{B}C = AC$. These relations are easy to derive using Boolean algebra, but they are *obvious by inspection* of the Karnaugh map.

The concept can be extended to groupings of four adjacent squares as shown in Fig. 8.9b, where the labels are omitted from the squares. The 4-square cluster outlined in color is independent of both **B** and **C** and the 4-square in-line group is independent of both **A** and **B**; that is, $f_3 = \overline{A}\,\overline{B}C + \overline{A}BC + ABC + A\overline{B}C = C$. Enlarging groups by overlapping simplifies the terms. Note that the map is "continuous" in that the last column on the right is "adjacent" to the first column on the left. The 4-square group in Fig. 8.9c is just equal to $\overline{B}$.

The standard labeling scheme for Karnaugh maps (Fig. 8.9c) is convenient for mapping from a truth table. Each square in the map corresponds to a row in the truth table. A specific logic function is *mapped* by placing a **1** in each square for which the function is **1**. When this has been done, possible simplifications are easily recognized.

EXAMPLE 10

Map the vote-taker function and simplify the circuit realization, if possible.

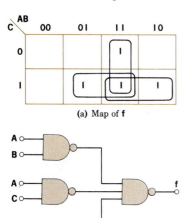

(a) Map of **f**

(b) Minimal realization

Figure 8.10   Simplification by Karnaugh mapping.

From the truth table of Fig. 8.8a, 1s are placed in the squares corresponding to rows in the truth table for which the function is **1**, representing

$$f = \overline{A}BC + A\overline{B}C + AB\overline{C} + ABC$$

All the 1s can be included in three overlapping 2-square groups; ∴ the complete function can be represented by

$$f = AB + AC + BC$$

There is no simpler expression for this function.

By using DeMorgan's theorem, any "sum of products" can be converted to a "**NAND**ed product of **NAND**s." Here

$$f = \overline{\overline{AB} \cdot \overline{AC} \cdot \overline{BC}}$$

which can be synthesized using **NAND** gates only. Note that in the circuit realization of Fig. 8.10b, the number of gate leads has been reduced to 9 input + 4 output = 13 from the 21 in Fig. 8.8b. This vote taker could be realized in a single SSI chip.

Mapping in Four Variables

The Karnaugh technique is even more valuable in simplifying functions in four variables. (For more than four variables, other techniques are usually more convenient.) As indicated in Fig. 8.11, 2-square groups are independent of one variable, 4-square groups are independent of two variables, and 8-square groups are independent of three variables. Note that the standard labeling scheme provides that adjacent rows differ by only one complement bar and the bottom row is *adjacent* to the top row. The four corner squares form a combination that is a little difficult to visualize.

There are detailed rules[†] for finding the minimal expression from a Karnaugh

[†] See pp. 69 ff. of the Peatman reference at the end of Ch. 7.

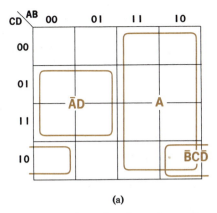 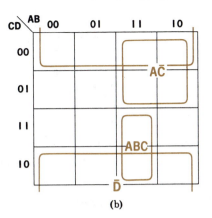

(a)   (b)

Figure 8.11   Examples of grouping on four-variable maps.

map, but some general guidelines will suffice for our purposes:

Include all 1s in groups of eight, four, two, or one.
Groups may overlap; larger groups result in simpler terms.
Of the possible combination of terms, select the simplest.

The technique is illustrated in Example 11.

EXAMPLE 11

Map the function

$$f = \overline{AB}(\overline{C} + D) + \overline{A}CD + AB\overline{C}\,\overline{D} + A\overline{B}\,\overline{C}D$$

and obtain a minimal sum-of-products expression.

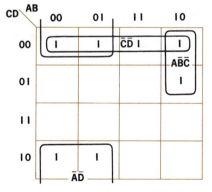

Figure 8.12   Four-variable simplification.

As an alternative to forming the truth table, let us map the function by considering the factors individually.

$\overline{AB}$ limits the first term to the **00, 01,** and **10** columns and $(\overline{C} + D)$ corresponds to the first row, implying three 1s as shown in Fig. 8.12.

The $\overline{A}CD$ term is independent of **B**, implying a 2-square group in the lower left-hand corner.

The four-variable terms imply 1s in the **1100** and **1001** squares.

All the 1s can be included in the two 4-square and one 2-square groups encircled. Therefore,

$$f = \overline{A}\,\overline{D} + \overline{C}D + A\overline{B}\,\overline{C}$$

Other expressions are possible, but none will include fewer, simpler terms.

Two additional comments should be made. In some circuits, certain combinations of inputs never occur; such *don't care* combinations may be mapped as **X**s and considered as either **0**s or **1**s, whichever provides the greatest simplification. In other circuits, the simplest realization results from implementing $\bar{f}$ as a sum of products and then inverting to obtain **f**.

Practice Problem 8-5

Map the function $f = A\bar{B}CD + \bar{A}BC + \bar{A}\bar{B}C + BCD$ and find the minimal sum-of-products form.

Answer: $f = \bar{A}C + CD$.

The emphasis in this section is on synthesizing logic circuits from optimum arrangements of basic gates. This is the proper approach in designing complex circuits for mass production or in designing custom circuits requiring only a few gates. As we shall see, other approaches using standard IC packages are better where the cost of design time is an important factor.

REGISTERS

In addition to the logic circuits that process data, digital systems must include memory devices to store data and results. We know that a flip-flop can store or "remember" one digit of a binary number, one bit. A *register* is an array of flip-flops that can temporarily store data or information in digital form. For example, the 16-bit registers in a microprocessor, composed of sixteen flip-flops in parallel, can handle 2-byte instructions or 16-digit numbers. A great variety of registers is available in IC form.

Shift Registers

The sophistication of the response of the JK flip-flop makes it useful in computer applications. The *serial shift register* of Fig. 8.13 consists of four trailing-edge-triggered JK flip-flops connected so that $J \neq K$. (See Fig. 7.26.) At the trailing edge of each clock pulse, **Q** follows **J** in each flip-flop of the 4-bit register. The data are

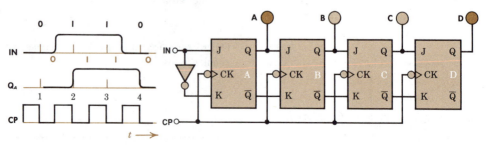

Figure 8.13 Entering **0110** into a 4-bit serial shift register.

entered serially, that is, one bit at a time, and shifted right through the register at each clock pulse.

Table 8-5 shows how **0110** would be placed in the register. We begin with the least significant bit. With a **0** at **IN = J$_A$** and **K ≠ J**, at the trailing edge of the first clock pulse (CP1), **Q$_A$** follows **J$_A$** and the LSB is transferred to the output of flip-flop A. During the next clock cycle, **J$_B$ = Q$_A$ = 0** and the second bit, a **1**, is applied to **IN = J$_A$**. At CP2, the **0** is transferred to **Q$_B$** (i.e., shifted one position to the right) and the **1** is transferred to **Q$_A$**. After four clock pulses, the 4-bit number is stored in the register and **0110** is available at parallel outputs **ABCD**. One application of such a register is as a *serial-to-parallel converter*, changing serial data to parallel form for processing all bits simultaneously. There is a single input line, but four lines are required for the parallel data output.

Table 8-5 Serial Shift Register

CP	IN	Q$_A$	Q$_B$	Q$_C$	Q$_D$
1	0 ⟶ 0				
2	1 ⟶ 1	0			
3	1 ⟶ 1	1	0		
4	0 ⟶ 0	1	1	0	

The shift register of Fig. 8.14 consists of D flip-flops with **CLEAR** and **PRESET** capabilities. It is similar to MSI TTL units available commercially. The symbols indicate that the flip-flops are cleared to **0** if **CLR** goes **LOW** while **PR** is inactive (**HIGH**) (clearing is independent of the clock level). On the positive-going edge of the clock signal, the input at **D** is transferred to **Q**. (See Fig. 7.25.)

Because both inputs and outputs are accessible, this unit can be used as a general-purpose register. It can function as a 4-bit storage register (serial or parallel), as a serial-to-parallel converter (**SERIAL INPUT** to **Q$_A$ Q$_B$ Q$_C$ Q$_D$ OUTPUT**), or as a parallel-to-serial converter. For the last function, all stages are cleared to **0**, and the data to be loaded are applied to the **ABCD INPUTS**. A **HIGH** signal to **PRESET ENABLE** is **NAND**ed with any **1** input to send **PR LOW** setting **Q** to **1** in that stage (independent of the clock level). At the next upward transition of the clock pulse (unless the clock is inhibited), the data are shifted to the right; the value of **Q$_D$** is output and the value of **Q$_C$** is transferred to **Q$_D$**, etc. As we shall see, shifting bits one place to the right is equivalent to division by 2 in binary notation, and the shift operation is useful in computation.

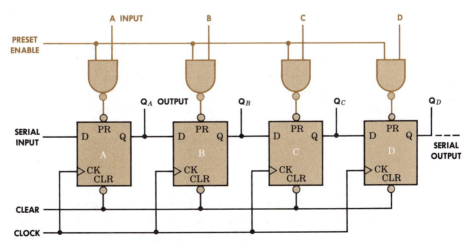

Figure 8.14 A general-purpose 4-bit shift register.

A Practical Register—The 74173

The 74173 TTL unit is a versatile 4-bit register incorporating D flip-flops and three-state outputs for use in "bus-organized" systems (Fig. 8.28). The "totem pole" outputs (R_4, T_4, D, and T_3 in Fig. 7.13) are capable of driving the bus lines directly. (Rated drive current is 16 mA.) As shown by the "pinout" in Fig. 8.15, the four data inputs (D) and the four data outputs (Q) use eight pins in the 16-pin package. Power supply (V_{CC}, GND), direct **CLEAR**, and **CLOCK** account for four pins. The remaining pins provide versatility.

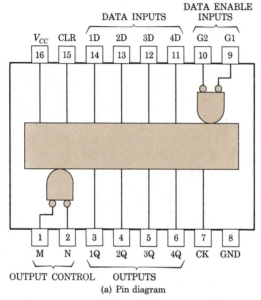

(a) Pin diagram

INPUTS					OUTPUT Q
CLEAR	CLOCK	DATA ENABLE G1	DATA ENABLE G2	DATA D	
H	X	X	X	X	L
L	L	X	X	X	Q_0
L	↑	H	X	X	Q_0
L	↑	X	H	X	Q_0
L	↑	L	L	L	L
L	↑	L	L	H	H

H = high level (steady state)
L = low level (steady state)
↑ = low–to–high–level transition
X = irrelevant (any input including transitions)
Q_0 = the level of **Q** before the indicated steady–state input conditions were established.

(b) Function table

Figure 8.15 The 74173 4-bit register.

Access to the output control is provided by pins M and N. For **M + N = 0**, normal logic states are available; for **M + N = 1**, the three-state outputs are effectively disconnected from the bus. This means that we can disconnect the unit by sending a signal from either of two components of the system. Disabling of the outputs is independent of the clock level.

The entry of data into the flip-flops is also controlled. As indicated in the function table, when both **DATA ENABLE** inputs are **LOW** (**L**), data at the D inputs are loaded into their respective flip-flops on the next positive transition of the **CLOCK**.

These registers are used as bus buffer registers between data sources with inadequate drive or lacking three-state capability and the system bus. The typical propagation delay time is 23 ns, and the typical power consumption is 250 mW. They can accept clock frequencies up to 25 MHz.

COUNTERS

Flip-flops can be connected to function as electronic counters to count random events or to divide a frequency or to measure a parameter such as time, distance, or speed. Counters are used to keep track of operations in digital computers and in instrumentation. The JK master-slave flip-flop is widely used in counter design, but T and D flip-flops are also useful.

Divide-by-*n* Circuits

A divide-by-*n* counter produces one output pulse for *n* input pulses: *n* is called the "modulo" of the counter. As shown in Fig. 7.30, a toggle flip-flop divides by two; for T held **HIGH**, the output **Q** is the number of CK inputs divided by two. Two D flip-flops can be used in the divide-by-four circuit of Fig. 8.16. With Q_A and Q_B

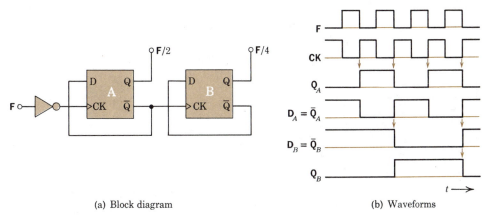

(a) Block diagram   (b) Waveforms

Figure 8.16   A divide-by-4 circuit using D flip-flops.

cleared, when clock signal **F** goes **LOW**, **CK** goes **HIGH** and $D_A = \overline{Q}_A = 1$ is transferred to Q_A. The next time **F** goes **LOW**, Q_A changes to **LOW**; two events at **F** complete one cycle of Q_A or $Q_A = F/2$. When Q_A goes **LOW**, $\overline{Q}_A = CK_B$ goes **HIGH** and $D_B = \overline{Q}_B = 1$ is transferred to Q_B; four events at **F** complete one cycle of Q_B or $Q_B = F/4$.

Binary Ripple Counter

The counter of Fig. 8.17 consists of three JK flip-flops in cascade. With J and K held **HIGH** (the connections to +5 V are not shown), the flip-flops toggle at each downward transition of the pulse at CK. The lowest order bit Q_A changes state after each input pulse. The next bit Q_B changes state whenever Q_A goes **LOW** since Q_A supplies CK_B. Similarly, Q_C changes state whenever Q_B goes **LOW**. This is an *asyn-*

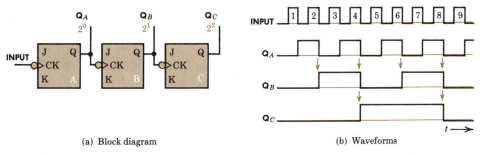

(a) Block diagram   (b) Waveforms

Figure 8.17   A three-stage binary ripple counter.

chronous, *binary*, *modulo-8*, *ripple* counter; *asynchronous* because all flip-flops do not change at the same time; *binary* because it follows the binary number sequence with bit values of 2^0, 2^1, and 2^2; *modulo-8* because it counts through 8 distinct states; *ripple* because the changes in state ripple through the stages. The ripple effect is seen at the fourth input pulse where Q_A goes **LOW**, causing Q_B to go **LOW**, causing Q_C to go **HIGH**.

EXAMPLE 12

Design a modulo-5 binary counter using T flip-flops with **CLEAR** capability.

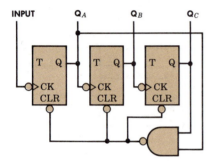

Figure 8.18   A modulo-5 counter.

Three stages are required to count beyond 4. A modulo-5 counter must count up to 4 and then, on the fifth pulse, clear all flip-flops to 0. The sequence of states is

Count	Q_C	Q_B	Q_A	
0	0	0	0	
1	0	0	1	
2	0	1	0	
3	0	1	1	
4	1	0	0	
5	1	0	1	(Unstable)
	0	0	0	(Stable)

At the count of 5, the 1s at Q_A and Q_C can be **NAND**ed to generate a **CLEAR** signal as shown in Fig. 8.18.

Decade Counters

Counters that communicate with humans usually display results in the decimal system. However, flip-flops count in binary; therefore, binary numbers must be *coded* in decimal. To count to 10 in the 8421 code, four flip-flops are required. The usual way to get ten distinct states is to modify a 4-bit binary counter so that it skips the last six states. It counts normally from 0 to 9, and then feedback logic is provided so that at the next count the decimal sequence is reset to zero. (See Exercise 31.)

Synchronous Counters

One disadvantage of ripple counters is the slow speed of operation caused by the long time required for changes in state to ripple through the flip-flops. Another problem is that the cumulative delays can cause temporary state combinations (and voltage spikes called *glitches*) that result in false counts. Both of these difficulties are avoided in *synchronous* counters in which all flip-flops change state at the same instant.

In the synchronous counter of Fig. 8.19, JK master-slave units are connected as T (toggle) flip-flops. At each count, flip-flop A toggles, and the other flip-flops are clocked. The master-slave characteristic causes flip-flop B to toggle on the next count after Q_A becomes **1**, as called for in the truth table. Similarly, the **AND** gate causes

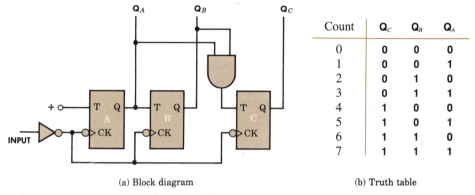

Count	Q_C	Q_B	Q_A
0	0	0	0
1	0	0	1
2	0	1	0
3	0	1	1
4	1	0	0
5	1	0	1
6	1	1	0
7	1	1	1

(a) Block diagram   (b) Truth table

Figure 8.19 A synchronous modulo-8 counter.

flip-flop C to toggle on the next count after Q_A **AND** $Q_B = 1$, as called for in the truth table. In general, synchronous counters are faster and more trouble-free.

A Practical Counter—The 74163

The 74163 (Fig. 8.21 on page 223) is a popular 4-bit counter using TTL technology. The technical features include:

4-bit binary—counts 0 to 15 and then resets; both P and T enable inputs must be **HIGH** to count.

Carry output—permits cascading 74163 counters for higher counting capacity. The carry pin (15) provides a positive pulse that can enable successive stages.

Synchronous operation—clocks flip-flops simultaneously so that, when enabled, outputs change simultaneously.

Buffered clock input—triggers the four JK master-slave flip-flops on the rising edge of the clock pulse.

Programmable—allows presetting the outputs. When **LOAD** goes **LOW**, the values at the **DATA INPUTS** are set into the **OUTPUTS** after the next clock pulse.

Synchronous clear—when **CLEAR** is **LOW**, sets all flip-flop outputs to **LOW** after the next clock pulse (regardless of inputs). This permits modifying the count length by decoding the maximum count desired using an external **NAND** gate connected to the **CLEAR** input.

The circuits are designed for operation in ambient temperatures from 0 to 70° C; typical power consumption is 325 mW.

Figure 8.21 illustrates the kind of information provided by a manufacturer of digital devices; with our background we should appreciate the capabilities of the 74163 and be able to use it in designing counter circuits. In step 1 of the illustration, **CLEAR** is sent **LOW** and on the next upward **CLOCK** transition the **OUTPUTS** are set to **0**. (Following the functional diagram, we see that a **0** at **CLEAR** puts a **0** on each J and a **1** on each K so that each Q goes to **0** at the next upward transition of **CLOCK**.) In step 2, while **LOAD** is **LOW**, the DCBA = 1100 → 12D previously entered as **DATA** is loaded into $Q_D Q_C Q_B Q_A$ at the next **CLOCK** transition.

Practice Problem 8-6

Follow the functional block diagram during step 2 of the 74163 sequence illustrated in Fig. 8.20d to determine the values at J and K of flip-flop C and the behavior of Q_C.

Answer: J = 1, K = 0, Q_C goes to 1.

In step 3 of the 74163 illustration, Q_A toggles at each **CLOCK** transition and Q_B toggles at every other transition to count to 15, then 0, 1, etc. At 15, a **1** appears at **CARRY OUT**, which is logically equal to $Q_A \cdot Q_B \cdot Q_C \cdot Q_D \cdot$ **T**, to **ENABLE** the next stage in cascaded counters. In step 4, a command to inhibit further counting sends **ENABLE P LOW** and counting stops. The flexibility of the 74163 is illustrated in Example 13.

EXAMPLE 13

Basically, the 74163 is a divide-by-16 counter with the input at CK and CY OUT going **HIGH** once in 16 clock cycles.

Design a simple divide-by-12 counter using a 74163. Show the pin diagram and the external connections required.

We could shorten the count length by decoding the outputs on the eleventh count (**1011**), using a 3-input **NAND** gate to drive the **CLEAR LOW** on the next count as explained earlier. This approach is illustrated in the truth table at left below.

Count	DCBA	CLR		Count	DCBA	OUT
0	0000	1		0	1000	1
1	0001	1		1	1001	1
2	0010	1		2	1010	1
3	0011	1		3	1011	1
4	0100	1		4	1100	1
5	0101	1		5	1101	1
6	0110	1		6	1110	1
7	0111	1		7	1111	1
8	1000	1		8	0100	0
9	1001	1		9	0101	0
10	1010	1		10	0110	0
11	1011	0		11	0111	0
0	0000	1		0	1000	1

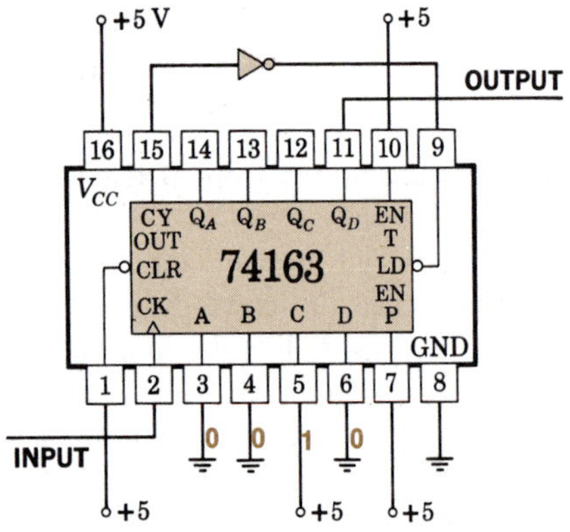

Figure 8.20   A divide-by-12 counter.

Another approach is shown in Fig. 8.20. To divide by 12, we prepare to load $16 - 12 = 4$ into the counter by setting **DATA INPUT C HIGH** and tying the other **DATA INPUTS LOW**. This makes input **DCBA = 0100** → 4D.

In output state **1111**, the **CARRY** is inverted to set **LOAD LOW**; after the next clock pulse, **DCBA = 0100** is loaded into the **OUTPUTS**. The counting sequence is shown in the truth table on the right above. On the 12th count, the output at Q_D goes **HIGH** in state 8D = **1000B**. This signal is used for the counter **OUTPUT**.

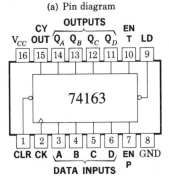

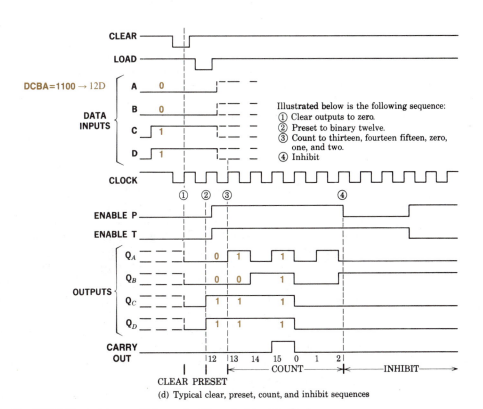

Figure 8.21 The SN74163 synchronous 4-bit counter (Texas Instruments, Inc.).

READ-AND-WRITE MEMORY

In a digital computer, instructions and numbers are stored in the *memory*, an organized arrangement of elements called *memory cells*, each capable of storing one bit of information. In *Read-And-Write Memory* (RAM), the computer can store ("write") data at any selected location ("address") and, at any subsequent time, retrieve ("read") the data. In *Read-Only-Memory* (ROM), data are initially and permanently stored (by the manufacturer or the user); the computer can read the data at any address, but it cannot alter the stored bits.

In Fig. 8.22a, the k address lines can designate $2^k = m$ words whose n bits are carried on the n parallel input and output lines. The 64 bits of the 4×16 ROM in

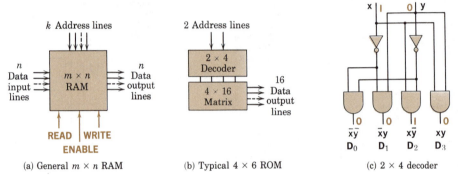

(a) General $m \times n$ RAM   (b) Typical 4×6 ROM   (c) 2×4 decoder

Figure 8.22 Examples of IC memories.

Fig. 8.22b are stored in 64 memory cells arranged in $2^k = 2^2 = 4$ words of 16 bits each available at the 16 output lines. The address is coded as a k-bit binary number; a *decoder* translates the coded address and specifies one of the 2^k words. In the 2×4 decoder of Fig. 8.22c, the address $xy = 10$ yields a **1** at D_2 ($x\bar{y}$) and specifies that word 2 is to be read, that is, connected to the 16 output lines.

RAM

The Read-And-Write Memory of Fig. 8.22a includes an $m \times n$ matrix of memory cells. Each cell consists of a binary storage element and the associated control logic. In the simple cell shown in Fig. 8.23a, the cell is selected when **S** goes **HIGH**. With **R/W̄ HIGH**, the output is enabled and the value of **Q** is read out on the output data line. With **R/W̄ LOW**, the input is enabled to write in the input data value.

The organization of memory cells into an elementary RAM with a capacity of two 3-bit words is shown in Fig. 8.23b. The decoder activates one of the two word-select lines in accordance with the address. In the **WRITE** operation (**R/W̄ LOW**), 3 bits of data are transferred from the input lines to the selected word. In the **READ** operation, the 3 bits of the selected word are transferred to the output lines via the **OR** gates. The outputs of the unselected words are all **LOW**.

Because the words in memory can be accessed in any order, this is a *random-access memory*. (Originally RAM meant random-access memory, but now it is interpreted as read-and-write memory.) In the 2×3 RAM shown, address selection is

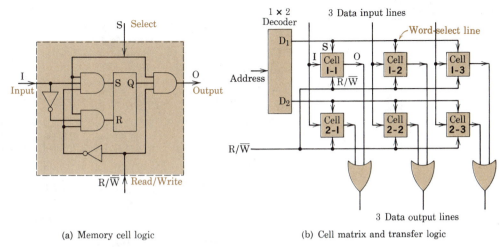

(a) Memory cell logic

(b) Cell matrix and transfer logic

Figure 8.23 Functional representation of a 2 × 3 RAM.

linear; that is, addressing is performed by activating one word-select line. In large memories, selection is *coincident*; that is, each cell is accessed by addressing an X select line to select the row and a Y select line to select the column. The intersection of the X and Y lines identifies one cell in a two-dimensional matrix. For example, a 1024-bit (1K) RAM could consist of 1024 memory cells arranged in a 32 × 32 matrix on the silicon chip. For operation as a 256 × 4 RAM, the 32 columns could be organized into 8 groups of 4 columns representing eight 4-bit words in each of 32 rows. The 32 row-select lines would be addressed by a 5-bit address ($2^5 = 32$), and the 8 column-select lines would be addressed by a 3-bit address ($2^3 = 8$).

Practice Problem 8-7

(a) In Fig. 8.23a, S(Select) = **1**, R/$\overline{\text{W}}$ = **0**, and I = **1**; predict the values of Q and output O.

(b) For S(Select) = **1**, R/$\overline{\text{W}}$ = **1**, and Q = **1**, predict the output.

Answers: (a) Q = **1**, O = **0**; (b) **1**.

MOS RAM

Monolithic semiconductor memory dominates the market for small, fast memory applications. The high packing density and low power consumption of MOS devices have led to their wide use; storage of 256K bits on a single IC chip is common.

The MOS memory cell of Fig. 8.24 contains 6 n-channel enhancement-mode MOSFETs. Transistor T_1 is the *driver* and T_3 the *load* for an inverter (see Fig. 7.2c, p. 170) that is cross-coupled to a second inverter T_2-T_4 to form the storage flip-flop (Fig. 7.21). When T_2 is **ON** and T_1 is **OFF**, output **Q** is logic **1**. With the row-select line

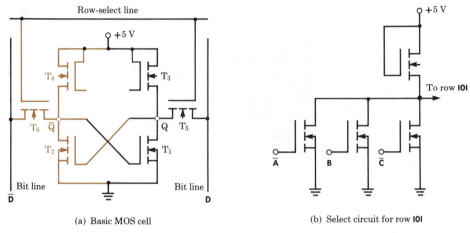

(a) Basic MOS cell

(b) Select circuit for row **101**

Figure 8.24   Typical MOS memory cell and row-select circuit.

LOW, transistors T_5 and T_6 are **OFF**, and the cell is isolated from the bit lines. In a **READ** operation, when row-select goes **HIGH**, T_5 and T_6 provide coupling and **Q** appears on bit line **D**. A *sense amplifier* (similar to that shown in Fig. 8.25b) connected to the bit line provides buffering, and the proper logic level appears on the data output line. In a **WRITE** operation, the selected row of cells is connected to the bit lines, and **Q** is **SET** or **RESET** by a **1** placed on bit line **D** or **D̄** by the *write amplifier*. In this *static* RAM with *nondestructive* read-out, the state of the flip-flop persists as long as power is supplied to the chip, and it is not altered by the **READ** operation.

The row-select circuit of Fig. 8.24b is one of eight similar circuits that decode 3-bit addresses to select a row of MOS memory cells. The three select input pins are connected to buffer-inverters that generate **A**, **Ā**, **B**, **B̄**, **C**, and **C̄**. The circuit is a **NOR** gate that performs the **AND** function on the complements of the inputs (see the truth table in Fig. 7.6a). For the address **101**, the **NOR** gate inputs are **0**, **0**, **0**; the driver transistor is **OFF**; the output is **HIGH**; and only one row-select line will go **HIGH**.

EXAMPLE 14

A 64-bit RAM organized to provide sixteen 4-bit words is to use circuitry similar to Fig. 8.23. The row-select circuits are similar to those in Fig. 8.24b. Specify the number of address lines, the number of **NOR** gate inputs in the row-select circuits, and the decoder connections for the fourth row.

To address 16 words requires 4 address lines since $2^k = m$ or $2^4 = 16$. Assuming linear selection, there will be 16 rows, and 4-input **NOR** gates are required to decode the addresses. Assuming the first 4 rows are designated **0000**, **0001**, **0010**, and **0011**, the fourth row-select inputs must be **AB̄CD̄** so that all drivers are **OFF**.

For another address **0010**, the last transistor will be **ON**, the output will be **LOW**, and the fourth row will not be active.

Bipolar RAM

Bipolar read-and-write memories are extremely fast, but they are less compact and less energy efficient than MOS RAMs. They are TTL compatible and are used as small "scratch-pad" memories for data being processed.

The RAM cell of Fig. 8.25a employs two cross-coupled triple-emitter transistors

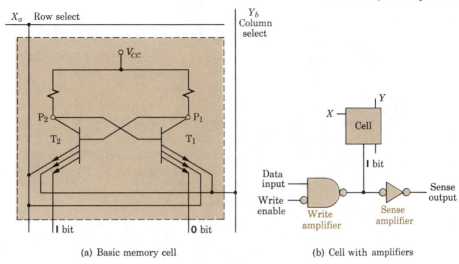

(a) Basic memory cell   (b) Cell with amplifiers

Figure 8.25   Bipolar RAM with coincident selection.

in a flip-flop. One emitter of each transistor is connected to a bit line to provide for data transfer and storage. The other emitters are connected to address lines for coincident selection. The address lines are normally **LOW**, and the currents from all the conducting transistors in the matrix flow out along these lines.

Assuming T_2 is **ON**, V_{P2} is **LOW**, T_1 is **OFF**, and V_{P1} is **HIGH**. (For example, with $V_{CC} = 3.5$ V and $V_X = V_Y = 0.3$ V, $V_{BE2} \cong 0.7$ V and $V_{CE(\text{sat})} = 0.3$ V, $V_{P1} = V_X + V_{BE2} = 0.3 + 0.7 = 1.0$ V. Then $V_{P2} = V_X + V_{CE(\text{sat})} = 0.3 + 0.3 = 0.6$ V; T_1 cannot be forward biased.) To address cell a-b, the corresponding address lines X_a and Y_b are taken **HIGH**. (All cells except the one addressed have at least one address line—X or Y—**LOW** and are unaffected.) As X_a and Y_b go **HIGH**, the connected emitters rise in potential, and the current in T_2 is transferred to the **1**-bit line (held at 0.5 V, say, slightly above logic **0**). This current activates the sense amplifier (Fig. 8.25),[†] and a logic **1** appears at the output. The **READ** operation has been performed.

The **WRITE** operation proceeds as follows. The particular address lines are taken **HIGH**, **WRITE ENABLE** goes **LOW**, and a **1**, say, is input to the write amplifier. The amplifier output takes the **1**-bit line **LOW**, and T_2 is turned **ON** (assuming it was **OFF**). The unaddressed cells on the same bit line are not affected since at least one emitter in each transistor is held **LOW** by an inactive address line. After **WRITE ENABLE** goes **HIGH**, the bit line returns to normal (0.5 V, say). When the address lines return to normal (0.3 V), the emitter current of T_2 is diverted to an address line, ready for any **READ** operation.

READ-ONLY MEMORY

In Read-Only Memory, binary data is physically and permanently stored by defining the state of the constituent memory cells. A set of input signals on the address lines is decoded to access a given set of cells whose states then appear on the output lines.

[†] The sense amplifier is a transistor switch that is turned on when the **1**-bit line current enters the base.

The function performed can be described in three ways. If the set of inputs is identified as an address and the output is the word (data or instructions) stored there, this is a "memory." If the set of inputs represents data coded in one form and the outputs represent the same data coded in another form, this is a "code converter." If the input is identified as a set of binary variables (a binary *function*) and the corresponding output is identified as a related binary function, this is a "combinational logic circuit" that can replace a network of logic gates.

The same bipolar and MOS technologies are used in IC ROM as in RAM. In general, ROM is simpler than RAM since fewer control elements are necessary (Fig. 8.22b) and no provision is made for changing cell states. The essential components are an address decoder, a matrix of memory cells, and appropriate buffers.

ROM Cells

The MOS memory cell shown in Fig. 8.26a consists of a single enhancement-mode transistor connected to word and bit lines plus a passive load R (actually another MOSFET). If this cell is selected, the word line goes **HIGH**, the conductivity of the

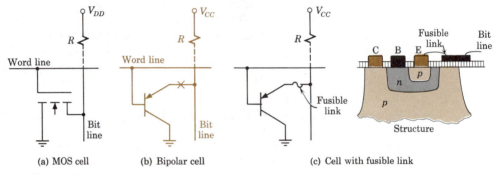

(a) MOS cell   (b) Bipolar cell   (c) Cell with fusible link

Figure 8.26 Elementary ROM cells.

n channel is enhanced, drain current flows, and the bit line is pulled **LOW**; a **0** bit is output. If a **1** is to be stored at this location in memory, this MOSFET is made inoperative by proper masking during chip fabrication. Instead of the normal thin oxide layer between gate and channel, a thick oxide layer is deposited, and the threshold voltage for conducting is excessively high. When the word line is raised, there is no pull-down, and a **1** is output. The finished matrix consists of $m \times n$ similar transistors with a pattern of **1**s corresponding to the thick-oxide cells. This pattern is specified by the user ordering "custom-programmed" ROMs.

The bipolar memory cell of Fig. 8.26b operates in an analogous way. If this cell is selected, the word line goes **LOW**, the emitter-base junction is forward biased, and the bit line is pulled **LOW**; a **0** bit is output. If a **1** is to be stored at this location, the emitter connection at **X** is not completed during fabrication. The specifications of the customer are translated into a pattern of **0**s and **1**s represented by complete and incomplete transistors.

PROMs

The process just described is called *mask programming*; it can be done only by the manufacturer. Also available are *field programmable* memories called PROMs.

The PROM cell of Fig. 8.26c contains a special conducting segment that can be melted by a high-current pulse after packaging. The PROM is fabricated with all **0**s; the user "programs" the unit by electrically changing appropriate **0**s to **1**s. This is an irreversible process; once fused, the links are destroyed. PROMs are economical in small quantities; the high cost of programming masks is justifiable only in large runs.

EPROMs

Erasable programmable read-only memories, called EPROMs for obvious reasons, can be programmed and erased repeatedly. They are particularly useful in the development of new products where the stored information is changed as the development proceeds. One common type employs MOS devices in which the silicon gate element "floats" in an insulating oxide layer. With the proper design, a high-voltage pulse can cause avalanche breakdown and injection of avalanche-created electrons across the very thin oxide layer into the floating gate. After the programming pulse is over, the trapped negative charge enhances the channel conductivity and creates a MOSFET representing, after buffering, a stored **1**. These EPROMs are packaged with transparent quartz covers. Irradiation of the floating gates with ultraviolet light for a few minutes dissipates the trapped charge and the memory is "erased."

A Practical EPROM—The 2732

The 2732 is a 4096-word by 8-bit erasable and electrically programmable ROM suited for digital system development where experimentation with memory pattern is involved. The pinout and functional diagram for the 2732A (Intel's 250-ns access time version of the 2732) is shown in Fig. 8.27. On the 24-pin package, there are 2 power supply terminals, 12 address inputs ($A_0 - A_{11}$), 8 data outputs ($O_0 - O_7$), a chip enable pin ($\overline{CE}$), and a control/program pin ($\overline{OE}/V_{PP}$).

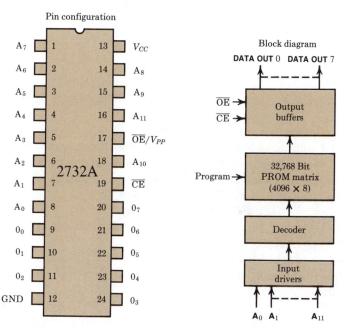

Figure 8.27   The Intel high-speed NMOS 2732A EPROM.

As fabricated, all 32,768 bits are in the **1** state. Information is stored by selectively programming **0**s in the appropriate locations. In the programming process, each word is addressed in sequence, and a high-current pulse is applied to store the desired pattern in the 8 bits simultaneously. In practice, the complete bit pattern, called a "program," is loaded into a "PROM programmer." The microcomputer-controlled programmer runs through the addressing-pulsing sequence 32 times and verifies the stored program by checking against the loaded program—all in less than 2 minutes. To erase the program, the chip under the transparent quartz lid is exposed to high-intensity ultraviolet light for 15 to 20 minutes.

INFORMATION PROCESSING

In the typical digital system there are several sources of information and several destinations. Providing a transfer path for every possible combination of source and destination is usually not feasible.

Routing and Transfer

The practical approach is to provide a means for selecting the desired source and transferring its information to a common path or *bus* and providing a means for connecting the desired destination to the bus (from *omnibus*: a conveyance for all.) This approach (Fig. 8.28) is called "multiplexing." A multiplexer or *data selector*

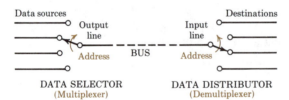

Figure 8.28 A bus-organized information transfer system.

selects the desired source and places its information on the bus; a demultiplexer or *data distributor* transfers information on the bus to the selected destination.

Decoder/Demultiplexer/Data Distributor. The decoder of Fig. 8.22c translates the ROM address and places one of four 16-bit words on the ROM output lines. In general, decoders convert binary information from one coded form to another. The unit that controls the numerals displayed by a digital clock is a "BCD-to-seven-segment" decoder; decimal numbers coded in binary form are converted to provide readout of the corresponding seven-segment decimal figure. Also available are "BCD-to-decimal" decoders that drive one of ten indicator lamps in accordance with the BCD input address.

As shown in Fig. 8.29, the same unit can serve as a decoder or as a data distributor, depending on how the terminals are interpreted. When enabled by **E** going **LOW**, the decoder places a **0** on the **OUTPUT** line corresponding to the **INPUT** code; all other output lines remain **HIGH**. In Fig. 8.29c, **DATA** from the bus is applied to the **E** terminal of the demultiplexer and appears on the **DESTINATION** line selected by the **ADDRESS**.

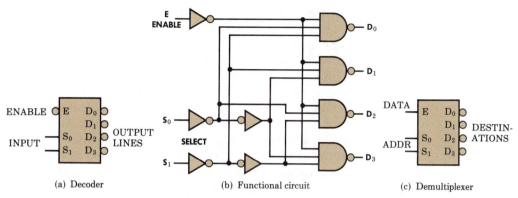

(a) Decoder   (b) Functional circuit   (c) Demultiplexer

Figure 8.29   A 2 × 4 line decoder/demultiplexer.

A data distributor has one input line and many output lines. The TTL 75154 is a 4 × 16 line decoder/demultiplexer. In the 24-pin package, there are four address lines, sixteen output lines, an enable input, a data input, and two power supply leads. It can decode a 4-bit address to send one of the sixteen output lines **LOW**, or it can distribute input data to the addressed output line.

Practice Problem 8-8

In Fig. 8.29b, if the inputs are $E = 0$, $S_0 = 0$, and $S_1 = 1$, what are the outputs? If E is connected to a data bus and $S_0 S_1$ specifies a particular device to receive the data, what function is performed by this unit?

Answers: $D_3 D_2 D_1 D_0 = $ **1011**; data distributor.

Multiplexer/Data Selector.   A multiplexer has many input lines and one output line. The multiplexer of Fig. 8.30 has four possible information sources; it chooses the one source whose address is on the select terminals. The information on the chosen

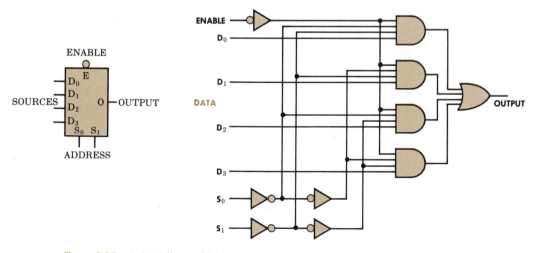

Figure 8.30   A 4 × 1 line multiplexer.

input line is transferred to the output line when the device is enabled. This is a "data selector" in that it selects a data source and transfers its data to the output. The TTL 74150 selects one of sixteen data sources; the 74151 selects one of eight. The TTL 74153 is a dual 4 × 1 line multiplexer that can route 2-bit data from one of four sources to a 2-bit bus.

The basic logic circuit design approach outlined earlier proceeds from a function statement to a truth table and then to a logic expression that can be manipulated into the simplest form using Boolean algebra or Karnaugh mapping. As another approach, a ROM can replace a network of logic gates if the address input is identified as a set of binary variables and the stored words as the corresponding output function. In complex problems, the great saving in design time may justify the high cost of a ROM. In many problems of moderate complexity, the best approach is to save design time by using a data selector as a decision-making circuit. As shown in Example 15, logic circuit design is reduced to simply generating the truth table by placing **1**s and **0**s on the data selector inputs as required. Because the inputs can be easily changed, the data selector becomes, in effect, an EPROM.

EXAMPLE 15

Design a vote-taker by using a 74151 1-of-8 data selector (Fig. 8.31).

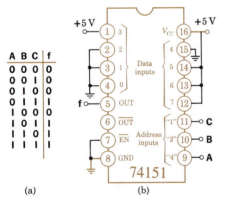

A	B	C	f
0	0	0	0
0	0	1	0
0	1	0	0
0	1	1	1
1	0	0	0
1	0	1	1
1	1	0	1
1	1	1	1

(a)   (b)

Figure 8.31 Data selector logic design.

The 74151 selects one of eight inputs and transfers the data on the selected input (or its complement) to the output. For normal operation, the enable pin must be **LOW**.

To synthesize a logic function, variables **A**, **B**, and **C** are applied to the address inputs. The data inputs, corresponding to rows of the truth table, are connected **LOW** or **HIGH** as required by the truth table of the vote-taker. For **ABC = 011** → 3D, Data input 3 is connected to logic **1** → 5 V, etc. With **ENABLE** held **LOW**, an input of **011** on the address lines will result in f = 1; action will be taken in accord with the majority vote.

Note: A 1-of-8 data selector can generate any logic function of up to four variables by using the "folding" technique described on p. 141 of the Lancaster reference at the end of Chapter 7.

Computer Organization

An electronic digital computer is a vast assemblage of simple logic gates and memory devices organized to perform complex calculations at high speed by automatically processing information in the form of electrical pulses. The information consists of *data* and *instructions*; a complete set of instructions is called a *program*.

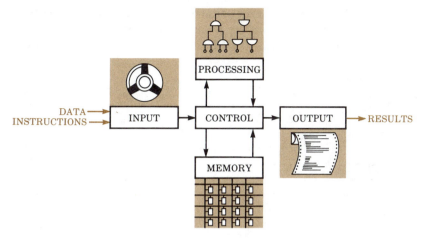

Figure 8.32 Functional components of a general-purpose digital computer.

The functional organization of a computer is illustrated in Fig. 8.32. Although the designs of individual computer components vary widely, all general-purpose computers follow the same general plan.

Input

Instructions and data are entered into the machine through an input device that reads data on punched cards, magnetic tape, or magnetic disks and converts it to electrical waveforms. All communication from the operator to the machine is through the input device. Most commonly this is a magnetic tape reader or a disk drive.

Memory

Each instruction or datum is "stored" in the computer memory until it is "retrieved" for use. Flip-flops are used for temporary storage of data and instructions in active registers. The working computer memory usually consists of an array of ICs capable of storing millions of bits of information.

Control

The flow of instructions, data, and results is governed by the control unit, which sets the various logic gates, feeds the numerical data, and provides the clock pulses that regulate the speed of all operations. Instructions are received from the input device and stored in one part of the memory; data are stored in another part. In accordance with the program, the control unit provides for a desired sequence of arithmetic operations and retrieves the numerical data from memory.

The operation of the control unit is illustrated in the elementary example of Fig. 8.33. Here stored numbers **A** and **B** are to be operated on in accordance with instructions in the form of a 4-bit word. The instruction is drawn from memory and

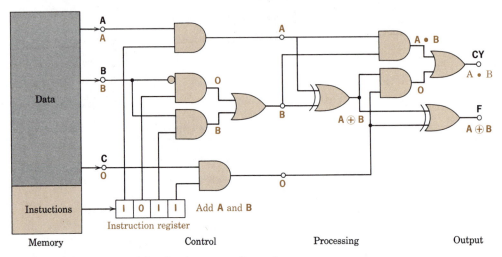

Figure 8.33   Control of data flow by program instructions.

placed in a register associated with a logic circuit of five gates. The instruction **1011**, as shown, sets the logic gates so that **A**, **B**, and **0** are available for processing. Other instructions set the logic gates for different operations. (See Exercises 43 and 44.)

Processing

The processing or *arithmetic* unit consists of logic gates arranged to perform addition, complementing, and incrementing, and the associated registers for temporary storage of data or results. One basic operation in computation is the addition of two binary numbers. In adding a column, the first number is taken from memory and placed in a register, the second number is added, a third is then added to the sum of the first two, and so on. In *parallel addition*, there is a separate adder for each bit and all orders are added in one operation. In *serial* addition, numbers are added one bit at a time by a single full-adder; the numbers and the sum are stored in shift registers. In Fig. 8.33, the processing logic is instructed to "Add **A** and **B**" and output the **SUM** and **CARRY**.

Output

The computer presents its results to the operator in display, print, or graphical form by means of an output device. Cathode-ray tubes are used for temporary display of results or plots. Modern line printers can convert electrical signals to typed characters at the rate of 20 lines per second. Electrostatic printing permits even higher speeds. Results on magnetic tape or disks are readily stored for reuse. Plotters are available for graphical presentation of calculations where this is desirable.

SUMMARY

- Conversions from binary to decimal and vice versa follow simple rules.
  Binary arithmetic follows the same rules as decimal arithmetic.
  Two's complement notation simplifies arithmetic with negative numbers.

- Boolean algebra is used to describe and to manipulate binary relations.
  Boolean theorems are summarized in Tables 8-2, -3, and -4.
  DeMorgan's theorems are particularly helpful in logic circuit synthesis.

- To analyze a logic circuit, trace through the circuit from input to output, noting the function obtained at each step.
  To synthesize a logic circuit, use Boolean algebra or Karnaugh mapping to simplify the function, and realize it with standard elements.

- Useful components and their logic functions include:

  $$\textbf{Exclusive–OR gate:} \quad \mathbf{f = (A + B)\overline{AB}} \quad (\text{or } \mathbf{f = A\overline{B} + \overline{A}B})$$

  $$\text{Half-adder:} \quad \mathbf{S = (A + B)\overline{AB}} \quad \mathbf{C = AB}$$

  $$\text{Equality comparator:} \quad \mathbf{f = AB + \overline{A}\,\overline{B}}$$

- A register is an array of flip-flops that can store binary numbers.
  In general, data may be entered and extracted in serial or parallel form.
  Practical IC registers may provide **CLEAR**, **PRESET**, and **ENABLE** capability.

- An array of flip-flops may be connected to function as a digital counter.
  Synchronous counters are faster but more expensive than ripple counters.
  Practical IC counters provide versatile operation at low cost.

- In Read-And-Write Memory (RAM), the computer can store data at any location and retrieve it at any subsequent time; k lines can address 2^k words.
  RAM units consists of a matrix of memory cells and an address decoder.
  MOS devices offer high density at low power; bipolar devices are faster.

- In Read-Only Memory (ROM), data are permanently stored as cell states.
  ROMs can act as addressable memory, or code converter, or decision logic.
  PROMs are field-programmable ROMs; EPROMs are erasable PROMs.

- In a bus-organized system, a multiplexer selects a source and puts its data on the bus; a demultiplexer transfers the data to a selected destination.

- A digital computer is a complex array of logic and memory elements organized to perform binary calculations at high speed.
  Computers include input, memory, control, processing, and output sections.

REVIEW QUESTIONS

1. Why is the binary system useful in electronic data processing?
2. How can decimal numbers be converted to binary? Binary to decimal? To octal?
3. Explain "2's complement" notation. What are its advantages?
4. What is a "binary-coded decimal"? What is decimal 476 as a BCD in the 8421 code?
5. What is Boolean algebra and why is it useful in digital computers?
6. Which of the Boolean theorems in **0** and **1** contradict ordinary algebra?
7. Explain the association and distribution rules for Boolean algebra.
8. State DeMorgan's theorems in words. When are they useful?
9. How can a logic circuit be analyzed to obtain a logic statement?
10. How can a logic statement be synthesized to create a logic circuit?
11. What function is performed by a half-adder? An **Exclusive–OR** gate?
12. Explain how binary numbers are added in a parallel adder.
13. Explain how we can design a **NAND** circuit from a truth table.
14. Explain the principle behind the numbering of columns in a Karnaugh map.
15. How can a shift register serve as a serial-to-parallel converter?
16. What is meant by a "bus-organized" system?
17. Explain the operation of a binary ripple counter.
18. Distinguish between RAM and ROM and PROM.
19. How many address lines are required to address a 1-kilobyte memory?
20. Explain the **WRITE** operation in a MOS RAM.
21. Explain the **READ** operation in a bipolar ROM.
22. Differentiate between a multiplexer and a decoder.

EXERCISES

1. (a) Write in binary the following decimals: 4, 12, 23, 31.
   (b) Write in decimal the following binaries: **00011, 00101, 01010, 10101**.
2. (a) Write in binary the following decimals: 10, 14, 21, 39, 75.
   (b) Write in decimal the following binaries: **00111, 01110, 11010, 101011**.
   (c) Convert decimals 0.875 and 0.65 to binaries.
3. Given four numbers: (a) 165_{10}, (b) **11001101B**, (c) **207Q**, and (d) **D8$_{16}$**. Convert each to equivalent numbers in three different notations.
4. Repeat Exercise 3 for: (a) 85_{10}, (b) **11100101B**, (c) **62Q**, and (d) **6F$_{16}$**.
5. Repeat Exercise 3 for: (a) 103_{10}, (b) **01101110B**, (c) **207Q**, and (d) **A5$_{16}$**.
6. (a) Convert the number **161Q** to binary, decimal, hex, and BCD.
   (b) Convert the number **5F$_{16}$** to binary, decimal, octal, and BCD.
7. Convert the BCD **10010011** to decimal, binary, and octal.
8. (a) Using signed 2's complement notation, express as 8-bit words the decimal numbers 25, 121, -17, and -96.
   (b) Show in binary notation the arithmetic operations: $121 - 25$, $25 + (-17)$, and $25 - 96$.
   (c) Write a brief explanation of how subtraction is accomplished using 2's complement notation.
9. (a) Using signed 2's complement notation, express as 8-bit words the decimal numbers 37, 92, -30, and -48.
   (b) Show in binary notation the arithmetic operations: $92 - 37$, $37 + (-30)$, and $37 - 48$.
   (c) Write a brief explanation of how subtraction is accomplished using 2's complement notation.
10. In general, what is the 2's complement of the 2's complement of binary number **A**?
11. Use truth tables to prove the following Boolean theorems:
    (a) **A + 1 = 1**
    (b) **A + B = B + A**
    (c) **AB + AC = A(B + C)**
    (d) **A(BC) = (AB)C**
12. Use Boolean theorems to prove the following identities:
    (a) **A + A̅B = A + B**
    (b) **ABC + ABC̅ = AB**

(c) $A(\overline{A} + B) = AB$

(d) $(A + B)(\overline{A} + C) = AC + \overline{A}B$

(e) $(A + C)(A + D)(B + C)(B + D) = AB + CD$

13. When writing equations for "Programmed Array Logic" circuits, complicated expressions must be broken down into simple "sums-of-products" (like Eq. 8-6).

(a) Write the following expression as a sum-of-products. Show the Boolean theorems used during each step of simplification.

$$[(A \cdot B \cdot \overline{C}) \cdot (A \cdot \overline{B} \cdot C) + \overline{A}\,\overline{B}] \cdot \overline{D}$$

(b) Invert the result from part (a), and factor it into a sum-of-products, showing theorems used.

14. The function $f = A + B$ is to be realized using only **NAND** gates. Use DeMorgan's theorems to express f in terms of $\overline{C \cdot D}$ where C and D can be expressed in terms of A and B. Draw the necessary logic circuit and check by constructing the truth table.

15. Analyze the logic circuit of Fig. 8.34 and determine f in terms of A and B. Simplify using Boolean algebra and check your result with a truth table.

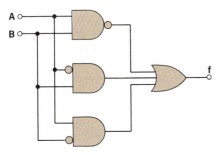

Figure 8.34

16. (a) What single gate is equivalent to the circuit of Fig. 8.35a? Check your answer using Boolean algebra.

(b) Repeat part (a) for the circuit of Fig. 8.35b.

(c) What theorem is illustrated in parts (a) and (b)?

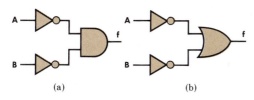

(a)   (b)

Figure 8.35

17. The 8212 I/O (input/output) port consists of 8 data latches with noninverting three-state buffers as shown in Fig. 8.36. When this port (Device) is to be used (Selected) in the output Mode, the microprocessor places 8 bits of data on the **DI** lines and sends device select signal **DS** = 1 and control signals **M** = 1 and **S** (Strobe) = 1. Assuming **S** = 1 at all times:

(a) Construct a truth table showing **M**, **DS**, **CK**, and **E**.

(b) Explain the operation of this "output latch" when **M** = 1.

(c) Explain the operation of this "gated buffer" when **M** = 0.

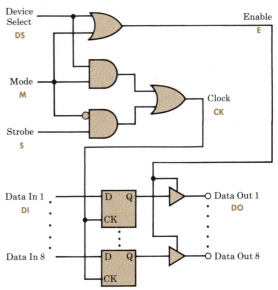

Figure 8.36 I/O port.

18. To provide a comparison of the most common logic operations, construct a table with inputs **A** and **B** and outputs **AND**, **NAND**, **OR**, **NOR**, **Exclusive–OR**, and **Equality Comparator**.

19. Synthesize logic circuits to realize the following functions as written:

(a) $\overline{\overline{A + B} + A \cdot B + \overline{A + B}}$

(b) $\overline{(A \cdot B + A \cdot B) \cdot \overline{A + B}}$

(c) $\overline{(A + B) + \overline{AB} \cdot \overline{AB}}$

(d) $\overline{A + B \cdot \overline{AB}} + AB$

20. Given the logic function

$$f = \overline{AB} + \overline{A}\,\overline{B} + \overline{A}B$$

(a) Assuming the complements are available, simplify the function using DeMorgan's theorem and synthesize it using the basic gates.

(b) Assuming the complements are not available, simplify the function and synthesize it from five **NAND** gates.

21. Using DeMorgan's theorem, convert the vote-taker expression (Eq. 8-6) to a **NAND** expression and synthesize it using **NAND** gates only.

22. Map the following functions and find the minimal sum-of-products form:
   (a) $ABC + AB\overline{C} + A\overline{B}\,\overline{C}$
   (b) $A\overline{B} + ABC + \overline{A}\,\overline{B}$
   (c) $(A + \overline{C})(\overline{B} + \overline{C})$
   (d) $ABC + BC + A\overline{B}C$

23. Evaluate the combination of the four corner squares of a four-variable Karnaugh map.

24. Map the following functions and find the minimal sum-of-products form:
   (a) $ABC\overline{D} + A\overline{B}C + \overline{B}\,\overline{C}$
   (b) $AB + \overline{A}\,\overline{B}CD + A\overline{B}C$
   (c) $\overline{A}(C + D) + A\overline{B}C + A\overline{B}\,\overline{C}\,\overline{D}$
   (d) $AB\,\overline{C}\,\overline{D} + \overline{A}\,\overline{B}\,\overline{C}\,\overline{D} + \overline{B}\,\overline{C}D + \overline{B}C$

25. Decimal numbers coded in binary are to drive the display of Fig. 8.37. (*Note*: The anodes of all LED segments are connected to +5 V; a segment is lit when its cathode is taken **LOW**, that is, logic **0**.) You are to design a "BCD-to-seven segment" code converter.
   (a) Draw up a truth table showing decimal numbers, their binary codes, and the state of each segment.
   (b) How many decoder input and output lines are required?
   (c) Segment *b* should be lit for certain input numbers. Write the logic expression for segment *b* in terms of input variables **DCBA** (**A** is LSB).
   (d) Simplify the logic expression and synthesize it using basic logic gates.

Figure 8.37 Seven-segment decimal display.

26. The array of delay flip-flops in Fig. 8.38 serves as a three-bit register.
   (a) For **LOAD HIGH**, what are the D values?
   (b) For **LOAD LOW**, what are the D values?

(c) Describe the operation of this register.
(d) For the input waveforms shown, draw the waveform of Q_A, the first bit of output.

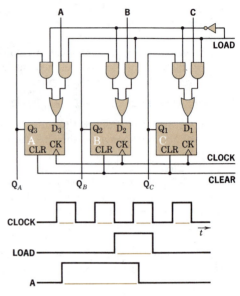

Figure 8.38   Three-bit register.

27. A 4-bit "serial-to-parallel converter" receives data one bit at a time (LSB first) and then transmits the four bits simultaneously.
   (a) Using leading-edge-triggered D flip-flops with active-low clear, design the converter; that is, draw a labeled connection diagram using standard flip-flop symbols.
   (b) Assuming that converter outputs Q_A (LSB), Q_B, Q_C, and Q_D are initially **0**, show the output waveforms on the timing diagram of Fig. 8.39.

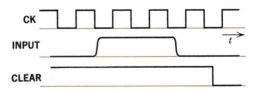

Figure 8.39

28. Design (i.e., draw a schematic diagram of) a binary counter that can count up to decimal 100. Include a provision to reset the counter to zero after 100 is reached.

29. The "ring counter" of Fig. 8.40 is used in a computer to activate (**ENABLE**) several devices in turn.
  (a) For input waveforms **X** and **CK** as shown, draw the output waveforms.
  (b) In a truth table, show the output "data word" $Q_D Q_C Q_B Q_A$ after each downward transition of **CK**. Describe the operation of this counter.

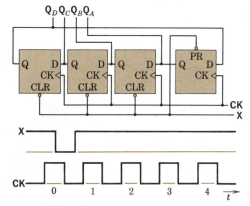

$Q_D Q_C Q_B Q_A$

Figure 8.40 Ring counter.

30. (a) Define the operation of a leading-edge-triggered D (delay) flip-flop.
  (b) Construct a truth table for a synchronous divide-by-3 counter using two D flip-flops. Show "state," Q_1, Q_2, D_1, and D_2.
  (c) Design the counter circuit so that Q_2 goes **HIGH** once in every 3 states.

31. Design an 8421 BCD ripple counter as follows:
  (a) Draw a block diagram of a 4-stage binary ripple counter using T flip-flops *ABCD* (**T** held **HIGH**).
  (b) Show the truth table of flip-flop outputs for a decade counter that counts normally to decimal 9 and then resets to **0000**. How does it differ from the truth table for a binary counter in the tenth row?
  (c) Modify the circuit to accomplish the following: On the eighth count, the change in state of the *D* (MSB) flip-flop is to disable the input to the *B* flip-flop so it will *not* change on count 10. (See part (d).)
  (d) Modify the circuit so that on the tenth count the *D* flip-flop will be reset to **0** by the output of the *A* flip-flop, without affecting the use of Q_C to toggle *D*.
  (e) Check the operation of your decade counter by drawing **CK** and flip-flop waveforms.

32. In a certain application, you need a circuit that will start and stop counting clock pulses on command.
  (a) Start by designing a 3-bit *synchronous* counter using positive edge-triggered JK flip-flops. Label the bits Q_A, Q_B, and Q_C, with Q_C the most significant bit.
  (b) Now modify the circuit using **AND** gates so that it counts whenever the **COUNT ENABLE** lead is **HIGH**, stops counting when **CTE** goes **LOW**, and resumes counting from where it stopped when **CTE** goes **HIGH** again. Label bits as before.

33. Refer to the divide-by-12 counter of Fig. 8.21. Draw the waveforms of **INPUT** (clock), $Q_D Q_C Q_B Q_A$, **CARRY**, **LOAD**, and **OUTPUT** for one complete cycle of the counter. (See typical waveforms in Fig. 8.20.)

34. Using 74163 counters, design a counter that will count up to 100 and then reset to zero.
  (a) What is the counting range of one 74163? Two?
  (b) Describe in words how your counter is to work.
  (c) Draw a schematic diagram of your counter.

35. Assuming that each byte in RAM has a unique address:
  (a) How many bytes can be specified by an address of 8 bits? Of 16 bits? Of *n* bits?
  (b) How many memory cells are required in the RAMs of part (a)?

36. An array of eight memory cells is arranged in two rows and four columns. You are to design the addressing system consisting of a row decoder and a column decoder. Assume a row or column is selected when driven **HIGH** (logic **1**).
  (a) Rows **0** and **1** are selected by row-select (RS) values **0** and **1** and columns 0 to 3 are selected similarly. Compose the truth tables for **RS** and for CS_1 and CS_2 (column select).
  (b) Show the eight memory cells (lettered *A* through *D* and *E* through *H*). Design the decoders using **AND** and **NOT** gates only and show the decoder circuits on your diagram.
  (c) Specify the address values that will select cell *F*.
  (d) How many row-select and column-select lines would be required to address 1024 cells arranged in 16 rows and 64 columns?

37. The BCD-to-seven segment code conversion to drive the LED display in Exercise 25 is to be performed by a ROM.

(a) Draw up a truth table showing decimal numbers, their binary codes, and the state of each segment.

(b) Describe how a ROM could be used to perform the decoding function; specify the number of address and data lines. Draw a conclusion regarding the relative difficulty of ROM vs logic gate design.

38. Show how a ROM could be used to realize:
(a) The vote-taker of Fig. 8.8.
(b) The multiplexer of Fig. 8.30.
(c) The decoder of Fig. 8.29.

39. Design (draw the circuit of) an 8-bit parallel-to-serial data converter using a data selector and a clocked binary counter.

40. Using a 74151 data selector:
(a) Design the sum circuit of the 1-bit full-adder of Fig. 8.7.
(b) Design a simulated 3-variable **Exclusive–OR** gate.

41. (a) Analyze the behavior of the circuit of Fig. 8.41 and draw up a truth table showing the output for values of the control lines **AB**.
(b) Describe in words the function performed by this circuit.
(c) Where in a computer would such a device be useful?

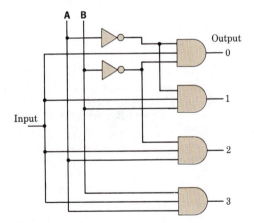

Figure 8.41

42. Four-bit registers A and B and a data bus are interconnected as shown in Fig. 8.42.
(a) What operation is accomplished when $S_1 S_2$ is **01**?

(b) What is the name and function of this computer component?
(c) Where in a computer would such a component be used? For what purpose?

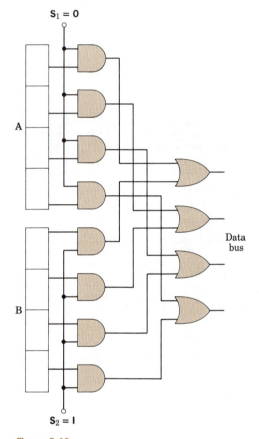

Figure 8.42

43. Refer to Fig. 8.33 and interpret the logic (ignore the **CY**) instruction:
(a) **1110**.
(b) **1010**.

44. Refer to Fig. 8.33 and interpret the arithmetic instruction: **0101** with **C = 1**.
(a) Write expressions for **F** and **CY**.
(b) Form the truth table for **B**, $\overline{\textbf{B}}$, **F**, **CY**, and the 1's and 2's complements of **B**.
(c) Interpret the instruction as an arithmetic operation.

PROBLEMS

1. An electric light is to be controlled by three switches. The light is to be **ON** whenever switches A and B are in the same position; when A and B are in different positions, the light is to be controlled by switch C.
   (a) Draw up a truth table for this situation.
   (b) Represent the light function **f** in terms of **A**, **B**, and **C**.
   (c) Simplify the function and design a practical switching circuit.

2. One way of checking for errors in data processing is to "code" each word by attaching an extra bit so that the number of **1**s in each "coded" word is *even*, that is, the words all have *even parity*. Assuming that the probability of *two* errors in one word is *negligible*, the coded word can be checked for even parity at various points in the process. Assuming **Exclusive–OR** gates are available, design a *parity checker* for coded 4-bit words such that the output is **0** for even parity.

3. In a pocket calculator, each character of the 15-digit display consists of 7 light-emitting diodes arranged as shown in Fig. 8.37. For example, decimal "**1**" is displayed when segments *c* and *d* are lit.
   (a) Determine how each decimal from 0 to 9 could be represented and draw the corresponding segment display.
   (b) Draw up the truth table showing the state of each segment for the decimal numbers. (Assume "lit" equals logic **0**.)
   (c) If a four-stage binary counter similar to Fig. 8.17 is used to drive the display, segment *b* should be lit for certain combinations of outputs from flip-flops A through D. Write and simplify the logic expression governing segment *b*.

4. The "comparator," widely used in computers, has an output of **0** unless the inputs are identical. Synthesize, using 4-input **NAND** gates only, a circuit that "compares" the 2-digit binary numbers A_1A_2 and B_1B_2.

5. Devise a "half-subtractor."
   (a) Construct the truth table providing **A**, **B**, **Borrow**, and **Difference** columns.
   (b) Synthesize the circuit using only the five basic gates.
   (c) Synthesize the circuit assuming **Exclusive–OR** gates are available.

6. Analyze Fig. 8.33 with instruction **1101** with **C = 1**.
   (a) Determine the outputs and form a truth table.
   (b) Form a truth table for **A** plus **B** and **A** minus **B**.
   (c) What arithmetic operation is performed?

7. In the RAM memory cell shown in Fig. 8.43, $V_{CE(\text{sat})} = 0.3$ V and $V_{BE} \cong 0.7$ V. The word line can be switched from **LOW** (0.3 V) to **HIGH** (4 V). Currents in the data bit lines can be sensed. Assume initially that T_1 is **ON**.
   (a) With the word line **LOW**, what is voltage V_{P1}? V_{P2}? Is T_2 **ON** or **OFF**?
   (b) What is current I_D?
   (c) If now the word line is switched to **HIGH**, what is I_D? What memory function has been performed? What is the effect on T_1 of then returning the word line to **LOW**?
   (d) Now the word line and bit line D are simultaneously raised to **HIGH**. What is the effect on T_1? On T_2?
   (e) After returning word and bit lines to **LOW**, what memory function has been performed?

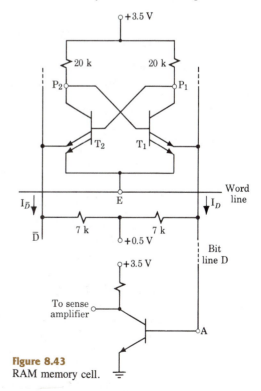

Figure 8.43
RAM memory cell.

Microprocessors

A Four-Bit Processor

A Stored-Program Computer

Microprocessor Programming

Practical Microprocessors

A Minimum Computer System

In 1969 Ted Hoff, an Intel Corporation engineer, was handed a difficult problem in LSI design. A Japanese calculator company wanted Intel to produce chips for an extremely complex logic design. Using his background in high-density memories, Hoff came up with a bold new approach—put a tiny central processing unit (CPU) on one chip, store a program of instructions in memory on another, and use a third to move data in and out of the CPU. The resulting crude computer turned out to be too slow for its intended application, and it was rejected; but the idea was revolutionary, and it has changed the world in significant ways.

The age of electronic computers began in 1946 with ENIAC (Electronic Numerical Integrator And Computer), a 30-ton, 140-kW, 18,000-vacuum tube monster that could add two 12-digit numbers in 200 μs. In the 1950s, vacuum tubes were replaced by transistors which, in turn, were replaced by integrated circuits. The first microprocessor chip, a fourth-generation computer component about the size of three letters on this page and costing only a few dollars, just about matched the computing power of ENIAC. Its invention may well be the greatest technological development of our time.

The microprocessor is more than just a new device with applications in such diverse areas as automobile ignition and pollution control, traffic lights, microwave ovens, cash registers, entertainment electronics, medical diagnosis, and "smart" instruments. Its computing power, small size, and low cost have completely changed our approach to the design of all kinds of machines used in work and play, at home and in business, in the factory and in the laboratory. In particular, it has changed our philosophy of design in electronics; instead of creating an elaborate network of components to solve a data processing problem, we take a cheap, mass-produced "brain" with a flexible "memory" and "tell" it how to solve our problem.

In this brief introduction to a complex subject, we begin with an elementary processor and learn how it operates and how it can be programmed. We gradually increase its complexity by adding capability and flexibility a little at a time until we are using a computer that is functionally equivalent to a commercial unit. With this background, we study a popular microprocessor and learn how to build and program a basic microcomputer.

A FOUR-BIT PROCESSOR

Let us start with a very simple data processor and see how it is organized to execute instructions. Later we shall add complexity in order to obtain more versatility.

An Elementary Processor

Some of the essential elements of a processor are shown in Fig. 9.1. Instructions in the form of 4-bit words placed in the *Instruction Register* cause the *Controller* to perform operations on the *Accumulator* register in response to signals received from the *Timer*. The Controller is a set of logic gates that guides the flow of binary data in response to the combination of bits in the Instruction Register (IR). (See Fig. 8.33.) The Accumulator is at the heart of the processor; arithmetic and logic operations are performed by the Controller on the contents of the Accumulator. Binary data represented by the *Input* switches can be transferred, on command, to the Accumulator; and the contents of the Accumulator can be transferred, on command, to the *Output* light display.

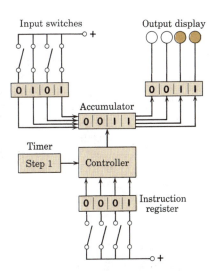

Figure 9.1 An elementary processor (after Cohn and Melsa).

The operation of the processor is illustrated in Fig. 9.2. Initially (Step 0), the Accumulator contains the binary word remaining after the last previous operation (say, **1111**), all lights in the display are **OFF** (**0000**), and the input switches are set for **0101**, for example. Note that each register always contains some 4-bit word; a register can never be "empty." The Timer is initially set at Step 0, and we assume that at each step of the Timer the Controller will execute the instruction in the IR. The first instruction is a 4-bit word placed in the IR; in this case, $2^4 = 16$ different instructions are possible by setting the four instruction switches. Let us define the instruction **0100** (in binary

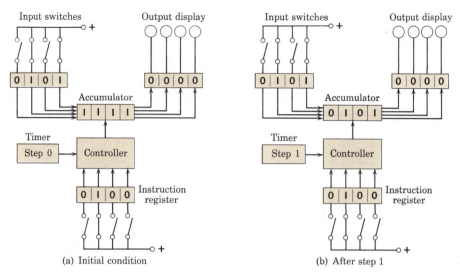

Figure 9.2 Operation of an elementary processor.

code) as **INPUT TO A**, meaning: "The data at the Input is moved to register A (the Accumulator)." At the first signal from the Timer (Step 1), the Controller "decodes" the instruction and enables the input to the A register; the Input word **0101** is transferred to the Accumulator (Fig. 9.2b).

To display the Input word at the Output, a second step is required. Let us define the instruction **0101** as **OUTPUT FROM A**, meaning: "The content of register A is moved to the Output." In Fig. 9.3a, instruction **0101** has been placed in the IR by resetting the instruction switches. At the next signal from the Timer (Step 2), the Controller decodes the instruction to enable the output of the A register, and the word **0101** is displayed at the Output (Fig. 9.3b). Note that the content of the Accumulator

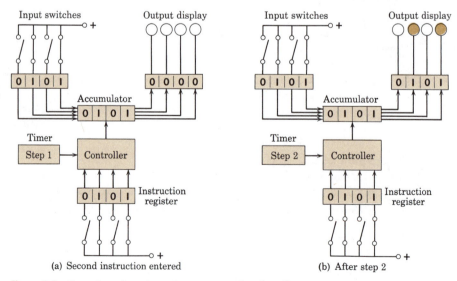

Figure 9.3 Operation of an elementary processor (continued).

is unaffected by executing this instruction; this is a "nondestructive readout" of information. Note also that the same binary word (**0101**) represents "data" in the Accumulator and an "instruction" in the IR.

In addition to transferring data in and out, the Controller is capable of manipulating the content of the Accumulator. One useful instruction (designated **0001** here) is **COMPLEMENT A**, meaning: "The content of register A is complemented; each zero bit becomes **1**, and each **1** becomes **0**." This instruction is useful in performing arithmetic operations.

Another useful instruction (designated **0010**) is **ROTATE A LEFT**, meaning: "Each bit in the Accumulator is shifted left one position; the low-order bit is set to the value shifted out of the high-order bit position." Similarly, the instruction **0011** can be defined as **ROTATE A RIGHT**, meaning: "Each bit in the Accumulator is shifted right one position; the high-order bit is set to the value shifted out of the low-order bit position." These instructions cause the Accumulator to act as a shift register with an end-around loop.

A Simple Program

In any microcomputer, the Controller must be told, in great detail, just what to do; in other words, it must be *programmed*. The programmer decides which tasks are to be performed in what order and defines the sequence in a written program. The program is a detailed list of specific instructions, selected from a fairly small repertoire of allowable instructions, each corresponding to an elementary step in the overall operation.

Suppose, for example, that: "The binary number at the Input is to be multiplied by two and the result displayed at the Output." Since in binary numbers a shift of one position left is equivalent to a multiplication by two, the programmer breaks the desired operation down into the following sequence and adds some explanatory "comment."

Step	CODE	INSTRUCTION	Comment
1	0100	**INPUT TO A**	Move data at Input to Accumulator.
2	0010	**ROTATE A LEFT**	Multiply content of A by two.
3	0101	**OUTPUT FROM A**	Move result in A to Output.

The execution of this program is illustrated in Fig. 9.4 where Input and Output are represented by registers. Here it is assumed that the Input number (limited to **0111** in this example) is **0101**, the initial content of the Accumulator is **xxxx**, and the initial setting of the Output is **yyyy** (irrelevant). The first instruction is placed in the IR and then, during Step 1, the Input number is moved to the Accumulator for manipulation. The second instruction is placed in the IR and then, during Step 2, the content of A is shifted left one position. The third instruction is placed in the IR and then, during Step 3, the Input number multiplied by two is transferred to the Output. Although this program is very simple, the concepts involved are powerful and, when expanded and extrapolated, are capable of sophisticated information processing.

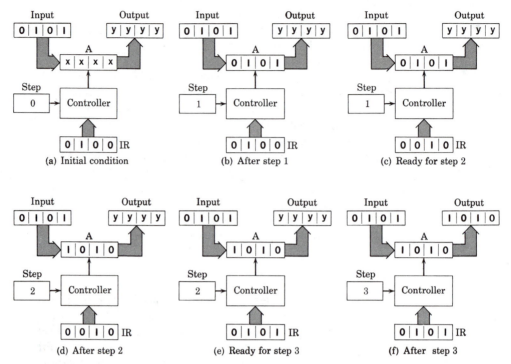

Figure 9.4   Step-by-step execution of a simple program.

A More Versatile Processor

In order to solve more challenging problems, we now increase the capability of our elementary processor by four significant additions.

Multiple I/O Ports. In general a processor can receive information or data from more than one input device and send results to more than one output device. Each device is connected to a separate "port" of the processor identified by number. Henceforth, our input/output (I/O) transfer instructions must include a second 4-bit word specifying the particular port involved. For example, let us define the two-word instruction **0100–0100** as **INPUT TO A FROM PORT 4**. Similarly, to transfer data from the Accumulator to Output Port 7, we will use **OUTPUT FROM A TO PORT 7** (**0101–0111**). Henceforth, our Controller will know that following the "operation code" ordering an I/O transfer, the next word identifies (in binary code) the specific port.

A Bidirectional Data Bus. To facilitate transfer of Accumulator data to and from multiple ports, we add a "bidirectional data bus," a common path (four lines in parallel) for the transmission of 4-bit words in either direction. The use of three-state buffers permits the physical connection of many devices to the common bus. (See Fig. 7.15.)

Some General-Purpose Registers. As shown in Fig. 9.5, our more versatile (but still simplified) processor incorporates a data bus permitting information flow between

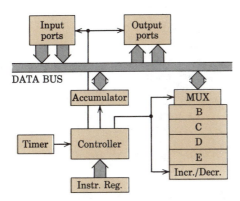

Figure 9.5   A more versatile processor.

the Accumulator and I/O ports and also between the Accumulator and four general purpose RAM registers via a multiplexer (MUX). Registers B, C, D, and E are convenient locations for the storage of data used in computation or control. The content of each register can be incremented or decremented. Unlike the A register, however, these registers are not capable of data manipulation; for example, it is not possible to complement the content of the B register directly. Furthermore, only the A register can communicate with the I/O ports.

More Instructions. To take advantage of the additional registers, we expand the capability of the Controller to respond to instructions placed in the IR. The Instruction Set of Table 9-1 has been expanded to include three new types of instruction.

The **INCREMENT/DECREMENT** instructions are useful in counting; for example, we can use register B (designated by the code number **0000**) as a counter that will be incremented after each in a series of external events or internal program steps. The two-word instruction would be **0110–0000**.

The **LOAD** instructions permit us to transfer information between the Accumulator and one of the general-purpose registers. For example, to transfer the content of the Accumulator to register D for temporary storage, we use the two-word instruction **LOAD D FROM A**, coded as **1010–0010**. The second word of such a **LOAD** instruction is the code designation of the register specified as source or destination.

The **LOAD DATA** (sometimes called "move immediate") instructions allow us to load data directly into a register. For example, in counting we can start by loading the desired number of events (say, **0101**) into a register (say, B) and then count down to zero by decrementing the register. To accomplish the operation **LOAD B** with **0101**, we need an operation code, a register specification, and the data; the necessary three-word instruction is coded as **1101–0000–0101**.

The logic instructions **AND**, **OR**, and **XOR** permit bit-by-bit operations on the content of the Accumulator and a register. In effect, four simultaneous logic operations are performed with one instruction; the results are stored in the Accumulator, and the other register is unaffected. One simple application is in selecting the low-

Table 9-1   Instruction Set for a 4-Bit Processor

CODE	INSTRUCTION	Meaning
0000	HALT	The processor is stopped; registers are unaffected.
0001	COMPLEMENT A	The content of A is complemented bit-by-bit.
0010	ROTATE A LEFT	Each bit in A is shifted left one position; low-order bit takes the value shifted out of high-order position.
0011	ROTATE A RIGHT	Each bit in A is shifted right one position; high-order bit takes the value shifted out of low-order position.
0100 pppp	INPUT TO A FROM p	The data at the specified port p is moved to the Accumulator.
0101 pppp	OUTPUT FROM A TO p	The content of A is moved to the specified port p without changing the content of A.
0110 rrrr	INCREMENT r	The content of register r is incremented by one. [Register codes: A(0111), B(0000), C(0001), D(0010), E(0011)]
0111 rrrr	DECREMENT r	The content of register r is decremented by one.
1000 rrrr	ADD TO A, r	Add to the content of A the content of register r without changing the content of r.
1001 rrrr	LOAD A FROM r	Transfer to the Accumulator the content of r without changing the content of r.
1010 rrrr	LOAD r FROM A	Transfer to register r the content of A without changing the content of A.
1011 rrrr	LOAD r FROM MEM	Transfer to r the content of the memory location whose address is in register E without changing MEM.
1100 rrrr	LOAD MEM FROM r	Transfer content of r to the memory location whose address is in register E without changing content of r.
1101 rrrr dddd	LOAD r WITH data	Load into register r the data d specified in the third word.
1110 LLRR	LOGIC OP A WITH R	Perform logic operation LL on content of A with content of register RR without changing content of RR. [Logic codes: OR(00), XOR(01), AND(11)] [Register codes: B(00), C(01), D(10), E(11)]
1111 cccc addr	JUMP IF cc TO addr	If condition specified in second word is met, jump to program instruction whose address is given in third word; if condition not met, continue sequentially. [Condition codes: UNcond(0000), Not Zero(1000), Zero(1001), No Carry(1010), Carry(1011), Positive(1110), Negative(1111)]

order bit, say, from a 4-bit number. A **0001** "mask" is moved into a register and **AND**ed with **A**; the result in A contains only the low-order bit. The instruction would be **1110–1110** to **AND** (**LL = 11**) the content of **D(RR = 10)** with **A**.

Practice Problem 9-1

Assuming register C contains **0110** and the Accumulator contains **1010**, predict their contents after the instruction **1110-0101** is executed.

Answer: 0110, 1100.

To provide some computation capability, the arithmetic **ADD** operation is included. By using this instruction, the content of a specified register is added to the content of the Accumulator, and the result is placed in A. As we have seen (p. 204), subtraction can be accomplished by **ADD**ing the 2's complement of the subtrahend. (*Note*: Table 9-1 contains other instructions that will be used after we learn how programs are stored in memory.)

EXAMPLE 1

Show the instructions necessary to subtract decimal 5 = **0101B** from the content of A.

CODE	INSTRUCTION	Comment
1010 0001	**LOAD** **C FROM A**	Temporarily store content of Accumulator in register C.
1101 0111 0101	**LOAD** **A WITH** **0101**	Move immediately decimal number 5 into Accumulator.
0001 0110 0111	**COMPLEMENT A** **INCREMENT** **A**	Form the 2's complement of decimal 5 in the Accumulator.
1000 0001	**ADD TO A,** **C**	Add initial content of Accumulator to 2's complement of subtrahend.

A STORED-PROGRAM COMPUTER

The processor becomes truly useful when we provide it with a memory into which we can write the program. The processor fetches the instructions from memory, decodes them to arrange the processing circuitry, executes the instructions, and then makes the results available at output ports. Using modern LSI technology, the instruction decoder, controller, accumulator, auxiliary registers, timing circuits, and I/O buffers can be fabricated on a single chip and mass produced at low cost. We obtain a customized digital system when we provide the mass-produced processor with a tailor-made program stored in memory.

Program Storage

A memory is a set of locations where binary information can be stored. For example, a 64-bit memory could be organized into sixteen 4-bit words with each word location identified by a 4-bit address. The programmer stores the program instructions at successive locations in memory; this simplifies addressing because the next address is just the present address incremented by one.

In Fig. 9.6, a 16-word memory is shown with 4-bit address and data buses. The program used in Example 1 has been written into locations 0–9. If this were a field-

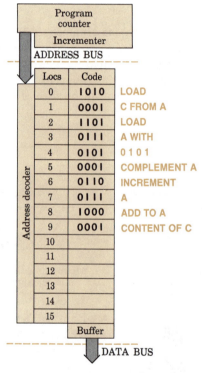

Locs	Code	
0	1010	LOAD
1	0001	C FROM A
2	1101	LOAD
3	0111	A WITH
4	0101	0 1 0 1
5	0001	COMPLEMENT A
6	0110	INCREMENT
7	0111	A
8	1000	ADD TO A
9	0001	CONTENT OF C
10		
11		
12		
13		
14		
15		

Figure 9.6 A program stored in ROM.

programmable memory (see PROM, p. 228) fabricated with all **0**s, the programmer would address each location and electrically change the **0**s to **1**s where appropriate.

The Program Counter

The processor includes a register that contains the address of the next instruction. The processor updates this Program Counter (PC) by adding **1** to the counter content each time it fetches an instruction. We say that the PC always "points" to the next instruction. To execute the given program, the PC would be set to **0** and then incremented through nine addresses.

Timing

The activities of the processor are cyclical: fetching an instruction, transferring data, performing an operation, writing the result, fetching the next instruction, and so on. To execute an instruction in a few microseconds requires precise timing governed by a clock. In general, an interval called a "state" (equal to several clock periods) is necessary for the completion of a specific processing activity, and there are several states in a "machine cycle." The time required to fetch and completely execute an instruction is called an "instruction cycle" and includes one or more machine cycles. The typical processor can transmit only one address during a machine cycle. The first machine cycle is always an instruction fetch; if execution requires another reference to memory or to an addressable I/O device, a second machine cycle is required.

EXAMPLE 2

Assuming the initial content of A is **1111**, show the content of the IR, A, C, and PC registers for the first two instruction cycles of the program in Example 1.

Locs	CODE	Instruction
0	1010	LOAD C
1	0001	FROM A
2	1101	LOAD
3	0111	A WITH
4	0101	0101

Instruction Cycle	IR	A	C	PC
	xxxx	1111	yyyy	0000
1	1010	1111	yyyy	0001
	0001	1111	1111	0010
	1101	1111	1111	0011
2	0111	1111	1111	0100
	0101	0101	1111	0101

Practice Problem 9-2

Continue Example 2 for the next three instruction cycles and predict the final content of the IR, A, C, and PC registers.

Answer: 0001, 1010, 1111, 1010.

A Basic Microcomputer

A stored-program computer is shown in Fig. 9.7. If the CPU components within the dashed line are fabricated on a single IC chip, this is a *microprocessor* (μP). The combination of microprocessor, IC memory devices, and I/O ports constitutes a *microcomputer* (μC).

In this more sophisticated system, a Program Counter, Incrementer, Address Bus, and external Memory are provided. Data and address buses are terminated in pins. Timing is performed by a Clock, and Timing and Control circuitry is separated from the Arithmetic/Logic Unit (ALU).

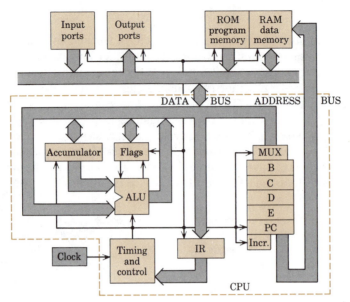

Figure 9.7 A basic microcomputer.

The ALU

In accordance with Control signals, the ALU performs arithmetic and logical operations on binary data. All ALUs contain an Adder that will combine the content of the Accumulator with a word on the internal data bus and place the result back on the data bus. The ALU also contains hardware that will efficiently perform incrementing, left or right bit shifts, and logic operations such as **AND**, **OR**, and **XOR**.

The ability of the ALU to "raise flags" signaling certain key conditions that occur in data processing greatly increases the versatility of the μP. A "conditional instruction" specifies an operation to be performed "only if" a certain condition has been satisfied; for example, if an addition results in a carry, the μP is diverted to a sequence of instructions to handle that contingency. Conditional instructions are extremely valuable in a stored-program computer because they provide decision-making capability.

In Fig. 9.7, the Flag register is set by the ALU in accordance with the result of the last instruction. Most μPs provide Zero, Carry, and Sign flags that indicate whether the result of the last instruction is zero (or not zero), generates a carry (or no carry), or is positive (or negative). For example, if the last instruction is an **ADD** and a carry results, "the Carry flag is set" meaning that a **1** is placed in the Carry bit of the flag register. During a conditional instruction, the status of the specified flag is tested to determine if the "branch" in the program is to be followed.

JUMP Instructions

So far we have considered only *sequential* programs that run through a series of instructions in order. **JUMP** instructions allow us to add great variety by altering the sequence in which stored instructions are executed. Because the Program Counter tells the μP where to get the next instruction, a **JUMP** instruction must change the content of the PC register. As shown in the Instruction Set of Table 9-1, the three-word **JUMP IF cc TO addr** instruction means: "If the condition specified in the second word is met, control is passed to the program instruction whose address is given in the third word." In effect, the binary word **addr** is forced into the PC. Unconditional jumps are useful in "looping" when we wish the μP to execute a part of the program over and over. Conditional jumps provide "intelligence"—the ability to make decisions and implement different programs depending on circumstances.

Practice Problem 9-3

Write a program segment to: Add the number in register C to the number in register B; if there is a carry, get the next instruction at location **1010** in program memory.

Answer: 1001, 0000, 1000, 0001, 1111, 1011, 1010.

Data Memory

The μC of Fig. 9.7 can address either the ROM Program Memory or a RAM Data Memory. The RAM memory can function like a series of registers into which data can be written and from which data can be retrieved when needed. Our instruction set provides a flexible memory addressing scheme in which the address of the location to be accessed is stored internally in register E. To load the content of the Accumulator into memory location 6, the address is first placed in E by using the instruction **LOAD E WITH 0110**. Then the content of A is stored by using **LOAD MEM FROM A**. The **LOAD MEM** operation code (**1100**) transfers data from the specified register (A, here) to the memory location (**0110**) whose *address* is stored in E. This "indirect addressing" scheme is particularly convenient where we need to address the same memory location repeatedly or where we wish to address locations sequentially. To store or retrieve data words sequentially, the E register is simply incremented.

EXAMPLE 3

Write a program segment, starting at location **0000**, to do the following: Decrement and test the content of register B. If the content of B = **0**, store the content of C in data memory location **0110**; if not, decrement B again and repeat.

LOCS	CODE	INSTRUCTION	Comment
0000	1101	LOAD	Load into register E the
0001	0011	E WITH	specified memory
0010	0110	0110	location.
0011	0111	DECREMENT	Count by decrementing
0100	0000	B	register B.
0101	1111	JUMP	If decrement of B sets Zero
0110	1001	IF ZERO	flag, jump to "store C"
0111	1011	TO 1011	instruction.
1000	1111	JUMP	Loop back and decrement
1001	0000	UNCOND	again.
1010	0011	TO 0011	
1011	1100	LOAD MEM	Store content of C in
1100	0001	FROM C	memory at location
			whose address is in E.

Input/Output Operations

In performing its information processing function, the μP must "talk" to many devices: keyboards, sensors, disks, cathode-ray tubes, printers, and control mechanisms. Between each device and the digital system there must be an "interface" unit that transcribes parallel binary data to and from the "language" used by the device. In Fig. 9.8, we assume that each I/O device provides its own interface so that the I/O ports process binary data that is compatible with the μP signals.

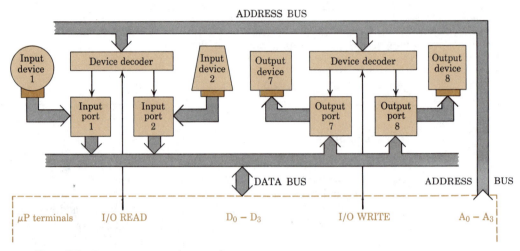

Figure 9.8 Input/output control connections.

A μP using the architecture shown in Fig. 9.8 would execute an **INPUT TO A FROM p** instruction by placing the port address **pppp** on the address bus and sending an **I/O READ** signal to the input *device decoder*. The 4×16 decoder ($2^4 = 16$) generates a *device select* signal to **ENABLE** the one addressed input port to place its data on the data bus. The μP addresses an I/O port and reads the content of the port buffer just as it addresses a memory location and reads its content.

MICROPROCESSOR PROGRAMMING

The ALU is capable of performing a very few operations at very high speed. Our Instruction Set indicates that the ALU in combination with various registers and data processors can be controlled by binary operation codes to perform 16 unique instructions. Following these instructions, our μP can transfer data inside the CPU, perform logical and arithmetic manipulations, make decisions based on manipulation results, and move data into and out of the μP.

The μP cannot, by itself, perform simple multiplication or figure percentage. It must be told in unambiguous terms, in complete detail, just what to do and in just what order. The physical components, connections, and logic circuits that decode and execute the operation codes constitute the computer *hardware*. The lists of specific instructions, selected from those allowed by the μP manufacturer and organized to control computer operations, are *software*.

Programming Technique

Programming is the art of directing the processor hardware to perform the elementary steps in solving a computation or control problem in an effective way. In writing a program (in general), the programmer:

- Defines the problem to be solved.
- Determines a feasible solution plan.
- Draws up a flow chart identifying the sequence of operations.
- Writes the program instruction by instruction.
- Checks out the program to see that it solves the problem.

Flow Charts

A flow chart is a block diagram representing the structure of the program. The essential symbols are rectangles (with one input and one output), used for processes, and diamonds (with two or more outputs), used at decision points. Lines connecting the blocks indicate the flow of information. The flow chart is a valuable aid in writing a program and explaining it to someone else. Flow charts visually break large, complex programs into smaller, logical units and make it easier to make changes and corrections. With the program structure clearly defined in English, the flow chart can be converted directly into a set of instructions, using any programming language.

Programming technique can be illustrated using Example 3. In general terms, a delay is to be introduced into a μP-controlled operation.

Define the Problem. The programmer converts this to: "N instruction cycles after a certain event, the content of register C is to be stored in a specified memory location."

Determine a Solution Plan. One possibility is: "Use register B as a counter and assume an appropriate count has been loaded into B. Load the specified address (**0110**) into register E. Decrement and test register B. If the content of B = **0**, store the content of C in data memory location **0110**; if not, decrement B again and repeat the test."

Draw a Flow Chart. The solution plan is displayed graphically in Fig. 9.9. The process rectangles contain brief English expressions for the operations performed; the decision diamond contains the question: "Does Counter B = **0**?" The two outputs are labeled "Yes" and "No" for convenience. The other labels outside the blocks are also for convenience in identifying points in the program.

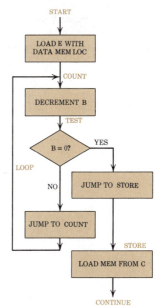

Figure 9.9 Flow chart for Example 3.

Write the Program. This can be done directly from the flow chart; the results should agree with Example 3. The decision alternatives, the repeating loop, and the **JUMP TO STORE** command, all seen clearly in the flow chart, are more obscure in the written program. The more elaborate flow chart of Example 4 includes two methods of counting and two types of jump commands. The next step would be to write the program.

EXAMPLE 4

A processor is to read in 3 integers from input port 4, store them in external RAM locations starting at **1000**, add the 3 integers, and send their sum to output port 5. Outline a solution plan and draw a flow chart. (Disregard overflows and memory limitations.)

Solution Plan

Use register D to store sums.
Preset counter B to $16 - 3 = 13$ and count up until a carry appears.
Clear counter C, count up and back down to zero.
Use register E as a memory pointer.

Bring in integers one at a time and store them in sequential RAM locations.
Do 3 integers, counting with register B.
Fetch integers from memory one at a time and add them.
Do three integers, counting with register C.
Output the sum.

Flow Chart

The solution plan leads to the flow chart of Fig. 9.10.

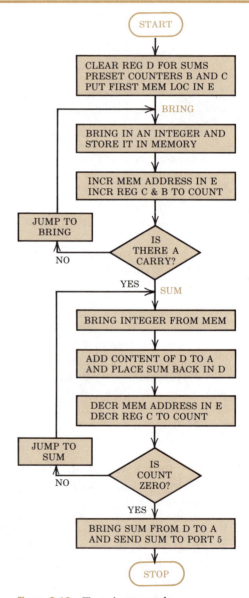

Figure 9.10 Flow chart example.

Check the Program. The smallest error may make a program completely ineffective. The identification and elimination of all errors is called "debugging." Working through the flow chart and checking the written program instruction by instruction are essential steps in debugging, but the only true check is an error-free execution by the processor.

After the program has been written and debugged, it must be physically entered into ROM program memory by electrically setting **1**s and **0**s into the ROM registers. This could be done manually for Example 3 by entering the first 4-bit location (**0000**) into the address lines of a PROM and then introducing high currents in the Code pattern (**1101**). The second location would then be addressed and the bit pattern **0011** entered, and so on until 13 memory locations were filled. The program of instructions understandable to humans would then reside in ROM in a form understandable to the machine, that is, in *machine language*.

Program Development

Looking ahead to the use of real μPs with instruction sets of a hundred or so 8-bit instructions in practical applications employing programs requiring a thousand or so memory locations, we see some disadvantages of manual programming in machine language. We are going to need help in storing an error-free program in a reasonable time and, fortunately, it is available. The principal programming aids are:

Text Editor. A text editor is a computer program that allows the programmer to make changes, correct errors, and make substitutions easily in preparing, and displaying in human readable form, an error-free program.

Assembler. A symbolic assembler is a program that automatically translates instructions in symbolic form (**DECR B**) into machine language (**0111–0000**); it also keeps track of memory locations for the programmer.

System Monitor. The system monitor is a computer program that facilitates the overall μP programming process by performing intricate supervisory operations in response to simple instructions.

Development System. Text Editor, Assembler, and Monitor programs are controlled by a small computer called a Development System (Fig. 9.11). With a teletypewriter (TTY) keyboard for entering·instructions and commands and a cathode-ray tube (CRT) for displaying them, the development of programs for new μP-based products is simplified.

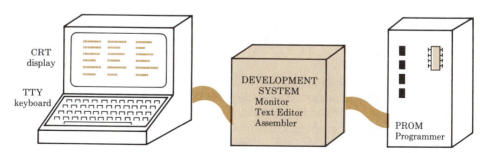

Figure 9.11 A microprocessor development system with peripherals.

PROM Programmer. This is a computer-controlled development tool that will read the final "software" program and automatically load it into a PROM.

Programming Languages

The set of instructions that directs any computer can be written in *machine language*, in *assembly language*, or in a *higher-level language*.

In machine language, which we have been using up to now, the "words" are sets of binary digits (CODE) that can be stored in memory or moved in and out of registers along buses. Binary instruction words control sets of switches according to the architecture of the computer. In a given μP, all communication must be in the machine language fixed by the manufacturer in the fabrication process.

High-level languages such as ALGOL (algebraic language) or FORTRAN (formula translation language) simplify programming sophisticated data processing because a single symbol such as $\sqrt{}$ can represent a complex sequence of machine steps. Such high-level languages are translated into machine language by a *compiler*, a sophisticated computer program itself. High-level languages make programming easy, but they are inefficient in their use of program memory that is usually "costly" in a μC.

An assembly language uses short symbolic phrases that are understandable to people to represent specific binary instructions that are understood by the machine. The phrases, made up of alphanumeric symbols, are called *mnemonics* (memory aids). Because it is derived from English, assembly language simplifies the programmer's job, reduces the number of errors, and makes errors easier to find. Assembly language programs are efficient in their use of memory and, because they require fewer instructions, they execute faster than those written in high-level language. Assembly language is translated into machine language by an Assembler, essentially a computer-based dictionary. For a given μP, any user can devise his or her own set of mnemonics and the necessary Assembler.

Assembly Language Programming

The translation dictionary embodied in an Assembler is similar to the Code-Instruction columns of our Instruction Set for a 4-Bit Processor, but our instructions are much too long for efficient handling by a computer. If we are willing to use brief but unambiguous abbreviations and a strictly prescribed arrangement of symbols (syntax), we can greatly simplify assembling the program. For example, if we reserve "**IN**" for input operations, and if we agree to follow "**IN A**" by a comma and then the code number of the port, the instruction **INPUT TO A FROM port 2** becomes "**IN A,2.**" The role of the Assembler is indicated in Fig. 9.12.

The Assembler program includes a fixed dictionary of mnemonics and their machine language translations into binary code (Table 9-2). It also can "build" an address dictionary wherein **LABEL**s assigned by the programmer to certain instructions are identified by their addresses in memory. For example, in a **JUMP** instruction, the address of the next instruction must be specified. In the flow chart of Fig. 9.9, there is a jump to the address of the **DECREMENT B** instruction, but that address in memory is not known until after the program is assembled. However, if that instruction carries

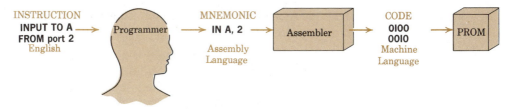

Figure 9.12   The role of the Assembler in programming.

the **LABEL** "**COUNT**," the programmer can refer to it by **LABEL** and the Assembler will straighten it out after assembly is complete. (Labels contain no more than five characters.)

In assembly language syntax, each instruction occupies a single line. First comes the **LABEL** (if needed), then the **MNEMONIC** (always present), then any Comment the programmer finds useful in checking the program or explaining it to someone else. Sometimes the entity (data, complement, address) on which the instruction operates is identified separately as the **OPERAND** (2 in the **IN A,2** instruction); here we shall include operation and operand in the **MNEMONIC**.

The Assembler is well trained; for example, if the first entry in an instruction line is a character (in ASCII code), it is identified as the first character of a **LABEL**. If the first entry is a space, the Assembler assumes there is no **LABEL** and treats the first character as the beginning of a **MNEMONIC**. The beginning of a Comment is signaled by a special character (a semicolon here). In creating the machine language program, the Assembler completely ignores the Comment, although the Comment is carried in the display and will be printed in any printout of the assembled program.

Table 9-2   Instruction Mnemonics

CODE	MNEMONIC	Instruction	CODE	MNEMONIC	Instruction
0000	HALT	Stop processing	1010 rrrr	LOAD r,A	Load r from A
0001	COMPL A	Complement A			
0010	ROTL A	Rotate A left	1011 rrrr	LOAD r,M	Load r from MEM
0011	ROTR A	Rotate A right	1100 rrrr	LOAD M,r	Load MEM from r
0100 pppp	IN A,p	Input to A from p			
			1101 rrrr dddd	LDIM r,d	Load r with data
0101 pppp	OUT p,A	Output to p from A			
0110 rrrr	INCR r	Increment r	1110 00RR	OR R,A	OR R with A
0111 rrrr	DECR r	Decrement r	1110 01RR	XOR R,A	XOR R with A
1000 rrrr	ADD A, r	Add to A, r	1110 11RR	AND R,A	AND R with A
1001 rrrr	LOAD A, r	Load A from r	1111 cccc addr	JP cc,addr	JUMP if cc to addr

To provide the Assembler with needed information that it cannot deduce, special *assembler directives* are entered as mnemonics:

ORG addr—The program is to originate at address **addr** in program memory. (This information is needed to locate the labels.)

END   —There are no more executable instructions forthcoming. (This tells the Assembler its task is completed.)

EXAMPLE 5

Assuming the program is to be stored in memory starting at location **0000**, show the program of Example 3 in assembly language syntax as it would be displayed after assembly.

LOCS	CODE	LABEL	MNEMONICS	Comment
			ORG 0000	;Start program at loc **0000**
0000	1101, 0011, 0110		LDIM E,0110	;Load E with spec mem loc **0110**
0011	0111, 0000	COUNT	DECR B	;Count by decrementing reg B
0101	1111, 1001, 1011		JP Z,STORE	;If decrement of B sets Zero flag, ;jump to store C instruction
1000	1111, 0000, 0011		JP UN,COUNT	;Loop back and decrement again
1011	1100, 0001	STORE	LOAD M,C	;Store C in mem at specified addr
			END	;No more instructions

In Example 5, note the following:

- Each instruction occupies a single line. (Comments may run over.)
- Only the first location for a multiword instruction is shown.
- The Assembler will "define" **COUNT** as **0011** and **STORE** as **1011**. If a new instruction is inserted between **JP Z,STORE** and **JP UN,COUNT**, the Assembler will automatically redefine **STORE**.
- Ordinarily no Comment is required for **ORG** and **END** directives.

Practice Problem 9-4

The program of Example 5 is modified by first storing the number 5 in register B. Determine the new Assembler definitions of **COUNT** and **STORE**.

Answer: 0110 and **1110**.

PRACTICAL MICROPROCESSORS

The basic function of a μP is to perform arithmetic, logic, and control operations on data obtained from input devices; the results are made available to various output devices, all in accordance with a stored program of instructions. The 4-bit processor that we have studied in detail can perform all the basic functions, but it is severely limited in the size of data words handled (4-bit) and the number of instructions (16).

Practical Limitations

Ideally, a μP of negligible size and cost should accept unlimited amounts of data in any form, respond to programs of any length composed of instructions of infinite variety, deal with any number of peripheral devices, and perform all operations instantaneously without consumption of power. Although available μPs approximate these ideal characteristics, there are practical limitations within which μP manufacturers must work.

A major constraint on the design of LSI components is the limited number of terminals available for external connection. Because most industrial testers will not accept components with more than 40 (or 42) pins, common μPs are limited to 40 pins. The number of internal elements, now 10 to 20,000 transistors on a single chip, continues to climb, but the total number of terminals for all purposes is fixed. With a minimum of 2 pins for power supply and 2 for clocking, say, only 36 pins are available for data, address, and control buses (unless a multiplexing arrangement is used). The most common result of this limitation is a design providing 8 bidirectional data lines, 16 address lines, and 12 control lines (Fig. 9.13). A device with 8 data lines is said to have a "word size" of 8 bits or one byte. (See Example 6.)

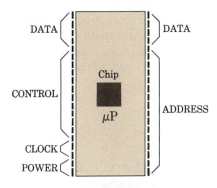

Figure 9.13   A 40-pin DIP package.

Because instructions are carried over the data bus, the instruction code of an 8-bit processor would allow a maximum of $2^8 = 256$ different instructions. For practical reasons, all bit combinations are not convenient, and the number of unique instructions is usually only one-third to one-half of the total possible.

EXAMPLE 6

(a) The Intel 8080 is an "8-bit microprocessor with 16 address lines." What is the range of signed integers that can be processed?

(b) How many unique locations can be addressed?

(a) In signed 2's complement notation, 8-bit positive integers range from **00000000** to **01111111** or from 0 to 127D and negative integers range from **10000000** to **11111111** or from -128 to -1D. Therefore, integers from -128 to $+127$D can be processed.

(b) Because $2^{16} = 65{,}536$, the number of locations possible is 65,536.

The complexity of the operations performed by a μP depends on the density of circuit elements because this limits the number of internal registers and the extent of "invisible" internal microprograms that control ALU operations. In comparison to bipolar devices, MOS technology provides higher density but lower speed. The newer NMOS technology used in the Intel 8085, Motorola 6800, and Zilog Z80, for example, is faster than p-channel MOS because of the greater mobility of the electrons employed in conduction. Clock rates of 5 MHz permit typical instruction cycles of less than 1 μs.

A complete digital system including processor (μP), program memory (ROM), data memory (RAM), and input/output ports (I/O) is called a *microcomputer* (μC). In the most common computer configuration, the CPU is on one chip and the other components are on one or more additional chips. The great virtue of microprocessors is that mass-produced, low-cost hardware can be used to solve diverse control problems through individualized programs.

It is now possible to put all essential elements (except quartz crystal and power supply) on a single LSI chip—the Intel 8048, for example. This is truly "a computer on a chip," and it is an amazingly versatile computer considering its small size and low cost. Furthermore, since the program is internal, the lines ordinarily reserved for addressing are available for external communication. On the other hand, the chip area devoted to memory and I/O is unavailable for computation, and the ROM portion must be individually masked during fabrication. As a result of these limitations, the principal application of one-chip microcomputers is in simple control devices (requiring small memories) that are produced in very large numbers (justifying the high mask costs).

The Z80

Microprocessor technology is changing rapidly and the devices described here are likely to become obsolete in a short time. However, at the time of its introduction (1976), the Zilog Z80 CPU was an 8-bit microprocessor of advanced design and it continues to provide great versatility and high speed in an elegantly simple configuration. N-channel depletion-mode MOS technology (Fig. 6.5) provides high element density and permits operation at 5 V with a 4-MHz clock at speeds up to eight million instructions per second. CMOS versions now provide 4-MHz operation with significantly lower power consumption.

Clock signal and power supply use 3 pins, and the data bus and 16 address lines use another 24 pins, leaving 13 pins of the 40-pin DIP (Dual In-line Package) for a sophisticated set of control signals. The instruction set includes 158 distinct instructions that provide flexibility and computing power rivaling some fairly large (and expensive) minicomputers. The combination of simplicity and versatility makes the Z80 attractive for our purpose.

Registers. The architecture is similar to that of the basic CPU shown in Fig. 9.7. As shown in Fig. 9.14, there is an 8-bit Accumulator (A) that holds the result of arithmetic and logic operations and an 8-bit Flag register (F) that indicates specific operating conditions. The six general purpose registers B, C, D, E, H, and L are

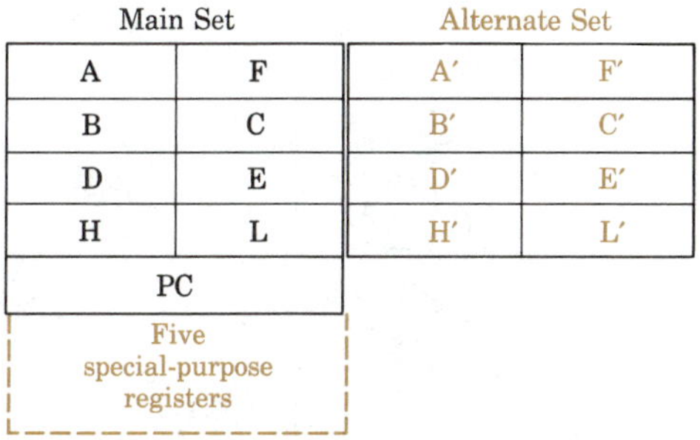

Main Set		Alternate Set	
A	F	A′	F′
B	C	B′	C′
D	E	D′	E′
H	L	H′	L′
PC			

Five
special-purpose
registers

Figure 9.14 Register set of the Z80.

arranged so that they can be used as individual 8-bit registers or as 16-bit register pairs BC, DE, and HL. The HL pair can be used to store the high (H) and low (L) bytes of a 16-bit address in RAM memory for a transfer-to-or-from-memory instruction. The Program Counter (PC) is a 16-bit register that contains the address of the next instruction. The other five special-purpose registers provide increased programming and operating convenience; their use requires more background than is justified in this introduction.

Alternate Registers. The alternate set of Accumulator, Flag, and general-purpose registers is particularly useful in systems where an I/O device can *interrupt* the program when it is ready to be serviced. For example, an output device such as a line printer may be slow in comparison to the μP controlling it. An efficient way to handle such a device is to supply it with one byte of information and then let the μP continue with other control or computation activities as defined by the program. When the device has finished printing a character, say, it sends an *interrupt* signal to the μP. At the next convenient time, the μP shifts operation to the alternate set of registers, goes through the appropriate *service routine* to supply to the printer the data for the next character, and then returns to its main program. Without this duplicate set, the contents of operating registers would have to be saved by proper storage in RAM so that the main program could continue where it left off at the interruption. Because the time-consuming "save" operation is avoided, the Z80 is known for its fast response to interrupts.

ALU. Arithmetic and logic instructions executed in the ALU include:

ADD	**COMPARE**	**INCREMENT**
SUBTRACT	**SHIFT LEFT**	**DECREMENT**
AND	**SHIFT RIGHT**	**SET BIT**
OR	**ROTATE LEFT**	**RESET BIT**
XOR	**ROTATE RIGHT**	**TEST BIT**

The **SUBTRACT, SHIFT,** and **ROTATE** operations are particularly useful in integer multiplication and division. The bit manipulations are useful in setting flags or external control conditions.

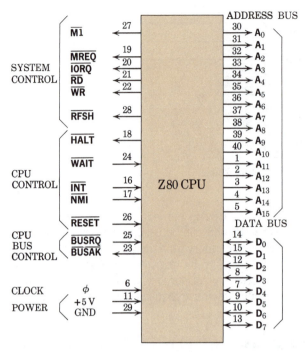

Figure 9.15 Z80 pin designation (Zilog, Inc.).

Pin Description. The Z80 "pinout" is shown in Fig. 9.15. Arrowheads differentiate among input, output, and bidirectional signals. A bar over a label indicates that the pin is active low. Included are the following:

Address Bus (A_0-A_{15} [MSB]) is a 16-bit, three-state bus. I/O addressing uses the 8 lower bits to select up to 256 input or 256 output ports.

Data Bus (D_0-D_7) is an 8-bit bidirectional, three-state bus.

Machine Cycle 1 ($\overline{M1}$) indicates that the CPU is in the op code fetch cycle.

Memory Request ($\overline{MREQ}$) indicates that the address bus holds a valid address for a memory read or memory write operation.

I/O Request ($\overline{IORQ}$) indicates that the address bus holds a valid address for an I/O read or write.

Read ($\overline{RD}$) indicates that the CPU wishes to read data. The device addressed uses this signal to put data on the data bus.

Write ($\overline{WR}$) indicates to memory or an I/O device that the data bus holds valid data to be stored.

Refresh ($\overline{RFSH}$) is used to refresh dynamic memories.

Wait ($\overline{WAIT}$) tells the CPU that the device addressed is not ready for a data transfer.

Halt ($\overline{HALT}$) indicates that the CPU is "stopped" (ignoring program instructions) and is awaiting an interrupt.

Interrupt Request ($\overline{INT}$) accepts the signal generated by an I/O device requiring service.

NonMaskable Interrupt (**NMI**) is a high priority interrupt to save program status in case of an "error," for example, a power failure; it forces the CPU to restart in a designated location.

Reset (**RESET**) forces the PC to zero and gets the CPU ready to start at location zero.

Bus Request (**BUSRQ**) asks the CPU to disconnect from data and address buses so that external devices can control these buses temporarily.

Bus Acknowledge (**BUSAK**) indicates that the CPU three-state data, address, and control elements are set to the high-impedance state, that is, disconnected.

EXAMPLE 7

Draw up a function table showing the Z80 output signal levels used in controlling read/write operations on memory and I/O devices.

The outputs are active low; therefore:

MREQ	IORQ	RD	WR	Function
LOW	HIGH	LOW	HIGH	Read from memory.
LOW	HIGH	HIGH	LOW	Write into memory.
HIGH	LOW	LOW	HIGH	Read from I/O.
HIGH	LOW	HIGH	LOW	Write into I/O.

Timing. In executing instructions, the Z80 steps through a series of elementary operations corresponding to *machine cycles*. Depending on its complexity, each machine cycle takes from three to six clock cycles, and complete instructions are executed in one to six machine cycles. (See Table 9-4 on p. 267.) Figure 9.16 shows how machine cycles make up a typical instruction cycle. The first machine cycle (**M1**) is always an instruction fetch during which the next instruction code is transferred to the instruction register. During T_1, **M1** goes **LOW**, the content of PC is placed on the address bus, then **MREQ** and **RD** go **LOW** to select the program memory (as in Example 7). During T_2, the memory responds by placing the instruction on the data bus, and

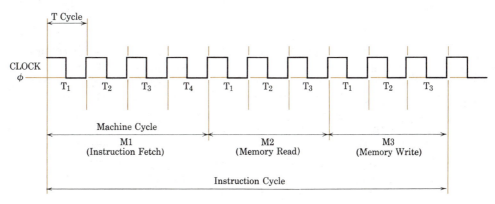

Figure 9.16 Example of Z80 timing.

then the μP loads the instruction into the IR. During T_3 and T_4, the PC is incremented, and the current instruction is decoded and executed (if no further reference to memory is required). A memory read or write generally requires three clock periods unless "wait states" are requested by the memory device with a $\overline{\text{WAIT}}$ signal. Other operations are handled in essentially the same way.

The Z80 Instruction Set

The Z80, like the Motorola 6800, evolved from the Intel 8080, a very popular 8-bit microprocessor. The Z80 can execute 158 instructions, including all 78 of the 8080 and all 16 of our 4-bit processor. For compatibility, the machine language codes for the 78 common instructions are identical in the Z80 and 8080; the mnemonics are different, however, and therefore different assemblers are required. For comparison, some examples are shown in Table 9-3.

Note that the concept of mnemonic and code developed for our simple processor is found in current microprocessors. The availability of 8 bits permits one-word instructions where two words were required with 4 bits. The **n** designates an 8-bit number. The **SSS** designates one of eight general-purpose "source" registers; a **DDD** code designates one of eight "destination" registers.

Table 9-3   Instruction Code Examples

Our 4-Bit Processor		Zilog Z80		Intel 8080	
MNEMONIC	CODE	MNEMONIC	CODE	MNEMONIC	CODE
HALT	0000	HALT	01110110	HLT	01110110
LOAD A,r	1001 rrrr	LD A,r	01111rrr	MOV A,r$_2$	01111sss
IN A,p	0100 pppp	IN A,(n)	11011011 ←n→	IN port	11011011 pppppppp

Note also that 8-bit codes are confusing and errors are easily made in working with them. For that reason, hexadecimal notation is used to represent instruction, data, and address codes. To convert a binary number to "hex," represent each 4-bit group by its hex equivalent. For example, the **HALT** code becomes **0111 0110 → 76**, and the address **0101101010001111** becomes **0101 1010 1000 1111 → 5A8F** in hex.

Table 9-4 lists the mnemonics and codes for a basic set of Z80 instructions. The full list of 158 instructions, detailed timing diagrams, and suggestions for use of the more sophisticated Z80 capabilities are included in the *Z80-CPU Technical Manual* available from the manufacturer. For our purposes the abridged instruction set is more than adequate. It will permit writing a variety of programs to illustrate practical (but limited) application of microprocessors.

Table 9-4   Abridged Z80 Instruction Set

Four-Bit Processor		Z80				
Instruction	MNEMONIC	MNEMONIC	CODE[1]	Bytes	Clock Periods	Flags[2]
Stop processor	HALT	HALT	76	1	4	
*No operation		NOP	00	1	4	
Complement A	COMPL A	CPL	2F	1	4	
Rotate A left	ROTL A	RLCA	07	1	4	$C \leftarrow$ prev MSB
Rotate A right	ROTR A	RRCA	0F	1	4	$C \leftarrow$ prev LSB
Input to A from p	IN A,p	IN A,(n)	DB n	2	10	
Output to p from A	OUT p,A	OUT (n),A	D3 n	2	11	
Increment r	INCR r	INC r	00rrr100	1	4	Z,S ↕
Decrement r	DECR r	DEC r	00rrr101	1	4	Z,S ↕
Add to A,r	ADD A,r	ADD A,r	10000rrr	1	4	C,Z,S ↕
Load A from r	LOAD A,r	LD A,r	01111rrr	1	4	
Load r from A	LOAD r,A	LD r,A	01rrr111	1	4	
*Load r from r'		LD r,r'	01rrrr'r'	1	4	
Load r from MEM	LOAD r,M	LD r,(HL)	01rrr110	1	7	
Load MEM from r	LOAD M,r	LD (HL),r	01110rrr	1	7	
Load r with data	LDIM r,d	LD r,n	00rrr110 ←n→	2	7	
*Load HL with addr		LD HL,nn	21nn	3	10	
OR R with A	OR R,A	OR r	10110rrr	1	4	$C \leftarrow 0$; Z,S ↕
XOR R with A	XOR R,A	XOR r	10101rrr	1	4	$C \leftarrow 0$; Z,S ↕
AND R with A	AND R, A	AND r	10100rrr	1	4	$C \leftarrow 0$; Z,S ↕
*Jump uncond to addr	JP UN addr	JP nn	C3nn	3	10	
Jump if cc to addr	JP cc,addr	JP cc,nn	11ccc010 nn	3	10	
*Compare A with MEM [A − (HL)]		CP (HL)	BE	1	7	Z ↕ ; C set if A < (HL)

Notes:
* New or modified instruction.
[1] In hex where convenient; otherwise binary.
[2] C = Carry; Z = Zero; S = Sign = 1 (negative) if MSB = 1.
↕ = Flag is affected by the result of the operation.
n = 8-bit number.
Numbers in parentheses are "addresses."
The Z80 uses the lower 8 bits of a 16-bit address first.

Jump Conditions				Registers	
cc	ccc	Condition		r	rrr
NZ	000	Not Zero		B	000
Z	001	Zero		C	001
NC	010	No Carry		D	010
C	011	Carry		E	011
P	110	Sign Positive		H	100
M	111	Sign Negative		L	101
				A	111

A common application of microprocessors is processing a set of data. Figure 9.17 shows the flow chart for a typical program using the HL register pair of the Z80 as a memory location pointer. The actual program as displayed by a commercial assembler is reproduced in Table 9-5 on page 269.

EXAMPLE 8

Assuming the program of Example 5 is to be stored in a 1 K (1024-byte) PROM starting at location **256D = 100000000B = 0100H**, show the PROM contents and the program in Z80 assembly language syntax. (*Note*: The Z80 uses the lower 8 bits of a 16-bit address first.) How long would the Z80 take to execute this 5-instruction segment?

The programmer writes a specific mnemonic for each instruction, clearly identifying numbers in hex. The assembler converts the mnemonics to machine language and displays the code in hex, one byte per 8-bit location. After the assembler defines **COUNT** and **STORE**, the display is as shown in black below.

LOCS	CODE	LABEL	MNEMONICS	Bytes	Periods
			ORG 0100H		
0100	21 06 00		LD HL,0006H	3	10
0103	05	COUNT	DEC B	1	4
0104	CA 0A 01		JP Z,STORE	3	10
0107	C3 03 01		JP COUNT	3	10
010A	71	STORE	LD (HL),C	1	7
			END	11	Total 41

At 4 MHz, the 41 clock periods take $41 \div 4 \times 10^6 \cong 10 \ \mu s$.

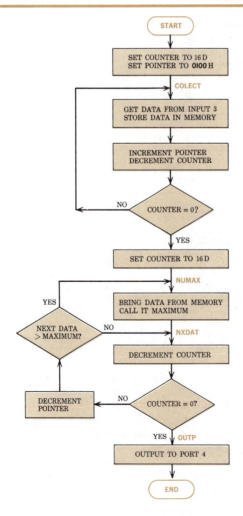

Figure 9.17 Flow chart for the program displayed in Table 9-5.

Table 9-5 Printout by a Commercial Assembler of a Z80 Program

DISPLAY LINE NO.		0001	;	CROMEMCO CDOS Z80 ASSEMBLER VERSION 02.12
		0002	;	
		0003	;	PROGRAMMING EXAMPLE
		0004	;	A COMMON APPLICATION OF MICROPROCESSORS IS PROCESSING A
		0005	;	SET OF DATA. THIS IS DONE BY COLLECTING THE DATA, STORING IT IN
		0006	;	AN "ARRAY", MANIPULATING THE DATA, AND OUTPUTTING IT. IN THE Z80
		0007	;	THIS IS FACILITATED BY USING REGISTER PAIR HL AS A MEMORY
		0008	;	LOCATION POINTER.
		0009	;	
		0010	;	PROCESSING A SET OF DATA
		0011	;	
		0012	;	WRITE A PROGRAM TO COLLECT SIXTEEN 8-BIT UNSIGNED DATA
		0013	;	NUMBERS (PRESUMED TO BE AVAILABLE AT PORT 3H), STORE THEM IN
		0014	;	MEMORY (STARTING AT LOCATION 0100H), SELECT THE LARGEST, AND
		0015	;	SEND IT TO PORT 4H.

MEM LOCS	OBJECT CODE		LABEL	MNEMONIC	COMMENT
		0016	LABEL	MNEMONIC	COMMENT
		0017		ORG 0000H	; START THE PROGRAM AT 0000H
0000	0E10	0018	BEGIN:	LD C,16D	; SET THE COUNTER TO 16 DECIMAL
0002	210001	0019		LD HL,0100H	; SET REGISTER PAIR HL TO LOC 0100H
		0020	;		
0005	DB03	0021	COLECT:	IN A,(3H)	; BRING DATA INTO ACCUMULATOR
0007	77	0022		LD (HL),A	; PUT IN MEM LOC POINTED TO BY HL
0008	2C	0023		INC L	; INCREMENT HL TO POINT TO NEXT LOC
0009	0D	0024		DEC C	; DECREMENT COUNTER.
000A	C20500	0025		JP NZ,COLECT	; CONT COLLECT IF C NOT EQUAL ZERO
		0026	;		
000D	0E10	0027		LD C,16D	; SET COUNTER TO 16 DECIMAL
000F	7E	0028	NUMAX:	LD A,(HL)	; BRING IN PROVISIONAL MAXIMUM
0010	0D	0029	NXDAT:	DEC C	; DECREMENT COUNTER
0011	CA1C00	0030		JP Z,OUTP	; JUMP TO OUTPUT COMMAND IF C ZERO
0014	2D	0031		DEC L	; DECR POINTER TO NEXT LOCATION
0015	BE	0032		CP (HL)	; IS NEXT DATA GREATER THAN MAX?
		0033			; THE CARRY FLAG IS SET IF (A) < (HL)
0016	DA0F00	0034		JP C,NUMAX	; IT IS, SO REPLACE PROVISIONAL MAX
0019	C31000	0035		JP NXDAT	; NO, KEEP LOOKING
		0036	;		
001C	D304	0037	OUTP:	OUT (4H),A	; OUTPUT THE TESTED MAXIMUM
001E	76	0038		HALT	; ALL DONE SO STOP
001F	(0000)	0039		END BEGIN	; TELL THE ASSEMBLER TO STOP

ERRORS   0

CROSS REFERENCE LISTING

BEGIN	0018	0039	
COLECT	0021	0025	
NUMAX	0028	0034	
NXDAT	0029	0035	
OUTP	0037	0030	

Notes: Items in color have been added.
Some lines are left blank to make it easier to read.
BEGIN identifies this program segment.
ERRORS refer to syntax errors.
Items in "label dictionary" are listed at left.

A MINIMUM COMPUTER SYSTEM

In addition to the CPU, a working microcomputer must include clock signal, program memory, data memory, input port, output port, and power supply. For proper operation of the control system, some logic gates may also be required. We are familiar with all the components; now let us put them together in a practical Z80 microcomputer.

Clock

The only requirement is that the clock input be a square wave swinging from 0 to 5 V (nominal) at a frequency between 500 kHz and 4 MHz. At frequencies close to the maximum, the oscillator should be controlled by a quartz crystal to provide the necessary stability.

Program Memory

For flexibility let us choose a user-programmed memory such as the 2732 EPROM. (See p. 229.) This 4096-byte (4K) erasable/programmable memory (Fig. 9.18) is well suited to digital system experimentation. During programming, the $\overline{OE}/V_{PP}$ input is pulsed from a TTL level up to 21 V. In the read mode, a single 5-V power supply is required and all inputs are at TTL levels. Applying a **HIGH** signal to the $\overline{CE}$ input puts the 2732A in the standby mode and power is reduced by 75%.

The three-state data outputs of the 2732 allow us to physically connect the 8 Data Out pins to the data bus. When $\overline{CE}$ (Chip Enable) is taken **LOW**, this device is selected; when $\overline{OE}$ (Output Enable) is taken **LOW**, the outputs come out of the high-impedance state and place on the data bus the content of the location addressed by the 12 address lines A_0–A_{11}.

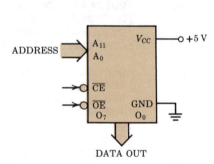

Figure 9.18 The 2732 EPROM as program memory.

Data Memory

To store data and the results of intermediate calculations we need some read-and-write memory. For simplicity we choose a *static* RAM whose contents remain stable as long as power is available over a *dynamic* RAM, which requires "refreshing" every millisecond or so but which is cheaper for large systems. The 2116 in Fig. 9.19 is an NMOS 2K (2048-word × 8-bit) static RAM that operates on a single 5-V supply and is TTL compatible.

During a **WRITE** cycle, the $\overline{OE}$ (Output Enable) signal is **HIGH** to place the Data pins in the input mode, signal $\overline{WE}$ (Write Enable) goes **LOW**, and the data is written into the location designated by A_0–A_{10} (Address Inputs). During a **READ** cycle, $\overline{OE}$ goes **LOW** to place the Data pins in the output mode, signal $\overline{WE}$ goes **HIGH**, and the data at the designated address is made available.

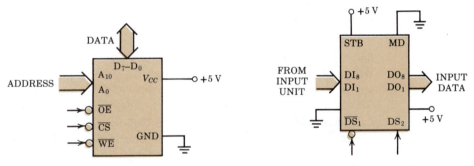

Figure 9.19   The 2116 static RAM as data memory.   **Figure 9.20**   The 8212 as an input port.

Input Port

The simplest input port is a register into which an input unit places data that, at the appropriate instant, is transferred to the data bus. The 8212 is a versatile input/output port consisting of an 8-bit data-latch register (Fig. 7.23) with three-state output buffers and control logic for operating in various modes. With the Mode terminal (MD) **LOW** and the Strobe terminal (STB) **HIGH** (Fig. 9.20), the device operates as a "gated buffer." The outputs are in the high-impedance state until the device is *selected* by taking $\overline{DS}_1$ **LOW AND** DS_2 **HIGH**. While the device is selected by the appropriate signal from the CPU, the input unit data are transferred to the data bus.

Output Latch

A simple buffer is not satisfactory as an output port because the CPU operates so fast that output data are valid for only a few microseconds. In Fig. 9.21, with **MD HIGH** (output mode), the output buffers are enabled, and the device select signal ($\overline{DS}_1 \cdot DS_2$) provides the **CLOCK** signal that latches the data bus values into the output register of the 8212. While the 8212 is "selected," **CK = DS_2** is **HIGH**, and the output **Q** of each flip-flop follows the input **DI** value. When the 8212 is "unselected," **CK** goes **LOW** and **Q** remains at the latched value. (See Exercise 17 in Chapter 8). The output unit can take its time processing or displaying the output data, which remain constant until the CPU again selects the output 8212.

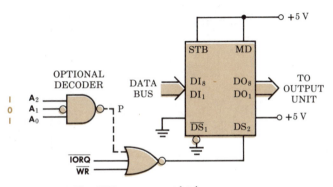

Figure 9.21   The 8212 as an output latch.

Port Selection

The Z80 control outputs can be used to "notify" the input or output port that it has been selected. When the CPU is ready to write data at an output port, $\overline{\text{IORQ}}$ and $\overline{\text{WR}}$ go **LOW**. The output of the **NOR** gate in Fig. 9.21 goes **HIGH** supplying $DS_2 = 1$ and selecting the output latch.

In a more flexible system with, say, eight output ports, port selection can be handled by **NAND** gate decoders. In the **OUT (n),A** instruction, the port address **n** is placed on the three lower-order lines of the address bus. In Fig. 9.21, only address **101** will take point P low and select this device. Each of the other seven ports would have its own unique decoder.

Power Supply

The Z80 requires a single 5-V supply capable of supplying 1.1 W at 5 V $\pm$ 5%. The 8212s and TTL logic gates also operate at 5 V and require less than 1 W each. The 2732 EPROM needs about 0.75 W in ROM operation and only 0.2 W in standby. The 2116 requires 0.2 W when active and only 0.025 W on standby. These power supply requirements are easily met.

A Practical Microcomputer

The complete microcomputer shown in Fig. 9.22 is simple in layout and connection. Only a fraction of the capability of the components is used here and only part of the terminals are directly involved. In general, output terminals can be left open,

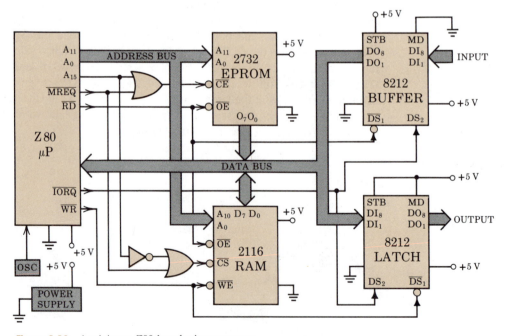

Figure 9.22 A minimum Z80-based microcomputer.

but unused input terminals should be connected high (+5 V) or low (ground) depending on their characteristics.

In accordance with a 4-kilobyte program stored in the 2732 EPROM, the Z80 CPU will input 8-bit data through the 8212 buffer, use the 2116 2K RAM to process it following customized instructions, and make it available at the 8212 output latch for display or control. With a few dollars worth of chips, a few hours of construction time, and some imaginative programming, the designer can create a fast and powerful computer.

THE ULTIMATE PERSONAL COMPUTER

The relentless progress in memory density, microprocessor speed, and user friendliness led Alan Kay of Apple Computer to this set of "specifications": Pack millions of bits of memory capacity into a clipboard-size pad, fitting in a briefcase. Fill it with millions of bytes of software instructions but leave plenty of additional memory available so that people could fill it with everything from shopping lists to stock portfolios. Add multiple ultrafast processors, so that the machine would respond instantly to even the most esoteric command or request. Endow it with communications abilities so complete that the machine or its user could link to any other machine or user in the world. And make it cheap enough so that even if you couldn't think of anything to do with it, you would buy one anyway.[†]

SUMMARY

- In a processor, binary instruction words placed in the Instruction Register are interpreted by the Controller, which then performs arithmetic and logic operations on the content of the Accumulator in response to signals from the Timer.
  The Controller has a limited repertoire of allowable instructions.
  A program is a detailed list of specific instructions, each corresponding to an elementary step in the overall operation.

- For greater versatility, the processor can be provided with a bidirectional Data Bus, multiple I/O ports, general-purpose registers, more instructions, a Memory in which the program is stored, and a Program Counter that always contains the address of the next instruction.

- Conditional instructions, contingent upon certain key conditions that occur in data processing, provide decision-making capability when flags are raised.
  Jump instructions add variety by altering the sequence of instructions.
  Indirect addressing provides flexibility in storing or retrieving data from RAM.
  I/O ports are addressed and read or written into just as memory locations are.

- Programming is the art of directing the processor to perform the elementary steps in solving a computation or control problem in an effective way.
  Flow charts visually break large, complex programs into smaller logical units that are easier to follow and easier to change and correct.
  Program development is simplified by a Development System computer controlling Editor, Assembler, and System Monitor programs.

[†] Forbes, p. 92, January 27, 1986.

- Programs can be written in machine, assembly, or higher-level language. Within a μP, all communication is in machine language fixed at manufacture. Assembly language uses short symbolic mnemonics understandable to people to represent specific binary instructions that are understood by the machine. Assembly language instructions are translated into machine language by an Assembler that can also build a dictionary of Labels and their addresses in memory.

- A practical 40-pin, 8-bit μP may have 8 bidirectional data lines, 16 address lines, and 12 control lines with 4 pins available for power and clocking. A complete digital system including processor, program memory, data memory, input/output ports, and clock, is called a microcomputer. Clock rates of 5 MHz permit typical instruction cycles of less than 1 μs.

REVIEW QUESTIONS

1. What is the role of the Instruction Register? Controller? Accumulator? Timer?
2. How does a processor know that a binary "word" is an instruction and not a number?
3. What is an instruction? A program?
4. Why are three-state buffers used in processor registers?
5. Describe two ways of using a register as a counter?
6. How can a mass-produced processor be used in a customized application?
7. Why are instructions stored sequentially?
8. Explain the phrase: "The PC always points to the next instruction."
9. What is an instruction fetch? Why does it come first?
10. List three different functions of the ALU.
11. Why are conditional instructions useful?
12. Explain "indirect addressing"; "software."
13. Why are flow charts useful?
14. Differentiate between machine, assembly, and higher-level languages.
15. Differentiate between assembler and compiler.
16. How does an assembler treat Labels?
17. How are assembler directives used?
18. How does a microcomputer differ from a microprocessor?
19. Write a one-sentence technical description of the Z80.
20. Define the 13 Z80 control signals.
21. How does the Z80 execute a **CPL** instruction? An **IN A,(n)** instruction?
22. Explain the following Z80 instructions: **RLCA**; **ADD A,r**; **LD r,r'**; **LD r,(HL)**; **CP (HL)**.
23. List the components of a minimum Z80-based computer and explain the function of each.

EXERCISES

1. A simple controller for the elementary processor of Fig. 9.2 is shown in Fig. 9.23. Complete the wiring diagram so that the proper registers are

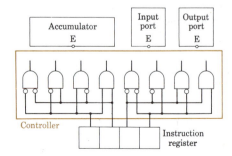

Figure 9.23

ENABLED (with a **1**) when codes **0100** and **0101** are placed in the Instruction Register.

2. The initial condition of an elementary processor (Fig. 9.4a) is defined by: Step = **0**, Input = **1110**, A = **1011**, Output = **0110**. The "program" consists of Step 1: **0011**, Step 2: **0001**, and Step 3: **0101**. Following Fig. 9.4, show the step-by-step execution of this program.

3. Repeat Exercise 2 for the 4-step program whose instruction codes are: **0001**, **0101**, **0100**, **0011**.

4. Write instructions (from Table 9-1) to perform the following operations. Show Code, Instruction, and Comment as in Example 1.
   (a) Store data at input port 2 in register D.
   (b) Complement the content of register C.

(c) Clear register C for use as a counter.

(d) Simulate an IC "quad inverter" by inverting each of 4 bits at port 1 and outputting them at port 2.

5. The Controller of our 4-bit processor needs an Instruction Decoder that will output a **1** to certain devices for each instruction code in the table below. Using **NOT** and **AND** gates, design and draw the logic circuit for such a decoder. Label the input and output terminals.

Instruction Code				Terminal Activated
I_4	I_3	I_2	I_1	
0	0	0	0	HALT
0	0	0	1	COMPLEMENT A
0	0	1	0	ROTATE A LEFT
1	1	1	1	JUMP

6. Write a program with comments for the 4-bit processor that:

(a) Increments register B without using the **INCR r** instruction.

(b) Decrements register B without using the **DECR r** instruction and justify including unique **INCR** and **DECR** instructions.

7. Write a program for the 4-bit processor that forms the 2's complement of the content of the A register without using the **COMPL A** instruction. (*Hint*: Use **XOR**.)

8. Showing Code, Instruction (Table 9-1), and Comment, write a program segment to:

(a) Multiply by 3 the number at port 4.

(b) Reduce the number in register C by the number at port 5.

9. Repeat Exercise 4 for the following:

(a) Input a number from port 3, add it to the number stored in register C, and output the sum to port 2.

(b) Subtract the number in register B from the number in register C and output the difference to port 4.

10. Using two- and three-word instructions complicates controller operation.

(a) Explain why the complication occurs.

(b) Suggest how a more sophisticated controller could be implemented.

11. Write instructions (from Table 9-1) to perform the following operations. Show Code, Instruction, and Comment as in Example 1.

(a) "Mask off," that is, set to zero, the two LSB of a 4-bit word in the accumulator.

(b) Compare the contents of registers C and A;

if they are equal, jump to program memory address **1100**. (*Hint*: Use **XOR**.)

(c) Add the numbers stored in data memory locations **1010** and **1011**, and store the sum in register E. (*Hint*: Use **LOAD r FROM MEM**.)

12. Decode the following program, add comments, and describe the function performed. Show the content of B and PC at the beginning of each instruction cycle.

LOCS	Code	LOCS	Code
0000	1101	0110	1001
0001	0000	0111	1011
0010	0011	1000	1111
0011	0111	1001	0000
0100	0000	1010	0011
0101	1111	1011	0000

13. (a) The programmer, by mistake, writes **0000** in location **1010** of Exercise 12. What is the effect of this "bug"? Show the content of B and PC for 12 instruction cycles.

(b) Repeat part (a) for **0001** in location **1010**.

(c) Comment on the effect of small errors in conditional instructions.

14. Repeat Exercise 11 for the following:

(a) Complement the content of memory location **1111**.

(b) Fill data memory locations 0 to 15D with zeros.

(c) Substitute the MSB in the Accumulator for the MSB in register D.

(d) Subtract the positive integer at port 2 from the positive integer at port 1 and output the difference at port 3.

(e) Add the numbers stored in data memory locations **1000** and **1001**, and store the sum in register B. (*Hint*: Use **LOAD r FROM MEM**.)

15. Assuming the stored numbers in Exercise 14(e) are **0011** and **0100**, show the contents of the IR, A, E, B, and PC registers for each machine cycle of your addition "program." (See Example 2. Assume PC initially at **0000**.)

16. You are to write a program "routine" that will introduce a time delay of exactly 80 machine cycles. (Assume each word of instruction requires 1 machine cycle for execution.)

(a) Outline, in general terms, a feasible solution based on Table 9-2.

(b) Draw up a flow chart identifying the sequence of operations.

(c) Write the program segment in assembly language syntax, showing memory locations and contents as in Example 5.

17. Repeat Exercise 16 for a program segment that will add the positive integer at input port 1 to the positive integer at port 2, and send the carry, if any, to port 4. (Assume register C initially contains **0001**.)

18. Write a Z80 program in assembly language syntax to execute the instructions of Example 4 (Fig. 9.10).

19. Write a Z80 program segment that will take a number (assumed less than 26D) from port 1, multiply it by 10, and send the product to port 2. (Use the **ROTATE** instruction.)

20. In a nuclear reactor, an alarm is to be sounded whenever the temperature sensed at input port 1 exceeds 120° C. The alarm is activated by placing the code number **99H** on output port 4. Write an appropriate Z80 program in assembly language syntax; the first instruction is to be in program memory location **0010H**.

21. A "monostable multivibrator" (MM), when triggered, generates a single square pulse of given length. Prepare a flow chart and write a Z80 program to simulate an MM with a pulse length of approximately 1 second. Assume 1 clock period = 2 μs.

22. One hundred 8-bit numbers reside in data memory at consecutive locations starting at **0101H**. The code number **FA** is known to occur at least once; you are to find the first occurrence. Prepare a flow chart and write a Z80 program (starting at program memory location **0010H**) that will search for the first occurrence of the code number **FA** and display the lower 8 bits of its address at output port 5.

23. In one mode of operation, the Z80 can be "interrupted" and caused to perform a program segment called a "service routine," the final instruction of which is a **RETURN** to the main program.

The pressure (in psi) in a rocket engine under test is 3.5 times the reading of a strain gauge (interfaced to port 4). Periodically the pressure is tested. Any time the pressure exceeds 65 psi, the positive pressure difference in psi is to be recorded on a strip chart (interfaced to port 8). Prepare a flow chart and write an efficient program from the interrupt to **RETURN** in assembly language syntax. (Disregard overflows. Let the microprocessor do the calculations.)

24. When a certain computer is turned on, it first determines the size of the RAM memory (assumed to be < 256 bytes) that is available to the CPU. Prepare a flow chart and write a Z80 program that will test each successive byte of memory, starting at **1000H**, by storing the number **10101010B** and then recalling it to see whether the number was stored correctly. The address of the last contiguous byte of working memory is to be stored in register pair HL.

25. One hundred 8-bit unsigned integers are stored in RAM at consecutive locations starting at **0001H**. Write a Z80 program in assembly language syntax to add the numbers five at a time, send each sum to output port 3, and stop after processing all the numbers (20 sums). Include a flow chart.

26. Sixty-five signed numbers are stored in data memory locations **0100** to **0140H**. Prepare a flow chart and write a Z80 program that will instruct the processor to survey the list of numbers and tally the number of negative and positive entries in registers B and C, respectively.

27. (a) Using a decoder (Fig. 8.29) and appropriate logic, design a circuit that will select one of four output latches.

    (b) Use four additional 2 × 4 decoders in a circuit to select one of 16 output latches. (*Hint*: Use outputs of first decoder to enable the other decoders.)

10

Alternating Current Circuits

Phasors

Circuit Analysis

Frequency Response

Resonance

Information in digital form is processed by devices and circuits that respond to signals containing only two discrete levels that correspond to binary logic values. In contrast, much of the communication of information by wire and by radio is in the form of continuously varying voltages and currents ranging in frequency from a few cycles per second to billions of cycles per second. The analysis of such analog systems is usually in terms of sinusoidal quantities. Also, most electrical power is generated, transmitted, and utilized in the form of steady-state alternating currents. To round out our knowledge of circuit analysis, we need to learn how to predict the response of circuits to ac signals.

The first three chapters in this book provide the necessary foundation of basic laws and circuit principles. Building on that foundation, we now develop techniques and approaches that transform difficult problems into routine exercises. First we learn how to represent continuously changing voltages and currents by easily manipulated constant quantities. Then we study voltage and current relations and learn to analyze series and parallel combinations of R, L, and C. Then we consider circuits in which the frequency of the applied signal is varied and investigate the interesting phenomenon of resonance. Our goal is to gain the analytical tools needed in designing or predicting the behavior of the real amplifiers whose idealized models were used in the circuits of Chapter 3.

PHASORS

If a sinusoidal source supplies a linear circuit consisting of resistances, inductances, and capacitances, all voltages and currents will be sinusoids of the same frequency. To perform the addition and subtraction of voltages and currents required in the application of Kirchhoff's laws, we need a technique for combining quantities that are continuously changing. The convenient and general method that we shall use is based on complex algebra and an ingenious transformation of functions of time into constant quantities.

Complex Numbers

In Fig. 10.1, *real* numbers are plotted along the horizontal axis and *imaginary* numbers are plotted along the vertical axis.[†] The real and imaginary axes define the *complex plane*. The combination of a real number and an imaginary number defines a point in the complex plane and also defines a *complex number*. The complex number may be considered to be the point or the directed line segment to the point; both interpretations are useful. (Imaginary numbers and complex algebra are explained in the appendix.)

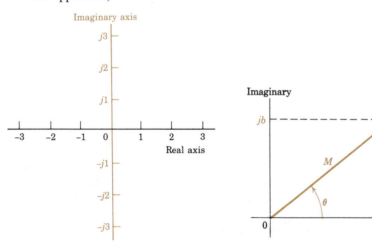

Figure 10.1 The complex plane.   **Figure 10.2** A complex number.

Given the complex number **W** of magnitude M and direction θ, in *rectangular* form,

$$\mathbf{W} = a + jb$$

or

$$\mathbf{W} = M(\cos \theta + j \sin \theta)$$

by inspection of Fig. 10.2. By Euler's theorem (see appendix p. 453),

$$\cos \theta + j \sin \theta = e^{j\theta} \tag{10-1}$$

Therefore,

$$\mathbf{W} = M e^{j\theta} \tag{10-2}$$

[†] Engineers use j instead of i for imaginary numbers to avoid confusion with the symbol for current. Charles P. Steinmetz, who first used combinations of real and imaginary numbers in circuit analysis, called them "general numbers." See *IEEE Spectrum*, April 1965.

This is called the *exponential* or *polar* form and can be written symbolically as

$$\mathbf{W} = M\underline{/\theta} \qquad (10\text{-}3)$$

which is read "magnitude M at angle θ." (In this book, complex numbers are always in boldface type.)

To convert complex numbers to *polar* form from rectangular,

$$M = \sqrt{a^2 + b^2} \qquad \theta = \arctan\frac{b}{a} \qquad (10\text{-}4)$$

To convert complex numbers to *rectangular* form from polar,

$$a = M\cos\theta \qquad b = M\sin\theta \qquad (10\text{-}5)$$

Practice Problem 10-1

Use your hand calculator in an efficient way (consult your Owner's Manual) to:

(a) Convert $V_1 = 5 + j6$ to polar form.
(b) Convert $V_2 = 10\underline{/30°}$ to rectangular form.
(c) Find $V_2 + V_3$ in polar form where $V_3 = 20\underline{/-45°}$.

Answers: (a) $7.81\underline{/50.19°}$; (b) $8.66 + j5.00$; (c) $24.57\underline{/-21.85°}$.

Representation of Sinusoids

The alternating currents and voltages carrying information in analog form are sinusoidal functions of time with various amplitudes, frequencies, and phase angles. (See Fig. 3.7.) To represent sinusoidal *signals*, we need to extend our concept of complex numbers to include *complex variables*.

The complex constant $\mathbf{W} = M\,e^{j\theta}$ is represented by a fixed radial line. If the line rotates at an angular velocity ω, as shown in Fig. 10.3, $\mathbf{W}$ is a complex function of time and

$$\mathbf{W}(t) = M\,e^{j(\omega t + \theta)} \qquad (10\text{-}6)$$

The projection of this line on the real axis is

$$W_{\text{real}} = f_1(t) = M\cos(\omega t + \theta)$$

and the projection on the imaginary axis is

$$W_{\text{imag}} = f_2(t) = M\sin(\omega t + \theta)$$

Either component could be used to represent a sinusoidal quantity; in this book, we shall always work with the real components.

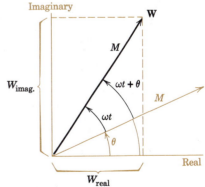

Figure 10.3 A complex function of time.

Phasor Representation

If an instantaneous voltage is described by a sinusoidal function of time such as

$$v(t) = V_m\cos(\omega t + \theta) = \sqrt{2}V\cos(\omega t + \theta) \qquad (10\text{-}7)$$

where V_m is the amplitude and V is the effective value (Eq. 3-27), then $v(t)$ can be interpreted as the "real part of" a complex function or

$$v(t) = \text{Re } \{V_m e^{j(\omega t + \theta)}\} = \text{Re } \{(V e^{j\theta})(\sqrt{2}\, e^{j\omega t})\} \tag{10-8}$$

In the second form of Eq. 10-8, the complex function in braces is separated into two parts; the first is a complex constant, and the second is a function of time that implies rotation in the complex plane and includes the conversion factor relating effective to maximum values (Eq. 3-27). The first part we define as the *phasor* **V** where

$$\mathbf{V} = V e^{j\theta} = V\underline{/\theta} \tag{10-9}$$

The phasor $\mathbf{V} = V e^{j\theta}$ is called a *transform* of the voltage $v(t)$. It is obtained by transforming a function of time into a complex constant that retains the essential information: effective value and phase angle. The term $e^{j\omega t}$ indicates rotation at angular velocity ω, but this is the same for all voltages and currents associated with a given source and therefore this term may be put to one side until needed.

EXAMPLE 1

(a) Write the equation of the current shown in Fig. 10.4a as a function of time and represent the current by a phasor.

The current reaches a positive maximum of 10 A at $\pi/6$ rad or 30° before $\omega t = 0$; therefore, the phase angle θ is $+\pi/6$ rad and the equation is

$$i = 10 \cos \left(\omega t + \frac{\pi}{6} \right) A$$

This could also be expressed as

$$i(t) = \text{Re } \{10 e^{j(\omega t + \pi/6)}\} = \text{Re } \{7.07\, e^{j(\pi/6)}\sqrt{2}\, e^{j\omega t}\} A$$

By comparison with the defining Eqs. 10-8 and 10-9, the current phasor in Fig. 10.4b is

$$\mathbf{I} = 7.07\, e^{j(\pi/6)} A \qquad \text{or} \qquad 7.07\underline{/30°}\, A$$

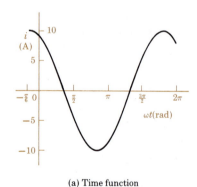

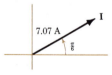

(a) Time function   (b) Phasor

Figure 10.4   Phasor representation.

(b) Represent, by a phasor, the current $i = 20\sqrt{2} \sin(\omega t + \pi/6)$ A.

Our definition of phasor is based on the cosine function as the real part of the complex function of time. Because $\cos(x - \pi/2) = \sin x$, we first write

$$i = 20\sqrt{2} \cos \left(\omega t + \frac{\pi}{6} - \frac{\pi}{2} \right) = 20\sqrt{2} \cos \left(\omega t - \frac{\pi}{3} \right) A$$

Then the current phasor is $\mathbf{I} = 20 e^{j(-\pi/3)} A = 20\underline{/-60°}$ A.

Phasor Diagrams

Electrical engineering deals with practical problems involving measurable currents and voltages that can exist in and across real circuit components such as those in your TV set. In solving circuit problems, however, we choose to deal with abstractions rather than physical currents and voltages. Learning to read and write and interpret such abstractions is an essential part of an engineering education. Working with the abstract concept that we call a phasor will provide some good experience in this important activity.

The phasor has much in common with the vector that is so valuable in solving problems in mechanics. As you may recall, vector representation of forces permits the use of graphical methods for the composition and resolution of forces. Also, and this may be more important, the use of vector representation—in free-body diagrams, for example—provides a picture of the physical relationships that is not provided by the algebraic equations alone. In the same way, *phasor diagrams* can be used for graphical solution, as a quick check on the algebraic solution, and to gain new insight into voltage and current relations.

EXAMPLE 2

Given $v_1 = 150\sqrt{2} \cos (377t - \pi/6)$ V and $\mathbf{V}_2 = 200\underline{/+60°}$ V, find $v = v_1 + v_2$.

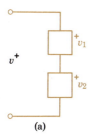

(a)

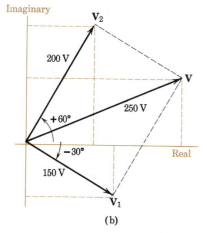

(b)

Figure 10.5   Use of phasor diagrams.

The expression for v_1 is a mathematical abstraction that carries a great deal of information: a voltage varies sinusoidally with time, completing $377/2\pi = 60$ Hz and reaching a positive maximum value of $150\sqrt{2}$ V when $t = (\pi/6)/377 = 0.00139$ s. To work with sinusoid v_1, we first *transform* the time function to phasor $\mathbf{V}_1 = 150\underline{/-30°}$ V.

The expression for $\mathbf{V}_2$ is "an abstraction of an abstraction"; part of the information is specified and part is only implied. The sinusoidal character is implied by the boldface phasor notation. The effective value is clearly 200 V and the phase angle is 60°. Since v_1 and v_2 are associated, we assume that they have the same angular frequency or $\omega_2 = 377$ rad/s. Reading "between the lines," we conclude that $v_2 = 200\sqrt{2} \cos (377t + \pi/3)$ V.

The phasor diagram of Fig. 10.5b shows the complex quantities $\mathbf{V}_1$ and $\mathbf{V}_2$ plotted to scale. Using the parallelogram law for adding vectors, the construction is as shown. Scaling off the resultant, it appears that $\mathbf{V}$ has a magnitude of about 250 V at an angle of approximately 23°.

Ignoring the direct calculation capability of your hand calculator for this example, we write (Eq. 10-5)

$$\mathbf{V}_1 = 150 \cos (-30°) + j150 \sin (-30°) = 130 - j75 \text{ V}$$

$$\mathbf{V}_2 = 200 \cos 60° + j200 \sin 60° = 100 + j173 \text{ V}$$

By the rules for equality and addition (see pp. 453 and 454),

$$\mathbf{V} = \mathbf{V}_1 + \mathbf{V}_2 = 230 + j98 = 250\underline{/23.1°} \text{ V}$$

After the inverse transformation on $\mathbf{V}$,

$$v = v_1 + v_2 = 250\sqrt{2} \cos (377t + 23.1°) \text{ V}$$

Looking at Fig. 10.5, we see that in Example 2 the algebraic addition involves the same components as the graphical addition, and we can conclude that phasor diagrams can be used for graphical calculations with the phasors treated just like vectors. A second virtue of the phasor diagram is that a quick sketch only approximately to scale provides a good check on algebraic calculations. Mistakes in sign, decimal point, or angle are easily spotted.

Practice Problem 10-2

Two currents into a node are $i_1 = 10\sqrt{2} \cos(\omega t + 135°)$ and $i_2 = 6\sqrt{2} \cos(\omega t - 60°)$.

(a) Represent the two currents by phasors.
(b) Sketch the current phasors approximately to scale and estimate the phasor sum.
(c) Evaluate $I_1 + I_2$ and compare with your estimate.

Answers: (a) $10\underline{/135°}$, $6\underline{/-60°}$; (c) $4.48\underline{/155.27°}$.

CIRCUIT RESPONSE TO SINUSOIDS

Although an "alternating" current can be nonsinusoidal, unless otherwise stated we assume that an alternating current or an "ac voltage" is sinusoidal. Signals may be generated by sinusoidal current sources or voltage sources. When sinusoidal signals are applied to ideal resistance, inductance, or capacitance elements, or to any series or parallel combination of these elements, the response is also sinusoidal. The response of circuit elements to sinusoidal voltages or currents can be obtained by considering the defining element equations.

Element Impedance

In general, the ratio of a voltage phasor to the corresponding current phasor is a measure of the opposition to the flow of current called the *impedance* **Z**.

When a current $i = \sqrt{2}I \cos \omega t$ represented by phasor $I\underline{/0°}$ flows through a resistance R, the voltage is given by

$$v_R = Ri = \sqrt{2}RI \cos \omega t = \sqrt{2}V_R \cos \omega t \qquad (10\text{-}10)$$

represented by phasor $V_R\underline{/0°}$. In this case the opposition to current flow is

$$\mathbf{Z}_R = \frac{V_R\underline{/0°}}{I\underline{/0°}} = \frac{RI\underline{/0°}}{I\underline{/0°}} = R\underline{/0°} \qquad (10\text{-}11)$$

The quantity R is called the *ac resistance*[†] and is measured in ohms.

When the current $i = \sqrt{2}I \cos \omega t$ flows through an inductance L, the voltage is given by

$$v_L = L\frac{di}{dt} = \sqrt{2}\omega LI(-\sin \omega t) = \sqrt{2}\omega LI \cos(\omega t + 90°) \qquad (10\text{-}12)$$

[†]The ac resistance of a physical resistor may be different from the *dc resistance*.

represented by phasor $V_L\underline{/90°}$. The opposition to current flow is

$$\mathbf{Z}_L = \frac{V_L\underline{/90°}}{I\underline{/0°}} = \frac{\omega L I\underline{/90°}}{I\underline{/0°}} = \omega L\underline{/90°} = j\omega L \qquad (10\text{-}13)$$

The magnitude of the impedance in this case is called the *inductive reactance* $X_L = \omega L$ in ohms.

When a voltage $v = \sqrt{2}V\cos\omega t$ appears across a capacitance, the current is given by

$$i_C = C\frac{dv}{dt} = \sqrt{2}\omega CV(-\sin\omega t) = \sqrt{2}\omega CV\cos(\omega t + 90°) \qquad (10\text{-}14)$$

represented by phasor $I_C\underline{/90°}$. The opposition to current flow is

$$\mathbf{Z}_C = \frac{V\underline{/0°}}{I_C\underline{/90°}} = \frac{V\underline{/0°}}{\omega CV\underline{/90°}} = \frac{1}{\omega C}\underline{/-90°} = -j\frac{1}{\omega C} \qquad (10\text{-}15)$$

The magnitude of the impedance in this case is called the *capacitive reactance*. We write $X_C = -1/\omega C$ in ohms, where the minus sign indicates the 180° phase difference between X_L and X_C.

Circuit Laws in Terms of Phasors

The element equations define the voltage-current characteristics of the passive elements shown in Fig. 10.6. The sign convention indicates that for the assumed

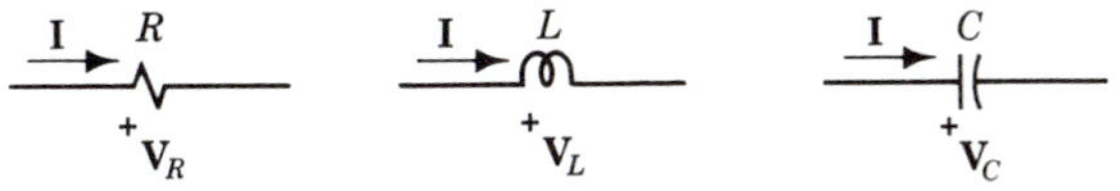

$$V_R = RI \qquad V_L = j\omega L I \qquad V_C = -j\frac{1}{\omega C}I = \frac{1}{j\omega C}I$$

$$I_R = \frac{1}{R}V_R \qquad I_L = \frac{1}{j\omega L}V_L \qquad I_C = j\omega C\,V_C$$

Figure 10.6 Element equations in terms of phasors.

reference direction of positive current, the corresponding reference polarity for voltage is as shown. Such a convention is necessary because the actual current and voltage are both changing sign periodically, but not simultaneously.

When a current $i(t) = \sqrt{2}\,I\cos(\omega t + \theta)$ represented by phasor $I\underline{/\theta}$ flows through a resistance R (Fig. 10.7, p. 284), the voltage is

$$V_R = RI = R\underline{/0°} \cdot I\underline{/\theta} = RI\underline{/\theta}$$

Therefore, the voltage and current phasors for R are drawn at the same angle. The voltage and current are shown as functions of time; the resistance voltage and current waves reach maximum and zero values simultaneously. We say

The voltage across a resistance is in phase with the current through it.

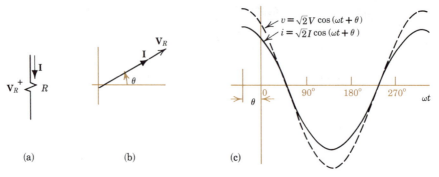

Figure 10.7 Current-voltage relations for a resistance.

For an inductance (Fig. 10.8),

$$\mathbf{V}_L = j\omega L\,\mathbf{I} = \omega L\,\underline{/90^\circ} \cdot I\,\underline{/\theta} = \omega LI\,\underline{/\theta + 90^\circ}$$

Therefore, the voltage phasor for L is 90° *ahead* of the current phasor. When the inductance voltage is shown as a function of time, it is clear that the voltage wave reaches its maximum and zero values at an *earlier* time than the current wave. We say

The voltage across an inductive reactance leads the current through it by 90°.

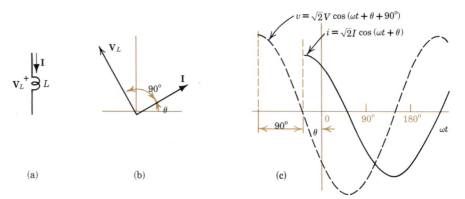

Figure 10.8 Current-voltage relations for an inductance.

For a capacitance (Fig. 10.9),

$$\mathbf{V}_C = (1/j\omega C)\,\mathbf{I} = (1/\omega C)\underline{/-90^\circ} \cdot I\,\underline{/\theta} = (1/\omega C)I\,\underline{/\theta - 90^\circ}$$

Therefore, the voltage phasor is 90° *behind* the current phasor. When the capacitance voltage is shown as a function of time, it is clear that the voltage wave reaches its maximum and zero values at a *later* time than the current wave. We say

The voltage across a capacitive reactance lags the current through it by 90°.

These phrases emphasize that the *voltage across an element* is specified with respect to the *current through that element* and not with respect to any other current. Obviously we could say "The *current* through an inductance *lags* the *voltage* across it by 90°" and "The *current* through a capacitance *leads* the *voltage* across it by 90°."

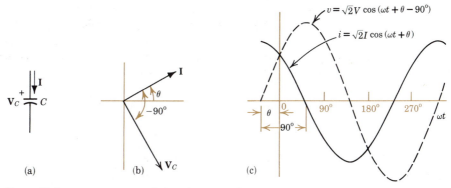

Figure 10.9   Current-voltage relations for a capacitance.

In Fig. 10.10a, the current $i(t)$ represented by phasor $I\underline{/\theta}$ flows through three circuit elements connected in series, and we are interested in the total voltage across the series combination. Kirchoff's voltage law is defined for instantaneous voltages, and we know that the total voltage $v(t)$ is just equal to $v_R(t) + v_L(t) + v_C(t)$. By definition, instantaneous voltages are the real parts of the complex numbers we call voltage phasors. By the rule of equality (see p. 453), if two complex numbers are equal their real parts are equal. Therefore, the instantaneous values of a voltage phasor are always just equal to the sum of the instantaneous values of its component phasors.

Assuming that all sources in a linear circuit are sinusoids of the same frequency, all steady-state voltages and currents are sinusoids of that same frequency. Kirchhoff's voltage law $\Sigma v = 0 = v_1 + v_2 + \cdots + v_k$ becomes, in terms of phasors,

$$\Sigma \mathbf{V} = 0 = \mathbf{V}_1 + \mathbf{V}_2 + \cdots + \mathbf{V}_k \qquad (10\text{-}16a)$$

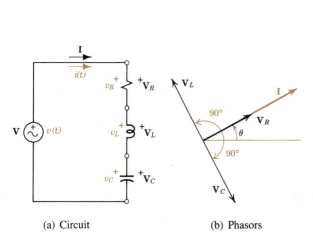

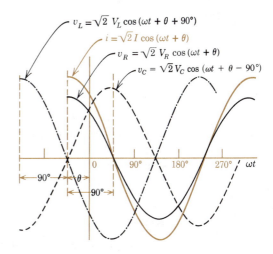

(a) Circuit   (b) Phasors   (c) Time functions for R, L, and C

Figure 10.10   Voltage-current relations in ac circuits.

In other words, for steady-state sinusoidal functions, Kirchhoff's voltage law holds for voltage phasors and $\Sigma\mathbf{V} = 0$ around any closed loop. In Fig. 10.10b total phasor voltage $\mathbf{V}$ could be found as the phasor sum $\mathbf{V}_R + \mathbf{V}_L + \mathbf{V}_C$. (See Exercise 14.) Following a similar line of reasoning, Kirchhoff's current law holds for current phasors and

$$\Sigma\mathbf{I} = 0 = \mathbf{I}_1 + \mathbf{I}_2 + \cdots + \mathbf{I}_K \qquad (10\text{-}16\text{b})$$

into any node.

EXAMPLE 3

Voltage $v = 120\sqrt{2}\cos(1000t + 90°)$ V is applied to the circuit of Fig. 10.11a where $R = 15\,\Omega$, $C = 83.3\ \mu\text{F}$, and $L = 30$ mH. Find $i(t)$.

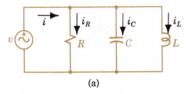

(a)

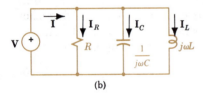

(b)

Figure 10.11 Response to a sinusoid.

The first step is to transform $v(t)$ into phasor $\mathbf{V} = 120\underline{/90°}$ V. Then

$$\mathbf{I}_R = \frac{1}{R}\mathbf{V} = \frac{1}{15}120\underline{/90°} = 8\underline{/90°} = 0 + j8\ \text{A}$$

$$\mathbf{I}_C = j\omega C\mathbf{V} = (0.0833\underline{/90°})(120\underline{/90°})$$
$$= 10\underline{/180°} = -10 + j0\ \text{A}$$

$$\mathbf{I}_L = \frac{1}{j\omega L}\mathbf{V} = \frac{120\underline{/90°}}{30\underline{/90°}} = 4\underline{/0°} = 4 + j0\ \text{A}$$

By Kirchhoff's current law, $\Sigma\mathbf{I} = 0$ or

$$\mathbf{I} = \mathbf{I}_R + \mathbf{I}_C + \mathbf{I}_L = (0 - 10 + 4) + j(8 + 0 + 0)$$
$$= -6 + j8 = 10\underline{/127°}\ \text{A}$$

The final step is to transform phasor $\mathbf{I}$ into

$$i(t) = 10\sqrt{2}\cos(1000t + 127°)\ \text{A}$$

AC CIRCUIT ANALYSIS

Impedance $\mathbf{Z}$ is defined for sinusoidal waveforms. With this restriction, impedances can be combined in series and parallel just as resistances are. When we take advantage of this concept, differential equations are replaced by algebraic equations; steady-state ac circuit problems become only slightly more difficult to solve than corresponding dc circuit problems.

Series Circuits

For the series combination of R and L in Fig. 10.12a, $\Sigma\mathbf{V} = 0$ or

$$\mathbf{V} = \mathbf{V}_R + \mathbf{V}_L = \mathbf{I}(R + j0) + \mathbf{I}(0 + j\omega L) = \mathbf{I}(R + j\omega L) = \mathbf{IZ}$$

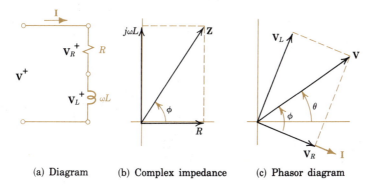

(a) Diagram   (b) Complex impedance   (c) Phasor diagram

Figure 10.12   Series *RL* circuit analysis.

Therefore,

$$\mathbf{Z} = R + jX_L = R + j\omega L = Z\underline{/\phi} = \sqrt{R^2 + (\omega L)^2}\,\underline{/\arctan\,\omega L/R}$$

and we see that the equivalent impedance of elements in series is just the sum of the individual impedances, which is the expression of Eq. 2-25 for ac circuits.

Assuming that **V** is given as $V\underline{/\theta}$, where $V = V_{\text{eff}} = V_m/\sqrt{2}$,

$$\mathbf{I} = \frac{\mathbf{V}}{\mathbf{Z}} = \frac{V\underline{/\theta}}{Z\underline{/\phi}} = \frac{V}{Z}\underline{/\theta - \phi} = I\underline{/\theta - \phi} \qquad (10\text{-}17)$$

As shown in Fig. 10.12c, the total voltage phasor leads the current phasor[†] by *phase angle* ϕ, something less than 90°. The voltage across the resistance is in phase with the current, and the voltage across the inductive reactance is 90° ahead of the current. These are components of the total applied voltage, and the phasor sum of $\mathbf{V}_R$ and $\mathbf{V}_L$ is equal to **V**.

For the series combination of *R* and *C* (Fig. 10.13a),

$$\mathbf{Z} = R + jX_C = R + j\left(\frac{-1}{\omega C}\right) = Z\underline{/\phi} = \sqrt{R^2 + \left(\frac{1}{\omega C}\right)^2}\,\underline{/\arctan\frac{-1/\omega C}{R}}$$

$$(10\text{-}18)$$

where ϕ is a negative angle. Assuming that **I** is given as $I\underline{/\theta} = (I_m/\sqrt{2})\underline{/\theta}$,

$$\mathbf{V} = \mathbf{ZI} = (Z\underline{/\phi})(I\underline{/\theta}) = ZI\underline{/\phi + \theta} = V\underline{/\phi + \theta} \qquad (10\text{-}19)$$

As shown in Fig. 10.13c, the voltage across the resistance *R* is in phase with the current **I**. The voltage across the capacitive reactance X_C lags behind the current by 90°. The resulting total voltage $\mathbf{V} = \mathbf{V}_R + \mathbf{V}_C$ lags behind the current by an angle less than 90°.

[†] In drawing phasor diagrams, all voltages are drawn to one scale and all currents are drawn to one scale (which may be quite different from the voltage scale).

EXAMPLE 4

In the circuit of Fig. 10.13a below, $V = 20\underline{/45°}$ V, $R = 5$ Ω, and reactance $X_C = -1/\omega C = -8.66$ Ω. Find I, V_R, and V_C.

First, we calculate $\mathbf{Z}$.
By Eq. 10-18,

$$\mathbf{Z} = R + j\left(\frac{-1}{\omega C}\right) = 5 - j8.66 = 10\underline{/-60°}\ \Omega$$

Then we calculate $\mathbf{I}$.
By Eq. 10-3,

$$\mathbf{I} = \frac{\mathbf{V}}{\mathbf{Z}} = \frac{20\underline{/45°}}{10\underline{/-60°}} = 2\underline{/105°}\ A$$

Hence

$$\mathbf{V}_R = R\mathbf{I} = (5\underline{/0°})(2\underline{/105°}) = 10\underline{/105°}\ V$$

and

$$\mathbf{V}_C = \mathbf{Z}_C\mathbf{I} = (8.66\underline{/-90°})(2\underline{/105°})$$
$$= 17.32\underline{/15°}\ V$$

These results are plotted in Fig. 10.13c.

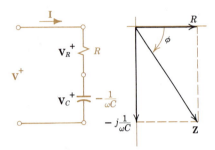

(a) Diagram   (b) Complex impedance

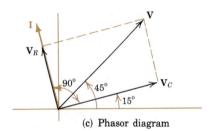

(c) Phasor diagram

Figure 10.13 Series RC circuit analysis.

Assuming ω is known, the phasors in Example 4 could be transformed into time functions, but this is seldom necessary. For the reasons discussed in Chapter 3, the quantities of interest in steady-state circuit analysis are the effective phasors, and ammeters and voltmeters are calibrated to read these values directly.

Practice Problem 10-3

Use your calculator in an efficient way to:

(a) Find the impedance at 2000 Hz of a coil of 100-mH inductance and 400-Ω resistance.
(b) Find the capacitance required in a series R-C circuit to provide an impedance $\mathbf{Z} = 100\underline{/-80°}$ Ω at 5000 Hz.
(c) Find the frequency at which the impedance of a series combination of $R = 200$ Ω and $L = 10$ mH will have a phase angle of 45°.

Answers: (a) $1319\underline{/72.34°}$ Ω; (b) 0.32 μF; (c) 3183 Hz.

Complex Impedance

In general, phasor voltage and phasor current are related by a *complex impedance* $\mathbf{Z} = Z(j\omega)$ or

$$\mathbf{V} = \mathbf{ZI} \qquad (10\text{-}20)$$

which is sometimes referred to as the Ohm's law of sinusoidal circuits. For sinusoidal signals, any two-terminal network is completely defined by the impedance $\mathbf{Z} = \mathbf{V}/\mathbf{I}$ at the terminals (Fig. 10.9). In general, $\mathbf{Z}$ is complex and can be written as

$$\mathbf{Z} = \frac{\mathbf{V}}{\mathbf{I}} = Z\underline{/\phi_Z} = R + jX \qquad (10\text{-}21)$$

where

$$Z = \sqrt{R^2 + X^2} \qquad \text{and} \qquad \phi_Z = \arctan X/R$$

The real part of $\mathbf{Z}$ is the *ac resistance* in ohms and is designated R; in the series *RLC* circuit of Fig. 10.14c, the real part of $\mathbf{Z}$ is the resistance of a particular element, but

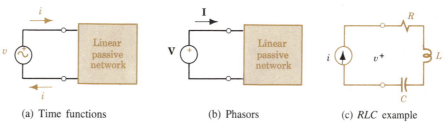

(a) Time functions   (b) Phasors   (c) *RLC* example

Figure 10.14   Representation of a two-terminal network.

in general this is *not* true. The imaginary part of $\mathbf{Z}$ is the *reactance* in ohms and is designated X; for the series *RLC* circuit,

$$X = X_L + X_C = \omega L - 1/\omega C$$

A positive X is associated with a predominantly inductive circuit.

Impedance $\mathbf{Z}$ is measured in ohms when phasors $\mathbf{V}$ and $\mathbf{I}$ are in volts and amperes, respectively. Note that $\mathbf{Z}$ is a complex quantity, like a phasor, but $\mathbf{Z}$ is *not* a phasor. The term phasor is reserved for quantities representing sinusoidal varying functions of time. The chief virtue of complex impedance is that although it expresses the relation between two time-varying quantities, $\mathbf{Z}$ itself is *not* a function of time.

Whenever we deal with sinusoidal voltages or currents, we think of phasors $\mathbf{V}$ or $\mathbf{I}$ containing the important information—effective value and phase angle. To find the response of a circuit, we think of complex impedance $\mathbf{Z}$. In these terms, the rules for series and parallel combinations of circuit elements and the principles of voltage and current division are applicable to a wide variety of alternating current problems. (See Example 5 on p. 290.)

The *v-i* relations for resistance, inductance, and capacitance are $v_R = Ri$, $v_L = L\,di/dt$, and $i_C = C\,dv/dt$. Using these relations to solve Example 5 would lead to a set of simultaneous differential equations. By using phasors, complex impedances, and circuit principles, we can solve a complicated problem in a straightforward way.

EXAMPLE 5

A voltage $v = 12\sqrt{2}\cos 5000t$ V is applied to the circuit of Fig. 10.15. Find the individual and combined impedances and the current $i(t)$.

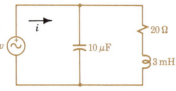

(a) Circuit elements

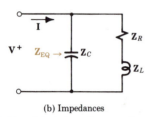

(b) Impedances

Figure 10.15 Calculation of current in an ac circuit.

$\mathbf{Z}_R = R = 20 = 20 + j0 = 20\underline{/0°}\,\Omega$

$\mathbf{Z}_L = j\omega L = j5000 \times 0.003 = j15\,\Omega = 15\underline{/90°}\,\Omega$

$\mathbf{Z}_C = -j\dfrac{1}{\omega C} = \dfrac{-j}{5000 \times 10^{-5}} = -j20 = 20\underline{/-90°}\,\Omega$

$\mathbf{Z}_{RL} = \mathbf{Z}_R + \mathbf{Z}_L = 20 + j15 = 25\underline{/37°}\,\Omega$

$\mathbf{Z}_{EQ} = \dfrac{\mathbf{Z}_{RL} \cdot \mathbf{Z}_C}{\mathbf{Z}_{RL} + \mathbf{Z}_C} = \dfrac{25\underline{/37°} \cdot 20\underline{/-90°}}{20 + j15 - j20} = \dfrac{500\underline{/-53°}}{20 - j5}$

$= \dfrac{500\underline{/-53°}}{20.6\underline{/-14°}} = 24.3\underline{/-39°} = 18.9 - j15.3\,\Omega$

(Note that R_{EQ} is *not* equal to R.)

$$\mathbf{I} = \dfrac{\mathbf{V}}{\mathbf{Z}} = \dfrac{12\underline{/0°}}{24.3\underline{/-39°}} = 0.49\underline{/39°}\ \text{A}$$

$i(t) = 0.49\sqrt{2}\cos(5000t + 39°)$ A

Practice Problem 10-4

Use current and voltage division theorems to find, in phasor form, the current in $\mathbf{Z}_C$ and the voltage across R in Fig. 10.15.

Answers: $0.59\underline{/90°}$ A, $9.60\underline{/-37°}$ V.

Complex Admittance

The steady-state sinusoidal response of any two-terminal network is completely defined by the impedance $\mathbf{Z} = \mathbf{V}/\mathbf{I}$ at the terminals. Another way of characterizing a network is in terms of the *admittance* $\mathbf{Y}$, defined as the ratio of phasor current to voltage or

$$\mathbf{Y} = \dfrac{\mathbf{I}}{\mathbf{V}} = \dfrac{1}{\mathbf{Z}}$$

where $\mathbf{Y}$ is the *complex admittance* in siemens. Specifically, for R, L, and C elements,

$$\mathbf{Y}_R = \dfrac{1}{R\underline{/0°}} = G\underline{/0°} \qquad \mathbf{Y}_L = \dfrac{1}{j\omega L} = \dfrac{1}{\omega L}\underline{/-90°} \qquad \mathbf{Y}_C = j\omega C = \omega C\underline{/90°}$$

In general, $\mathbf{Y}$ is complex and can be written as

$$\mathbf{Y} = Y\underline{/\phi_Y} = G + jB \qquad (10\text{-}22)$$

where

$$Y = \sqrt{G^2 + B^2} \qquad \text{and} \qquad \phi_Y = \arctan \frac{B}{G}$$

The real part of **Y** is the *conductance* in siemens and is designated G; in general, G is *not* the conductance of a particular element. The imaginary part of **Y** is called *susceptance* in siemens and is designated B. A positive B is associated with a predominantly capacitive circuit.

Parallel Circuits. The admittance approach is particularly useful in dealing with parallel circuits. For the parallel combination of Fig. 10.11b, $\Sigma \mathbf{I} = 0$ or

$$\mathbf{I} = \mathbf{I}_R + \mathbf{I}_C + \mathbf{I}_L = \mathbf{V}\mathbf{Y}_R + \mathbf{V}\mathbf{Y}_C + \mathbf{V}\mathbf{Y}_L = \mathbf{V}(\mathbf{Y}_R + \mathbf{Y}_C + \mathbf{Y}_L) = \mathbf{V}\mathbf{Y}_{EQ}$$

Therefore,

$$\mathbf{Y}_{EQ} = \mathbf{Y}_R + \mathbf{Y}_C + \mathbf{Y}_L$$

and we see that the equivalent admittance of elements in parallel is just the sum of the individual admittances, which is the expression of Eq. 2-26 for ac circuits.

Because **Z** and **Y** are complex quantities, they obey the rules of complex algebra and can be represented by plane vectors. In contrast to phasors, **Z** and **Y** can lie in only the first and fourth quadrants of the complex plane because R and G are always positive for passive networks. Application of the admittance approach to a more general circuit is shown in Example 6.

EXAMPLE 6

In Example 5 (Fig. 10.15), voltage $\mathbf{V} = 12\,\underline{/0°}$ V is applied to the parallel combination of $\mathbf{Z}_C = 20\,\underline{/-90°}$ Ω and $\mathbf{Z}_{RL} = 25\,\underline{/37°}$ Ω. Use admittances to calculate branch and total currents and show them on a phasor diagram.

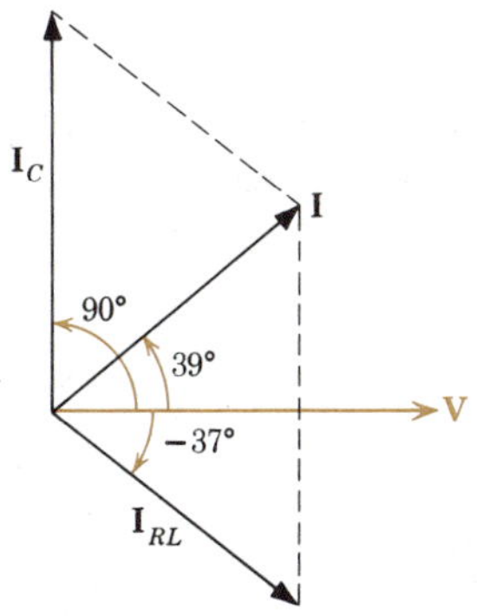

Figure 10.16   Phasor diagram of currents calculated from admittances.

The admittances are

$$\mathbf{Y}_C = \frac{1}{\mathbf{Z}_C} = \frac{1}{20\,\underline{/-90°}} = 0\,05\,\underline{/+90°} = 0 + j0.05 \text{ S}$$

$$\mathbf{Y}_{RL} = \frac{1}{\mathbf{Z}_{RL}} = \frac{1}{25\,\underline{/37°}} = 0.04\,\underline{/-37°} = 0.032 - j0.024 \text{ S}$$

$$\mathbf{Y}_{EQ} = \mathbf{Y}_C + \mathbf{Y}_{RL} = 0 + j0.05 + 0.032 - j0.024$$

$$= 0.032 + j0.026 = 0.041\,\underline{/39°} \text{ S}$$

The currents are

$$\mathbf{I}_C = \mathbf{Y}_C\mathbf{V} = 0.05\,\underline{/+90°} \times 12\,\underline{/0°} = 0.6\,\underline{/+90°} \text{ A}$$

$$\mathbf{I}_{RL} = \mathbf{Y}_{RL}\mathbf{V} = 0.04\,\underline{/-37°} \times 12\,\underline{/0°} = 0.48\,\underline{/-37°} \text{ A}$$

$$\mathbf{I} = \mathbf{I}_C + \mathbf{I}_{YL} = 0 + j0.6 + 0.384 - j0.288$$

$$= 0.384 + j0.312 = 0.49\,\underline{/39°} \text{ A}$$

Alternatively, the total current is

$$\mathbf{I} = \mathbf{Y}_{EQ}\mathbf{V} = 0.041\,\underline{/39°} \times 12\,\underline{/0°} = 0.49\,\underline{/39°} \text{ A}$$

The phasor diagram in Fig. 10.16 provides a visual check on the magnitudes and phases of the currents.

Practice Problem 10-5

Calculate phasor current $\mathbf{I}$ in Fig. 10.11b from the equivalent admittance of the three elements in parallel.

Answer: $\mathbf{I} = 10\underline{/127°}$ A.

General Circuit Analysis

In talking about circuit analysis in general, we need a term that includes both impedance and admittance. For this purpose we shall use the word *immittance*, derived from *im*pedance and ad*mittance*.

We are now ready to formulate a general procedure applicable to a variety of problems in steady-state sinusoidal circuits. While the exact sequence depends on the particular problem (see Example 7), the following steps are usually necessary:

1. Transform time functions to phasors and convert element values to complex immittances.
2. Combine immittances in series or parallel to simplify the circuit.
3. Determine the desired response in phasor form, using connection equations.
4. Draw a phasor diagram to check computations and to display results.
5. Transform phasor results to time functions if required.

AC Network Theorems

The theorems applied in Chapter 2 to two-terminal resistive networks are easily extended to include general ac one-ports.

Equivalence. Two passive one-ports are equivalent if they have the same input impedance or the same input admittance; for sinusoidal excitation and response, for example, this means the same $\mathbf{Z}$ or $\mathbf{Y}$. Ordinarily, two real networks are equivalent at one frequency only.

Network Reduction. Replacing a complicated network with a relatively simple equivalent is advantageous in network analysis. Already we have used the fact that in passive circuits impedances and admittances are readily combined. For impedances in series,

$$\mathbf{Z}_{EQ} = \mathbf{Z}_1 + \mathbf{Z}_2 + \cdots + \mathbf{Z}_n$$

and for admittances in parallel,

$$\mathbf{Y}_{EQ} = \mathbf{Y}_1 + \mathbf{Y}_2 + \cdots + \mathbf{Y}_n$$

Dividers. The useful voltage divider and current divider become, for ac circuits,

$$\mathbf{V}_2 = \frac{\mathbf{Z}_2}{\mathbf{Z}_1 + \mathbf{Z}_2}\mathbf{V} = \frac{\mathbf{Z}_2}{\mathbf{Z}_{EQ}}\mathbf{V} \quad \text{and} \quad \mathbf{I}_2 = \frac{\mathbf{Y}_2}{\mathbf{Y}_1 + \mathbf{Y}_2}\mathbf{I} = \frac{\mathbf{Y}_2}{\mathbf{Y}_{EQ}}\mathbf{I} \quad (10\text{-}23)$$

EXAMPLE 7

The circuit of Fig. 10.17a represents a load consisting of C, R, and L supplied by a generator over a transmission line approximated by L_T. For a desired load voltage $v = 28.3\cos(5000t + 45°)$ V, determine the current $i(t)$ and the branch currents.

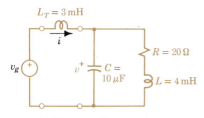

$L_T = 3\,\text{mH}$
$R = 20\,\Omega$
$C = 10\,\mu\text{F}$
$L = 4\,\text{mH}$

(a) Element values

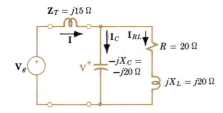

$\mathbf{Z}_T = j15\,\Omega$
$R = 20\,\Omega$
$-jX_C = -j20\,\Omega$
$jX_L = j20\,\Omega$

(b) Complex impedances

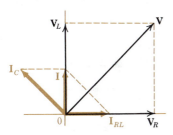

(c) Current phasors

Figure 10.17 General circuit analysis.

In phasor form, $v(t)$ becomes $\mathbf{V} = 20\underline{/45°}$ V.

$$\mathbf{Z}_T = j\omega L_T = j5000 \times 3 \times 10^{-3} = j15\,\Omega$$

$$\mathbf{Y}_C = j\omega C = j5000 \times 10^{-5} = j0.05\,\text{S}$$

$$\mathbf{Z}_C = 1/\mathbf{Y}_C = -j20\,\Omega$$

$$\mathbf{Z}_L = j\omega L = j5000 \times 4 \times 10^{-3} = j20\,\Omega$$

The diagram of Fig. 10.17b displays these values.

Since $\mathbf{I} = \mathbf{YV}$, we need the admittance of the parallel circuit constituting the load.

$$\mathbf{Y}_{RL} = \frac{1}{\mathbf{Z}_{RL}} = \frac{1}{20 + j20} = \frac{1}{20\sqrt{2}\underline{/45°}}$$

$$= 0.025\sqrt{2}\underline{/-45°} = 0.025 - j0.025\,\text{S}$$

$$\mathbf{Y} = \mathbf{Y}_C + \mathbf{Y}_{RL} = j0.05 + (0.025 - j0.025)$$

$$= 0.025 + j0.025 = 0.025\sqrt{2}\underline{/+45°}\,\text{S}$$

$$\mathbf{I} = \mathbf{YV} = (0.025\sqrt{2}\underline{/45°})(20\underline{/45°}) = 0.7\underline{/90°}\,\text{A}$$

$$i(t) = 0.7\sqrt{2}\cos(5000t + 90°)\,\text{A}$$

The phasor diagram should be drawn as the calculations proceed, to provide a continuous check as well as a final display of results. In addition to the given V and the calculated I, we know

$$\mathbf{I}_C = \mathbf{Y}_C\mathbf{V} = (0.05\underline{/90°})(20\underline{/45°}) = 1\underline{/135°}\,\text{A}$$

$$\mathbf{I}_{RL} = \mathbf{Y}_{RL}\mathbf{V} = (0.035\underline{/-45°})(20\underline{/45°}) = 0.7\underline{/0°}\,\text{A}$$

$$\mathbf{V}_R = \mathbf{R}\mathbf{I}_{RL} = (20\underline{/0°})(0.7\underline{/0°}) = 14\underline{/0°}\,\text{V}$$

$$\mathbf{V}_L = \mathbf{Z}_L\mathbf{I}_{RL} = (20\underline{/90°})(0.7\underline{/0°}) = 14\underline{/90°}\,\text{V}$$

These are shown on the phasor diagram of Fig. 10.17c. The construction lines verify the connection equations, $\mathbf{I} = \mathbf{I}_C + \mathbf{I}_{RL}$ and $\mathbf{V} = \mathbf{V}_R + \mathbf{V}_L$. Other checks are also available; $\mathbf{I}_C$ leads $\mathbf{V}$ by 90°, $\mathbf{V}_R$ is in phase with $\mathbf{I}_{RL}$, and $\mathbf{V}_L$ leads $\mathbf{I}_{RL}$ by 90°.

To determine the required generator voltage, you could write the loop equation

$$\Sigma\mathbf{V} = 0 = \mathbf{V}_g - \mathbf{I}\mathbf{Z}_T - \mathbf{V}$$

and solve for $\mathbf{V}_g$. A phasor diagram would confirm that $\mathbf{V}_g = \mathbf{V}_T + \mathbf{V}$ and provide a check on your answer. (See Practice Problem 10-6.)

When a particular voltage or current is desired, using these relations is preferable to writing and solving loop or node equations.

Superposition. The concept of linearity and the theorem of superposition are inseparable; the first is essential to the second, and the second defines the first. We say that function f is linear in x if

$$f(kx) = kf(x) \qquad \text{and} \qquad f(x_1 + x_2) = f(x_1) + f(x_2) \qquad (10\text{-}24)$$

These defining equations paraphrase the superposition theorem (p. 46), which is particularly useful in predicting the response of a network to a complex signal.

If cause and effect are linearly related, the total effect of several signals acting simultaneously is equal to the sum of the effects of the individual signals acting one at a time.

EXAMPLE 8

The signal shown in Fig. 10.18a is applied to a filter circuit. Predict the output.

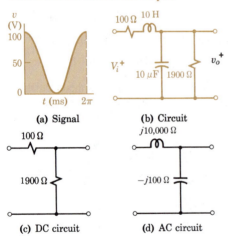

(a) Signal

(b) Circuit

(c) DC circuit

(d) AC circuit

Figure 10.18 Superposition application.

The signal can be described as

$$v_i = 50 + 50 \cos 1000t = V_i + V_{mi} \cos 1000t \text{ V}$$

For the dc component, the voltage-divider output is

$$V_o = \frac{R_2}{R_1 + R_2} V_i = \frac{1900}{100 + 1900} 50 = 47.5 \text{ V}$$

For the ac component, the resistances are negligible in comparison to the reactances; therefore, the output amplitude is

$$V_{mo} \cong \frac{Z_C}{Z_L + Z_C} V_{mi} = \frac{-j100}{j10,000 - j100} 50 = -0.5 \text{ V}$$

Using superposition, the total output is

$$v_o = 47.5 - 0.5 \cos 1000t \text{ V}$$

Active Networks

Active networks containing ac energy sources can be represented by simpler equivalents using Thévenin's theorem or Norton's theorem.

Insofar as a load is concerned, any one-port network of linear elements and energy sources can be replaced by a series combination of an ideal voltage source and a linear impedance or a parallel combination of an ideal current source and a linear admittance.

Here the Thévenin equivalent is a phasor voltage $\mathbf{V}_T$ in series with a complex impedance $\mathbf{Z}_T$; the Norton equivalent is a phasor current $\mathbf{I}_N$ in parallel with a complex admittance $\mathbf{Y}_N$. The parameters are: $\mathbf{V}_T = \mathbf{V}_{OC}$, $\mathbf{Z}_T = \mathbf{V}_{OC}/\mathbf{I}_{SC}$, $\mathbf{I}_N = \mathbf{I}_{SC}$, and $\mathbf{Y}_N = \mathbf{I}_{SC}/\mathbf{V}_{OC} = 1/\mathbf{Z}_T$. You can interchange the equivalents whenever it is convenient.

What approach should you use in a given problem? First, consider the simplest methods: series or parallel combination, voltage or current division, source transformation. The Thévenin and Norton theorems are powerful, especially in drawing general conclusions about a circuit. Avoid writing simultaneous equations if possible, but use Kirchhoff's laws and phasor diagrams to check your results.

Practice Problem 10-6

In the circuit of Example 7, $\mathbf{V}_g = 14.6\underline{/76°}$ V. Replace the generator and transmission line by a Norton equivalent and use current division to check the value of $\mathbf{I}_C$.

FREQUENCY RESPONSE

The different responses of inductance and capacitance to sinusoidal signals can be exploited in communication circuits. Because $X_L = \omega L$ is directly proportional to frequency, whereas $X_C = -1/\omega C$ is of opposite sign and inversely proportional to frequency, inductive and capacitive elements can be combined in networks that are *frequency selective*. For example, in a radio receiver a parallel combination of L and C is used to select a single signal from the myriad radio waves that are intercepted by the antenna. The extreme degree of frequency selectivity achievable is one of the key properties of electric circuits.

Simple Filters

The simple *RL* circuit of Fig. 10.19a is often employed to discriminate between input signals of different frequencies. For an input voltage of constant amplitude V and varying frequency $\omega = 2\pi f$, the impedance is $\mathbf{Z} = R + j\omega L$. At low frequencies, the current is limited by the resistance alone and $I_{lo} \cong V/R$. At high frequencies, the ωL

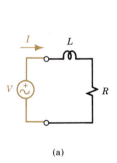

(a)

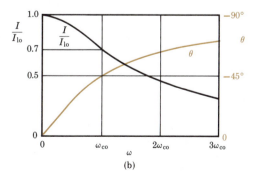

(b)

Figure 10.19   A frequency-selective circuit.

term predominates and $I_{hi} \cong V/\omega L$ or current is inversely proportional to frequency. In general,

$$\mathbf{I}(\omega) = \frac{\mathbf{V}}{\mathbf{Z}} = \frac{\mathbf{V}}{R + j\omega L} = \frac{V/R}{1 + j\omega L/R} \qquad (10\text{-}25)$$

or, since $\mathbf{I}_{lo} = \mathbf{V}/R$,

$$\frac{\mathbf{I}}{\mathbf{I}_{lo}} = \frac{1}{1 + j\omega L/R} = \frac{1}{\sqrt{1 + (\omega L/R)^2}} \underline{/-\tan^{-1}(\omega L/R)} = \frac{I}{I_{lo}}\underline{/\theta} \qquad (10\text{-}26)$$

The variation of magnitude and phase angle with frequency is shown in Fig. 10.19b. At the intermediate frequency

$$\omega_{co} = \frac{R}{L} \qquad (10\text{-}27)$$

resistive and reactive components of impedance are equal and

$$\frac{\mathbf{I}}{\mathbf{I}_{lo}} = \frac{1}{1 + j1} = \frac{1}{\sqrt{2}}\underline{/-45°} = 0.707\underline{/-45°} \qquad (10\text{-}28)$$

At this "70% current point," the power delivered to the circuit (I^2R) drops to one-half the maximum value. This *half-power frequency* is an easily calculated measure of the frequency response. Because signals of frequencies below ω_{co} are passed on to the resistance, this *RL* circuit is said to be a "low-pass filter" with a "cutoff frequency" of ω_{co}. (See Exercise 10.)

The simple *RC* circuit of Fig. 10.20a can serve as a high-pass filter because at high frequencies the current is limited by the resistance alone and $I_{hi} = V/R$. In general,

$$\mathbf{I} = \frac{\mathbf{V}}{\mathbf{Z}} = \frac{\mathbf{V}}{R - j1/\omega C} = \frac{V/R}{1 - j1/\omega CR} \qquad (10\text{-}29)$$

or, in dimensionless form,

$$\frac{\mathbf{I}}{\mathbf{I}_{hi}} = \frac{1}{1 - j1/\omega CR} = \frac{1}{\sqrt{1 + (1/\omega CR)^2}}\underline{/\tan^{-1}(1/\omega CR)} = \frac{I}{I_{hi}}\underline{/\theta} \qquad (10\text{-}30)$$

as shown in Fig. 10.20b. Here the cutoff or half-power frequency is defined by $(1/\omega CR) = 1$ or

$$\omega_{co} = \frac{1}{RC} \qquad (10\text{-}31)$$

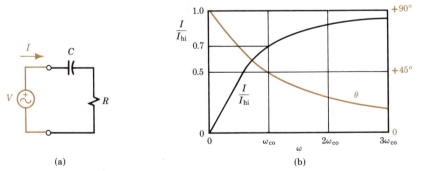

(a)　　　　　　　　　　　　　　(b)

Figure 10.20 A simple high-pass filter.

In the high-pass filter of Fig. 10.20a, specify C so that the cutoff frequency is 500 rad/s for $R = 5000\ \Omega$. What is the half-power frequency in hertz?

Answers: $0.4\ \mu F$, 79.6 Hz.

In the typical application of the circuit of Fig. 10.20a, the filter output is the voltage across R, which is directly proportional to I, and V_{out} as a function of ω has the same shape as I/I_{hi} in Fig. 10.20b and the same cutoff frequency. If R and C are interchanged and the output taken across C, the cutoff frequency ω_{co} will be the same, but now the output voltage will be high at low frequencies because $\mathbf{Z}_C = -j/\omega C$ is large at low frequencies. In fact, the variation of V_{out} as a function of ω will follow the curve of I/I_{lo} of Fig. 10.19b, and this RC combination acts as a low-pass filter.

L-Section Filters

Since the RL circuit of Fig. 10.19a is a low-pass filter, it could be used in a power supply (see p. 84) to reduce the ripple while passing the dc components of the rectified wave without opposition. However, inductors in the sizes required for low ripple are relatively expensive; for a given investment, the combination of inductance and capacitance shown in Fig. 10.21b gives better results. Such a two-port is called an L-*section* because of its geometry.

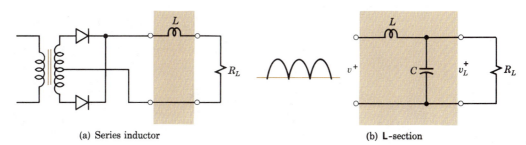

(a) Series inductor   (b) L-section

Figure 10.21   Filters employing inductors.

If the inductance L is greater than a certain critical value[†] L_c where

$$L_c \cong \frac{R_L}{6\pi f} \cong \frac{R_L}{1000} \tag{10-32}$$

for 60-Hz operation, the flywheel effect of the inductance keeps the current flowing continuously (instead of in short pulses), and the output voltage changes only slightly with changes in load resistance R_L. Predicting the filtering action under this condition is a straightforward application of ac circuit analysis.

As represented in Fig. 10.21b, the output of a full-wave rectifier consists of a series of half-sinewaves. Such a periodic wave can be represented by a series (the

[†] See Millman and Halkias, *Electronic Devices and Circuits*, McGraw-Hill Book Co., New York, 1967.

Fourier series) of sinusoidal components of various frequencies and amplitudes. In this case, the series is

$$v = \frac{2}{\pi}V_m(1 - \tfrac{2}{3}\cos 2\omega t - \tfrac{2}{15}\cos 4\omega t - \tfrac{2}{35}\cos 6\omega t - \cdots) \qquad (10\text{-}33)$$

The first term, a constant, represents the dc component of relative magnitude 1. The second term is a second harmonic of relative magnitude $\frac{2}{3}$ and frequency 2ω. The succeeding terms in this infinite series are smaller and, furthermore, they are easier to filter out. Our analysis is concerned with the second harmonic term because, if that is filtered out effectively, all the higher harmonics are removed even more effectively. (See Exercise 17.)

A common problem is to determine the ripple factor in the voltage across R_L for given values of L and C. These values are not critical, and it is convenient to make simplifying assumptions that reduce the work involved. In the circuit of Fig. 10.22, let us assume that:

1. The input voltage to the filter is $v = (2/\pi)V_m - (4/3\pi)V_m \cos 2\omega t$, representing the dc and second harmonic terms of the rectified voltage.
2. The resistance of inductor L is negligibly small and, therefore, the full dc component of voltage appears across R_L.
3. The reactance $1/2\omega C$ is small compared to R_L (as it must be for good filtering) and, for the parallel combination, $Z_{RC} = Z_{23} \cong -1/2\omega C$.
4. The reactance of the inductor $2\omega L$ is large compared to Z_{RC} (as it must be for good filtering), and $Z_{13} \cong (2\omega L - 1/2\omega C) \cong 2\omega L$.

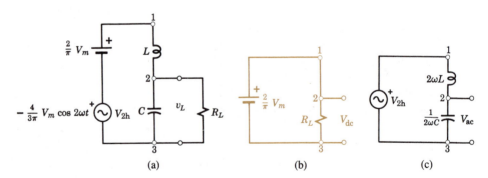

(a)   (b)   (c)

Figure 10.22 Analysis of an L-section filter by superposition.

Because the filter is linear, the principle of superposition is applicable and the circuit of Fig. 10.22a can be replaced by an equivalent combination of two simpler circuits. For the effective values shown on the voltage divider of Fig. 10.22c (based on assumptions 3 and 4),

$$\frac{V_{ac}}{V_{2h}} = \frac{Z_{23}}{Z_{13}} \cong \frac{1/2\omega C}{2\omega L} \cong \frac{1}{4\omega^2 LC} \qquad (10\text{-}34)$$

or the output is

$$V_{ac} = \frac{1}{4\omega^2 LC}V_{2h} = \frac{1}{4\omega^2 LC}\frac{4V_m}{3\pi\sqrt{2}} \qquad (10\text{-}35)$$

The desired result of rectification and filtering is direct current. As a measure of the effectiveness of the process, we define the *ripple factor r* where

$$r = \frac{I_{ac}}{I_{dc}} = \frac{V_{ac}}{V_{dc}} = \frac{\text{rms value of ac components}}{\text{dc component}} \qquad (10\text{-}36)$$

For the L-section filter, the second harmonic is the only significant ripple component, so the ripple factor is

$$r = \frac{V_{ac}}{V_{dc}} = \frac{4V_m}{4\omega^2 LC(3\pi\sqrt{2})} \cdot \frac{\pi}{2V_m} = \frac{\sqrt{2}/3}{4\omega^2 LC} = \frac{0.47}{4\omega^2 LC} \qquad (10\text{-}37)$$

The ω^2 factor in the denominator of Eq. 10-35 justifies our assumption that higher harmonics can be neglected. Example 9 illustrates ripple calculation.

EXAMPLE 9

A load requiring 50 mA at 20 V ($R_L = 400\,\Omega$) is to be supplied from a full-wave rectifier with an L-section filter consisting of $L = 20$ H and $C = 40\ \mu\text{F}$. Check the assumptions made in deriving Eq. 10-37 and determine the ripple factor.

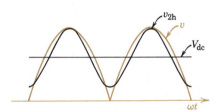

Figure 10.23 Calculating ripple factor from the Fourier components of the filter input.

By Eq. 10-32, $L_c = 400/1000 = 0.4$ H. Therefore, $L > L_c$ and the ac analysis is applicable.

Assuming 60-Hz operation, for the second harmonic,

$$X_L = 2\omega L = 2(2\pi\,60)20 = 15{,}080\ \Omega$$

$$X_C = -\frac{1}{2\omega C} = -\frac{1}{2(2\pi 60)40 \times 10^{-6}} = -33\ \Omega$$

and assumptions 3 and 4 appear to be justified.

The rms value of the second harmonic input component in Fig. 10.23 is, by Eq. 10.33,

$$V_{2h} = \frac{2}{3\sqrt{2}}V_{dc} \cong 9.4\ \text{V}$$

The filter reduces this to (Eq. 10-34)

$$V_{ac} \cong \frac{1/2\omega C}{2\omega L}V_{2h} = \frac{33}{15{,}080}9.4 = 0.021\ \text{V}$$

The ripple factor is

$$r = \frac{V_{ac}}{V_{dc}} = \frac{0.021}{20} \cong 0.001 \cong 0.1\%$$

The same result is obtained directly from Eq. 10-37.

To reduce the ripple factor further, a second L-section can be added before the load resistor. For two identical sections, the denominator of Eq. 10-37 becomes $(4\omega^2 LC)^2$.

RESONANCE

In a series *RLC* circuit, the impedances to sinusoidal signals presented by the inductive and capacitive elements are of opposite sign and they vary with frequency in inverse fashion. Inductive reactance ωL is directly proportional to frequency and capacitive reactance $-1/\omega C$ is inversely proportional to frequency. If an inductance and a capacitance are connected in series, there is always a frequency at which the two reactances just cancel. A practical circuit for achieving this effect consists of a coil in series with a capacitor.

Series Resonance

For a series *RLC* circuit the impedance is

$$\mathbf{Z} = R + j\omega L + \frac{1}{j\omega C} = R + j\left(\omega L - \frac{1}{\omega C}\right) \tag{10-38}$$

$$\mathbf{Z} = \sqrt{R^2 + \left(\omega L - \frac{1}{\omega C}\right)^2} \bigg/ \tan^{-1}\frac{\omega L - 1/\omega C}{R} \tag{10-39}$$

The frequency at which the reactances just cancel and the impedance is a pure resistance is the *resonant frequency* ω_0.

In Fig. 10.24, where $\theta_Z = 0$ and $(\omega_0 L - 1/\omega_0 C) = 0$, the resonant frequency is

$$\omega_0 = \frac{1}{\sqrt{LC}} \, \text{rad/s} \quad \text{or} \quad f_0 = \frac{1}{2\pi\sqrt{LC}} \, \text{Hz} \tag{10-40}$$

By definition, at series resonance the impedance $Z_0 = R$ and $\theta_Z = 0$. At frequencies lower than ω_0, the capacitive reactance term predominates; the impedance increases

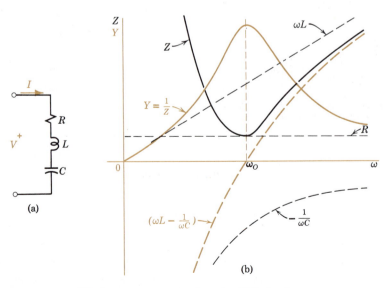

Figure 10.24 The frequency response of a series *RLC* circuit.

rapidly as frequency decreases, and θ_Z approaches $-90°$. At frequencies higher than ω_0, the inductive reactance term predominates; the impedance increases rapidly with frequency, and θ_Z approaches $+90°$. (Can you visualize the curve of θ_Z as a function of ω?)

Practice Problem 10-8

Calculate the resonant frequency in radians per second for a series combination of $R = 5\ \Omega$, $L = 8$ mH, and $C = 0.05\ \mu$F, and calculate X_L and X_C at this frequency.

Answers: 50 krad/s, 400 Ω, $-400\ \Omega$.

Also shown in Fig. 10.24 is the variation with frequency of admittance $Y = 1/Z$. The admittance is high at resonance, where $Y_0 = 1/R$, and it decreases rapidly as frequency is changed from ω_0. Because $I = YV$, the variation of current with a constant amplitude voltage has the same shape as the admittance curve. If we apply voltages of various frequencies to a series RLC circuit, one voltage of frequency ω_0 will be favored and a relatively large current will flow in response to this particular voltage. The circuit is *frequency selective*.

EXAMPLE 10

The variable capacitor behind a radio dial provides a capacitance of 1 nF when the dial setting is "80" on the AM band. Calculate the inductance of the associated coil.

A dial setting of "80" means a frequency of 800 kHz. The coil-capacitor combination provides resonance at this frequency; therefore, $\omega_0 L = 1/\omega_0 C$ and

$$L = \frac{1}{\omega_0^2 C} = \frac{1}{(2\pi f_0)^2 C}$$

$$= \frac{1}{(2\pi \times 8 \times 10^5)^2 \times 1 \times 10^{-9}} = 39.6\ \mu\text{H}$$

Phasor Diagram Interpretation

Additional information about the phenomenon of resonance is revealed by phasor diagrams drawn for various frequencies. Because current is common to each element in this series circuit, phasor $\mathbf{I}$ (rms value) is drawn as a horizontal reference in Fig. 10.25. The voltage $\mathbf{V}_R$ across the resistance is in phase with the current, and voltages $\mathbf{V}_L$ and $\mathbf{V}_C$ across the inductance and capacitance, respectively, lead and lag the current. For $\omega = \omega_0$, $\mathbf{V}_L$ and $\mathbf{V}_C$ just cancel (Why?), and the voltage across the resistance is exactly equal to the applied voltage $\mathbf{V}$. For $\omega < \omega_0$, $\mathbf{V}_C$ is greater than $\mathbf{V}_L$ and the sum of the three voltages is $\mathbf{V}$. In comparison to the case for $\omega = \omega_0$, a larger $\mathbf{V}$ is required for the same current or the admittance is lower; θ_Z is negative indicating a leading power factor. For $\omega > \omega_0$, $\mathbf{V}_L$ is greater than $\mathbf{V}_C$ and the current lags $\mathbf{V}$ (the sum of the three component voltages) by angle θ_Z.

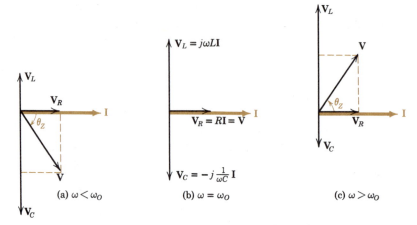

Figure 10.25 Phasor diagrams for a series *RLC* circuit.

For a given current **I**, the impedance **Z** is directly proportional to the voltage **V**. Therefore, the voltage diagrams of Fig. 10.25 can be transformed into impedance diagrams. For $\omega < \omega_O$, the impedance is predominantly capacitive. For $\omega = \omega_O$, the impedance is a pure resistance and has a minimum value. For $\omega > \omega_O$, the impedance is predominantly inductive and increases with frequency.

EXAMPLE 11

A generator supplies a variable frequency voltage of constant amplitude = 10 V (rms) to the circuit of Fig. 10.26 where $R = 5\ \Omega$, $L = 4$ mH, and $C = 0.1\ \mu F$. The frequency is to be varied until a maximum rms current flows. Predict the maximum current, the frequency at which it occurs, and the resulting voltages across the inductance and capacitance.

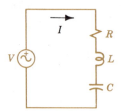

Figure 10.26 A series resonant circuit.

For a series *RLC* circuit, maximum current corresponds to maximum admittance, which occurs at the resonant frequency. At $\omega = \omega_O$, $\omega_O L = 1/\omega_O C$, and $Y_O = 1/Z_O = 1/R$. Therefore,

and

$$I = VY = \frac{V}{R} = \frac{10}{5} = 2\ \text{A}$$

$$\omega_O = \frac{1}{\sqrt{LC}} = \frac{1}{\sqrt{4 \times 10^{-3} \times 10^{-7}}}$$

$$= 5 \times 10^4\ \text{rad/s}$$

Then

$$V_L = \omega LI = 5 \times 10^4 \times 4 \times 10^{-3} \times 2 = 400\ \text{V}$$

and

$$V_C = \frac{I}{\omega C} = \frac{2}{5 \times 10^4 \times 10^{-7}} = 400\ \text{V}$$

As expected, V_L is just equal to V_C and the phasor diagram is similar to that in Fig. 10.25b.

Quality Factor Q

How can the voltage across one series element be greater than the voltage across all three, as in Example 11? The answer is related to the fact that *L* and *C* are energy-storage elements and high *instantaneous* voltages are possible. As indicated in Fig. 10.25b, phasors V_L and V_C are 180° out of phase; instantaneous voltages are also

180° out of phase, and a high positive voltage across L is cancelled by a high negative voltage across C. (See Exercise 39). The voltages do exist, however, and they can be measured with ordinary voltmeters. The ability to develop high element voltages is a useful property of resonant circuits, and we need a convenient measure of this property.

At resonance, $\omega_0 L = 1/\omega_0 C$ and $I = I_0 = V/R$. Therefore, the reactance voltages are

$$V_L = \omega_0 L I_0 = \omega_0 L \frac{V}{R} = \frac{\omega_0 L}{R} V \tag{10-41}$$

and

$$V_C = \frac{I_0}{\omega_0 C} = \frac{V}{\omega_0 CR} = \frac{\omega_0 L}{R} V = V_L \tag{10-42}$$

By definition, in a series circuit

$$\frac{\omega_0 L}{R} = \frac{1}{\omega_0 CR} = Q \tag{10-43}$$

where Q is the *quality factor*, a dimensionless ratio.

In a practical circuit, R is essentially the resistance of the coil since practical capacitors have very low losses in comparison to practical inductors. Hence, Q is a measure of the energy-storage property (L) in relation to the energy dissipation property (R) of a coil or a circuit. A well-designed coil may have a Q of several hundred, and an RLC circuit employing such a coil will have essentially the same Q since the capacitor contributes very little to R. In practical coils, the losses (and therefore R in the circuit model) increase with frequency, and $Q = \omega L/R$ is fairly constant over a limited range of frequencies.

Equation 10-41 indicates that for a given value of current there is a "resonant rise in voltage" across the reactive elements equal to Q times the applied voltage. In addition to indicating the magnitude of this resonant rise, Q is a measure of the frequency selectivity of the circuit. A circuit with high Q discriminates sharply; a low-Q circuit is relatively unselective. The solid admittance curve of Fig. 10.27a is redrawn from Fig. 10.24; for the same L and C, the admittance curve near resonance depends on R. Increasing the resistance by 50% reduces Y_0 to $\frac{2}{3}$ of its original value and the selectivity is reduced. Decreasing the resistance to $\frac{2}{3}$ of its original value (and thereby increasing Q by a factor of $\frac{3}{2}$) increases Y_0 and increases selectivity.

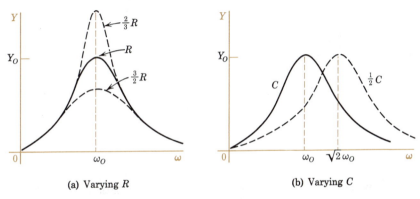

(a) Varying R (b) Varying C

Figure 10.27   The effect of circuit parameters on frequency response.

Practice Problem 10-9

An inductor of 20-mH inductance and 6.28-Ω resistance is used in a series circuit resonant at 5000 Hz.

(a) Specify the necessary value of capacitance.
(b) What total applied voltage will produce a voltage of 5 V across the capacitance at resonance?
(c) If the capacitance is one-half the value specified in part (a), what is the resonant frequency? The Q of the new circuit?

Answers: (a) 0.0507 μF; (b) 0.05 V; (c) 7071 Hz, 100.

The effect of changing one reactive element while keeping the other and the resistance constant is indicated by Fig. 10.27b. Although ω_0 is shifted, the general shape of the response curve is still the same. A *general resonance equation* that describes series *RLC* circuits with various parameters and various resonant frequencies would be very valuable. Is it possible that such a general equation also describes parallel resonant circuits? Let us investigate that possibility.

Parallel Resonance

Figure 10.28 shows a parallel *GCL* circuit. For this circuit the admittance is

$$\mathbf{Y} = G + j\omega C + \frac{1}{j\omega L} = G + j\left(\omega C - \frac{1}{\omega L}\right) \tag{10-44}$$

$$\mathbf{Y} = \sqrt{G^2 + \left(\omega C - \frac{1}{\omega L}\right)^2} \Big/ \tan^{-1}\frac{\omega C - (1/\omega L)}{G} \tag{10-45}$$

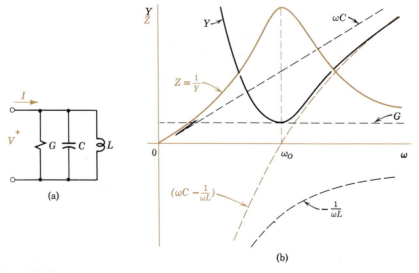

(a)

(b)

Figure 10.28 The frequency response of a parallel *GCL* circuit.

The frequency at which the susceptances just cancel and the admittance is a pure conductance is the resonant frequency ω_0. Where $\theta_Y = 0$ and $(\omega_0 C - 1/\omega_0 L) = 0$, the resonant frequency is

$$\omega_0 = \frac{1}{\sqrt{CL}} \text{ rad/s} \quad \text{or} \quad f_0 = \frac{1}{2\pi\sqrt{CL}} \text{ Hz} \tag{10-46}$$

By definition, at parallel resonance the admittance $Y_0 = G$ and $\theta_Y = 0$. At frequencies lower than ω_0, the inductive susceptance term predominates; the admittance increases rapidly as frequency decreases and θ_Y approaches $-90°$. At frequencies higher than ω_0, the capacitive susceptance term predominates; the admittance increases rapidly with frequency and θ_Y approaches $+90°$.

The phasor diagrams for a parallel GCL circuit are shown in Fig. 10.29. In this case there is a "resonant rise in current" in the reactive elements. At resonance, $V = V_0 = I/G$. Therefore, the reactance currents are

$$I_C = \omega_0 C V_0 = \omega_0 C \frac{I}{G} = \frac{\omega_0 C}{G} I = Q_P I \tag{10-47}$$

and

$$I_L = \frac{V_0}{\omega_0 L} = \frac{I}{\omega_0 LG} = \frac{\omega_0 C}{G} I = Q_P I \tag{10-48}$$

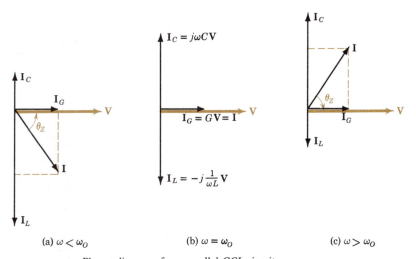

(a) $\omega < \omega_0$ (b) $\omega = \omega_0$ (c) $\omega > \omega_0$

Figure 10.29   Phasor diagrams for a parallel GCL circuit.

The expressions for Q in a parallel circuit are quite different from those in a series circuit; in fact, if the same circuit elements are reconnected, $Q_P = 1/Q_S$. But basically the phenomena described by Q are the same in both circuits. In essence, quality factor is a measure of the energy-storage property of any circuit in relation to its energy dissipation property. Note that for a reactive *component* such as a coil, $Q = \omega L/R$, a function of frequency; for a resonant *circuit*, however, Q is evaluated at ω_0 and is a constant. (See Example 12 on p. 306.)

EXAMPLE 12

A typical radio receiver employs parallel resonant circuits in selecting the desired station. Using the values from Example 10 ($C = 1$ nF and $L = 39.6$ μH), the resonant frequency is again 800 kHz and the dial setting is "80." If the *GCL* linear model has a Q of 100, what is the value of G?

If two signals of the same amplitude (50 μA rms) but different frequencies (f_1 at "80" and f_2 at "85") are introduced from the antenna, what are the respective voltages developed across the tuned circuit?

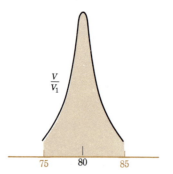

Figure 10.30   Frequency selectivity.

For a Q of 100, Eq. 10-47 yields

$$G = \frac{\omega_0 C}{Q_P} = \frac{2\pi \times 8 \times 10^5 \times 10^{-9}}{100} = 50.3 \ \mu S$$

and the equivalent resistance is

$$R = \frac{1}{G} \cong 20{,}000 \ \Omega$$

At the resonant frequency f_1,

$$V_1 = \frac{I_1}{Y_O} = \frac{I_1}{G} = \frac{50 \times 10^{-6}}{50.3 \times 10^{-6}} \cong 1 \ V$$

At frequency $f_2 = 850$ kHz, $\omega = 2\pi \times 8.5 \times 10^5$ rad/s,

$$Y_2 = \sqrt{G^2 + (\omega C - 1/\omega L)^2}$$
$$= \sqrt{(50.3)^2 + (5341 - 4728)^2} = 615 \ \mu S$$

and

$$V_2 = \frac{I_2}{Y_2} = \frac{50 \times 10^{-6}}{615 \times 10^{-6}} \cong 0.08 \ V$$

For the same impressed current, the voltage response for the desired signal is more than 12 times as great as that for the undesired signal (Fig. 10.30), and the power developed in G is more than 150 times as great since $P = V^2 G$.

Normalized Response

The frequency selectivity demonstrated in Example 12 is characteristic of all resonant circuits, and the response curves of all such circuits have the same general shape. Now we wish to derive a simple but general equation that describes series and parallel circuits, resonant at high and low frequencies, with large and small energy dissipation. To eliminate the effect of specific parameter values, we use dimensionless ratios. For illustration we choose to work with the admittance of a series *RLC* circuit, but the result has a more general interpretation. For the series circuit of Fig. 10.31a,

$$\mathbf{Z} = R + j\left(\omega L - \frac{1}{\omega C}\right) = \frac{1}{\mathbf{Y}}$$

and, at resonance,

$$\mathbf{Z}_O = R = \frac{1}{\mathbf{Y}_O}$$

The dimensionless ratio of the admittance at any frequency ω to that at the resonant frequency ω_O is

$$\frac{\mathbf{Y}}{\mathbf{Y}_O} = \frac{\mathbf{Z}_O}{\mathbf{Z}} = \frac{R}{R + j\left(\omega L - \dfrac{1}{\omega C}\right)} = \frac{1}{1 + j\left(\dfrac{\omega L}{R} - \dfrac{1}{\omega C R}\right)} \qquad (10\text{-}49)$$

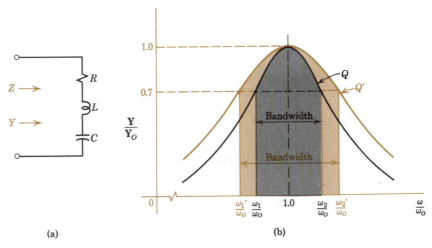

(a)

(b)

Figure 10.31 The normalized response of a series *RLC* circuit.

Introducing the factor ω_0/ω_0 and letting the dimensionless ratio $\omega_0 L/R = Q = 1/\omega_0 CR$, the admittance ratio for a series *RLC* circuit is

$$\frac{\mathbf{Y}}{\mathbf{Y}_o} = \frac{1}{1 + j\left(\dfrac{\omega L}{R} \cdot \dfrac{\omega_0}{\omega_0} - \dfrac{1}{\omega CR} \cdot \dfrac{\omega_0}{\omega_0}\right)} = \frac{1}{1 + jQ\left(\dfrac{\omega}{\omega_0} - \dfrac{\omega_0}{\omega}\right)} \qquad (10\text{-}50)$$

This simple equation also describes the impedance ratio $\mathbf{Z}/\mathbf{Z}_o$ for a parallel *GCL* circuit; the even simpler reciprocal equation describes the impedance ratio for a series circuit and the admittance ratio for a parallel circuit.

Bandwidth

The curves in Fig. 10.31b have been normalized; all variables are dimensionless ratios. The difference in the two curves shown is due to a difference in selectivity. A convenient quantitative measure of selectivity is defined by letting the imaginary term in the denominator of Eq. 10-50 be equal to $\pm j1$. Where

$$\left(\frac{\omega}{\omega_0} - \frac{\omega_0}{\omega}\right) = \pm\frac{1}{Q}$$

then

$$\frac{\mathbf{Y}}{\mathbf{Y}_o} = \frac{1}{1 \pm j1} = \frac{1}{\sqrt{2}}\underline{/\mp 45^\circ} \qquad (10\text{-}51)$$

Imposing this condition defines two frequencies, ω_1 and ω_2, such that

$$\frac{\omega_1}{\omega_0} - \frac{\omega_0}{\omega_1} = -\frac{1}{Q} \quad \text{and} \quad \frac{\omega_2}{\omega_0} - \frac{\omega_0}{\omega_2} = +\frac{1}{Q} \qquad (10\text{-}52)$$

Frequencies ω_1 and ω_2 are called the lower and upper "70% points," because at these frequencies the magnitude Y/Y_o is $1/\sqrt{2} = 0.707$. For a given applied voltage, current is proportional to admittance and these are also called "70% current points." More generally, these are *half-power frequencies* because power is proportional to the square of the current. (Do ω_1 and ω_2 also designate half-power frequencies in parallel circuits? Yes, because Eq. 10-50 also describes the relative impedance Z/Z_o that defines the frequency response of a parallel resonant circuit.)

The frequency range between the half-power points, $\omega_2 - \omega_1$, is called the *bandwidth*, a convenient measure of the selectivity of the circuit. Solving Eqs. 10-52 and selecting the consistent roots, we obtain

$$\omega_1 = \omega_0 \sqrt{1 + \left(\frac{1}{2Q}\right)^2} - \frac{\omega_0}{2Q} \quad \text{and} \quad \omega_2 = \omega_0 \sqrt{1 + \left(\frac{1}{2Q}\right)^2} + \frac{\omega_0}{2Q} \quad (10\text{-}53)$$

Subtracting the first equation from the second, the bandwidth is

$$\omega_2 - \omega_1 = \frac{\omega_0}{Q} \quad \text{or} \quad f_2 - f_1 = \frac{f_0}{Q} \quad (10\text{-}54)$$

Note that the bandwidth is inversely proportional to the quality factor Q, a very convenient result. (See Example 13.)

EXAMPLE 13

A series circuit is to be resonant at 800 kHz. Specify the value of C required for a given inductor ($L = 40\ \mu H$ and $R = 4.02\ \Omega$) and predict the bandwidth. Assume that the capacitor is ideal, i.e., introduces no resistance.

For the specified resonant frequency,

$$C = \frac{1}{\omega_0^2 L} = \frac{1}{(2\pi 8 \times 10^5)^2 \times 4 \times 10^{-5}} = 0.99\ \text{nF}$$

and the Q of the circuit or the inductor is

$$Q = \frac{\omega_0 L}{R} = \frac{2\pi 8 \times 10^5 \times 4 \times 10^{-5}}{4.02} = 50$$

Then the bandwidth is, by Eq. 10-54,

$$f_2 - f_1 = \frac{f_0}{Q} = \frac{800 \times 10^3}{50} = 16\ \text{kHz}$$

In the typical case where $Q \geq 10$, the factor $\sqrt{1 + (1/2Q)^2} \cong 1$ and

$$\omega_1 = \omega_0 - \frac{\omega_0}{2Q} \quad \text{and} \quad \omega_2 = \omega_0 + \frac{\omega_0}{2Q} \quad (10\text{-}55)$$

In other words, the resonance curve is symmetric and the half-power points are equidistant from the resonant frequency, or

$$\omega_2 - \omega_0 = \omega_0 - \omega_1 = \frac{\omega_0}{2Q} \quad (10\text{-}56)$$

This simple relation makes it easy to sketch the response curve of a series or parallel resonant circuit by plotting three points.

Practice Problem 10-10

A parallel circuit with a Q of 60 has a maximum impedance of 10,000 Ω at 240 kHz. Calculate the frequencies at which the impedance drops to 70.7% of the maximum value and sketch the impedance as a function of frequency.

Answers: $f_1 = 238$ kHz, $f_2 = 242$ kHz, $f_O = 240$ kHz.

Practical Resonant Circuits

In focusing attention on principles, we neglected some practical aspects of resonant circuits. The practical parallel circuit consists of an inductor in parallel with a capacitor, and the only resistance is usually that due to losses in the inductor. The linear model of Fig. 10.32 is the appropriate representation of the actual circuit; if Q is 20 or more, the general equations (Eqs. 10-43, 46, 47, 48 and 56) hold. Example 14 illustrates practical circuit design.

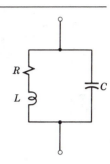

Figure 10.32 Model of a practical parallel circuit.

EXAMPLE 14

Design a parallel circuit to be resonant at 800 kHz with a bandwidth of 32 kHz. Use the inductor from Example 13 with $L = 40$ μH and $R = 4.02$ Ω and predict the impedance at resonance.

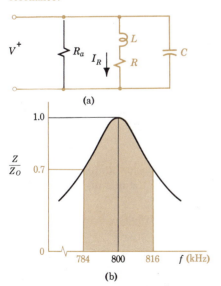

(a)

(b)

Figure 10.33 Design of a parallel resonant circuit.

The desired quality factor is

$$Q = \frac{f_O}{f_2 - f_1} = \frac{800 \times 10^3}{32 \times 10^3} = 25$$

Because $Q > 20$, Eq. 10-46 holds for this practical circuit and $C = 1/\omega_O^2 L = 0.99$ nF as before. From Example 13, the Q of the inductor is 50. Therefore, to double the bandwidth we must halve the Q and double the losses. One possibility is to add another 4.02-Ω resistance in series with R.

Another possibility is shown in Fig. 10.33a. To double the losses, add R_a so that

$$P_a = \frac{V^2}{R_a} = I_R^2 R \cong \left(\frac{V}{\omega_O L}\right)^2 R$$

or

$$R_a = \frac{(\omega_O L)^2}{R} \cdot \frac{R}{R} = Q^2 R = 50^2 \times 4.02 = 10,050\ \Omega$$

The parallel combination of inductor, capacitance, and resistance can be replaced by the model of Fig. 10.28a where (Eqs. 10-44 and 10-48),

$$G = Y_O = \frac{1}{Z_O} = \frac{1}{Q\omega_O L}$$

Therefore,

$$Z_O = Q\omega_O L = 25 \times 2\pi 8 \times 10^5 \times 4 \times 10^{-5} = 5025\ \Omega$$

The emphasis in this discussion is on a variable frequency ω. In many cases the circuit is "tuned" by varying a capacitor or, less frequently, an inductor. This corresponds to varying ω_O in Eq. 10-50, and the same general resonance behavior is observed.

The specific values of the circuit parameters are usually determined by economic considerations. At audiofrequencies (20 to 20,000 Hz), inductors with iron cores are used to obtain inductances in the order of henrys. At radio broadcast frequencies (above 500,000 Hz), air-core inductors with inductances in the order of microhenrys are resonated with variable air-dielectric capacitors with capacitances in the order of nanofarads. Reasonable values of resistance and quality factor are obtained in properly designed inductors using copper wire. At higher frequencies, undesired resonance may occur because of the inductance of a short lead (a small fraction of a microhenry) and the capacitance between two adjacent conductors (a few picofarads).

SUMMARY

■ A sinusoidal function of time $a = A_m \cos(\omega t + \theta)$ can be interpreted as the real part of the complex function $A_m e^{j(\omega t + \theta)} = A e^{j\theta} \cdot \sqrt{2} e^{j\omega t}$.
The complex constant $A e^{j\theta}$ is defined as phasor $\mathbf{A}$, the transform of $a(t)$.
Phasor calculations follow the rules of complex algebra.

■ Phasor voltage and current are related by the complex impedance $\mathbf{Z} = Z(j\omega)$ or the complex admittance $\mathbf{Y} = Y(j\omega)$.

$$\mathbf{V} = \mathbf{ZI} \quad \text{where} \quad Z_R(j\omega) = R \quad Z_L(j\omega) = j\omega L \quad Z_C(j\omega) = 1/j\omega C$$

$$\mathbf{I} = \mathbf{YV} \quad \text{where} \quad Y_R(j\omega) = G \quad Y_L(j\omega) = 1/j\omega L \quad Y_C(j\omega) = j\omega C$$

For phasors, Kirchhoff's laws are written: $\Sigma \mathbf{V} = 0$ and $\Sigma \mathbf{I} = 0$.

■ The sinusoidal response of a two-terminal network is completely defined by

$$\mathbf{Z} = Z\underline{/\phi_Z} = R + jX \quad \text{or} \quad \mathbf{Y} = Y\underline{/\phi_Y} = G + jB$$

where $\phi_Z = \tan^{-1}(X/R)$ $\phi_Y = \tan^{-1}(B/G)$
    R = ac resistance in ohms G = ac conductance in siemens
    X = reactance in ohms B = susceptance in siemens
    $= \omega L$ or $-1/\omega C$ $= \omega C$ or $-1/\omega L$

■ To determine the forced response to sinusoids:
1. Transform time functions to phasors and evaluate complex immittances.
2. Combine immittances in series or parallel to simplify the circuit.
3. Determine the desired response in phasor form.
4. Draw a phasor diagram to check values and display results.
5. Transform phasors to time functions if required.

■ The frequency response of simple filters is characterized by ω_{co}, the cutoff or half-power or 70% current frequency.
In L-section filters, the inductor opposes current changes and the capacitor absorbs voltage peaks; ac circuit analysis shows that the ripple is

$$r = \frac{V_{ac}}{V_{dc}} \cong \frac{0.47}{4\omega^2 LC} \quad \text{(with full-wave rectification)}$$

- A circuit containing inductance and capacitance is in resonance if the terminal voltage and current are in phase. At the resonant frequency the impedance and admittance are purely real.

  For series RLC or parallel GCL circuits, $\omega_o = 1/\sqrt{LC}$ or $f_o = 1/2\pi\sqrt{LC}$.

- The frequency selectivity of a resonant circuit is determined by bandwidth.

$$BW = \omega_2 - \omega_1 = \frac{\omega_o}{Q} \qquad \text{where} \qquad \omega_1 \text{ and } \omega_2 \text{ are half-power frequencies}$$

$$\text{For } Q \geq 10, \; \omega_1 = \omega_o - \frac{\omega_o}{2Q} \qquad \text{and} \qquad \omega_2 = \omega_o + \frac{\omega_o}{2Q}$$

- For resonant circuits:

$$\text{Series:} \quad Q = \frac{\omega_o L}{R} = \frac{1}{\omega_o CR} \qquad \text{and} \qquad V_L = V_C = QV \text{ at } \omega_o$$

$$\text{Parallel:} \quad Q = \frac{\omega_o C}{G} = \frac{1}{\omega_o LG} \qquad \text{and} \qquad I_L = I_C = QI \text{ at } \omega_o$$

$$\text{Series } \frac{\mathbf{Y}}{\mathbf{Y}_o} = \text{parallel } \frac{\mathbf{Z}}{\mathbf{Z}_o} = \frac{1}{1 + jQ\left(\dfrac{\omega}{\omega_o} - \dfrac{\omega_o}{\omega}\right)}$$

REVIEW QUESTIONS

1. Write out in words a definition of a phasor.
2. How can a phasor, a constant quantity, represent a variable function of time?
3. Given $i_1 = 10 \cos (1000t + \pi/2)$ and $i_2 = 5 \cos 2000t$, is the phasor method applicable to finding $i_1 + i_2$? Explain.
4. Describe the quantity designated by the code symbols $\mathbf{I} = 20\underline{/\pi/4}$, $\omega = 60$.
5. Sketch three phasors on the complex plane and show the graphical construction to obtain $\mathbf{V}_1 + \mathbf{V}_2 - \mathbf{V}_3$.
6. Write an equation for $i(t)$ in terms of $\mathbf{I}$.
7. How can we neglect the imaginary part of the exponential function used in phasor representation?
8. What is the distinction between a phasor and a complex quantity?
9. Outline in words the reasoning that leads to the relation $\Sigma\mathbf{I} = 0$ into any node.
10. Define impedance and admittance.
11. Define reactance, susceptance, and immittance.
12. Given $\mathbf{I} = 10\underline{/150°}$ and $\mathbf{V} = 200\underline{/-150°}$, is the associated $\mathbf{Z}$ inductive or capacitive? What is the phase angle ϕ?
13. Explain in words the two equations used to define linearity.

14. Under what circumstances is Thévenin's theorem useful?
15. Is the same power dissipated in an active circuit and the Norton equivalent?
16. How are voltage and current sources "removed"?
17. Given a circuit containing R, L, and C elements, outline the general procedure for determining the current in one element in response to a sinusoidal input.
18. Explain how reactive elements can be connected to remove high-frequency components of an input signal.
19. What actually happens when you turn the tuning dial of a radio?
20. Above ω_o, is a series RLC circuit capacitive or inductive? Below ω_o?
21. How can the voltage across one series element in an RLC circuit be greater than the total voltage across all three?
22. Why is Q called the "quality factor"?
23. Name two important virtues of resonant circuits.
24. Do ω_1 and ω_2 defined by Eq. 10-52 designate half-power frequencies in parallel circuits?

EXERCISES

1. Sketch, approximately to scale, the following functions:
   (a) $v_1 = 10 \cos (628t - \pi/3)$ V.
   (b) $v_2 = 30 \sin (628t + 60°)$ V.
   (c) $i = 2 \cos (377t + \pi/4)$ A.
2. Represent the time functions in Exercise 1 by phasors in polar and rectangular form.
3. Express as a function of time, first in general and then in specific terms, each of the following sinusoidal signals:
   (a) A current having a period of 30 ms and passing through a positive maximum of 10 A at $t = 5$ ms.
   (b) A voltage having a period of 40 ms and passing through a negative maximum of -120 V at $t = 10$ ms.
   (c) A voltage having a frequency of 150 Hz and an amplitude of 20 V and passing through zero with positive slope at $t = 1$ ms.
   (d) A current reaching a positive maximum of 5 A at $t = 3$ μs and the next negative maximum at $t = 8$ μs.
   (e) A sinusoidal voltage that passes through a negative maximum of -2 V at $t = 5$ μs and the next positive maximum at $t = 15$ μs.
4. Represent the signals in Exercise 3 by phasors in polar and rectangular form.
5. Given $\mathbf{A} = 3 + j4$, $\mathbf{B} = 2e^{j(\pi/6)}$, and $\mathbf{C} = 4\underline{/-90°}$, sketch the three quantities on the complex plane and find:
   (a) $\mathbf{AB}$, (b) $\mathbf{B} - \mathbf{C}$, (c) $\mathbf{C}/\mathbf{A}$, (d) $\mathbf{CA}^*$,
   (e) $\mathbf{B} + \mathbf{AC}$, (f) $(\mathbf{B} + \mathbf{C})/\mathbf{A}$, (g) $\mathbf{B}^2$,
   (h) $\sqrt{\mathbf{AC}}$, (i) $(\mathbf{C}/\mathbf{B})^{1/3}$.
6. Write the time functions represented by:
   (a) $\mathbf{I} = 3 + j4$.
   (b) $\mathbf{V} = 2e^{-j(\pi/6)}$.
   (c) $\mathbf{I} = 4\underline{/-90°}$.
7. Given $\mathbf{A} = 6e^{j(\pi/4)}$, $\mathbf{B} = 0 - j2$, and $\mathbf{C} = 3\underline{/-60°}$, sketch the three quantities on the complex plane and find:
   (a) $\mathbf{AB}$, (b) $\mathbf{B} - \mathbf{C}$; (c) $\mathbf{A}/\mathbf{C}$, (d) $(\mathbf{B} + \mathbf{C})\mathbf{A}$,
   (e) $\mathbf{CA}^*$, (f) $\mathbf{B} + \mathbf{AC}$; (g) $\mathbf{B}^2$, (h) $\sqrt{\mathbf{AC}}$,
   (i) $(\mathbf{C}/\mathbf{B})^{1/3}$.
8. Find:
   (a) The fourth roots of (-1).
   (b) The cube roots of $(-1 + j\sqrt{3})$.

9. In Fig. 10.34, $v_1 = 8\sqrt{2} \cos (\omega t - 30°)$ V and $v_2 = 10\sqrt{2} \cos (\omega t + 45°)$ V. Estimate $\mathbf{V}$ graphically from a phasor diagram drawn to scale and then calculate $\mathbf{V}$ from the component phasors. Determine $v(t)$.

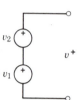

Figure 10.34

10. In Fig. 10.35, $i_1 = 4 \cos (\omega t + 30°)$ A and $i_2 = 5 \cos (\omega t - 30°)$ A.
    (a) Determine $\mathbf{I}$ graphically from a phasor diagram drawn to scale and check by an analytical phasor solution and determine $i(t)$.
    (b) Expand the trigonometric expressions and determine $i(t)$ using trigonometric identities. Compare the difficulty of the two methods.

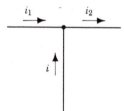

Figure 10.35

11. For a current $i = 0.5\sqrt{2} \cos (2000t + 30°)$ A, find the phasor voltage across:
    (a) Resistance $R = 5$ Ω.
    (b) Inductance $L = 10$ mH.
    (c) Capacitance $C = 20$ μF.
12. For a voltage $v = 12\sqrt{2} (5000t - 45°)$ V, find the phasor current in:
    (a) Resistance $R = 3$ kΩ.
    (b) Inductance $L = 2$ mH.
    (c) Capacitance $C = 8$ μF.
13. A current $i = 2\sqrt{2} \cos (2000t - \pi/4)$ A flows through a series combination of $R = 20\,\Omega$ and $L = 10$ mH.
    (a) Determine the voltage across each element.
    (b) Sketch the two voltages, $v = f(\omega t)$, approximately to scale and estimate the amplitude

and phase angle of $v_R + v_L$.

(c) Perform the addition of voltages in phasor form and compare the sum to the estimate in part (b).

14. A current $i(t) = 2\sqrt{2}\cos(5000t + 30°)$ mA flows in a series combination of $R = 2309\ \Omega$ and $L = 0.8$ H. Determine total voltage $v(t)$:

(a) By finding the individual phasor voltages and then adding them before transforming **V** to $v(t)$.

(b) By finding the total impedance and using **V = ZI**.

15. In Fig. 10.36, $v(t) = 12\sqrt{2}\cos(500t + 60°)$ mA, $R = 200\ \Omega$, and $C = 20\ \mu$F. Calculate the element impedances, the individual phasor currents, the total phasor current, and $i(t)$.

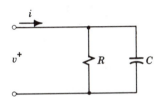

Figure 10.36

16. In Fig. 10.37, $R = 2$ kΩ, $C = 0.5\ \mu$F, $L = 2$ H, and $i_R = 10\sqrt{2}\cos(1000t + 90°)$ mA. Use Kirchhoff's laws for phasors to calculate **V**, $\mathbf{I}_L$, and **I**. Show $\mathbf{I}_R$, $\mathbf{V}_R$, $\mathbf{V}_C$, **V**, $\mathbf{I}_L$, and **I** on a phasor diagram.

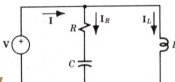

Figure 10.37

17. In Fig. 10.37, $R = 2\ \Omega$, $L = 0.2$ H, and $i_R = 10\sqrt{2}\cos(10t + 45°)$ A.

(a) On a phasor diagram, show $\mathbf{I}_R$ and $\mathbf{V}_R$ and draw a dashed arrow to show the direction of $\mathbf{V}_C$.

(b) If the phasor $\mathbf{V} = V\underline{/0°}$, show **V** and $\mathbf{V}_C$ on the diagram.

(c) Show $\mathbf{V}_L$ and determine $v(t)$.

18. When a voltage $v = 20\sqrt{2}\cos 120t$ V is applied across an impedance, the current is $i = 4\sqrt{2}\cos(120t + 37°)$ A; find the real and imaginary components of the complex impedance **Z**.

19. A voltage $v(t) = 20\sqrt{2}\cos(1000t - 50°)$ V is applied across a parallel combination of $R = 2$ kΩ and $C = 0.5\ \mu$F. Determine total current $i(t)$:

(a) By finding the individual phasor currents and then adding them before transforming **I** to $i(t)$.

(b) By finding the total admittance and using **I = YV**.

20. In Fig. 10.36, $R = 2$ kΩ and $C = 2.5\ \mu$F.

(a) Determine the two components of a *series* circuit having the same terminal admittance at $\omega = 1000$ rad/s.

(b) Repeat part (a) for $\omega = 500$ rad/s.

21. A two-terminal network consists of $R = 300\ \Omega$ and $C = 2\ \mu$F in series.

(a) Determine the two components of a *parallel* network having the same terminal impedance at $\omega = 1000$ rad/s.

(b) Repeat part (a) for $\omega = 2000$ rad/s.

22. An inductor of 8-Ω reactance and 6-Ω resistance is connected in parallel with a capacitor of 8-Ω reactance across a 120-V, 60-Hz line. Calculate the total admittance, total current, and branch currents in phasor form. Show $\mathbf{I}_C$, $\mathbf{I}_L$, and **I** on a phasor diagram using $V\underline{/0°}$ as a reference.

23. In Fig. 10.37, $L = 0.2$ H, $R = 3$ kΩ, and $C = 10$ nF.

(a) Determine the two components of a series circuit having the same terminal impedance at $\omega = 20$ krad/s.

(b) Repeat for $\omega = 10$ krad/s.

24. In the circuit of Fig. 10.38, $v = 5\sqrt{2}\cos 5t$ V, $R = 2\ \Omega$, $L = 1$ H, and $C = 0.05$ F.

(a) Find the input admittance $Y(j\omega)$.

(b) Determine $\mathbf{I}_R$, **I**, and $\mathbf{I}_L$ and show them with **V** and $\mathbf{V}_C$ on a phasor diagram.

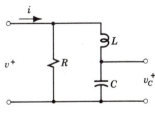

Figure 10.38

25. For the circuit of Exercise 24, use the voltage-divider concept to determine $v_C(t)$.

26. In Fig. 10.39, $L_1 = L = 2$ mH, $C = 500$ μF, and $R = 2\,\Omega$.
(a) For $v = 10\sqrt{2}\cos(1000t + 90°)$ V, determine the input impedance.
(b) Calculate currents $\mathbf{I}$, $\mathbf{I}_R$, and $\mathbf{I}_L$ and show them on a phasor diagram along with $\mathbf{V}$, $\mathbf{V}_{L_1}$, and $\mathbf{V}_C$.

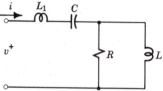

Figure 10.39

27. In the circuit of Exercise 26, L_1 is changed to 3 mH. Use node voltages to find $\mathbf{V}_R$ and $\mathbf{I}_R$.

28. In the circuit of Exercise 27, you are interested in only the current $\mathbf{I}_R$. Use Thévenin's theorem to isolate R and calculate $\mathbf{I}_R$.

29. In Fig. 10.40, the current in the 12-Ω inductive reactance is $12\underline{/90°}$ A. Use network theorems to find the current in L.

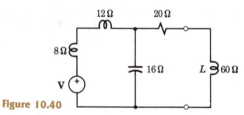

Figure 10.40

30. In Fig. 10.40, the voltage across the capacitive reactance is $20\underline{/-56.75°}$ V. Use network theorems to find the voltage across L.

31. In Fig. 10.40, the reactances are given in ohms for $\omega = 1000$ rad/s. Use superposition to find the steady-state current in L if:
(a) $\mathbf{V}$ is replaced by $v = 120 + 120\sqrt{2}\cos 1000t$ V.
(b) $\mathbf{V}$ is replaced by $v = 120\sqrt{2}\cos 1000t + 120\cos 4000t$ V.

32. For a load resistance $R = 20$ kΩ (Fig. 10.19), design a simple RL low-pass filter with a cutoff frequency at $f = 200$ Hz.

33. For a load resistance $R = 10$ kΩ (Fig. 10.20a), design a simple RC high-pass filter with a half-power frequency of 2000 Hz. For an input $V = 5$ V at 1000 Hz, what is the output voltage V_R?

34. Given a load resistance R, an inductance L, and a capacitance C:
(a) Arrange L, C, and R in two different ways to make a high-pass filter.
(b) Arrange L, C, and R in two different ways to make a low-pass filter.

35. The circuit of Fig. 10.41 is one type of filter. The input consists of four sinusoidal components with $\omega_1 = 10$, $\omega_2 = 100$, $\omega_3 = 1000$, and $\omega_4 = 10{,}000$ rad/s.
(a) Derive a general expression for the transfer ratio V_o/V_i.
(b) For $C = 10$ μF, $L = 1$ H, and $R = 10$ kΩ, evaluate the ratio V_o/V_i for each component and draw a conclusion.

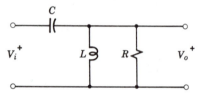

Figure 10.41

36. The output of a 60-Hz full-wave rectifier is applied to the filter in Fig. 10.21. At 60 Hz (NOTE!), the filter reactances are $X_L = 2000$ Ω and $X_C = 200$ Ω. The dc voltage is 300 V across $R_L = 2$ kΩ.
(a) Predict the ac ripple voltage (rms) in the output.
(b) Predict the rms value of the next higher harmonic in the output.

37. Design a 60-Hz full-wave rectifier circuit to supply 200 V at 50 mA with less than 1% ripple factor. Use an L-section filter with $C = 10$ μF. Assuming that the rectifier output voltage is represented by the first two terms of the Fourier series:
(a) Calculate R_L and the required transformer secondary voltage (rms), stating any necessary assumptions.
(b) Specify the required value of inductance L. (Is your approach valid?)
(c) Determine the factor by which the filter has reduced the magnitude of the ac voltage component.
(d) Predict, without detailed calculation, the ac voltage across R_L if two L-sections in cascade are used.

38. Examine the dial of a radio and determine the frequency range of the AM broadcast band. Calculate the range of capacitance needed with a 100-μH inductance to tune over the band.

39. A 10-nF capacitor is connected in series with a coil of 2-mH inductance and 5-Ω resistance.
  (a) Determine the resonant frequency.
  (b) If a voltage of 2 V is to appear across the capacitor at resonance, what total voltage must be applied to the series combination?
  (c) How can the voltage across one series component be larger than the total applied voltage? Sketch v_C and v_L as functions of ωt.

40. A signal source supplies $i = 20\sqrt{2}\cos 2000t$ mA to a series RLC circuit and meters arranged to read V_{total} and V_C. C is adjusted until the total voltage across R, L, and C is a minimum, 2 V; the voltage across C alone is observed to be 40 V.
  (a) Draw the circuit of this "Q meter."
  (b) Calculate the Q of the circuit.
  (c) Calculate C, L, and R.

41. In Fig. 10.42, $G = 200\ \mu S$, $C = 0.5\ \mu F$, and $L = 5$ mH.
  (a) Determine the resonant frequency ω_O.
  (b) For $I = 10$ mA at ω_O, determine I_G, I_C, and I_L and show them on a phasor diagram with voltage V.

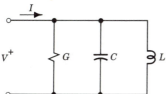

Figure 10.42

42. Design a series RLC circuit for a resonant frequency of 200 krad/s and a bandwidth of 5 krad/s. The available coil has an inductance of 2.5 mH and a Q of 50.

43. In Fig. 10.43, the vertical scale is current in milliamperes.
  (a) Design a circuit with the specified current response to an input voltage of constant amplitude $V = 62.8$ V.
  (b) Draw and label your circuit.

44. For the parallel circuit of Fig. 10.32:
  (a) Derive an expression for admittance as a function of ω.
  (b) For $R = 25\ \Omega$ and $L = 0.05$ mH, specify C for resonance at 10^7 rad/s.
  (c) For your circuit, estimate the Q, estimate ω_O following the suggestion on p. 309, compare to the specified ω_O, and draw a conclusion.

45. Design a parallel RLC circuit to provide an impedance at resonance of 50,000 Ω, a resonant frequency of 2 MHz, and a Q of 80. Replace the parallel RL combination by an equivalent series combination $R_S L_S$ as in the practical circuit of Fig. 10.32.

46. In Fig. 10.43, the vertical scale is voltage in volts.
  (a) Design a circuit to give the voltage response shown for an input current $I = 1$ mA.
  (b) Draw the circuit as you would construct it in the lab, labeling the components.

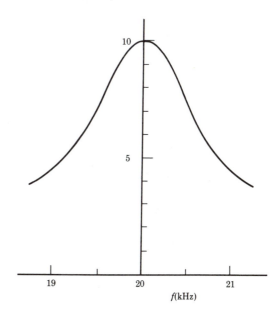

Figure 10.43

11

Operational Amplifiers

Inverting Circuits

Noninverting Circuits

Nonlinear Applications

Practical Considerations

The Analog Computer

We learned in Chapter 3 that the name *operational amplifier* was first applied to the amplifiers employed in analog computers to perform mathematical operations such as summing and integration. With sophisticated integrated-circuit (IC) amplifiers available for less than a dollar, the design of signal processing equipment has been radically altered. For linear or analog systems, the IC op amp plays the same basic building-block role as do the IC logic and memory elements in digital systems. In digital information processing systems, we use op amps to convert physical signals from and to the real world. Combinations of op amps and digital devices are widely used in instrumentation and control.

Our purpose here is to learn how these versatile amplifiers work and how we can use them in practical circuits. We start by reviewing ideal operational amplifiers and then look at some more circuits and applications. Next we consider some practical limitations and design aspects. Finally we look at the electronic analog computer to see how we can arrange operational amplifiers to simulate physical systems.

Ideal Operational Amplifiers

An op amp is a direct-coupled, high-gain voltage amplifier designed to amplify signals over a wide frequency range. Typically, it has two input terminals and one output terminal and a gain of at least 10^5, and it is represented by the symbol in Fig.

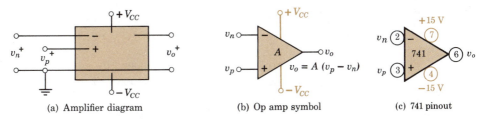

(a) Amplifier diagram

(a) Amplifier diagram   (b) Op amp symbol   (c) 741 pinout

Figure 11.1   Operational amplifier diagram, symbol, and pinout.

11.1b. It is basically a *differential amplifier* responding to the difference in the voltages applied to the positive and negative input terminals. (Single-input op amps correspond to the special case where the + input is grounded.) Normally we use an op amp with external feedback networks that determine the function performed.

The characteristics of the ideal op amp are as follows:

Voltage gain $A = \infty$
Output voltage $v_o = 0$ when $v_n = v_p$
Bandwidth $BW = \infty$
Input impedance $Z_i = \infty$
Output impedance $Z_o = 0$

Although these are extreme specifications, commercial units approach the ideal so closely that we can design many practical circuits assuming that these characteristics are available. One practical limitation that cannot be ignored is that, for linear operation, the output voltage v_o cannot exceed $\pm V_{CC}$. (See the linear region in Fig. 7.9b).

INVERTING CIRCUIT APPLICATIONS

In most cases, impedances Z_i and Z_o can be considered to be pure resistances R_i and R_o as shown in the model of Fig. 11.2a. In the amplifier circuit of Fig. 11.2b, we

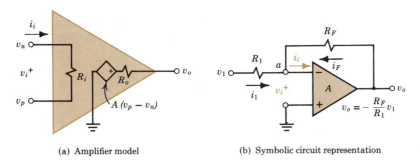

(a) Amplifier model   (b) Symbolic circuit representation

Figure 11.2   The basic inverting amplifier circuit.

apply the input signal to the negative terminal and ground the positive terminal. The input voltage v_1 is applied in series with resistance R_1, and the output voltage v_o is fed back through resistance R_F. Because the signal voltage is inverted, the feedback current is of opposite sign and i_F tends to cancel the input current i_1, leaving only a very small difference i_i. Another interpretation is that the part of v_o fed back tends to cancel the effect of v_1, leaving only a very small v_i; we say that "the high gain A drives v_i to zero."

For an ideal op amp, closely approximated by a commercial unit, $v_i = 0$ and $i_i = 0$; therefore, the sum of the currents into node a is

$$i_1 + i_F = \frac{v_1}{R_1} + \frac{v_o}{R_F} = 0 \tag{11-1}$$

and the gain of the inverting amplifier circuit is

$$\frac{v_o}{v_1} = A_F = -\frac{R_F}{R_1} \tag{11-2}$$

This basic relation has a variety of interesting interpretations.

Practice Problem 11-1

In the circuit of Fig. 11.2, $R_i = 1\ M\Omega$, $R_o \cong 0$, $A = 2.5 \times 10^5$, $R_1 = 10\ k\Omega$, and $R_F = 1\ M\Omega$. For $v_o = -5$ V, calculate $v_a = v_i$, v_1, i_i, and $i_1 = -i_F$. In an ideal op amp, $v_i/v_1 \ll 1$ and $i_i/i_1 \ll 1$; on that basis, can this op amp be considered ideal?

Answers: 0.02 mV, 50 mV, 0.00002 μA, 5 μA; yes.

Low-Pass Filter

A *passive filter* is a frequency-selective network consisting of passive resistors, inductors, and capacitors. (See Fig. 10.19.) Amplifiers can be used to realize *active filters* that avoid the use of bulky and expensive inductors and have other advantages. Low-cost op amps with desirable input and output properties are widely used in active filters.

As a simple example, consider Fig. 11.3a. This is a form of the general inverting circuit with impedance elements Z_F and Z_1 in the feedback network. Here V_1 is a

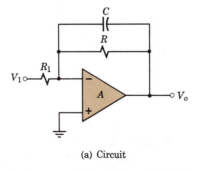

(a) Circuit

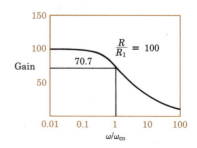

(b) Frequency response

Figure 11.3   A simple low-pass filter.

sinusoidal signal of variable frequency or a combination of signals of various frequencies. The voltage gain (a function of frequency here) is

$$\mathbf{A}_F = \frac{\mathbf{V}_o}{\mathbf{V}_1} = -\frac{\mathbf{Z}_F}{\mathbf{Z}_1} = -\frac{\dfrac{1}{(1/R) + j\omega C}}{R_1} = -\frac{R/R_1}{1 + j\omega RC} \qquad (11\text{-}3)$$

The cutoff or half-power frequency (see p. 296) is defined by $\omega_{co} RC = 1$, or

$$\omega_{co} = \frac{1}{RC} \qquad (11\text{-}4)$$

The voltage gain decreases rapidly above ω_{co} and this serves as a *low-pass filter*.

The response curve of Fig. 11.3b, drawn for $R/R_1 = 100$, does not exhibit a sharp cutoff. More complicated circuits designed by more sophisticated methods and using several op amps can provide sharp high-, low-, or band-pass filtering.[†] In designing active filters for high-frequency applications, the frequency response of the op amp itself must be taken into account.

Digital-to-Analog Converter

To translate a digital number to an analog signal, we need a digital-to-analog converter (DAC). Most commonly, the digital input is a binary word and the analog output is a voltage or current. For example, the 4-bit number

$$\mathbf{1100} = 1 \times 2^3 + 1 \times 2^2 + 0 \times 2^1 + 0 \times 2^0 = 8 + 4 + 0 + 0 = 12D \qquad (11\text{-}5)$$

could be translated to an output signal amplitude of 12 V. The conversion involves a weighted sum corresponding to the output of the op amp summing circuit of Fig. 3.23.

Practice Problem 11-2

Design a digital-to-analog converter to convert the binary number stored in a 4-bit register into a decimal-equivalent voltage. Assume the register provides inputs at precisely 0 V and +4 V. Let $R_F = 20$ kΩ.

Answers: R_1(MSB) $= 10$ kΩ, $R_2 = 20$ kΩ, $R_3 = 40$ kΩ, $R_4 = 80$ kΩ.

For accurate conversion, the inputs to a summing-circuit DAC would have to be precisely known voltages, a condition that is *not* required in digital systems. The various weighting resistances would also have to be precisely formed, a very difficult requirement in an IC 16-bit converter. In practice, the R-$2R$ *ladder* of Fig. 11.4a is far superior.

To predict the behavior of the R-$2R$ DAC, we replace the ladder by its Thévenin equivalent. With all voltage sources removed, we note that the resistance to

[†] See p. 151 of reference 5 at the end of this chapter.

READ

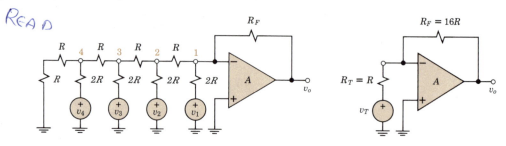

Figure 11.4 The R-$2R$ ladder op amp DAC.

the left of node 4 is $R + R = 2R$. Then we see that the equivalent resistance to the left of node 3 is

$$R + 2R \| 2R = 2R$$

and the same for nodes 2 and 1! By voltage division, the contribution of source v_1 is $v_1(2R/4R) = v_1/2$. After redrawing the circuit (see Exercise 12), the contribution of source v_2 is seen to be $v_2(2R/4R)(2R/4R) = v_2/4$. The complete Thévenin equivalent is

$$v_T = \frac{v_1}{2} + \frac{v_2}{4} + \frac{v_3}{8} + \frac{v_4}{16} \qquad \text{and} \qquad R_T = R \qquad (11\text{-}6)$$

As shown in Fig. 11.4b, for $R_F = 16R$ the output voltage is

$$v_o = -(8\,v_1 + 4\,v_2 + 2\,v_3 + v_4) \qquad (11\text{-}7)$$

Therefore, this circuit provides the desired conversion using only two resistance values. Furthermore, only the resistance *ratios* must be precise (easy to provide in ICs), and it turns out that the resistance "seen" by each binary source is the same (desirable for uniform behavior). The practical result is that high-precision DACs are available at low cost.

Current/Voltage Converters

An example of op amp versatility is in driving a low-resistance load (a coil, for example) with a voltage source of high internal resistance. The *voltage-to-current converter* shown in Fig. 3.25 performs this function effectively if the load R_o can "float" with neither side grounded. If one side of the load must be grounded, the remainder of the circuit (including the op amp power supply) must be isolated from ground.

A good constant-current source for a grounded load can be made from an op amp and an external transistor. In Fig. 11.5a, reference voltage V_1 is obtained from a voltage divider. Feedback drives v_i to zero so that $V_E = V_1$; if V_E tends to fall below V_1, the forward bias on the base-emitter (np) junction decreases and emitter current I_E drops, bringing V_E back to V_1. In effect, a voltage $V_S - V_1 = V_S \cdot R_2/(R_1 + R_2)$ establishes an emitter current in the *pnp* transistor and current

$$I_L = I_C \cong I_E = \frac{V_S \cdot R_2/(R_1 + R_2)}{R} = V_S \frac{R_2}{R(R_1 + R_2)} \qquad (11\text{-}8)$$

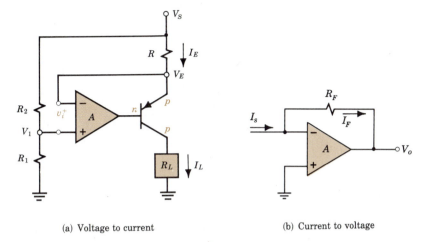

(a) Voltage to current   (b) Current to voltage

Figure 11.5 Voltage-current converter circuits.

is maintained in a changing load R_L. The voltage source V_S has been converted into a current source I_L.

The inverse function is performed by the circuit of Fig. 11.5b. Here a given current I_s is introduced into the $-$ terminal of the op amp, setting up an equal current in the feedback resistor R_F. Because the amplifier input voltage v_i is practically zero, the output voltage is

$$V_o = -I_F R_F = -I_s R_F \tag{11-9}$$

and the source current has been converted into a voltage. The *current-to-voltage converter* is useful where a current is to be measured without introducing an undesired resistance $(R_{oF} \cong 0)$ or where a current source, such as a photocell, has a high shunt resistance.

EXAMPLE 1

A photodiode generates 0.2 μA of current per μW of radiant power for a constant reverse-bias voltage. Design a circuit to measure radiant power with a voltmeter.

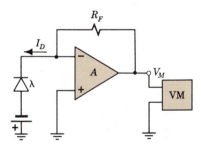

Figure 11.6 Photodiode light meter.

The photocurrent I_D (see p. 139) could be allowed to flow through a resistance to develop a measurable voltage, but this voltage would reduce the reverse bias and alter the photodiode characteristic.

Instead, we use the circuit of Fig. 11.6. Since $v_i \cong 0$, there is no bias voltage change with a change in I_D.

If $R_F = 5$ kΩ and the radiant power is 100 μW, the voltmeter will read

$$V_M = I_D R_F = (100 \times 0.2) \times 10^{-6} \times 5 \times 10^3 = 100 \text{ mV}$$

and the scale factor of the "light meter" is 1 μW/mV.

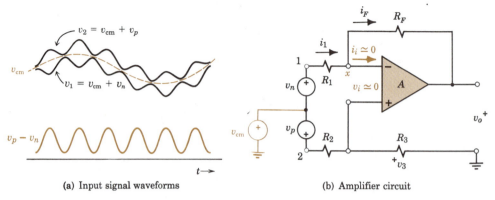

(a) Input signal waveforms   (b) Amplifier circuit

Figure 11.7   A direct-coupled differential amplifier.

Differential Amplifier

The *direct-coupled differential amplifier* shown in Fig. 11.7 is widely used in instrumentation. In the typical transducer, a physical variable such as a change in temperature or pressure is converted to an electrical signal that is then amplified to a useful level. The differential amplifier is advantageous because it discriminates against dc variations, drifts, and noise and responds only to the significant changes.

In Fig. 11.7, voltages v_1 and v_2 represent variables measured at two points in a complicated system—like the human body. We see that the measurements include a large *common-mode* voltage and a small *differential* voltage. The common-mode voltage has a dc component, perhaps due to drift of operating temperature, and a sinusoidal "hum" component picked up from power lines. The small differential voltage $v_p - v_n$ is the signal that we are interested in and wish to amplify.

The circuit model of signal sources and amplifier is shown in Fig. 11.7b. Assuming an ideal op amp, $v_i = 0$ and the potential at node x is equal to the voltage across R_3, or

$$v_3 = (v_{cm} + v_p)\frac{R_3}{R_2 + R_3} = v_x \qquad (11\text{-}10)$$

Because $i_i = 0$, the currents i_1 and i_F are equal, or

$$i_1 = \frac{v_{cm} + v_n - v_x}{R_1} = i_F = \frac{v_x - v_o}{R_F} \qquad (11\text{-}11)$$

Combining these equations and simplifying yields

$$v_o = v_{cm}\left(\frac{R_3}{R_2 + R_3} - \frac{R_F}{R_1} + \frac{R_F}{R_1} \cdot \frac{R_3}{R_2 + R_3}\right)$$

$$+ v_p\left(\frac{R_3}{R_2 + R_3} + \frac{R_F}{R_1} \cdot \frac{R_3}{R_2 + R_3}\right) - v_n\left(\frac{R_F}{R_1}\right) \qquad (11\text{-}12)$$

If we design the network so that $R_F/R_1 = R_3/R_2$, the coefficient of v_{cm} goes to zero, the *common-mode signal* is rejected, and the output is

$$v_o = -\frac{R_F}{R_1}(v_n - v_p) \qquad (11\text{-}13)$$

The ability of a differential amplifier to discriminate against signals that affect both inputs in common and to amplify the small differences in input is a valuable characteristic.

Practice Problem 11-3

In a direct-coupled differential amplifier, $R_1 = 10 \text{ k}\Omega$ and $R_2 = 20 \text{ k}\Omega$. Specify R_3 and R_F for common-mode signal rejection and a differential amplifier gain of 50. If the inputs are $v_1 = 5 + \sin 100t + 0.05 \sin 5000t$ V and $v_2 = 5 + \sin 100t - 0.03 \sin 5000t$ V, predict the output voltage.

Answers: 1 MΩ, 0.5 MΩ, 4 sin 5000t V.

NONINVERTING CIRCUIT APPLICATIONS

In Fig. 11.8, the input signal is applied to the noninverting + terminal, and a fraction of the output signal is fed back to the − terminal. Here R_1 and R_F constitute a voltage divider across the output voltage. For an ideal op amp with $v_i = 0$,

$$v_1 - \frac{R_1}{R_1 + R_F} v_o = v_i = 0 \tag{11-14}$$

and

$$A_F = \frac{v_o}{v_i} = \frac{R_1 + R_F}{R_1} \tag{11-15}$$

This basic noninverting amplifier has two distinctive characteristics. First, output signals are in phase with those at the input. Second, the input resistance is very high, approaching infinity in practical terms, and the output resistance is very low. This means that noninverting amplifiers do not "load" their sources and, in turn, they are not affected by their loads.

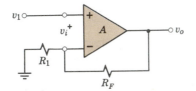

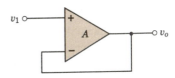

Figure 11.8   The noninverting amplifier circuit.   **Figure 11.9**   The voltage-follower circuit.

Voltage Follower

A useful special case of the noninverting circuit is shown in Fig. 11.9. Here $R_F = 0$ and $R_1 = \infty$ (open circuit). From Eq. 11-15, the circuit gain is now

$$A_F = \frac{v_o}{v_1} \cong \frac{R_1 + R_F}{R_1} = 1 \tag{11-16}$$

The output voltage is just equal to the input voltage and this is a *voltage follower*, so called because the potential at the v_o terminal "follows" the potential at the v_1 terminal.

EXAMPLE 2

In the circuit of Fig. 11.10a, $R_s = 1$ kΩ and $R_L = 10$ kΩ. For the op amp, $A = 10^5$, $R_i = 100$ kΩ, and $R_o = 100$ Ω. For $v_o = 10$ V, calculate v_s and v_o/v_s and estimate the input resistance of the circuit.

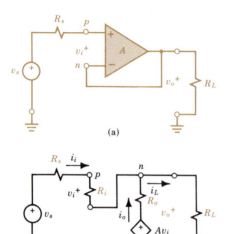

(a)

(b)

Figure 11.10 Voltage follower application.

The circuit is redrawn in Fig. 11.10b to show R_i in series with the input and R_o in series with Av_i as in Fig. 11.2a. For $v_o = 10$ V,

$$i_L = \frac{v_o}{R_L} = \frac{10}{10^4} = 10^{-3} \text{ A}$$

Expecting i_i to be very small, $i_o \cong i_L$ and we write

$$Av_i = v_o + i_o R_o \cong v_o + i_L R_o$$

$$= 10 + 10^{-3} \times 10^2 = 10.1 \text{ V}$$

$$\therefore v_i = (Av_i)/A = 10.1 \times 10^{-5} \text{ V}$$

Hence $i_i = v_i/R_i = v_i/10^5 = 1.01 \times 10^{-9}$ A and the assumption regarding i_i is justified. Then

$$v_s = v_o + i_i(R_s + R_i)$$

$$= 10 + 1.01 \times 10^{-9}(1.01 \times 10^5)$$

$$= 10.0001 \text{ V}$$

$$A_F = v_o/v_s = 10/10.0001 = 0.99999$$

This is indeed a voltage follower with unity gain. With feedback, the input resistance is

$$R_{iF} \cong \frac{v_s}{i_i} \cong \frac{10}{1.01 \times 10^{-9}} \cong 10^{10} \text{ Ω}$$

a very high value.

Unity-Gain Buffer

Example 2 demonstrates that the gain of a voltage follower is almost exactly 1; its usefulness lies in its ability to isolate a high-resistance source from a low-resistance load. To provide this isolation, the isolating network should have a very high input resistance and a very low output resistance. In general, such an isolating network is called a *buffer*. We cannot use the ideal op amp model to derive the general input and output characteristics of such a *unity-gain buffer* because they depend on the nonideal properties of the op amp. Instead we use the model shown in Fig. 11.11b that includes the finite values of R_i and R_o for the op amp, and we proceed without the simplifying assumptions made in Example 2.

Input Resistance. Writing loop equations for the rearranged circuit,

$$v_1 - R_i i_1 - R_o(i_1 - i_2) - Av_1 + Av_o = 0 \tag{11-17}$$

$$v_1 - R_i i_1 - R_L i_2 = 0 \tag{11-18}$$

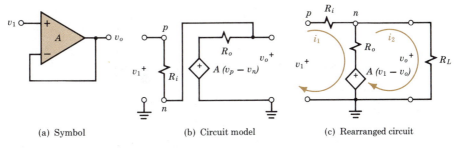

(a) Symbol   (b) Circuit model   (c) Rearranged circuit

Figure 11.11 Analysis of the unity-gain buffer circuit.

By inspection, $v_o = R_L i_2$ and, from Eq. 11-18, $i_2 = (v_1 - R_i i_1)/R_L$. Substituting these values in Eq. 11-17 yields

$$v_1 - R_i i_1 - R_o i_1 + R_o \frac{v_1 - R_i i_1}{R_L} - A v_1 + A R_L \frac{v_1 - R_i i_1}{R_L} = 0 \quad (11\text{-}19)$$

Solving, the input resistance with feedback is

$$R_{iF} = \frac{v_1}{i_1} = \frac{R_i + R_o + (R_o/R_L)R_i + A R_i}{1 + R_o/R_L}$$

$$= \frac{R_L R_i + R_L R_o + R_o R_i + A R_L R_i}{R_L + R_o} \quad (11\text{-}20)$$

For the practical case of A very large and $R_L \gg R_o$, this becomes

$$R_{iF} = \frac{(A+1)R_i R_L}{R_L} + \frac{R_o(R_L + R_i)}{R_L} \cong A R_i \quad (11\text{-}21)$$

A buffer using an op amp with $R_i = 100\ \text{k}\Omega$ and $A = 10^5$ would present an input resistance of about $10{,}000\ \text{M}\Omega$ (as in Example 2).

Output Resistance. Working from Fig. 11.11c, the output resistance can be obtained as v_{OC}/i_{SC}. For $i_2 = 0$,

$$v_{OC} = v_1 - R_i i_1 = v_1 - R_i \frac{v_1 - A v_1 + A v_{OC}}{R_i + R_o} \quad (11\text{-}22)$$

Solving,

$$v_{OC} = v_1 \frac{R_i + R_o + (A-1)R_i}{R_i + R_o + A R_i} = v_1 \frac{R_o + A R_i}{R_o + (1+A)R_i} \quad (11\text{-}23)$$

For the output shorted, $v_o = 0$ and

$$i_{SC} = \frac{v_1}{R_i} + \frac{A v_1}{R_o} = v_1 \frac{R_o + A R_i}{R_o R_i} \quad (11\text{-}24)$$

For the practical case of A very large and $R_i \gg R_o$, these yield

$$R_{oF} = \frac{v_{OC}}{i_{SC}} = \frac{R_o R_i}{R_o + (1+A)R_i} \cong \frac{R_o}{A} \quad (11\text{-}25)$$

A buffer using an op amp with $R_o = 100\ \Omega$ and $A = 10^5$ will present an output resistance of about $0.001\ \Omega$.

Practice Problem 11-4

A 741 op amp with typical characteristics (p. 463) is used as a unity-gain buffer. Estimate the input and output resistance of the buffer.

Answers: 400,000 MΩ, 0.38 mΩ.

EXAMPLE 3

An instrumentation transducer is characterized by a voltage $V_T = 5$ V in series with a resistance $R_T = 2000$ Ω. It operates an indicator characterized by an input resistance of 100 Ω. Predict the voltage and power delivered to the indicator with and without the use of a buffer.

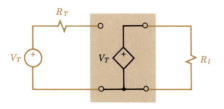

Figure 11.12   Buffer application.

Disregarding the buffer (Fig. 11.12), the voltage across the indicator would be

$$V_I = V_T \frac{R_I}{R_T + R_I} = 5\frac{100}{2100} = 0.238 \text{ V}$$

and the power delivered would be

$$P_I = V_I{}^2/R_I = (0.238)^2/100 = 0.566 \text{ mW}$$

With a unity-gain buffer, the output voltage follows the input. Since $R_i \rightarrow \infty$, no input current flows and, since $R_o \rightarrow 0$, there is no output voltage drop. Therefore, the model of the voltage follower is as shown. Now $V_I = V_T = 5$ V and the power is

$$P_I = V_I{}^2/R_I = 5^2/100 = 250 \text{ mW}$$

The power gain achieved by inserting the buffer is

$$\frac{P_2}{P_1} = \frac{250}{0.566} = 442$$

Voltage Regulator

The simple Zener diode voltage regulator of Fig. 5.18 is designed to minimize the effects of supply voltage fluctuation (V_1) and load current variation (I_L) on load voltage V_L. By using a high-gain amplifier and negative feedback, we can obtain much better regulation.

In designing a practical power supply, we choose a transformer to provide an ac voltage at the proper level, a diode rectifier to provide unidirectional current, a capacitor filter to develop an unregulated dc voltage, and we add a voltage regulator. In Fig. 11.13a, transistor T acts as a variable voltage dropping element "in series" with the load resistance R_L so that $V_{CE} = V_1 - V_L$. We use an op amp such as the 741 as a sensitive control of the *pass element* T. The Zener diode provides a constant reference voltage $V_Z = v_p$. The adjustable voltage divider R_1-R_2 takes a fraction $H = R_1/(R_1 + R_2)$ of the load voltage V_L and feeds it back to the inverting terminal

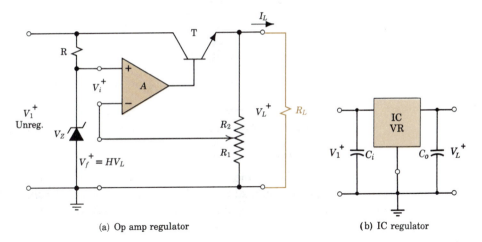

(a) Op amp regulator   (b) IC regulator

Figure 11.13   Voltage regulators with feedback.

so that $v_n = V_f = HV_L$. In effect, we compare V_f to the reference voltage V_Z; any difference is amplified and used to control the base current to transistor T. If for any reason the load voltage tends to drop, V_f decreases slightly, $V_i = v_p - v_n$ increases significantly, and the base current to T increases, increasing emitter current and stabilizing V_L.

The pass transistor must be able to dissipate the power $P_D = V_{CE} I_L$; an adequate "heat sink" is required. The variation of V_{BE} and V_Z with temperature is a principal limitation on the voltage stability of the circuit of Fig. 11.13a. Furthermore, this circuit provides no high-current protection; a short circuit may destroy T. These difficulties are overcome with the sophisticated circuitry available in IC form. Figure 11.13b shows a standard fixed-voltage monolithic regulator available at low cost. Capacitors C_i and C_o improve the operation by reducing the effect of transformer inductance at the input and minimizing the effect of sudden changes in load at the output. Such regulators will maintain specified voltages within 0.05% against wide variations in input voltage and load current.

NONLINEAR APPLICATIONS

The ideal op amp is a linear device in that the output is directly proportional to the input for all values of input voltage. There are several important nonlinear applications of op amps; one of the simplest is the *comparator*.

Comparator

In Fig. 11.14a, if the input voltage v_1 is larger than the reference voltage V_R, the output voltage v_o is positive. Because the gain is very large, only a small voltage difference will drive the amplifier into *saturation* because the maximum output swing is limited by the supply voltage. (See Fig. 7.9.) The transfer characteristic indicates that a small decrease in v_i (a fraction of a millivolt) will drive the op amp from positive saturation to negative saturation (tens of volts).

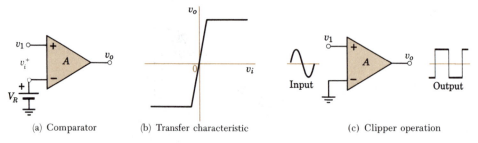

(a) Comparator   (b) Transfer characteristic   (c) Clipper operation

Figure 11.14 Nonlinear applications of an op amp.

If $V_R = 0$, this becomes a *zero-crossing comparator*. Such a comparator can be used to convert any alternating signal into a square wave by the clipping action shown in Fig. 11.14c. (A more sophisticated square-wave generator is shown in Fig. 11.17.)

Digital Voltmeter

The availability of stable, high-gain amplifiers whose characteristics are known precisely has contributed greatly to the art of instrumentation. By the use of amplifiers, the power required from the system under observation is reduced; the power needed to drive the indicating or recording instrument is provided by the amplifier. In contrast to the traditional moving-coil instrument, an electronic voltmeter may present an input impedance of millions of ohms and follow signal variations at megahertz frequencies.

Digital voltmeters, combining flexible digital controls and precise amplifiers, offer many advantages over other voltmeters: greater speed, higher accuracy and resolution, reduced operator error, and compatibility with other digital equipment. The widely used *dual-slope technique* is illustrated in Fig. 11.15. The unknown input

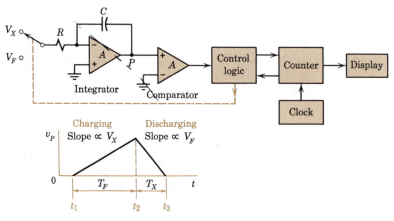

Figure 11.15 A digital voltmeter (DVM) using dual-slope integration.

voltage V_X is applied to the integrator at time t_1, and capacitor C charges at a rate proportional to V_X for a fixed time T_F. At time t_2 (after a fixed number of clock pulses has been counted), the control logic switches the integrator input to fixed reference voltage V_F and C discharges at a rate proportional to V_F. The counter is reset at t_2 and counts until the comparator indicates that the integrator output voltage v_P has returned to zero at t_3. To transfer the same charge $Q = IT$,

$$V_X T_F = V_F T_X \qquad \text{or} \qquad V_X = V_F(T_X/T_F) = kT_X$$

The count at t_3, proportional to the input voltage, is displayed as the measured voltage.

The integration process averages out noise and, by integrating over $1/60$ s, 60-Hz hum is almost eliminated. Three to five readings are taken per second. By using the appropriate *signal conditioner,* currents, resistances, or ac voltages can be measured by the same instrument. Such a digital multimeter (DMM) may incorporate automatic range and polarity selection so that measurement is reduced to selecting a variable and pressing a button.

Practice Problem 11-5

In the DVM of Fig. 11.15, $V_F = 10$ V and $T_F = 0.1$ s. Specify the frequency of the clock so that the decade counter will read V_X in millivolts.

Answer: 100 kHz.

Analog-to-Digital Converter

The analog data obtained in measurements on a physical system (temperature, pressure, displacement, voltage) can be converted into digital form for processing in various ways. One way is to use dual-slope integration with the counter of Fig. 11.15 providing a digital output. Another technique is illustrated in Fig. 11.16. Assuming that analog signal voltage V_i is available, the conversion is initiated by closing electronic switch S that charges capacitor C to V_i. The "sampled" input voltage is "held" after S is opened because the input resistance of the voltage follower is extremely high.

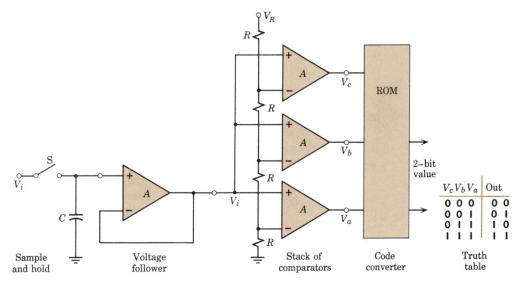

Figure 11.16   A parallel-comparator A/D converter.

Each comparator in the stack of three compares V_i to a "threshold" voltage derived from a precision voltage divider across the fixed reference V_R. Each comparator whose threshold voltage is below the analog signal voltage is driven into saturation (Fig. 11.14b), and the output is limited to the supply voltage. For example if $V_i > \frac{1}{4}V_R$, the first comparator saturates, providing an output $V_a = V_{CC}$ taken as logic **1**. If $\frac{2}{4}V_R < V_i < \frac{3}{4}V_R$, the first and second comparators saturate, providing an input code **011** to the ROM. The ROM code converter considers this input as an address and outputs the stored digital number **10**. In a practical A/D converter providing 5-bit digital output, $2^5 - 1 = 31$ comparators are required. High-speed IC converters operating on 20-ns cycles are available at low cost.

Square-Wave Generator

A simple and inexpensive *square-wave generator* can be constructed with one op amp and a pair of back-to-back Zener diodes. In Fig. 11.17a, a capacitor C is charged

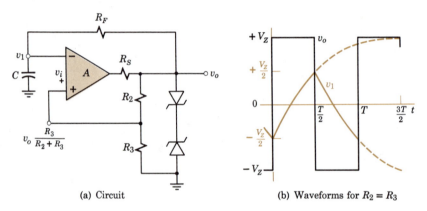

(a) Circuit   (b) Waveforms for $R_2 = R_3$

Figure 11.17   An elementary square-wave generator.

through resistor R_F from output voltage v_o limited to $+V_Z$ or $-V_Z$ by the diodes in combination with R_S. The op amp compares v_1 to $\frac{1}{2}V_Z$ obtained from the voltage divider where $R_2 = R_3$ (in this case). When $v_i = \frac{1}{2}V_Z - v_1$ changes sign, v_o changes sign. Half of v_o is fed back (positive feedback) to the noninverting terminal to drive the op amp into saturation.

To see how the circuit works, assume that $R_2 = R_3$ and that $V_Z = 10$ V. At $t = 0^-$, v_1 is approaching -5 V. At $t = 0$, v_1 reaches -5.01 V, say, and v_i goes positive, sending the op amp toward positive saturation but limited to $+10$ V. Because v_1 is the voltage on C, it cannot suddenly change and, at $t = 0^+$, $v_1 \cong -5$ V. Because $v_o = +10$ V, the voltage now tending to force current through R_F is $v_o - v_1 = 10 - (-5) = 15$ V. The capacitor voltage will rise exponentially, or

$$v_1 = 15(1 - e^{-t/R_F C}) - 5 \tag{11-26}$$

When v_1 exceeds $+\frac{1}{2}V_Z = +5$ V, the $+$ input terminal is less positive than the $-$ terminal, v_i changes sign, and v_o goes negative. Half of v_o is fed back to make v_i more negative, and v_o is driven rapidly and solidly to $-V_Z$.

In general, where $R_3/(R_2 + R_3) = H$ and $v_1 = -HV_Z$ at $t = 0$,

$$v_1 = (1 + H)V_Z(1 - e^{-t/R_FC}) - HV_Z \qquad (11\text{-}27)$$

for the first half-cycle. At $t = T/2$, $v_1 = +HV_Z$. Substituting these values in Eq. 11-27 and solving, the period is

$$T = 2R_FC \ln\frac{1 + H}{1 - H} \qquad (11\text{-}28)$$

Such a square-wave generator with symmetric Zener diodes and stable network elements will provide good square waves in the audiofrequency range.

EXAMPLE 4

Design a simple square-wave generator for operation at 200 Hz $(T = 0.005 \text{ s})$ with an amplitude of ± 10 V.

Two 10-V diodes (1N4104) are connected back-to-back as in Fig. 11.17a. Assuming op amp operation at ± 15 V, approximately $15 - 10 = 5$ V is to be dropped across R_S. Therefore, for $I_Z = 1$ mA, say,

$$R_S = V/I_Z = 5/0.001 = 5000 \ \Omega$$

Selecting $C = 0.1 \ \mu\text{F}$, $R_2 = R_3 = 1 \ \text{M}\Omega$, $H = 0.5$.

$$R_F = \frac{T}{2C \ln 3} = \frac{5 \times 10^{-3}}{2 \times 10^{-7} \times 1.1} = 22{,}750 \cong 22 \ \text{k}\Omega$$

Triangular-Wave Generator

Circuits containing a few operational amplifiers can be used to generate pulse trains, ramps, sawtooths, sine waves, or almost any other waveform of interest. For example, consider the *triangular-wave generator* in Fig. 11.18. In response to changes in v_i, the comparator switches between positive and negative saturation with its output v_1 clamped at $+V_Z$ or $-V_Z$. Assume that $v_1 = +V_Z$ at $t = 0$; the current

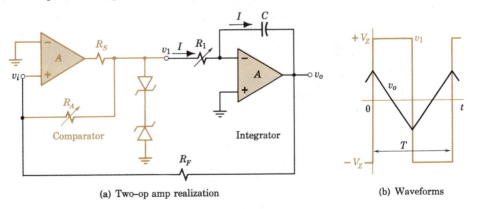

(a) Two–op amp realization

(b) Waveforms

Figure 11.18 An elementary triangular-wave generator.

flowing into the integrator is $I = V_Z/R_1$, which also serves to charge capacitor C. The integrator output v_o is just the capacitor voltage, or

$$v_o = V_0 - \frac{1}{C} \int_0^t I \, dt = V_0 - \frac{I}{C} t \tag{11-29}$$

This is a negative-slope *ramp* voltage. A part of voltage $v_o - v_1$ is fed back through R_F to the $+$ terminal of the comparator. When v_i changes sign to negative, the comparator switches to negative saturation, v_1 switches to $-V_Z$, and constant current I is reversed. This initiates a positive-going ramp and v_o is triangular. The amplitude can be controlled by R_A, which adjusts the feedback factor $H = R_A/(R_A + R_F)$, and the frequency can be adjusted by R_1, which controls the capacitor charging current.

PRACTICAL CONSIDERATIONS

The ideal op amp is characterized by infinite gain, infinite bandwidth, infinite input impedance, and zero output impedance. By the ingenious use of technological advances, design engineers provide at incredibly low cost commercial amplifiers that come close to the ideal.

A Four-Stage Op Amp

To provide the desired characteristics, several stages are required. In Fig. 11.19, the first stage is a *differential amplifier* with a double-ended output; this provides high

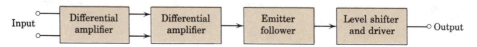

Figure 11.19 Block diagram of a four-stage operational amplifier.

input resistance, high gain for *difference* signals, and rejection of signals *common* to both terminals. The second stage is a single-output differential amplifier that provides more gain and more discrimination. The third stage is an *emitter follower*, a buffer to provide low output resistance and isolation of the amplifier from the load. The last stage is a combination *level shifter* and *driver* stage. The level shifting corrects for any dc offsets that have been introduced by bias networks or by component imbalances. The driver is a power amplifier to provide large output currents from a low output resistance.

The differential amplifier[†] is a versatile component of many electronic instruments and deserves further consideration. As shown in Fig. 11.20a, transistor T3 is a constant-current source where $I_{C3} = \beta I_{B3}$ and I_{B3} is obtained from a biasing network. The remainder of the circuit is a symmetric amplifier of matched resistors and transistors; in integrated circuit form, the corresponding elements are fabricated simultaneously and are nearly identical in all respects. Qualitatively speaking, any increase in v_P increases I_{B1}, which increases I_{C1} and lowers v_1. Because $I_{C1} + I_{C2} = I_{C3} =$

[†] Here we refer to an electronic circuit, a portion of an IC op amp; see the footnote on p. 77. The entire op amp can also function in a differential amplifier circuit (Fig. 11.7).

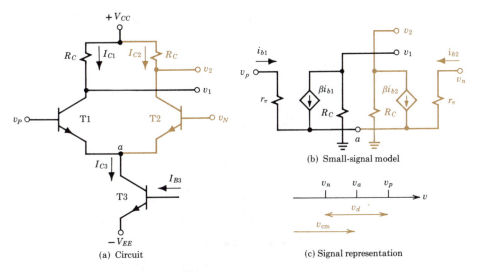

Figure 11.20   Analysis of a differential amplifier.

constant, an increase in I_{C1} results in a corresponding decrease in I_{C2} and v_2 goes up. The output $v_o = v_2 - v_1$ responds to the *difference* in v_P and v_N. If both v_P and v_N increase due to a *common* signal, T1 and T2 would have to respond in identical fashion, but I_{C1} and I_{C2} are unaffected because their sum must be constant. Therefore, $v_2 = v_1$ and there is no output.

For a quantitative analysis,[†] we consider the effect of *small changes* in v_P and v_N. Referring to Fig. 6.15a, we see that a small change in v_{BE} (from a steady value of, say, 0.7 V) produces a proportionate small change in base current Δi_B, which we designate i_{b1}. For such a "small signal,"

$$i_{b1} = \Delta i_{B1} = \frac{\Delta v_{BE}}{r_\pi} = \frac{v_p - v_a}{r_\pi}$$

where i_{b1}, v_p and v_a are signal values and r_π is an effective input resistance (the reciprocal of the slope of the i_B-v_{BE} curve). Referring to Fig. 6.15b, the resulting small change in collector current will be $i_{c1} = \beta i_{b1}$. Summing the currents into node a of Fig. 11.20b, we write

$$\Sigma i = (1 + \beta)i_{b1} + (1 + \beta)i_{b2} = (1 + \beta)\frac{v_p - v_a}{r_\pi} + (1 + \beta)\frac{v_n - v_a}{r_\pi} = 0$$

(11-30)

Hence, $(v_p - v_a) + (v_n - v_a) = 0$, which yields

$$v_a = \frac{v_p + v_n}{2}$$

(11-31)

or v_a takes on a value just halfway between v_p and v_n (Fig. 11.20c). Now let the input signals be described in terms of a common-mode signal v_{cm} and a difference signal v_d

---

[†] This analysis can be deferred until after the discussion of small-signal BJT models in Chapter 13.

so that $v_p = v_{cm} + v_d/2$ and $v_n = v_{cm} - v_d/2$. Using the small-signal models,

$$v_1 = -\beta \frac{v_p - v_a}{r_\pi} R_C = -\frac{\beta R_C}{r_\pi}(v_{cm} - v_a + v_d/2)$$

$$v_2 = -\beta \frac{v_n - v_a}{r_\pi} R_C = -\frac{\beta R_C}{r_\pi}(v_{cm} - v_a - v_d/2)$$

and the output voltage is

$$v_o = v_2 - v_1 = \frac{\beta R_C}{r_\pi} v_d = \frac{\beta R_C}{r_\pi}(v_p - v_n) \tag{11-32}$$

The common-mode signal is rejected and the difference signal is amplified.

Common-Mode Rejection Ratio

Ideally, a differential amplifier responds only to the difference voltage $v_p - v_n$ and not to the common-mode voltage v_{cm}. However, no practical op amp is perfectly balanced; an input v_{cm} always produces some output v_{ocm}. The ratio

$$A_{cm} = \frac{v_{ocm}}{v_{cm}} \tag{11-33}$$

is called the *common-mode gain*. The ratio of differential gain A to A_{cm} is called the *common-mode rejection ratio*, CMRR, where

$$\text{CMRR} = \frac{A}{A_{cm}} \tag{11-34}$$

In decibels (see p. 417), CMRR(dB) $= 20 \log$ CMRR. Typical values run from 50 to 100 dB.

EXAMPLE 5

A 741 op amp has a gain $A = 2 \times 10^5$ and a CMRR of 3.2×10^4. The input consists of a difference signal $v_p - v_n = 10 \ \mu V$ and a common-mode signal 100 times as large; that is, $v_p = 1005 \ \mu V$ and $v_n = 995 \ \mu V$. Determine A_{cm} and the output voltage.

By Eq. 11-34, the common-mode gain is

$$A_{cm} = \frac{A}{\text{CMRR}} = \frac{2 \times 10^5}{3.2 \times 10^4} = 6.3$$

By superposition, the output voltage is

$$v_o = A_d v_d + A_{cm} v_{cm}$$

$$= 2 \times 10^5 \times 10^{-5} + 6.3 \times 10^{-3}$$

$$= 2.0063 \ V$$

Even in this extreme case, imperfect CMRR introduces only a 0.3% error in the output.

Input Offset Voltage

When both input terminals are grounded, the ideal op amp develops zero output voltage. Under the same conditions, a practical amplifier will have a finite output because of inevitable small imbalances in the components. By definition, the *input offset voltage* V_{OS} is the dc input voltage required to reduce the output to zero. The offset voltage and its "drift" with changes in temperature and time is one of the important sources of error in op amp circuits. Op amps with appropriately low values of offset voltage must be selected or special nulling circuits must be employed. Many IC op amps provide connections to internal points for this purpose.

Bias Current Offset

The input terminals of a differential amplifier (Fig. 11.20a) carry the base bias currents for the transistors in the input stage. In an op amp circuit, these dc currents (typically a fraction of a μA) produce IR voltages that appear in the output as spurious signals. A practical op amp can be modeled by an ideal op amp with current generators in each lead. In an IC op amp, bias currents I_{B1} and I_{B2} will be very nearly equal and their effects can be balanced out by the insertion of a nulling resistor. Input offset current is important in amplifier circuits with large resistance values. Input offset voltage is important in circuits with high gain.

Frequency Response

The frequency response of an amplifier is defined by the *bandwidth*, the range of frequencies between the upper and lower 70.7% points, similar to the cutoff points of high- and low-pass filters. Whereas the ideal op amp has an infinite bandwidth, the gain A of a practical op amp begins to drop off as the frequency increases.

In many applications, however, the frequency response may be considered *flat* because the use of feedback greatly increases the bandwidth of the circuit. As we shall see in Chapter 15, with feedback the bandwidth of an amplifier is increased by the same factor as the gain is reduced, that is, by the factor A/A_F. For an op amp with $A = 10^5$ in an amplifier circuit where gain with feedback $A_F = 100$, the factor $A/A_F = 1000$. The bandwidth of the circuit will be 1000 times as great as the bandwidth of the op amp itself.

Stability

Actually, the high-frequency gain of an op amp is usually limited on purpose by the designer. At high frequencies, inherent capacitive effects introduce phase shift of the output signal with respect to the input. If this phase shift equals 180°, the feedback will be *positive*, and undesired oscillation may result.

One way to ensure that the amplifier is free from oscillation, or *stable*, is to include a low-pass filter in the output. This reduces the gain at high frequencies so that oscillation is prevented.

Slew Rate

The bandwidth determines the ability of an operational amplifier to follow rapidly changing small signals. In addition, there are limitations on the rate at which the output can follow large signals. This results from the fact that only finite currents are available within the amplifier to charge the various internal capacitances. The maximum rate at which the output voltage can swing from most positive to most negative is the *slew rate* ρ, typically a few volts per microsecond. For example, for an op amp with a slew rate of 1 V/μs operating at $V_{CC} = \pm 15$ V, the switching time will be

$$T_{SW} \cong \frac{\text{supply voltage}}{\text{slew rate}} = \frac{30}{1} = 30 \ \mu s$$

Slew rate may be a limitation on frequency response to large sinusoids, but it is of primary importance in switching applications such as the clipper in Fig. 11.14c.

THE ANALOG COMPUTER

The modern electronic analog computer is a precision instrument. Its basic purpose is to predict the behavior of a physical system that can be described by a set of algebraic or differential equations. The programming procedure is to arrange the operational amplifiers to perform the operations indicated in the describing equations and provide a means for displaying the solution.

In addition to the op amps (identified by function in Fig. 11.21), the practical computer includes an assortment of precision resistors and capacitors, a function

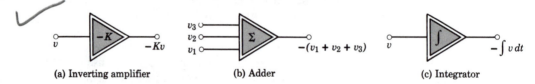

(a) Inverting amplifier   (b) Adder   (c) Integrator

Figure 11.21 Symbolic representation of operational amplifier functions.

generator to provide various inputs, means for introducing initial conditions, potentiometers for introducing adjustable constants, switches for controlling the operations, an oscilloscope or recorder for displaying the output, and a problem board for connecting the components in accordance with the program. In the hands of a skillful operator, the analog computer faithfully simulates the physical system, provides insight into the character of the system behavior, and permits the design engineer to evaluate the effect of changes in system parameters before an actual system is constructed.

Solution by Successive Integration

In one common application, the analog computer is used to solve linear integro-differential equations. To illustrate the approach, let us predict the behavior of the familiar physical system shown in Fig. 11.22. Assuming that mass M is constant, that

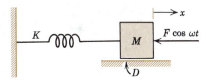

Figure 11.22 A physical system.

the spring is linear ($x = Kf$), and that friction force D is directly proportional to velocity u, the system is described by the linear differential equation

$$\Sigma f = 0 = -F \cos \omega t - M\frac{d^2x}{dt^2} - D\frac{dx}{dt} - \frac{1}{K}x \qquad (11\text{-}35)$$

and a set of initial conditions. The behavior of the system can be expressed in terms of displacement $x(t)$ or velocity $u(t)$ where $u = dx/dt$. We wish to display this behavior, so we proceed to program the computer to solve the equation.

The first step is to solve for the highest derivative. Anticipating the inversion present in the operational amplifier used, we write

$$\frac{d^2x}{dt^2} = -\left(\frac{F}{M}\cos \omega t + \frac{D}{M}\frac{dx}{dt} + \frac{1}{KM}x\right) \qquad (11\text{-}36)$$

To satisfy this equation, the mathematical operations required are addition, integration, inversion, and multiplication by constants. The required addition and the two integrations are shown in Fig. 11.23; in each operation there is an inversion.

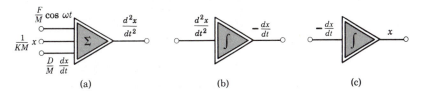

Figure 11.23 Operations needed in solving Eq. 11-36 by successive integration.

The next step is to arrange the computer elements to satisfy the equation. Knowing the required inputs to the adder, we can pick off the necessary signals (in the form of voltages) and introduce the indicated multiplication constants and inversions. Ignoring initial conditions for the moment, one possible program is outlined in Fig. 11.24.

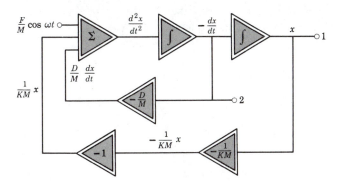

Figure 11.24 Analog computer program for Eq. 11-36.

Closing the circuit imposes the condition that the equation be satisfied. A properly synchronized cathode-ray tube connected at terminal 1 would display the displacement $x(t)$. The velocity $u(t)$ is available at terminal 2, but an inversion would be necessary to change the sign. Here six operational amplifiers are indicated; by shrewd use of amplifier capabilities, the solution can be accomplished with only three op amps.

EXAMPLE 6

For the physical system in Fig. 11.22, mass $M = 1$ kg, friction coefficient $D = 0.2$ N · s/m, and compliance $K = 2$ m/N. Devise an analog computer program using only three op amps to obtain the displacement and velocity for an applied force $f = F \cos \omega t$ N. (*Note:* Assume the op amps are single-input inverting; i.e., the $+$ terminal of each op amp is grounded.)

From Example 12 in Chapter 3, we know that one op amp can add and integrate the weighted sum. In this case, $M = 1$, $D/M = \frac{1}{5}$, and $1/KM = \frac{1}{2}$; after one integration, Eq. 11-36 becomes

$$\frac{dx}{dt} = -\int \left(F \cos \omega t + \frac{1}{5}\frac{dx}{dt} + \frac{1}{2}x \right) dt \quad (11\text{-}37)$$

A convenient reference is to let $RC = 1$ and specify resistances in megohms and capacitances in microfarads. For $C = 1$ μF and $R_1 = 1$ MΩ, a pure integration is performed; for $R_2 = 0.2$ MΩ, v_2 has a weighting of 5 in the result. On that basis, the appropriate values for Eq. 11-37 are shown in Fig. 11.25.

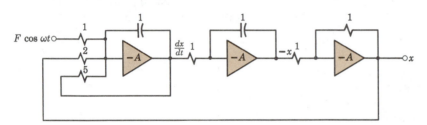

Figure 11.25 Analog computer program for solving Eq. 11-37. Resistance is in MΩ and capacitance in μF.

One interesting application of the analog computer is as a *function generator*. A desired time function can be obtained as a solution to another equation. For example (see Problem 7), with the proper initial conditions the function $y = Y \cos \omega t$ can be obtained as the solution of the equation

$$\frac{d^2y}{dt^2} + \omega^2 y = 0$$

SUMMARY

- The availability of low-cost high-performance op amps has radically altered the design of electronic signal processing equipment.

- The ideal op amp has infinite gain, bandwidth, input impedance, and common-mode rejection ratio and zero output impedance.
  Negative feedback is used to obtain the desired operating characteristics.

- Op amps are used in inverting, noninverting, and nonlinear modes in a great variety of applications.
  Active filters with improved characteristics can be realized with op amps.
  Digital signals can be converted to analog signals and vice versa.
  Voltage signals can be converted to current signals and vice versa.
  The unity-gain buffer is used to isolate a source from a load.
  High gain and negative feedback provide precise voltage regulation.
  A digital voltmeter can be made from an integrator and a comparator.
  Various waveforms can be generated by op amps operating nonlinearly.

- The differential amplifier is a symmetric two-input amplifier that provides high differential gain and a high common-mode rejection ratio.
  Practical limitations on op amp performance include input offset voltage, bias current offset, frequency response, stability, and slew rate.

- The analog computer is a flexible model of the system being studied; system behavior is represented by continuously varying quantities.
  Programming consists in arranging the operational amplifiers to perform efficiently the operations indicated in the equations describing the system.

REFERENCES

1. Roger Melen and Harry Garland, *Understanding IC Operational Amplifiers,* Howard W. Sams & Co., Indianapolis, 1971.
   An excellent nonmathematical introduction to IC op amps; how they are made, how they work, and how they are used. Practical applications.

2. Jacob Millman, *Microelectronics,* McGraw-Hill Book Co., New York, 1979, Chs. 15 and 16.
   A thorough treatment of op amp design and parameter measurement; discussion of many applications in analog and digital systems; description of some practical op amps available in IC form.

3. Jerry Eimbinder, ed., *Application Considerations for Linear Integrated Circuits,* Wiley-Interscience, New York, 1970.

   Practical advice for design engineers selecting and applying IC op amps. Specifications, general applications, and design considerations.

4. Jerald Graeme and Gene Tobey, eds., *Operational Amplifiers: Design and Application,* McGraw-Hill Book Co., New York, 1971.
   Thorough treatment of design and applications of IC op amps; prior circuit theory required. Appendix includes concise summary of basic concepts.

5. Paul Horowitz and Winfield Hill, *The Art of Electronics,* Cambridge University Press, Cambridge, 1980.
   Encyclopedic reference book emphasizing practical design techniques; linear and digital devices, their characteristics, and their application to instrumentation and signal processing.

REVIEW QUESTIONS

1. List the specifications of an ideal op amp.
2. Explain the operation of a low-pass filter.
3. Why must bit voltages be precise in a summing-circuit DAC?
4. In Fig. 11.5a, little current is drawn from the source; where does I_L come from?
5. Why is a differential amplifier used in the input stage of an op amp?
6. Distinguish between difference and common-mode signals.
7. What is a voltage follower? How does it work?
8. Explain the operation of an op amp voltage regulator.
9. What are the advantages of digital voltmeters?
10. Explain the operation of a dual-slope digital voltmeter.

11. What is the function of the ROM in the parallel-comparator ADC?
12. Explain the operation of the square-wave generator in Fig. 11.17.
13. Why does input bias current cause output voltage error?
14. Why is high-frequency gain intentionally reduced in an op amp?
15. What is the basic difference between digital and analog computers?
16. Why is the differentiator less useful than the integrator in computation?
17. Outline the procedure followed in programming an analog computer.
18. What would be needed to permit the solution of nonlinear equations?

EXERCISES

1. An op amp has the following low-frequency parameters: $R_i = 1$ MΩ, $R_o = 100$ Ω, and $A = 10^5$. Stating any necessary assumptions, design an amplifier circuit with:
   (a) $A_F = -200$, $R_{iF} = 5$ kΩ.
   (b) $A_F = -30$, $R_{iF} = 20$ kΩ.
2. The 741 of Fig. 11.1c is used in the circuit of Fig. 11.2b to provide a gain of -150. Specify R_1 and R_F if the maximum input current i_1 is to be ± 5 μA.
3. Use the model of Fig. 11.2a with $R_o = 0$ to derive an exact expression for voltage gain in the circuit of Fig. 11.2b.
   (a) Draw a convenient circuit diagram and use node voltages in your derivation.
   (b) Calculate the voltage gain and the error in Eq. 11-2 if $R_1 = 20$ kΩ $R_F = 1$ MΩ, $R_i = 10$ MΩ, and $A = 5 \times 10^4$.
   (c) Calculate the voltage gain if the actual op amp parameters are $R_i = 2$ MΩ and $A = 2 \times 10^4$. Draw a conclusion.
4. Repeat Exercise 1 for:
   (a) $A_F = +20$, $R_{iF} \geq 1$ MΩ.
   (b) $A_F = +50$, $R_{iF} \geq 1$ MΩ.
5. For inputs $v_1(t)$ and $v_2(t)$, devise a circuit to generate an output $v_o = +10 \int v_1\, dt - 5v_2$. The input resistance at each input terminal is to be ≥ 100 kΩ.

6. For the signals in Fig. 11.26, define v_o in terms of v_1 in words and mathematically. Design a circuit, i.e., specify X and Y in Fig. 11.27, so that $v_o(t)$ is obtained from $v_1(t)$.

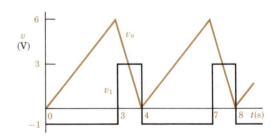

Figure 11.26

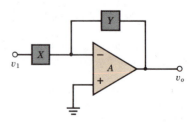

Figure 11.27

7. For the circuit of Fig. 11.27, where $X = L$ and $Y = R$, derive an expression for v_o in terms of v_1. What function is performed by the circuit?

8. Design a low-pass filter with a dc gain of 200 and a cutoff frequency of 1 kHz.

9. Design a low-pass filter with a dc gain of 1000 and a cutoff frequency of 100 Hz.

10. In Fig. 11.27, X consists of $R_1 = 10$ kΩ in series with $C_1 = 1$ μF and Y is $R_F = 100$ kΩ. If v_1 is a sinusoidal signal of variable frequency ω, derive an expression for V_o/V_1, plot $|V_o/V_1|$ versus ω, and identify the function performed.

11. Design a summing-circuit DAC that will convert the binary number stored in a 6-bit latch into a proportional *positive* voltage. The latch provides outputs at precisely 0 and +4 V; the maximum output voltage is to be approximately 8 V. Use ideal op amps with $R_F = 160$ kΩ.

12. To analyze the R-$2R$ DAC by superposition, redraw the circuit of Fig. 11.4a three times to isolate inputs v_2, v_3, and v_4. Calculate the individual contributions to the voltage at node 1.

13. Repeat Exercise 11 using an R-$2R$ circuit.

14. The current in a 100-Ω load R_L is to be proportional to $v_1 + 10v_2$, where v_1 and v_2 are sources with effective output resistances of over 10 kΩ. Using op amps of the type described in Exercise 1, design an effective circuit.

15. A voltmeter requires 1 mA to provide a full-scale deflection of 10 V. It is to be converted into an electronic voltmeter with an input resistance of at least 10 MΩ and ranges of 0 to 1 and 0 to 10 V.
    (a) Calculate the "input" resistance of the basic instrument.
    (b) Using an ideal op amp, design a circuit to provide the desired characteristics.
    (c) Predict the voltage indicated on the two voltmeters when connected, in turn, across a device represented by $V_T = 6$ V in series with $R_T = 500$ Ω. Draw a conclusion.

16. (a) Analyze the circuit of Fig. 11.28 and derive an expression for i_L in terms of the given quantities.

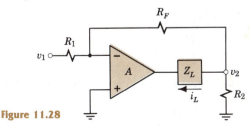

Figure 11.28

(b) A source characterized by $v_s = 2$ V in series with $R_s = 1$ kΩ is connected to the v_1 terminal. If $R_1 = 100$ kΩ, $R_F = 1$ MΩ, and $R_2 = 10$ kΩ, determine i_L. What function is performed by this circuit and what are its virtues?

17. In Fig. 11.28, the input combination of v_1 in series with R_1 is replaced by i_1 in parallel with R_1. Stating any necessary assumptions, derive an expression for i_L in terms of i_1. Evaluate the effect of R_1 and Z_L on the operation and define the function performed.

18. In the circuit of Fig. 11.29, $R_1 = R_2$. Use superposition to derive an expression for v_o in terms of the inputs. Draw the two partial circuits for considering v_a and v_b separately. Define in words the operation performed.

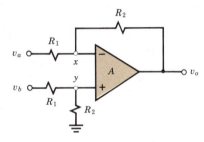

Figure 11.29

19. Starting from basic principles and stating any assumptions:
    (a) Express v_x and v_y in terms of v_a, v_b, and v_o in Fig. 11.29 and derive an expression for v_o.
    (b) Define the function of this circuit.
    (c) For $R_1 = R_2 = 50$ kΩ, $v_a = -3 + \sin \omega t$, and $v_b = \sin \omega t + 2 \sin 2\omega t$, predict v_o and explain the virtue of this circuit.

20. In the circuit of Fig. 11.29, the feedback resistor between v_o and x is relabeled $R_F = 200$ kΩ. Using superposition, design the remainder of the circuit to provide $v_o = 3 v_b - 5 v_a$.

21. An inferior op amp with $A = 2 \times 10^4$, $R_i = 50$ kΩ, and $R_o = 200$ Ω is used in the circuit of Fig. 11.10. Calculate A_F and R_{iF} and compare to the values in Example 3. Draw a conclusion.

22. Use one or more op amps of the type in Exercise 1 to design an amplifier circuit with $A_F = -20$, $R_{iF} \geq 1$ MΩ, and $R_o \leq 10$ Ω.

23. A sensor in an engine control system can be represented by an equivalent circuit with $V_S = 2$ V in series with $R_S = 2$ kΩ. It supplies a device characterized by $R_L = 100$ Ω.
    (a) Calculate V_L and P_L, the power supplied to the load device.
    (b) Use the op amp of Exercise 1 as a buffer, draw the circuit diagram, and calculate the new values of V_L and P_L. Draw a conclusion.

24. A cheap op amp with $A = 10^4$, $R_i = 10$ kΩ, and $R_o = 1$ kΩ is used in a unity-gain buffer. Predict R_{iF} and R_{oF}.

25. The op amp of Exercise 1 is used in the circuit of Fig. 11.5b. Derive a general expression for input resistance R_{iF} and evaluate it for the given op amp parameters and $R_F = 1$ kΩ.

26. The op amp circuit shown in Fig. 11.30 is suggested as a means for obtaining a steady output V_o for any value of V_s between 8 and 12 V, say.
    (a) Describe in words how this circuit might operate to provide a steady output voltage.
    (b) Stating any necessary assumptions, for $R = 6$ kΩ and $V_s = 10$ V, predict current I, voltage V_{CE}, and voltage V_o.
    (c) With V_s at 10 V, say, explain what would happen to the circuit variables if V_s increased to 12 V.

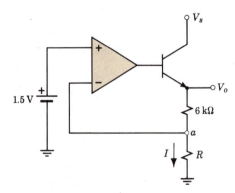

Figure 11.30

27. The circuit of Fig. 11.30 is to be used to supply a nearly constant voltage $V_o = 5$ V. Select a value of V_s and specify the value of R.

28. In the comparator circuit of Fig. 11.14a, $V_R = 2$ V.
    (a) If the op amp has the parameters of Exercise 1 and a supply voltage $V_{CC} = \pm 15$ V, what total change in v_1 will drive the op amp from negative to positive saturation?

    (b) For $v_1 = 6 \sin \omega t$ V, sketch the input and output waveforms.

29. For the safe operation of a reactor subsystem, if the temperature T_1 at point 1 ever exceeds the temperature T_2 at point 2, an alarm must sound. If the temperature sensors yield proportional voltages V_1 and V_2 and the alarm is activated at +5 V, design an op amp alarm system.

30. Design simple conditioner circuits to precede the digital voltmeter of Fig. 11.15 to allow measuring:
    (a) dc currents.
    (b) resistances.

31. A parallel-comparator ADC includes a 7-comparator stack across $V_R = 8$ V and a ROM code converter.
    (a) Draw a labeled schematic diagram and truth table.
    (b) Why is a sample-and-hold circuit necessary?
    (c) What is the digital output for $V_i = 0.9$ V? $V_i = 4.1$ V?

32. Draw the complete circuit model for two 10-V Zener diodes connected in series, back-to-back. Draw the composite i-v curve and explain why the only voltages allowable are ± 10 V.

33. (a) Derive Eq. 11-28 for the square-wave generator.
    (b) For convenience, you want to have the period $T = 2R_F C$; specify the value of H.

34. In Fig. 11.18a, $V_Z = 10$ V, $R_S = 5$ kΩ, $R_A = R_F = 100$ kΩ, $R_1 = 100$ kΩ, and $C = 0.1$ μF. Assume that at $t = 0$, $v_1 = V_Z$ and $v_o = 0$ V; at what time t will v_i change sign? Determine the amplitude and the frequency of the triangular wave.

35. For the differential amplifier of Fig. 11.20, $\beta = 99$, $r_\pi = 20$ kΩ, $R_{C1} = 100$ kΩ, and $R_{C2} = 110$ kΩ. Estimate the differential gain, the common-mode gain, and the CMRR.

36. Determine the CMRR for the differential amplifier of Fig. 11.7 if:
    (a) $R_1 = R_2 = 10$ kΩ and $R_F = R_3 = 100$ kΩ.
    (b) $R_1 = 10$ kΩ, $R_2 = 9$ kΩ, and $R_F = R_3 = 100$ kΩ.

37. Draw the block diagram of an analog computer program to solve:
$$\begin{cases} x - 3y = 5 \\ x + 2y = \sin \omega t \end{cases}$$

38. Draw the block diagram of an efficient analog computer program to solve:
$$\frac{d^2 y}{dt^2} + 2y = 5 \sin \omega t$$

39. Write the equation whose analog is shown in Fig. 11.31 (C in μF, R in MΩ).

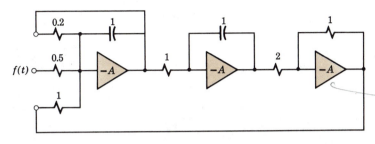

Figure 11.31

NOT ANALOGUE

PROBLEMS

1. The output of a flowmeter is $v = Kq$, where q is in cm^3/s and $K = 20$ mV $\cdot$ s/cm^3. The effective output resistance of the flowmeter is 2000 Ω. Design a circuit that will develop an output voltage $V_o = 10$ V (to trip a relay) after 200 cm^3 have passed the metering point.

2. Derive an expression for gain of the circuit of Fig. 11.32 as a function of frequency ω. For $R_1 = 10$ kΩ, $C_1 = 1$ μF, $C = 0.1$ μF, and $R = 100$ kΩ, determine the critical frequencies and sketch the gain as in Fig. 11.3b. What function is performed at low frequencies? At high frequencies?

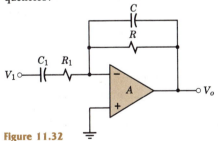

Figure 11.32

3. Derive a general expression for the output resistance of an op amp connected in the inverting circuit.

4. Derive a general expression for the input resistance of an op amp in the noninverting circuit.

5. An IC differential amplifier (Fig. 11.20) operates with $I_{C3} = 200$ μA. Assuming $\beta = 100$, calculate $r_\pi = dv_{be}/di_b$ by assuming that the emitter-base junction of a BJT behaves like the diode of Eq. 11-22. Estimate the input resistance $R_i = (v_p - v_n)/i_{b1}$ at room temperature.

6. Draw an efficient computer program to solve

$$2\frac{d^2y}{dt^2} + 4\frac{dy}{dt} + y = 2$$

provided $y = 3$ and $dy/dt = 1$ at $t = 0$.

7. Devise a computer program to solve the equation

$$\frac{d^2y}{dt^2} + \omega^2 y = 0$$

and specify the initial conditions so that the function generated is $y = A \cos \omega t$.

12

Large-Signal Amplifiers

Practical Amplifiers

Biasing Circuits

Power Amplifiers

Other Amplifiers

The two major functions performed by electronic devices are switching and amplifying. We have studied the use of diodes and transistors in switching and other nonlinear applications, and we have learned to use nearly ideal op amps in a great variety of special circuits. Using the characteristics of the transistors described in Chapter 6 and the circuit theory of Chapter 10, we are now ready to study the analysis and design of various types of amplifiers.

The design of electronic amplifiers to meet critical performance and cost specifications requires a great deal of knowledge and judgment. Electrical engineers will gain the detailed knowledge in subsequent courses and the judgment from practical experience. Other engineers and scientists will be more concerned with the use of existing amplifiers; if they do any designing (in connection with instrumentation, for example), it will be under circumstances where optimum performance is not essential and a simplified design procedure will give satisfactory performance.

In this chapter we discuss the factors that influence amplifier performance and present methods for designing simple amplifiers and predicting their performance. Our approach is first to look at the practical considerations in amplifier operation and see how to design circuits for maintaining the proper operating conditions, then to analyze the performance of one type of power amplifier, and finally to describe briefly some other important types of amplifiers. In subsequent chapters we shall learn to design amplifiers where frequency response and input-output characteristics must be considered.

PRACTICAL AMPLIFIERS

Amplifier Classification

We can classify amplifiers in terms of the number of stages, the size of the signals, the frequency range, or the bandwidth. A *single-stage* amplifier consists of one amplifying element and the associated circuitry; in general, several such elements are combined in a *multistage* amplifier. In a sound reproduction system, for example, the first stages are *small-signal* voltage (or current) amplifiers designed to amplify the output of a phonograph pickup, a few millivolts, up to a signal of several volts. The final stage is a *large-signal* or *power* amplifier that supplies sufficient power, several watts, to drive the loudspeaker.

Such amplifiers are called *audio* amplifiers if they amplify signals from, say, 30 to 15,000 Hz. In measuring structural vibrations, temperature variations, or the electrical currents generated within the human body, very low-frequency signals are encountered; to handle signals from zero frequency to a few cycles per second, *direct-coupled* amplifiers are used. In contrast, the *video* amplifier in a television receiver must amplify picture signals with components from 30 to 4,000,000 Hz.

A video amplifier is a *wide-band* amplifier that amplifies equally all frequencies over a broad frequency range. In contrast, a *radiofrequency* amplifier for the FM broadcast band (around 100 MHz) is *tuned* to select and amplify the signal from one station and to reject all others. In this chapter we are primarily interested in single-stage untuned audio power amplifiers.

Amplification and Distortion

The terms amplification and gain are used almost interchangeably. For sinusoidal signals or for a particular sinusoidal component of a periodic signal the voltage gain is

$$\mathbf{A}_V = \frac{\mathbf{V}_{\text{out}}}{\mathbf{V}_{\text{in}}} = A\,e^{j\theta} = A\,\underline{/\theta} \tag{12-1}$$

where $\mathbf{A}_V$ is the complex ratio of two voltage phasors. In a *linear amplifier*, A and θ are independent of signal amplitude and frequency, and the output signal is a replica of the input signal.

If there is *distortion* in the amplifier, the output is not a replica of the input. In Fig. 12.1, output is not proportional to input and there is *nonlinear* or *amplitude* distortion. In other words, A is not a simple constant. As a result of amplitude distortion, there is a frequency component in the output that is not present in the input. Analysis of the output reveals the presence of a "second harmonic," a component of

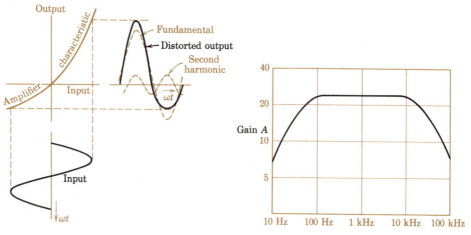

Figure 12.1   Amplitude distortion.

Figure 12.2   Frequency distortion.

twice the frequency of the "fundamental" input signal. Other forms of nonlinearity may introduce third-harmonic distortion. Amplitude distortion usually occurs when excessively large signals are applied to nonlinear elements such as transistors.

Practice Problem 12-1

The MOSFET of Fig. 6.7 is used as an amplifier with $R_L = 5$ kΩ. Predict the ac output voltage for an input signal $v_{GS} = 2 \sin 1000t$ V, and identify the type of distortion present.

Answer: $10 \sin 1000t - 1.25 \cos 2000t$ V; amplitude distortion.

At the other extreme is distortion due to *noise*, random signals unrelated to the input. If the input signal is too small, the output consists primarily of noise and is not a replica of the input. The "snow" that appears on a television screen when only a weak signal is available is a visual representation of noise. One source of noise is the random thermal motion of electrons in the amplifier circuit elements. The *shot effect* of individual electrons crossing a junction is another source of noise. Noise is of greatest importance in input stages where signal levels are small; any noise introduced there is amplified by all subsequent stages. The *dynamic range* of any amplifier is bounded at one end by the level at which signals are obscured by noise and at the other by the level at which amplitude distortion becomes excessive.

In Fig. 12.2, the *frequency-response curve* of an audio amplifier indicates that there is *frequency distortion*; all frequencies (within a finite band) are not amplified equally. In other words, for this amplifier gain A in Eq. 12-1 is a function of frequency. A signal consisting of a fundamental at 1 kHz, a tenth harmonic at 10 kHz, and a hundredth harmonic at 100 kHz would have a different waveform after amplification. No amplifier is completely free from frequency distortion.

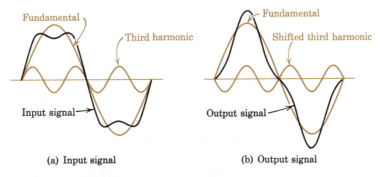

(a) Input signal   (b) Output signal

Figure 12.3 Phase distortion.

If θ is a function of frequency, the relative amplitudes of the signal components may be unchanged but the relative phase positions are shifted. As shown in Fig. 12.3, such *phase distortion* changes the shape of the output wave. The eye is sensitive to such distortion but the ear is not; a human ordinarily cannot distinguish between the two signals. On the other hand, the ear is quite sensitive to amplitude or frequency distortion.

Phase and frequency distortion are caused by circuit elements such as capacitive and inductive reactances that are frequency dependent. Some transistor parameters also are frequency dependent. In the design of untuned or wide-band amplifiers, special steps are taken to reduce the variation in gain with frequency.

Practical Considerations

Some of the practical considerations in amplifier design are illustrated in the simplified two-stage audio amplifier of Fig. 12.4.

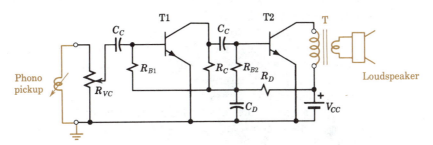

Figure 12.4 A simplified two-stage phonograph amplifier.

Biasing. For undistorted amplification, transistors must be operated in the linear portion of their characteristics. (Look again at Fig. 2.19.) Transistors are maintained at the proper operating point by dc power supplies and biasing networks. In portable units, batteries are used; in other units, the readily available 60-Hz current is rectified and filtered to provide the necessary direct current. In Fig. 12.4, the battery V_{CC} supplies the collector voltages to transistors T1 and T2 and, through resistors R_{B1}, R_{B2}, and R_D, the base-biasing currents.

Coupling. The voltage generated in the phonograph pickup is coupled to T1 by a combination of R_{VC} and C_C. Variable resistor R_{VC} provides volume control by voltage-divider action. Transistor T1 is coupled to the second stage by means of collector resistor R_C and coupling capacitor C_C. Such an RC coupling circuit develops a useful signal output across R_C and transfers it to the input of the next stage, but dc voltages and currents are blocked by C_C.

The loudspeaker is coupled to T2 by transformer T. Transformer coupling is more expensive than RC coupling, but it is more efficient and, as we shall see, it permits resistance matching for improved power transfer. The combination of R_D and C_D is a *decoupling filter* to prevent feedback of amplified signals to the low-level input stage. Such feedback is likely to occur when the common battery V_{CC} supplying several stages begins to age and develops an appreciable internal resistance, across which signal voltages may appear.

Load Impedance. If possible, the load impedance of an untuned amplifier is made purely resistive to minimize the variation in gain with frequency. In tuned amplifiers, the load impedance is usually a parallel resonant circuit. For voltage amplifiers, large values of R_C are desirable to develop large IR voltages; but large R_C requires large supply voltages, so the selected value is a compromise.

Input and Output Impedance. Ideally the input impedance of an amplifier stage should be high to minimize "loading" of the preceding stage, and the output impedance should be low for efficient power transfer. (See p. 76.) When several amplifier stages are connected in *cascade* so that the output of the first stage provides the input to the second stage, etc., the output characteristics of one stage are influenced by the input characteristics of the next. Here the signal output of T1 is determined by collector resistance R_C and also by biasing resistance R_{B2} and the input resistance of T2.

Unintentional Elements. Figure 12.4 is a wiring diagram and shows components, not circuit elements. Even a short, straight conductor can store a small amount of energy in the form of a magnetic field or an electric field (between the conductor and ground or a metal chassis). The magnetic field effect is important only at extremely high frequencies and will be neglected here. The electric field effect may be significant and can be represented by a *wiring capacitance*.

In the same way we represent the energy storage due to a difference in potential between the gate and source of a MOSFET by an *equivalent capacitance*. At a bipolar transistor junction there is a charge separation, and a potential difference appears across the depletion region (see Fig. 5.10); a change in junction potential causes a change in charge distribution ($q = Cv$), and this effect is represented by a *junction capacitance*. A circuit model used for analyzing amplifier performance at high frequencies must include these unintentional elements and, therefore, it will be quite different from the wiring diagram of Fig. 12.4.

BIASING CIRCUITS

The objective in biasing is to *establish* the proper operating point and to *maintain* it despite variations in temperature and variations among individual devices of the same type. Furthermore, this objective is to be achieved without adversely affecting the desired performance of the circuit. The biasing problem is difficult because of:

- The wide variations in device parameters expected in mass-produced transistors.
- The complicated interrelations among transistor variables.
- The inherent sensitivity of semiconductor devices to temperature.

Because of its importance, much attention has been given to this problem and as a result many ingenious circuits are available for the designer's use.

FET Biasing

Field-effect transistors are used primarily as "small-signal" amplifiers because the parabolic transfer characteristic introduces amplitude distortion if the signals are large. If the signal excursions along the transfer characteristic are small (Fig. 6.7), the biasing requirements are not critical.

DE MOSFET. Biasing a depletion-or-enhancement MOSFET is extremely simple because of its unique transfer characteristic that allows a quiescent operating point at $V_{GS} = 0$. The transfer characteristic of a DE MOSFET (Fig. 6.5b) is repeated in Fig. 12.5b. To obtain this characteristic, the drain-source channel must be biased by a dc

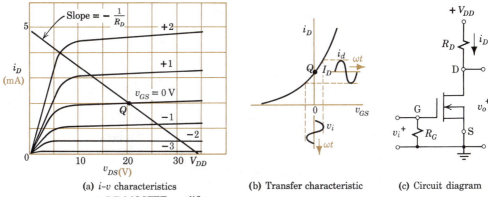

(a) i-v characteristics   (b) Transfer characteristic   (c) Circuit diagram

Figure 12.5   A DE MOSFET amplifier.

voltage that places v_{DS} in the "constant-current" region between turn-on and break-down where drain current i_D is nearly independent of v_{DS}. The amplifier circuit (Fig. 12.5c) contains a voltage supply V_{DD}, a drain resistor R_D across which signal voltages can be developed, and a gate resistor R_G that ties G to S so that $V_{GS} = 0$. R_G also allows any charge that might build up on the highly insulated gate to "leak" off.

As indicated by the transfer characteristic, a small-signal voltage applied to the input produces a signal current at the output. The output voltage $v_o = V_{DD} - i_D R_D$ contains a signal component that is an amplified replica of the input voltage.

Practice Problem 12-2

In the amplifier of Fig. 12.5, input voltage v_i varies from -1 to $+1$ V. Estimate, graphically, resistance R_D, variation in output voltage v_o, and the voltage gain v_o/v_i.

Answers: 7.1 kΩ, 14 V, 7.

JFET. Biasing a depletion-mode JFET is complicated by the fact that, for n-channel devices, positive values of v_{GS} are not permitted; ac signals must be super-imposed on a steady negative dc value V_{GS}. (For nomenclature, refer to Table 12-1.) A possible Q point is shown in Fig. 12.6a and b. A further complication is that V_{GS} for a given I_D varies from sample to sample and with temperature.

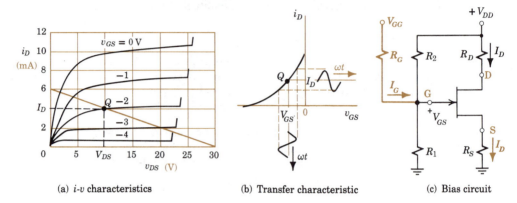

(a) $i\text{-}v$ characteristics   (b) Transfer characteristic   (c) Bias circuit

Figure 12.6 Biasing a depletion-mode JFET.

To provide the necessary negative value of V_{GS}, one possibility would be to use a negative source V_{GG}, but this would require a second battery. Another simple possibility is to place a resistor R_S between source and ground so that the source is at a potential $+I_D R_S$ with respect to ground; if then the gate is at ground potential, $V_{GS} = -I_D R_S$ as desired. A more complicated but better biasing scheme is shown in Fig. 12.6c, where R_1 and R_2 form a voltage divider to place the gate at the desired potential with respect to ground. As a first step in the analysis, replace V_{DD}, R_1, and R_2 by the Thévenin equivalents

$$V_{GG} = \frac{R_1}{R_1 + R_2} V_{DD} \quad \text{and} \quad R_G = \frac{R_1 R_2}{R_1 + R_2} \tag{12-2}$$

To facilitate design, these equations are solved for the circuit values

$$R_2 = R_G \frac{V_{DD}}{V_{GG}} \quad R_1 = \frac{R_G R_2}{R_2 - R_G} \tag{12-3}$$

Table 12-1 Transistor Nomenclature

	Field-Effect Transistors		Bipolar Transistors	
	Voltages	Currents	Voltages	Currents
Instantaneous total value	v_{GS}, v_{DS}	i_D	v_{EB}, v_{CB}	i_E, i_C
Instantaneous signal component	v_{gs}, v_{ds}	i_d	v_{eb}, v_{cb}	i_e, i_c
Quiescent or dc value	V_{GS}, V_{DS}	I_D	V_{EB}, V_{CB}	I_E, I_C
Effective (rms) value of signal	V_{gs}, V_{ds}	I_d	V_{eb}, V_{cb}	I_e, I_c
Supply voltage (magnitude)	V_{GG}, V_{DD}		V_{BB}, V_{CC}	

Summing voltages around the gate "loop" yields

$$\Sigma V = 0 = V_{GG} - I_G R_G - V_{GS} - I_D R_S \tag{12-4}$$

Disregarding the negligible gate current through a reverse-biased junction, the voltage equation reduces to

$$I_D R_S = V_{GG} - V_{GS} \quad \text{or} \quad R_S = \frac{V_{GG} - V_{GS}}{I_D} \tag{12-5}$$

For a specified Q point (I_D, V_{GS}) and chosen values of V_{GG} and R_G, the required values of R_S, R_1, and R_2 are easily calculated from Eqs. 12-3 and 5. Furthermore, by making the stable quantity V_{GG} large compared to the variable quantity $|V_{GS}|$, the effect of any shift in V_{GS} is reduced.

EXAMPLE 1

The JFET of Fig. 12.7 is to be operated at a quiescent point defined by $I_D = 4$ mA, $V_{DS} = 10$ V, and $V_{GS} = -2$ V. Design an appropriate biasing circuit with $V_{DD} = 30$ V.

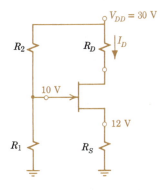

Figure 12.7   JFET bias design.

Using the voltage-divider circuit of Fig. 12.7, assume $V_{GG} = 10$ V so that $V_{GG} - V_{GS}$ is large compared to V_{GS} for stability, and assume $R_G = 10$ MΩ to keep the input resistance high. Then, by Eqs. 12-3 and 12-5,

$$R_2 = R_G \frac{V_{DD}}{V_{GG}} = 10\frac{30}{10} = 30 \text{ MΩ}$$

$$R_1 = \frac{R_G R_2}{R_2 - R_G} = \frac{10 \times 30}{30 - 10} = 15 \text{ MΩ}$$

$$R_S = \frac{V_{GG} - V_{GS}}{I_D} = \frac{10 - (-2)}{0.004} = 3 \text{ kΩ}$$

Summing voltages around the drain "loop" yields

$$\Sigma V = 0 = V_{DD} - I_D R_D - V_{DS} - I_D R_S$$

For $V_{DS} = 10$ V,

$$R_D = (V_{DD} - V_{DS} - I_D R_S)/I_D$$

$$= (30 - 10 - 12)/0.004 = 2 \text{ kΩ}$$

Note that a 20% shift in V_{GS} (to -2.4 V) would cause only a 3% change in $V_{GG} - V_{GS}$ and in I_D (Eq. 12-5).

Enhancement MOSFET. For an enhancement-only MOSFET (Fig. 6.3), normal operation requires a V_{GS} of the proper polarity to attract the holes or electrons necessary for conduction in the channel. If the voltage-divider bias circuit of Fig. 12.8b is used, V_{GG} is specified to provide the proper magnitude and polarity of $V_{GS} = V_{GG} - V_S = V_{GG} - I_D R_S$. (Some representative voltages are shown in color.)

The simpler bias circuit shown in Fig. 12.8c is satisfactory in many situations. Because the gate current is negligible, $V_{GS} = V_{DS}$ and operation is in the normal region for $|V_{GS}| > |V_T|$. (The design of such a bias circuit was illustrated in Example 1 of

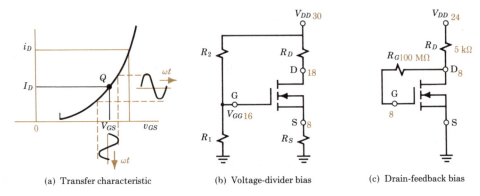

(a) Transfer characteristic   (b) Voltage-divider bias   (c) Drain-feedback bias

Figure 12.8   Biasing an n-channel enhancement-only MOSFET.

Chapter 6; the design values are shown in color in Fig. 12.8c.) In this *drain-feedback bias* circuit, any change in drain current due to changes in device or circuit parameters is fed back to the gate circuit in such a way as to compensate for the parameter change. For example, if I_D increases for any reason, V_{DS} decreases and the corresponding decrease in V_{GS} tends to decrease I_D. Feedback resistor R_G must be very large to minimize the degrading effect of *signal* current flowing from the output back to the input.

Practice Problem 12-3

Refer to the drain-feedback circuit in Example 1, page 150. For a particular sample of this E MOSFET, $K = 0.0003$ A/V^2 (50% higher than the manufacturer's specification). Assuming the other parameters are unchanged, predict the new value of drain current I_D and compare to that in Example 1. (*Hint:* Satisfy the drain-source loop equation and the transfer characteristic simultaneously.)

Answer: 3.33 mA; only 4% higher.

BJT Biasing

In normal operation of a bipolar transistor, the emitter-base junction is forward biased and the collector-base junction is reverse biased. Because in the common-emitter configuration these bias voltages have the same polarity, a single battery can supply both through an appropriate circuit. Here again, however, the variation in parameters among mass-produced transistors and the inherent sensitivity of semiconductors to temperature complicate the biasing problem, particularly in the large-signal amplifiers to be discussed in the next section. After demonstrating that the simplest solution is unsatisfactory, we shall analyze one commonly used bias circuit and outline an approximate design procedure.

Fixed-Current Bias. One possibility is to provide the desired dc base current from V_{CC} as in Fig. 12.9. For a forward-biased emitter-base junction, the voltage is a fraction of a volt and may be neglected in comparison to V_{CC}. On this basis, the quiescent base current is $I_B \cong V_{CC}/R_B$ and the operating point would be determined if the collector characteristics were known precisely.

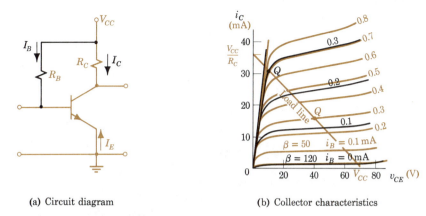

(a) Circuit diagram

(b) Collector characteristics

Figure 12.9   Fixed-current bias and effect of β change.

But the characteristics are not known precisely. For quiescent or dc values (see Table 12-1 on p. 350), Eq. 6.6 becomes

$$I_C = \frac{\alpha}{1 - \alpha} I_B + \frac{I_{CBO}}{1 - \alpha} = \beta I_B + (1 + \beta)I_{CBO} \qquad (12\text{-}6)$$

Even if I_B is fixed, collector current I_C may vary widely with the large variations in β and I_{CBO} expected among mass-produced transistors. As shown in Fig. 12.9b, a value $I_B = 0.3$ mA that places the quiescent point in the linear region of a transistor with $\beta = 50$ (characteristics in color) pushes operation into the saturation region of another transistor of the same type with $\beta = 120$. Also, I_C will vary widely with temperature changes because β increases nearly linearly with temperature and I_{CBO} increases exponentially with temperature.

Thermal Runaway.   The power loss in a transistor is primarily at the collector junction because the voltage there is high compared to the low voltage at the forward-biased emitter junction. If the collector current I_C increases, the power developed tends to raise the junction temperature. This causes an increase in I_{CBO} and β and a further increase in I_C, which tends to raise the temperature still higher. In a transistor operating at high temperature (because of high ambient temperature or high developed power), a regenerative heating cycle may occur that will result in *thermal runaway* and possibly destruction of the transistor.

For equilibrium, the power developed in the transistor is equal to the power (heat) dissipated to the surroundings. The ability to dissipate heat by conduction is proportional to the difference between junction and ambient temperatures and inversely proportional to the thermal resistance of the conducting path. For a constant ambient temperature, the ability to dissipate heat increases as the junction temperature goes up. Equilibrium will be upset and thermal runaway may occur when the increase in power to be dissipated is greater than the increase in dissipation ability due to the increase in junction temperature. One way to avoid this cumulative effect is to cool the collector junction; power transistors utilize the heat conducting capability of the metal chassis on which they are mounted or special heat-radiating fins. Another way is to use an effective biasing circuit.

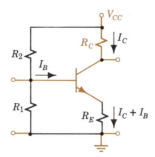

Figure 12.10 A self-biasing circuit.

Self Bias. The ingenious self-biasing circuit of Fig. 12.10 decreases the effect of changes in β or temperature on the quiescent operating point. Its operation is based on the fact that the critical variable to be stabilized is the collector current rather than the base current. The combination of R_1 and R_2 constitutes a voltage divider (as in Fig. 12.6c) to bring the base to the proper potential to forward bias the emitter junction. If I_C tends to increase, perhaps because of an increase in β due to a rise in temperature, the current $I_C + I_B$ in R_E increases, raising the potential of the emitter with respect to ground. This, in turn, reduces V_{BE}, the forward bias on the base-emitter junction, reduces the base current and, therefore, limits the increase in I_C. In other words, any increase in collector current is fed back to the base circuit and modifies the bias in such a way as to oppose a further increase in I_C.

Quantitative analysis of the circuit is simplified if the voltage divider (Fig. 12.11a) is replaced by its Thévenin equivalent where

$$V_{BB} = \frac{R_1}{R_1 + R_2}V_{CC} \quad \text{and} \quad R_B = \frac{R_1 R_2}{R_1 + R_2} \tag{12-7}$$

After replacing the transistor by a dc model derived from Eq. 12-6 (Fig. 12.11c), Kirchhoff's voltage law around the base loop yields

$$V_{BB} - I_B R_B - V_{BE} - (I_C + I_B)R_E = 0 \tag{12-8}$$

From Eq. 12-6, represented in the model by two current sources in parallel,

$$I_B = \frac{I_C}{\beta} - \frac{\beta + 1}{\beta}I_{CBO} \tag{12-9}$$

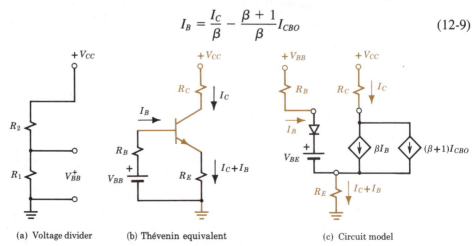

(a) Voltage divider   (b) Thévenin equivalent   (c) Circuit model

Figure 12.11 Analysis of the self-biasing circuit.

Substituting in Eq. 12-8 and solving,

$$I_C = \frac{V_{BB} - V_{BE} + \left(\dfrac{\beta + 1}{\beta}\right) I_{CBO}(R_B + R_E)}{R_E + \dfrac{R_B + R_E}{\beta}} \qquad (12\text{-}10)$$

This is a general equation for collector current in a self-biasing circuit where V_{BE}, β, and I_{CBO} are device parameters that are known imprecisely and that are sensitive to temperature change. Typically, $|V_{BE}|$ decreases 2.5 mV/°C, β may vary by 6:1 and increases linearly with temperature, and I_{CBO} may vary by 10:1 and approximately doubles for every 10° C increase above 25° C.

EXAMPLE 2

The β of individual specimens of a silicon transistor varies from 30 to 180. If V_{BE} may vary from 0.5 to 0.9 V and I_{CBO} may vary from 1 to 10 nA, predict the extreme variation in I_C in a self-biasing circuit where $R_1 = 10$ kΩ, $R_2 = 90$ kΩ, $R_E = 2$ kΩ, $R_C = 15$ kΩ, and $V_{CC} = 28$ V. Use the Thévenin equivalent circuit of Fig. 12.11b.

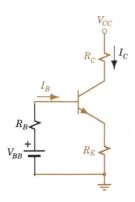

Figure 12.12 Stabilizing effect of a self-biasing circuit.

By Eqs. 12-7, the Thévenin equivalents are

$$R_B = \frac{R_1 R_2}{R_1 + R_2} = \frac{10 \times 90}{10 + 90} = 9 \text{ k}\Omega$$

$$V_{BB} = \frac{V_{CC} R_1}{R_1 + R_2} = \frac{28 \times 10}{10 + 90} = 2.8 \text{ V}$$

By Eq. 12-10, the collector current (Fig. 12.12) is

$$I_C = \frac{V_{BB} - V_{BE} + \left(\dfrac{\beta + 1}{\beta}\right) I_{CBO}(R_B + R_E)}{R_E + \dfrac{R_B + R_E}{\beta}}$$

For $\beta = 30$, $V_{BE} = 0.9$ V, and $I_{CBO} = 1$ nA (a "worst case"),

$$I_C = \frac{2.8 - 0.9 + \left(\dfrac{31}{30}\right) 10^{-9}(11{,}000)}{2000 + 11{,}000/30}$$

$$= \frac{1.9 + 0.00001}{2367} = 0.8 \text{ mA}$$

For $\beta = 180$, $V_{BE} = 0.5$ V, and $I_{CBO} = 10$ nA,

$$I_C = \frac{2.8 - 0.5 + \left(\dfrac{181}{180}\right) 10^{-8}(11{,}000)}{2000 + 11{,}000/180}$$

$$= \frac{2.3 + 0.0001}{2061} = 1.1 \text{ mA}$$

For these extreme variations, I_C shifts only 0.3 mA. Note that for a typical silicon transistor, $(1 + \beta) I_{CBO} = I_{CEO}$ is negligible in bias calculations.

Example 2 demonstrates that a properly designed self-biasing circuit is effective in stabilizing I_C despite variations in device parameters. The design of bias networks is a good illustration of an important engineering approach. Equation 12-10 reveals the general design criterion:

To stabilize operation, each term containing a variable parameter should be made "small" with respect to a circuit constant.

Because the terms are to be "small," rough approximations are acceptable in evaluating the variable terms; for example, for $\beta \geq 20$, $(\beta + 1)/\beta \cong 1$, and for $R_B \geq 5R_E$, $R_B + R_E \cong R_B$. A sizeable error in a small part becomes insignificant in the total term. With these approximations in mind, the specific design criteria are:

To make I_C independent of I_{CBO}, make

$$\left(\frac{1 + \beta}{\beta}\right) I_{CBO}(R_B + R_E) \cong I_{CBO} R_B \ll V_{BB} - V_{BE} \tag{12-11}$$

To make I_C independent of β, make

$$\frac{R_B + R_E}{\beta} \cong \frac{R_B}{\beta} \ll R_E \tag{12-12}$$

To make I_C independent of V_{BE}, make

$$V_{BE} \ll V_{BB} \tag{12-13}$$

The effect of temperature on the Q point can be predicted in terms of changes in V_{BE} and I_{CBO}. Assuming the circuit is designed to be independent of β (Eq. 12-12), the collector current is approximated by

$$I_C = \frac{V_{BB} - V_{BE} + I_{CBO} R_B}{R_E} \tag{12-14}$$

Assuming V_{BB}, R_B, and R_E are constant (actually $\pm 10\%$ variations are to be expected), the *change* in I_C will be

$$\Delta I_C = \frac{-\Delta V_{BE} + \Delta I_{CBO} R_B}{R_E} \tag{12-15}$$

The relative sensitivity of silicon and germanium to temperature changes is evaluated in Example 3.

As demonstrated in Example 3, silicon transistors are relatively unaffected by temperature changes and the small change in collector current is primarily due to ΔV_{BE}. Germanium transistors are sensitive to temperature changes and the principal factor is ΔI_{CBO}.

EXAMPLE 3

A silicon transistor with $V_{BE} = 0.7$ V and $I_{CBO} = 10$ nA and a germanium transistor with $V_{BE} = 0.3$ V and $I_{CBO} = 5$ μA (typical room temperature values) are used in the circuit of Example 2. Predict the effect on collector current of a 50° increase in operating temperature. (Use the typical rules given on page 355.)

The change in V_{BE} will be

$$\Delta V_{BE} = (-2.5 \text{ mV/°C})50°C \cong -0.1 \text{ V}$$

I_{CBO} will change by the factor

$$2^{\Delta T/10} = 2^{50/10} = 2^5 = 32$$

and the new values will be

$$\text{Si} : I_{CBO} = 32 \times 10 = 320 \text{ nA}$$

$$\text{Ge} : I_{CBO} = 32 \times 5 = 160 \text{ }\mu\text{A}$$

By Eq. 12-15, the *change* in I_C will be

$$\text{Si} : \Delta I_C = \frac{+0.1 + 310 \times 10^{-9} \times 9 \times 10^3}{2 \times 10^3}$$

$$= \frac{0.1 + 0.003}{2 \times 10^3} \cong 0.05 \text{ mA}$$

$$\text{Ge} : \Delta I_C = \frac{+0.1 + 155 \times 10^{-6} \times 9 \times 10^3}{2 \times 10^3}$$

$$= \frac{0.1 + 1.4}{2 \times 10^3} = 0.75 \text{ mA}$$

Bias Design, Approximate Method

In stabilizing against variations in β, Eq. 12-12 indicates that R_B should be small compared to βR_E. In practice, the ratio $\beta R_E/R_B$ is limited. R_B must not be too small because it appears directly across the input and tends to divert part of the signal current. Also, R_E must not be too large because part of the dc supply voltage V_{CC} appears across R_E (Fig. 12.12). As R_E is increased, less voltage is available for developing an output signal across R_C, and the collector circuit efficiency is reduced. (To prevent an undesirable ac voltage variation across R_E, it is customarily by-passed by a capacitor that offers low impedance at all signal frequencies. Coupling capacitors isolate the dc bias currents from the signal source and the following stage. See Fig. 12.16.)

As the designer, you must select values of R_1, R_2, and R_E to provide optimum performance, but the criteria are different in every situation and skill and experience are essential to an optimum design. If you do not expect large temperature changes and if less than optimum performance is acceptable, the bias circuit of Fig. 12.13 and the following procedure are satisfactory:

1. Select an appropriate nominal operating point (I_C, I_B, and V_{CE}) from the manufacturer's data (see Fig. 12.9b).

2. Arbitrarily assume that $V_E = I_E R_E \cong I_C R_E \cong 3$ V, say, and solve for R_E.

3. Select V_{CC} and R_C.
  (a) If V_{CC} is specified, $R_C \cong (V_{CC} - V_{CE} - 3)/I_C$.   (12-16)
  (b) If R_C is specified, $V_{CC} \cong 3 + V_{CE} + I_C R_C$.   (12-17)

4. Arbitrarily select R_B equal to $\beta_{min} R_E/10$.   (12-18)
  (For $\beta = 50$, a 10% change in β will cause only a 1% change in I_C.)

5. Calculate $V_{BB} = I_B R_B + V_{BE} + (I_B + I_C) R_E$.   (12-19)
  (Lacking other information, use $I_B = I_C/\beta$ and assume $V_{BE} \cong 0.7$ V for silicon or 0.3 V for germanium.)

6. Calculate R_2 and R_1 (see Eqs. 12-7) from

$$R_2 = R_B \frac{V_{CC}}{V_{BB}} \quad \text{and} \quad R_1 = \frac{R_2 R_B}{R_2 - R_B} \qquad (12\text{-}20)$$

The exact sequence in which these steps are taken can be modified to fit a given situation. (See Example 4, p. 361.)

Practice Problem 12-4

In the circuit of Fig. 12.10, the silicon transistor has a minimum $\beta = 70$ and a nominal β of 100. For $R_E = 2$ kΩ and $V_{CC} = 15$ V, specify R_C, R_1, and R_2 for quiescent operation at $I_C = 1.25$ mA and V_{CE} at 5 V.

Answers: 6 kΩ, 18.1 kΩ, 61.6 kΩ.

POWER AMPLIFIERS

To drive a loudspeaker or a recording instrument requires an appreciable amount of power. In analyzing or designing a *power amplifier*, we must consider large signals and the accompanying nonlinearity. Transistors capable of delivering high power output are expensive and, therefore, careful attention to efficient amplifier design is justified.

The basic design problem is to provide the desired power output with a stable circuit that uses the amplifying device efficiently and safely. Let us first consider the permissible operating region, see how operating point and load resistance are selected, derive expressions for power and efficiency, and then analyze a common type of audiofrequency amplifier that uses a power transistor.

Permissible Operating Region

An electronic amplifier must operate without introducing excessive distortion and without exceeding the voltage, current, and power limitations of the device. The bipolar junction transistor provides a large linear region and is commonly used where large signal excursions are required. The permissible operating region can be indicated on the output characteristics of the transistor as in Fig. 12.14. Line A represents the

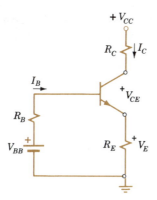

Figure 12.13 Approximate bias design.

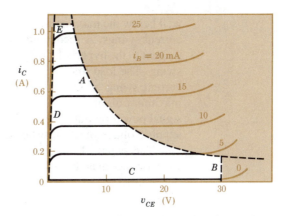

Figure 12.14 Permissible operating region of a transistor.

maximum allowable power dissipation P_D for the transistor; this is an hyperbola defined by $V_{CE}I_C = P_C = P_D$, where P_D is established by the manufacturer. Operation above this line may damage the device. Line B reflects the fact that at high collector voltages the avalanche effect causes a rapid increase in collector current and the curves become nonlinear.

For any electronic device there are regions of excessive nonlinearity that should be avoided. Line C bounds the region in which collector current is approaching zero and a further decrease in signal value (i_B) does not produce a corresponding decrease in output current. Line D bounds the saturation region in which a further increase in signal does not produce a corresponding increase in output current. Line E indicates an arbitrary limit within which the transistor manufacturer guarantees its specifications rather than a maximum allowable current.

Operating Point and Load Line

The purpose of the biasing arrangement is to locate and maintain operation in the permissible region. Within this limitation we wish to obtain maximum power output. In general, the load line should lie below the maximum dissipation curve, and its slope (determined by the load resistance) should reflect a compromise between large signals and low distortion.

The factors influencing the choice of Q point and load resistance are shown in Fig. 12.15, where R_L is assumed to be in the collector circuit. Quiescent point Q_1 is

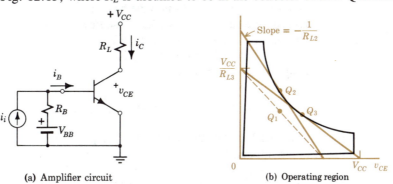

(a) Amplifier circuit (b) Operating region

Figure 12.15 Location of the load line for a simple amplifier.

well below the maximum dissipation curve and does not permit maximum voltage and current swings. A second possibility is Q_2 where the load line for $R_{L2} < R_{L1}$ is drawn tangent to the hyperbola; this permits the same voltage swing as Q_1 and a larger current swing. The load line for $R_{L3} > R_{L1}$, tangent to the hyperbola at Q_3, permits the same current swing as Q_1 and a larger voltage swing. Operation at any point along the maximum dissipation curve between Q_2 and Q_3 will permit approximately the same signal power output. The choice will depend on practical factors such as the available supply voltage or the resistance of the load.

The practical amplifier of Fig. 12.16a includes some additional complexities. Because the capacitances are open circuits for dc bias currents, the operating point and

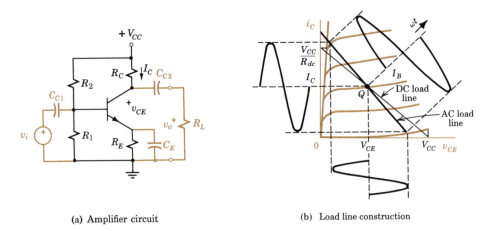

(a) Amplifier circuit

(b) Load line construction

Figure 12.16 Operation of a large-signal amplifier.

dc load line are determined from the equivalent dc circuit where $R_{dc} = R_C + R_E$. For maximum voltage swings, operation should be near the midpoint of the load line or $V_{CC} \cong 2V_{CE}$. Because the quiescent value $V_{CE} = V_{CC} - I_C R_{dc}$,

$$R_{dc} = R_C + R_E = \frac{V_{CC} - V_{CE}}{I_C} \cong \frac{V_{CE}}{I_C} \qquad (12\text{-}21)$$

a typical design condition.

The ac signals "see" a different circuit. $R_B = R_1 \| R_2$ is chosen large enough so that most of the signal current flows into the base. C_E is an ac short circuit; therefore, R_E is effectively removed. V_{CC} is a steady dc source and there can be no ac voltage across it; therefore the upper end of R_C is effectively grounded. C_{C2} is an ac short circuit, therefore R_C and R_L are effectively in parallel across the output voltage v_o, which is the signal component of v_{CE}. The *ac load line* defined by

$$R_{ac} = R_C \| R_L = \frac{R_C R_L}{R_C + R_L} \qquad (12\text{-}22)$$

is shown in Fig. 12.16b. For this circuit $R_{dc} > R_{ac}$ and the optimum location of the Q point is at the midpoint of the ac load line. (The ac load line defines the relation between signal values and device characteristics; the dc load line just identifies the supply voltage V_{CC} required for operation at Q. See Example 4.

EXAMPLE 4

The audiofrequency amplifier of Fig. 12.17 employs the silicon power transistor for which characteristics are shown. The manufacturer specifies $I_C(\text{max}) = 1$ A, $V_{CE}(\text{max}) = 50$ V, $P_D(\text{max}) = 5$ W (case at 25° C), and $\beta_{\text{min}} = 30$. Design the amplifier for maximum power to $R_L = 750$ Ω; an approximate design is satisfactory. (Assume capacitances are "large" and use "preferred values" of resistance; see p. 457.)

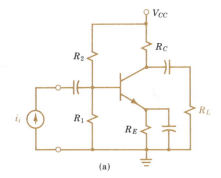

(a)

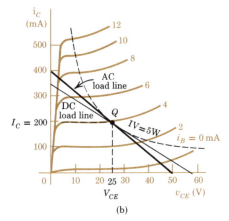

(b)

Figure 12.17   Amplifier design.

To select the operating point, we first sketch in the maximum dissipation line $I_C V_{CE} = 5$ W as shown. The ac load line should be tangent to the hyperbola and there is a range of slopes with nearly equal output powers. The high-voltage region looks to be slightly more linear than the high-current region, so we decide to use the full allowable voltage and let the current be limited by dissipation.

At $V_{CE} = V_{CEQ} = 50/2 = 25$ V and $I_C = I_{CEQ} = 0.2$ A,

$$P_C = V_{CE} I_C = 25 \times 0.2 = 5 \text{ W}$$

and the corresponding ac load resistance is

$$R_{\text{ac}} = \frac{V_{\text{pk}}}{I_{\text{pk}}} = \frac{V_{CE}}{I_C} = \frac{25}{0.2} = 125 \text{ Ω}$$

$$\therefore R_C = \frac{1}{\dfrac{1}{R_{\text{ac}}} - \dfrac{1}{R_L}} = \frac{1}{0.008 - 0.0013} = 150 \text{ Ω}$$

To design the bias circuit, we assume a voltage $V_E = 3$ V; the emitter resistance is

$$R_E \cong \frac{V_E}{I_C} = \frac{3}{0.2} = 15 \text{ Ω}$$

Then the necessary supply voltage is

$$V_{CC} = 3 + V_{CE} + I_C R_C = 3 + 25 + 30 = 58 \text{ V}$$

To minimize the effect of changes in β (Eq. 12-12), we let

$$R_B = \frac{\beta_{\text{min}} R_E}{10} = \frac{30 \times 15}{10} = 45 \text{ Ω}$$

Next, we read $I_B = 4$ mA, assume $V_{BE} = 0.7$ V for silicon, and by Eq. 12-19,

$$V_{BB} = I_B R_B + V_{BE} + (I_B + I_C) R_E$$

$$= 0.004 \times 45 + 0.7 + (0.204) 15$$

$$= 0.18 + 0.7 + 3.06 \cong 4 \text{ V}$$

To complete the bias design, by Eqs. 12-20,

$$R_2 = R_B \frac{V_{CC}}{V_{BB}} = 45 \frac{58}{4} = 652 \cong 680 \text{ Ω}$$

$$R_1 = \frac{R_2 R_B}{R_2 - R_B} = \frac{680 \times 45}{680 - 45} = 48 \cong 47 \text{ Ω}$$

Practice Problem 12-5

For the transistor of Fig. 6.16b operating as a large-signal amplifier with a Q point at $I_C = 5$ mA and $V_{CE} = 7.5$ V:

(a) Estimate the value of R_L for the given load line.
(b) Specify bias current I_B and collector dissipation P_C.
(c) For a swing in i_C from 0 to 9.5 mA, estimate the corresponding values of i_B and v_{CE}.

Answers: (a) 1500 Ω, (b) 0.125 mA, 37.5 mW; (c) 0 to 0.25 mA, 1 to 15 V.

Transformer Coupling

As illustrated in Example 4, the supply battery voltage V_{CC} must exceed the average collector voltage V_{CE} by the average voltage drop across the collector resistor $I_C R_C$. Furthermore, most of the signal power developed is dissipated in R_C instead of being transferred to R_L. The necessary supply voltage can be reduced and the power transfer improved by using a *transformer* to couple the output signal to the load. A transformer consists of two or more multiturn coils of high-conductivity wire wound on an iron core. The operation is defined by (Eqs. 3-32a and b)

$$\frac{V_2}{V_1} = \frac{N_2}{N_1} = \frac{I_1}{I_2} \tag{12-23}$$

If a resistance R_2 is connected to the output (*secondary* winding), the resistance "seen" at the input (*primary* winding) is

$$R_1 = \frac{V_1}{I_1} = \frac{V_2 \cdot N_1/N_2}{I_2 \cdot N_2/N_1} = \left(\frac{N_1}{N_2}\right)^2 \frac{V_2}{I_2} = \left(\frac{N_1}{N_2}\right)^2 R_2 \tag{12-24}$$

In other words, a transformer transforms resistance as well as voltage and current.

To couple an audiofrequency power amplifier to its load, a loudspeaker for example, the transformer has three important advantages:

1. Only a changing flux induces voltage and, therefore, direct currents are not "transformed." The transformer passes on the ac signal, but dc supply currents are kept out of the load. This is advantageous because some devices would be damaged or their characteristics would be adversely affected by direct currents.

2. The resistance of the primary winding is low and, therefore, the dc voltage drop across it is low. With transformer coupling, the supply voltage V_{CC} is approximately equal to V_{CE} instead of $V_{CE} + I_C R_C$. The dc power required is correspondingly reduced.

3. Maximum power transfer occurs when the resistance of a load is equal to the resistance of the source. (See Exercise 24.) The transformer transforms impedance and permits resistance matching for improved power transfer. For example, the low resistance of a loudspeaker, R_L, can be made to appear to a transistor as a much higher value $R_L' = R_L(N_1/N_2)^2$.

To gain these benefits, the output of an audiofrequency amplifier may be coupled to the load by a transformer. The performance of a well-designed transformer approaches the ideal over the major part of the audible range, but the output falls off at very low and very high frequencies because of "unintentional" inductance and capacitance inherent in the construction.

Power and Efficiency

The calculation of output power and efficiency can be illustrated for the case of the simplified transformer-coupled transistor amplifier of Fig. 12.18. Here the chosen

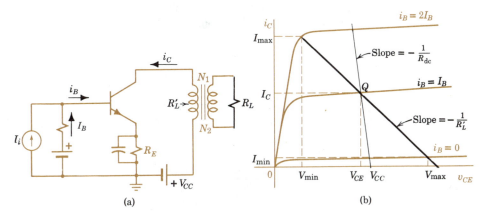

Figure 12.18 Simplified circuit of a transformer-coupled power amplifier.

Q point and desired signal swing define the ac load line whose slope is $-1/R'_L$. Once R'_L is determined, the turn ratio N_1/N_2 is selected to provide a match with the actual load resistance R_L. The dc load line has a slope of $-1/R_{dc}$; in this case R_{dc} is just R_E plus the small winding resistance of the transformer primary. Hence $I_C R_{dc} \ll V_{CE}$, and therefore $V_{CC} \cong V_{CE}$. (See Table of Transistor Nomenclature on p. 350.)

Assuming a sinusoidal signal current in a resistive load, the amplitude of the sinusoid is just one-half the difference between the maximum and minimum values of current, and the rms value is

$$I_c = \frac{I_{pk}}{\sqrt{2}} = \frac{1}{2\sqrt{2}}(I_{max} - I_{min}) \tag{12-25}$$

The output signal power is

$$P_o = I_c^2 R'_L = V_c I_c = \frac{V_c^2}{R'_L} \tag{12-26}$$

where V_c is the rms value of the signal voltage across R'_L.

With transformer coupling, the average dc power supplied by the collector battery (R_E is to be small and is neglected for the moment) is

$$P_{CC} = V_{CC} I_C \cong V_{CE} I_C \tag{12-27}$$

For sinusoidal signals, the average power dissipated in the transistor is

$$P_D = \frac{1}{T} \int_0^T .p \, dt = \frac{1}{2\pi} \int_0^{2\pi} v_{CE} \, i_C \, d(\omega t)$$

$$= \frac{1}{2\pi} \int_0^{2\pi} (V_{CE} - \sqrt{2} \, V_c \sin \omega t)(I_C + \sqrt{2} \, I_c \sin \omega t) \, d(\omega t)$$

$$= V_{CE} I_C - V_c I_c \qquad (12\text{-}28)$$

This interesting result indicates that the power dissipated is the difference between a constant input $V_{CE} I_C$ and a variable output $V_c I_c$. The losses to be dissipated are low when the signal output is high, and the power to be dissipated is maximum in the quiescent condition with no signal applied.

The efficiency of the output circuit is defined as signal power output over dc power input or

$$\text{Efficiency} = \frac{P_o}{P_I} = \frac{V_c I_c}{V_{CE} I_C} \qquad (12\text{-}29)$$

The theoretical limit for efficiency in this type of amplifier can be determined by considering the idealized current amplifier whose characteristics are shown in Fig.

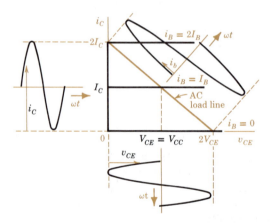

Figure 12.19 Operation of an idealized current amplifier.

12.19. A signal of amplitude equal to I_B produces a total current swing from $2I_C$ to zero and a total voltage swing from $2V_{CE}$ to zero. The ideal efficiency is

$$\frac{P_o}{P_I} = \frac{(V_{CE}/\sqrt{2})(I_C/\sqrt{2})}{V_{CE} I_C} \times 100 = 50\% \qquad (12\text{-}30)$$

At maximum signal input, the actual efficiency of transistor amplifiers is around 40 to 45%. (See Example 5.) The *average* efficiency with variable signal input is usually much lower.

EXAMPLE 5

Modify the design of Example 4 to provide transformer coupling for maximum power output to a load of 5 Ω, and predict the amplifier output and efficiency.

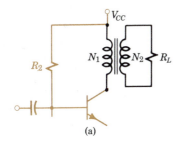

(a)

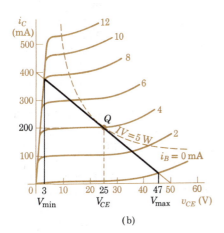

(b)

Figure 12.20 Amplifier performance.

The circuit modification is shown in Fig. 12.20a. The transformer should match the 5-Ω load to the 125-Ω load line and a commercial unit would be so labeled. The turn ratio should be (by Eq. 12-24)

$$N_1/N_2 = \sqrt{R_L'/R_L} = \sqrt{125/5} = 5$$

Now $I_C R_C \cong 0$ and the necessary supply voltage is

$$V_{CC} = V_{CE} + 3 = 25 + 3 = 28 \text{ V}$$

(R_1 and R_2 must be recalculated.)

To determine the amplifier performance, we note that the ac load line in Fig. 12.20b intersects the $i_B = 8$ mA line at $V_{min} \cong 3$ V, and the $i_B = 0$ line at $V_{max} = 47$ V; therefore the maximum undistorted voltage swing is

$$\frac{V_{max} - V_{min}}{2} = \frac{47 - 3}{2} = 22 \text{ V} = \sqrt{2}V_c$$

The power output is

$$P_o = \frac{V_c^2}{R_L'} = \frac{(22/\sqrt{2})^2}{125} = 1.94 \text{ W}$$

Neglecting R_E, the collector circuit efficiency is

$$\frac{P_o}{P_I} = \frac{1.94}{5} \times 100 \cong 39\%$$

If the emitter resistance loss is included, $V_{CC} = 28$ V and, assuming $I_E \cong I_C$, the efficiency is

$$\frac{P_o}{V_{CC}I_C} = \frac{1.94}{28 \times 0.2} \times 100 \cong 35\%$$

A small but significant amount of input power is required to "drive" the base of a large-signal amplifier. In Example 5, the rms value of base signal current is $4/\sqrt{2}$ mA and the necessary rms value of base-emitter signal voltage may be around $0.5/\sqrt{2}$ V. (The actual value can be predicted from I_C vs. V_{BE} characteristics provided by the manufacturer.) The required input power of approximately 1 mW is supplied by a small-signal amplifier of the type discussed in Chapter 14.

OTHER TYPES OF AMPLIFIERS

The emphasis in this chapter is on untuned power amplifiers. Some special forms of these amplifiers and some different types of amplifiers deserve mention.

Class B and C Operation

If corresponding values of i_B and i_C are obtained from the ac load line of Fig. 12.16b and plotted as in Fig. 12.21a, the *transfer characteristic* is obtained. For distortion-free operation of untuned amplifiers, the operating point is placed in the center of the linear part of the transfer characteristic and output current flows throughout the input-signal cycle. This condition is called *class A* operation and results in low distortion and also low efficiency because of the high value of dc current I_C with respect to the output signal current.

If the amplifier is biased to cutoff, output current flows only during the positive half-cycle and the output is badly distorted. The resulting *class B* operation is useful in a push-pull circuit (see Fig. 12.22). The second transistor supplies the other half-cycle of output current and the even harmonic distortion is cancelled. The average value of collector current is much lower and, therefore, the dc power input is less and the efficiency is higher.

If the amplifier is biased beyond cutoff, output current flows during only a small part of the cycle. With *class C* operation, the output is highly distorted, but with a tuned load impedance this presents no problem. (See Fig. 12.24.) The tuned circuit selects the fundamental component of the signal and rejects all others. A sinusoidal output is obtained and the efficiency is high because the average collector current is relatively low.

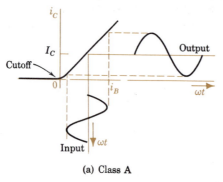

(a) Class A

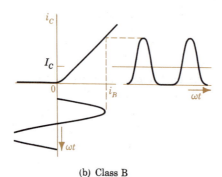

(b) Class B

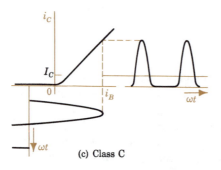

(c) Class C

Figure 12.21 Amplifier operation modes.

Push-Pull Amplifiers

Much of the distortion introduced in large-signal amplifiers can be eliminated by using two transistors in the *push-pull* circuit of Fig. 12.22. The input transformer T_1 receives a sinusoidal voltage from a low-level source. The signals applied to the two transistors are 180° out of phase, and the resulting collector currents are 180° out of phase. The output transformer T_2 delivers to the load a current that is proportional to

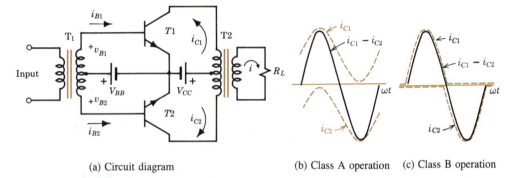

(a) Circuit diagram   (b) Class A operation   (c) Class B operation

Figure 12.22   Basic push-pull amplifier circuit.

the difference of the two collector currents. Any even-harmonic distortion components tend to cancel and the only distortion is that due to odd harmonics. Also the performance of transformer T_2 (a critical component) is improved because the dc components of i_{C1} and i_{C2} just cancel; magnetic core saturation and the accompanying nonlinearity are avoided.

The push-pull circuit is particularly useful for class B operation. Bias supply V_{BB} is set for the turn-on voltage, about 0.7 V for silicon transistors. (In practice, V_{BB} is obtained from a voltage divider across V_{CC} or from the voltage drop across a diode.) As v_{B1} goes positive, i_{B1} begins to flow and i_{C1} follows (Fig. 12.22c). Because v_{B2} is negative, transistor $T2$ is cut off. When v_{B1} goes negative, v_{B2} is positive and transistor $T2$ takes over. If the two transistors have identical characteristics, as they will have in integrated-circuit form, even harmonics are cancelled and the result is a nearly pure sinusoidal output.

Following reasoning similar to that in the derivation of Eq. 12-30, the ideal class B efficiency can be calculated. For two transistors, $P_I = 2V_{CC}I_C$ and $I_C = I_{dc} = I_{max}/\pi$. Therefore,

$$\text{Efficiency} = \frac{P_o}{P_I} = \frac{V_c I_c}{2V_{CC}I_C} = \frac{(V_{max}/\sqrt{2})(I_{max}/\sqrt{2})}{2V_{max}(I_{max}/\pi)}$$

$$= \frac{\pi}{4} = 0.785 \quad \text{or} \quad 78.5\%$$

Because *losses* in a transistor must not exceed the rated heat dissipation, two transistors in class B can supply nearly six times the power output of one similar transistor in class A. (See Exercise 33.)

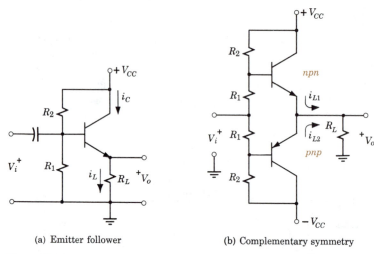

(a) Emitter follower   (b) Complementary symmetry

Figure 12.23 A class B complementary amplifier.

The expensive and bulky transformers can be eliminated by using an *npn* and a *pnp* transistor in the *complementary-symmetry* configuration. In Fig. 12.23a, the base-emitter voltage is nearly constant as long as the transistor is forward biased. If bias resistors R_1 and R_2 are chosen to place operation in the linear region, V_o will differ from V_i by a small constant value; in other words, the voltage at the emitter will *follow* the input voltage. (As we shall see in Chapter 15, the *emitter follower* has important advantages in small-signal amplifiers.)

If R_1 and R_2 are selected to bias the *npn* transistor to cutoff, V_o will follow V_i during the positive swing of the input signal and be zero during the negative half-cycle. While V_i is zero or negative, $i_C \cong i_L = 0$ and negligible power is dissipated. If a *pnp* transistor is added in the complementary connection of Fig. 12.23b, i_{L2} will flow during the negative half-cycle of V_i and the output of this push-pull amplifier will be linear. In IC versions of this circuit, resistances R_1 are replaced by diodes whose forward voltages "track" the base-emitter voltages of the transistors. The high efficiency of class B operation is obtained at low cost.

Tuned Amplifiers

The gain of a transistor amplifier depends on the load impedance. If a high-Q parallel resonant circuit is used for the load (Fig. 12.24), a very high resistance is presented at the resonant frequency and therefore the voltage gain is high. The frequency-response curve has the same shape as that of the resonant circuit. A narrow band of frequencies near resonance is amplified well, but signals removed from the resonant frequency are discriminated against.

The high selectivity of the load impedance eliminates nonlinear distortion; any harmonics in the input signal or in the collector current itself develop little voltage across the load impedance. As long as the collector current has a component at the resonant frequency, the output is nearly sinusoidal. With distortion eliminated, high efficiency is achieved by operating the transistor in a nonlinear region.

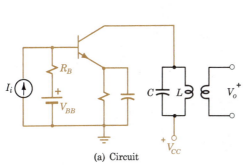

(a) Circuit

Figure 12.24 A tuned amplifier.

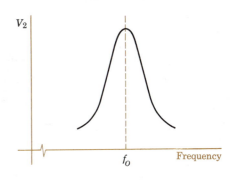

(b) Frequency response

SUMMARY

- An amplifier is a device for raising the level of a signal voltage, current, or power. For sinusoidal signals, the voltage gain is

$$\mathbf{A} = \frac{\mathbf{V}_{\text{out}}}{\mathbf{V}_{\text{in}}} = A\,e^{j\theta} = \mathbf{A}\,\underline{/\theta}$$

  If A and θ are constant, the amplification is linear and there is no distortion.

- In practical amplifiers, power supplies and biasing networks must maintain operation at the proper point, coupling circuits must transfer signals from one stage to the next without excessive discrimination, and load impedances must provide desired output without requiring excessive supply voltages.

- For FETs used as small-signal amplifiers, biasing is not critical and various circuits are available.
  For BJTs used as large-signal amplifiers, biasing by a combination of emitter resistor and voltage divider works satisfactorily.
  For noncritical design, an approximate bias procedure is available.

- The permissible operating region of a transistor is defined by maximum allowable current, voltage, power, and distortion.
  For large-signal class A operation, the midpoint of a properly located ac load line is the optimum quiescent point.

- Transformer coupling reduces the required supply voltage, isolates the signal output, permits impedance matching, and improves efficiency.
  In an idealized transformer-coupled, class A current amplifier:

$$\text{Input power (dc)} = P_I = V_{CC}I_C = V_{CE}I_C$$

$$\text{Output power (max)} = P_o = V_cI_c = \tfrac{1}{2}V_{CE}I_C = \tfrac{1}{2}P_I$$

- Class B and C amplifiers operate efficiently because output current is cut off during ineffective portions of the signal cycle.
  The push-pull amplifier circuit reduces distortion and improves efficiency.
  The tuned amplifier uses a resonant load circuit to obtain high selectivity and low distortion.

REFERENCES

1. Jacob Millman, *Microelectronics*, McGraw-Hill Book Co., New York, 1979, Chapters 11 and 18.
2. Charles Holt, *Electronic Circuits*, John Wiley and Sons, New York, 1978, Chapters 15 and 16.
3. Donald Schilling and Charles Belove, *Electronic Circuits: Discrete and Integrated*, McGraw-Hill Book Co., New York, 1979, Chapters 4 and 5.

REVIEW QUESTIONS

1. Distinguish between tuned and wide-band amplifiers and give an application of each (other than those in the text).
2. Distinguish between frequency, phase, and amplitude distortion. Explain the effect of each in a hi-fi system.
3. What determines the dynamic range of an amplifier?
4. What is the purpose of biasing in a MOSFET? In a BJT?
5. What are the relative merits of RC and transformer coupling?
6. Explain the practical considerations involved in the selection of values for: R_1, R_2, R_D, R_S in Fig. 12.8b.
7. Explain the operation of drain-feedback bias.
8. Why is fixed-current transistor bias unsatisfactory?
9. Explain thermal runaway.
10. Explain the function of each circuit element in Fig. 12.25.
11. Explain how the circuit of Fig. 12.25 stabilizes I_C if V_{CE} changes.
12. What are the practical limits on R_1 and R_3 in Fig. 12.25?
13. How are the quiescent point and load line related to the permissible operating region?

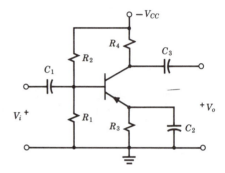

Figure 12.25

14. Sketch a set of transistor characteristics and indicate regions of high amplitude distortion.
15. Explain the statement: "A power transistor runs cool when the power output is high."
16. How does a transformer improve amplifier efficiency?
17. Why is impedance matching desirable?
18. Explain how push-pull amplifiers work. Where are they used?
19. How are "tuned amplifiers" tuned? Where are they used?
20. Explain why class B operation is more efficient than class A.

EXERCISES

1. Explain the causes of amplitude, frequency, and phase distortion. How would each be evidenced in a hi-fi system?
2. In Fig. 6.7, the input signal has a magnitude $V_m = 0.7$ V.
   (a) From the equation for output voltage v_L, evaluate the "distortion" component and express it as a percentage of the "fundamental" component. What type of distortion is this?
   (b) Reproduce the transfer characteristic. For an input signal $v_{GS} = 2 \sin \omega t$ V, determine graphically the output waveform and calculate the percent distortion.
   (c) How does percent distortion vary with signal size?

3. A MOSFET with the characteristics of Fig. 12.5a is used as a voltage amplifier with $V_{DD} = 30$ V and $R_D = 5$ kΩ.
   (a) Draw a "zero-bias" circuit and predict the quiescent values I_D and V_{DS}.
   (b) Sketch the transfer characteristic for a "zero-bias" circuit.
   (c) Estimate from the transfer characteristic the output signal current i_d for an input voltage $v_i = 0.5 \sin \omega t$ V.
   (d) Estimate the voltage gain of the amplifier.

4. Redraw the circuit of Fig. 12.5c modified by inserting resistance R_S between the S terminal and ground.
   (a) Derive an equation for i_D as $f(v_{DS})$ for the modified circuit.
   (b) The MOSFET of Fig. 12.5a is to operate at $I_D = 1$ mA and $V_{DS} = 15$ V; for $R_D = 5$ kΩ, specify R_S and V_{DD}. Check your answers by drawing the load line.

5. A JFET for which $I_{DSS} = 8$ mA and $V_p = -6$ V is used in the circuit of Fig. 12.26 where $V_{DD} = 32$ V, $V_{GG} = 10$ V, $R_D = 3$ kΩ, $R_S = 2.5$ kΩ, and $R_G = 10$ MΩ.
   (a) Estimate I_D by assuming $V_{GS} = 0$ and $I_G = 0$.
   (b) Using the estimated value of I_D, estimate V_{GS} by Eq. 6-1 and recalculate I_D from the circuit. Draw a conclusion about the method in part (a).

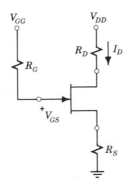

Figure 12.26

6. The bias circuit for a JFET is similar to Fig. 12.7 with $V_{DD} = 24$ V, $R_1 = 4$ MΩ, and $R_2 = 8$ MΩ. You are to complete the design to provide quiescent operation at $I_D = 2$ mA, $V_{DS} = 7$ V, and $V_{GS} = -3$ V.
   (a) Draw a circuit suitable for completing the bias design.
   (b) Use your circuit to specify R_S and R_D.

7. In the circuit of Fig. 12.7, $V_{DD} = 24$ V, $R_D = 3$ kΩ, $R_S = 5$ kΩ, $R_1 = 6$ MΩ, and $R_2 = 12$ MΩ.
   (a) Assume $V_{GS} = -2$ V and predict the quiescent values I_D and V_{DS}.
   (b) Assume V_{GS} is actually -1 V and repeat part (a).
   (c) Draw a conclusion about the stability of this bias scheme.

8. A JFET for which $I_{DSS} = 5$ mA and $V_{GS(OFF)} = -6$ V is to be operated at a Q point defined by $I_D = 3$ mA and $V_{DS} = 10$ V. Assuming $R_{in} \geq 10$ MΩ and $R_D = 3$ kΩ, design an appropriate bias circuit.

9. For the MOSFET of Fig. 12.8, the threshold voltage is 2 V and the drain current is 8 mA at a gate-source voltage of 6 V.
   (a) Sketch the transfer characteristic and predict the drain current at a quiescent point defined by $V_{GS} = 4$ V and $V_{DS} = 10$ V.
   (b) Design the bias circuit for $V_{DD} = 24$ V; i.e., given $R_1 = 1$ MΩ, specify R_2 and R_D.

10. The enhancement MOSFET of Fig. 6.3 is to be operated in the normal region at $I_D = 2$ mA. Draw a drain-feedback bias circuit and specify V_{DD} assuming $R_D = 7$ kΩ.

11. The data sheet of an enhancement MOSFET specifies $V_T = -3$ V and $I_D = -4$ mA for $V_{GS} = V_{DS} = -8$ V.
    (a) Determine V_{GS} for $I_D = -2$ mA.
    (b) Specify R_D for drain-feedback bias at $I_D = -2$ mA with $V_{DD} = 16$ V.

12. In the circuit of Fig. 12.9a, $I_B = 0.1$ mA, I_{CBO} for the germanium transistor is 5 μA, and β ranges from 50 to 200.
    (a) Predict the corresponding range of values of I_C.
    (b) If β = 150 and if I_{CBO} quadruples as a result of temperature increase, predict the change in I_C.

13. A silicon transistor with a nominal β = 100 is to operate with $I_C \cong 2$ mA and $V_{CE} = 6$ V in the circuit of Fig. 12.9.
    (a) For $V_{CC} = 12$ V, specify R_C and R_B.
    (b) If the β is actually 40, estimate the actual value of I_C for the value of R_B specified in part (a).

14. The transistor of Exercise 13 is to operate under the same quiescent conditions in the circuit of Fig. 12.10.
    (a) For $R_C = 3$ kΩ and $R_E = 1.5$ kΩ, specify V_{CC}.

(b) For $R_1 = 8.25$ kΩ and $R_2 = 22$ kΩ, calculate V_{BB} and R_B.

(c) If the β is actually 40, estimate the actual value of I_C.

(d) Compare the performance of the circuit with that in Exercise 13.

15. In Fig. 12.16a, $V_{CC} = 10$ V, $R_1 = 58$ kΩ, $R_2 = 142$ kΩ, $R_C = 8$ kΩ, and $R_E = 3.9$ kΩ. Drawing an appropriate circuit model and stating any assumptions, estimate dc current I_C and voltage V_{CE}.

16. In the self-biasing circuit (Fig. 12.10), $R_1 = 70$ kΩ, $R_2 = 140$ kΩ, $R_E = 3.9$ kΩ, and $V_{CC} = 15$ V. Estimate I_C and specify resistor R_C to establish the quiescent voltage across the silicon transistor (V_{CE}) at 5 V.

17. The one-stage audio amplifier shown in Fig. 12.25 employs a silicon transistor with a nominal β of 100 ($\beta_{min} = 70$).

(a) Replace the BJT by a suitable model, consider the effect of the capacitors at dc, and draw a new circuit appropriate for bias analysis.

(b) Use your circuit to explain how this bias scheme "stabilizes the collector current against changes in β and V_{BE}."

(c) Given $V_{CC} = 15$ V and $R_3 = 2$ kΩ, use your circuit to specify R_1, R_2, and R_4 for quiescent operation at $I_C = -1.25$ mA and V_{CE} at -5 V. State any assumptions.

18. A silicon transistor whose β may vary from 50 to 200 (nominal $\beta = 100$) is to be operated at $I_C = 2$ mA and $V_{CE} = 5$ V. For this amplifier application, $R_C = 2$ kΩ.

(a) Using the approximate procedure, design an appropriate bias network and draw a labeled circuit diagram.

(b) Determine the maximum variation in I_C expected for the given variation in β.

19. A germanium transistor with a nominal β of 60 is to be operated at $I_C = 0.5$ mA and $V_{CE} = 4$ V with the available $V_{CC} = 10$ V.

(a) Using the approximate procedure, design an appropriate self-biasing network and draw a labeled circuit diagram.

(b) If the actual β varies from 20 to 100, predict the maximum variation in I_C.

20. The data sheet for a 2N3114 (appendix A8) specifies a maximum power dissipation of 5 W with the case at 25° C and a "derating factor" of 28.6 mW/°C above 25° C to maintain a max-

imum junction temperature of 200° C.

(a) Explain the term "derating factor" in terms of energy transformations within the transistor and between the transistor and the metal case.

(b) If during operation the case temperature of the 2N3114 may rise to 75° C, estimate the allowable power dissipation.

21. A silicon transistor with a β of 50 and a maximum allowable collector dissipation of 10 W is to be used in the circuit of Fig. 12.15a. The maximum collector current is 2 A and the maximum collector voltage is 50 V.

(a) Making and stating any desirable simplifying assumptions, sketch the permissible operating region for this transistor.

(b) Draw two load lines, one for maximum voltage swings and one for maximum current swings. Estimate the value of load resistance R_L for each.

22. The collector dissipation for the transistor of Fig. 12.20b is to be limited to 3 W, maximum I_C is 0.5 A, and maximum V_{CE} is to be 60 V.

(a) Reproduce the characteristics and sketch in the permissible operating region.

(b) For operation at $V_{CE} = 20$ V, specify a suitable R_{ac} and the nominal I_B for maximum power output.

23. The silicon transistor in the circuit of Fig. 12.16a has a nominal β of 100; $V_{CC} = 15$ V, $R_1 = 10$ kΩ, and $R_2 = 30$ kΩ.

(a) Specify R_E and R_C to put the Q point at $V_{CE} = 6$ V and $I_C = 2$ mA.

(b) Sketch the i_C vs. v_{CE} characteristics and draw the dc load line.

(c) For $R_L = 3$ kΩ, draw the ac load line and estimate the output voltage v_o for an input current $i_b = 10 \sin \omega t$ μA.

24. A signal generator, consisting of a complicated network of linear passive and active elements, supplies a variable load resistance R_L.

(a) Use Thévenin's theorem to replace the generator by a combination of V_S and R_S.

(b) Express the power to the load as a function of V_S, R_S, and R_L.

(c) Prove the *maximum power transfer* theorem, which states that: "For maximum power transfer from a source to a load, the resistance of the load should be made equal to the Thévenin equivalent resistance of the source."

25. Specify the turn ratio of a transformer to provide impedance matching between a 3600-Ω source and:
    (a) A 16-Ω speaker.
    (b) Four 16-Ω speakers operated in parallel.

26. An amplifier uses the transistor of Exercise 21. For each of the two load lines in part (b), estimate the maximum possible signal power output in R_L and the efficiency.

27. Repeat Exercise 26 assuming that each R_L is replaced by an ideal transformer providing matched coupling to $R_L = 1000\ \Omega$. Specify the turn ratios of the transformers. Compare the results to those for Exercise 26, and draw a conclusion.

28. In Example 4 the Q point is at $V_{CE} = 25$ V and $I_{CE} = 200$ mA.
    (a) For no input signal, estimate the collector power supplied by the battery, the power dissipated in R_C and R_E, the power dissipated at the collector of the transistor, and the power delivered to R_L.
    (b) Repeat part (a) for an input signal $i_b = 2 \cos \omega t$ mA.
    (c) Define "efficiency" for this power amplifier and calculate it.
    (d) Compare the power output and efficiency to the results of Example 5.

29. The permissible operating region of a power transistor is defined by $P_D = 5$ W, $I_C(\text{max}) = 1$ A, $V_{CE}(\text{max}) = 100$ V, and $V_{CE}(\text{min}) = 2$ V.
    (a) Sketch the transistor characteristics and select an appropriate quiescent point for operation in the circuit of Fig. 12.15a.
    (b) Specify R_L for nearly maximum power output.
    (c) Calculate total dc power in, maximum signal power out, and overall efficiency.
    (d) If the actual load is 10 Ω, specify the appropriate transformer turn ratio and recalculate the overall efficiency of the amplifier.

30. The transistor of Fig. 12.14 is to operate with an effective load resistance of 45 Ω.
    (a) Select an optimum operating point and specify I_C and V_{CC}.
    (b) Show the Q point on a sketch of the characteristics and estimate the maximum symmetric swing in collector current and collector voltage.

(c) Estimate the rms value of input current required from the driver.
(d) Assuming that a transformer matches the load to $R_L' = 45\ \Omega$, calculate the dc power in, the maximum signal power out, and the overall efficiency.

31. The transistor of Fig. 12.14 is used in the circuit of Fig. 12.18 where R_L is a 4-Ω loudspeaker, $N_1/N_2 = 3$, and $R_E = 5\ \Omega$. The Q point is to be at $I_C = 0.4$ A and $V_{CE} = 10$ V.
    (a) Reproduce the characteristics and draw the load line.
    (b) Specify the battery voltage and estimate the input signal current (rms) required.
    (c) Estimate the maximum current and power delivered to the loudspeaker.

32. A transistor rated at $I_C(\text{max}) = 1.6$ A, $V_{CE}(\text{max}) = 50$ V, $V_{CE(\text{sat})} \cong 0$ V, $I_{CEO} \cong 0$, and $P_D(\text{max}) = 10$ W is used in the circuit of Fig. 12.15a.
    (a) Draw, approximately to scale, the permissible operating region.
    (b) For $R_L = 100\ \Omega$, select the "best" quiescent point and explain your reasoning.
    (c) Predict the average collector dissipation, the maximum power output to R_L, and the collector circuit efficiency.
    (d) If your Q point does not take full advantage of the capability of the transistor, rearrange the circuit to provide maximum possible power to the same 100-Ω load and repeat part (c).

33. For class B operation:
    (a) Show the operation of an amplifier on the idealized characteristics of Fig. 12.19.
    (b) Calculate the power output of a transistor rated at 10 W dissipation operating under ideal conditions in class A and in class B.
    (c) Assume practical operating efficiencies and demonstrate that two transistors in class B can supply nearly six times the signal power of one similar transistor in class A.

34. In the complementary amplifier of Fig. 12.23, $V_{CC} = 24$ V and $R_L = 8\ \Omega$. Estimate the maximum value of V_i for proper operation. Neglecting the effect of the constant base-emitter drop, estimate the average (dc) collector current and the signal power delivered to R_L for the maximum allowable signal input.

PROBLEMS

1. An inexperienced designer uses the transistor of Example 2 in the circuit of Fig. 12.9a with $V_{CC} = 20$ V and $R_C = 5$ kΩ.
   (a) Specify R_B for a nominal $I_C = 1$ mA.
   (b) Predict the extreme variation in I_C for this circuit, compare it to the variation in Example 2, and draw a conclusion regarding the effectiveness of this method.

2. A JFET (somewhat like Fig. 12.6a) and a BJT (somewhat like Fig. 12.9b) are connected in the *current-source bias* circuit of Fig. 12.27. Estimate I_D and V_{DS}. (*Hint:* Estimate the voltages at points 1, 2, and 3 in turn.) Evaluate the stability of this bias circuit.

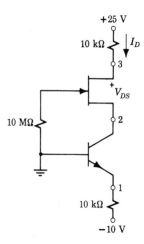

Figure 12.27

3. A 2N3114 transistor (Fig. A8) is to be operated as a single-stage audiofrequency amplifier at $I_C = 2$ mA and $V_{CE} = 40$ V with $R_C = 10$ kΩ. Draw an appropriate wiring diagram and, stating all assumptions, design the bias network.

4. The circuit of Fig. 12.28 is suggested as a simple means for obtaining improved stability.
   (a) Explain qualitatively what happens if I_C tends to rise as a result of an increase in I_{CBO} or β.
   (b) Derive an approximate expression for I_C in terms of I_{CBO} and β (somewhat like Eq. 12-10).
   (c) Under what conditions (i.e., for what relation between R_C and R_B) will I_C be insensitive to changes in β?

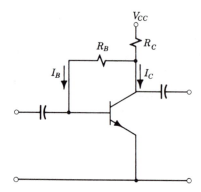

Figure 12.28

5. Prove that the point at which a load line is tangent to the hyperbola $P_D = V_{CE} I_{CE}$ is the midpoint of the load line.

6. A transformer-coupled audio amplifier is to supply 1 W to a 16-Ω loudspeaker. Available is a transistor whose permissible operating range is defined by $P_D = 4$ W, $V_{CE}(\text{max}) = 80$ V, $V_{CE}(\text{min}) = 2$ V at $I_C = 200$ mA, and $I_C(\text{max}) = 1$ A. Stating all assumptions, design the amplifier (draw the wiring diagram and specify the quiescent point, bias resistances, supply voltage, capacitor values, and transformer turn ratio) and estimate the total dc power requirement.

7. The circuit of Fig. 12.29 has two outputs labeled 1 and 2. Predict I_C and V_{CE} for $v_i = 0$. Predict and sketch the output voltages for $v_i = V_m \sin \omega t$. What function does this circuit perform? Where would it be useful?

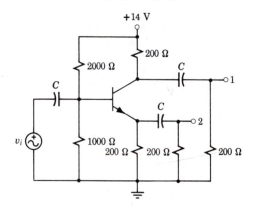

Figure 12.29

13

Small-Signal Models

Large and Small Signals

Diodes

Field-Effect Transistors

Bipolar Junction Transistors

The use of models to represent complicated phenomena or devices or systems is common practice throughout engineering and science. On the basis of experimental observations, the scientist may propose a tentative model that is then tested under various conditions. The engineer uses models to simplify the analysis of a known device or system or to predict the behavior of a proposed design. These models may be physical or mathematical in form; a linear model amenable to mathematical analysis is particularly desirable.

To predict the behavior of diodes and transistors in switching and biasing applications, we have used circuit models composed of ideal circuit elements. Working from characteristic curves, we devised relatively simple models to represent complicated physical devices under specified conditions. In an extension of that approach, we are now going to derive linear circuit models that are used to predict the behavior of semiconductor devices in practical applications over a wide range of frequencies. The character of the model depends on the accuracy required; more sophisticated models are used in computer-aided design for optimum performance. The character of the model also depends on the size of the signals expected.

LARGE AND SMALL SIGNALS

The circuit models derived in Chapters 5 and 7 are based on piecewise linearization of the characteristic curves and are useful for predicting the behavior of diodes and transistors under static conditions or where relatively large swings in voltage occur. A different approach is necessary if we are interested in the response to small signals superimposed upon dc values.

A Mathematical Model

Consider the circuit of Fig. 13.1 in which two sinusoidal signals $v_M(t)$ and $v_C(t)$ are superimposed on a bias voltage V_0 and the total voltage is applied to a diode circuit.

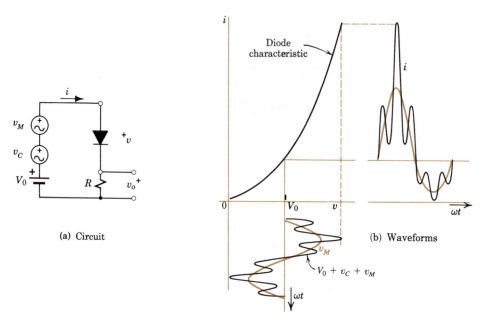

(a) Circuit

(b) Waveforms

Figure 13.1 Effect of nonlinearity on the sum of two signals.

Using the analytical technique of Chapter 2, we can represent a portion of the diode characteristic by a *mathematical model* described by the power series

$$i = a_0 + a_1 v + a_2 v^2 \tag{13-1}$$

If the voltage drop across R is relatively small, the full voltage

$$v = V_0 + v_M + v_C = V_0 + V_M \sin \omega_M t + V_C \sin \omega_C t \tag{13-2}$$

appears across the diode and

$$
\begin{aligned}
i = a_0 &+ a_1 V_0 + a_1 V_M \sin \omega_M t + a_1 V_C \sin \omega_C t \\
&+ a_2 V_0^2 + a_2 V_M^2 \sin^2 \omega_M t + a_2 V_C^2 \sin^2 \omega_C t \\
&+ 2a_2 V_0 V_M \sin \omega_M t + 2a_2 V_0 V_C \sin \omega_C t \\
&+ 2a_2 V_M V_C \sin \omega_M t \sin \omega_C t
\end{aligned}
$$

Substituting for $\sin^2 \omega t$ and $\sin \omega_M t \sin \omega_C t$ and collecting similar terms,

$$
\begin{aligned}
i = a_0 &+ a_1 V_0 + a_2 V_0^2 + \tfrac{1}{2} a_2 V_M^2 + \tfrac{1}{2} a_2 V_C^2 \\
&+ (a_1 + 2a_2 V_0) V_M \sin \omega_M t + (a_1 + 2a_2 V_0) V_C \sin \omega_C t \\
&- \tfrac{1}{2} a_2 V_M^2 \cos 2\omega_M t - \tfrac{1}{2} a_2 V_C^2 \cos 2\omega_C t \\
&+ a_2 V_M V_C \cos (\omega_C - \omega_M) t - a_2 V_M V_C \cos (\omega_C + \omega_M) t
\end{aligned} \tag{13-3}
$$

The output voltage v_o is directly proportional to i and contains the same eleven components. The first five terms constitute a dc component dependent on the amplitude of the input signals, as expected in a nonlinear device. The next two terms are replicas of the input signals resulting from the linear term of Eq. 13-1. Then there are two terms representing second harmonics generated by the nonlinearity. The last two terms are of particular interest in communication.

Practice Problem 13-1

A voltage $v = V_M \sin \omega_M t + V_C \sin \omega_C t$ is applied to a device whose transfer characteristic is $i = av^2$. Write out the expression for device current and identify the frequency components of current.

Answer: dc, $2\omega_M$, $2\omega_C$, $\omega_C - \omega_M$, $\omega_C + \omega_M$.

Amplitude Modulation

In a typical communication or control system, a high-frequency easily propagated signal serves as a *carrier* and information is superimposed on the carrier in the process called *modulation*. In amplitude modulation (AM), the amplitude of the carrier $v_C = V_C \sin \omega_C t$ is varied by the modulating signal $v_M = V_M \sin \omega_M t$. The equation of the amplitude-modulated wave is

$$v_{AM} = (1 + m \sin \omega_M t)V_C \sin \omega_C t \tag{13-4}$$

where m is the *degree of modulation* $= V_M/V_C$, approximately 80% in Fig. 13.2.

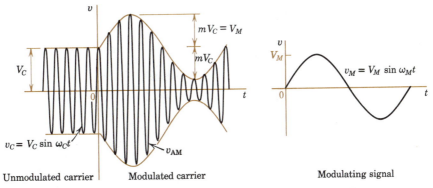

Unmodulated carrier   Modulated carrier   Modulating signal

Figure 13.2 A sinusoidal carrier amplitude-modulated by a sinusoidal signal.

Expanding Eq. 13-4,

$$\begin{aligned} v_{AM} &= V_C \sin \omega_C t + mV_C \sin \omega_M t \sin \omega_C t \\ &= V_C \sin \omega_C t + \tfrac{1}{2}mV_C \cos (\omega_C - \omega_M)t - \tfrac{1}{2}mV_C \cos (\omega_C + \omega_M)t \end{aligned} \tag{13-5}$$

This equation reveals that the amplitude-modulated wave consists of three sinusoidal components of constant amplitude. The carrier and the upper and lower *side frequencies* are shown in the frequency spectrum of Fig. 13.3a. In general the modulating signal is a complicated wave containing several components and the result is a carrier

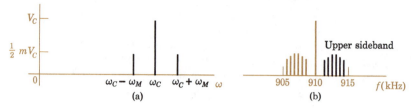

Figure 13.3 Frequency spectra of AM waves.

plus upper and lower *sidebands*. If the 910-kHz carrier of an AM station is modulated by frequencies from 100 to 4500 Hz, the frequency spectrum is as shown in Fig. 13.3b. The station is assigned a *channel* 10 kHz wide to accommodate side frequencies from 0 to 5000 Hz. The amplifiers in AM transmitters and receivers are designed to amplify well over a 10-kHz bandwidth and to discriminate against frequencies outside this range. In comparison, channels for frequency modulation (FM) broadcast are 200 kHz wide and permit lower noise and higher frequency response for greater fidelity. For satisfactory television transmission, sidebands approximately 6000 kHz wide are required.

EXAMPLE 1

A carrier signal $v_C = 100 \sin 10^7 t$ V is to be modulated by a signal $v_M = 100 \sin 10^4 t$ V, using a device with a characteristic defined by $i = v + 0.003v^2$ mA. The resulting current is filtered by a circuit that passes all frequencies from 9.9 to 10.1 Mrad/s and rejects all others. Describe the output signal.

Comparing the device characteristic to Eq. 13-1, we see that $a_0 = 0$, $a_1 = 1$, and $a_2 = 0.003$. Only the seventh, tenth, and eleventh components in Eq. 13-3 have frequencies within the passband of the filter.

The magnitudes of the carrier and side-frequency components are

$$a_1 V_C = 100 \text{ mA} \qquad \text{and} \qquad a_2 V_M V_C = 30 \text{ mA}$$

The output, an amplitude-modulated signal, is

$$i = 100 \sin 10^7 t - 30 \cos 1.001 \times 10^7 t$$
$$+ 30 \cos 0.999 \times 10^7 t \text{ mA}$$

Comparing the second term with the corresponding term in Eq. 13-5, we see that

$$\tfrac{1}{2} m I_C = \tfrac{1}{2} m \times 100 = 30 \qquad \text{or} \qquad m = \frac{60}{100}$$

and the degree of modulation is 60%.

Effect of Signal Size

As indicated in Example 1, a diode can be used as a *modulator*. The same nonlinear device will also serve as a *detector* to recover the information from the three radiofrequency signals or as a *mixer* to obtain the difference or "beat frequency" between two sinusoidal signals. (See Problems 1 and 3.) We know that the diode can also be used as a switch or as a rectifier. In all these functions, the signal excursions extend into the nonlinear regions of the diode characteristics. Obviously, the function actually performed is determined by the relative size of the coefficients of the various components in the output. These in turn depend on the factors a_0, a_1, and a_2 in the device characteristic, the bias voltage V_0, and the signal amplitudes V_M and V_C.

For linear applications, where the distorting effects of nonlinearity are to be avoided, we select the bias voltage to place operation on a portion of the characteristic where factor a_2 is relatively small. Then we make sure that the signal amplitudes are "small". Because the coefficients of the unwanted components vary as the square or the product of the signal amplitudes, any device can be considered to be linear if the signals are sufficiently small. By definition,

A "small signal" is one to which a given device responds in a linear fashion.

In electronics there are many situations where we obtain linear small-signal operation from devices whose gross behavior is highly nonlinear.

DIODES

If we are interested in the response of a diode to small signals superimposed upon dc values, a special circuit model is necessary.

Dynamic Resistance

As shown in Fig. 13.4, a small signal voltage v superimposed on a steady voltage V_{dc} produces a corresponding signal current i. The current-voltage relation for small

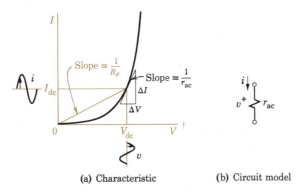

(a) Characteristic   (b) Circuit model

Figure 13.4 Dynamic resistance of a diode.

signals is defined by the slope of the curve at the operating point specified by V_{dc}. The *dynamic resistance* or *ac resistance* is defined as

$$r_{ac} = \frac{dV}{dI} \cong \frac{\Delta V}{\Delta I} \text{ in ohms} \tag{13-6}$$

In general, the dynamic resistance r_{ac} will differ appreciably from the static forward resistance defined as $R_F = V_{dc}/I_{dc}$.

The value of r_{ac} can be predicted for a theoretical diode for which

$$I = I_s(e^{eV/kT} - 1)$$

as in Eq. 5-19. Taking the derivative and assuming $I_s \ll I$,

$$\frac{dI}{dV} = \frac{e}{kT}I_s\,e^{eV/kT} = \frac{e}{kT}(I + I_s) \cong \frac{eI}{kT} \tag{13-7}$$

Hence, for this theoretical diode,

$$r_{ac} = \frac{dV}{dI} = \frac{v}{i} = \frac{kT}{eI} = \frac{kT/e}{I} \tag{13-8}$$

The dynamic resistance is directly proportional to absolute temperature T and inversely proportional to dc diode current I. At room temperature, $T \cong 300$ K, and $kT/e \cong 0.025$ V $= 25$ mV. Then

$$r_{ac} = \frac{0.025}{I} = \frac{25}{I(\text{in mA})}\,\Omega \tag{13-9}$$

This provides a simple means for estimating r_{ac} in practical diodes; however, the ohmic "body resistance" of the semiconductor (in series with the junction) may be the dominant factor at large currents.

Practice Problem 13-2

For a certain silicon diode, $I_s = 10$ nA. Estimate the dc forward resistance and small-signal ac resistance at room temperature for a current of 5 mA.

Answers: 66 Ω, 5 Ω.

FIELD-EFFECT TRANSISTORS

In the elementary amplifier of Fig. 13.5a, the input is a signal voltage v_i superimposed on the gate-source bias voltage V_{GG}. The signal voltage $v_{gs} = v_i$ modulates the width of the conducting channel and produces a signal component i_d of the drain current i_D. (For nomenclature, see Table 12-1 on p. 350.) The signal current develops across resistor R_L an amplified version of the input voltage.

The quantitative behavior of a small-signal amplifier could be determined graphically as we did in the case of large-signal BJT amplifiers, but this approach is not good for two reasons. First, the work is laborious and the results lack the generality needed in optimizing a design. Second, individual devices of a given type vary widely in their characteristics and the curves provided by the manufacturer are only general indications of how that type behaves on the average. A better approach is to devise a convenient model that will accurately predict device performance within limitations. If we limit our consideration to small signals superimposed on dc values, we are justified in assuming linear relations among signals in devices that are highly nonlinear

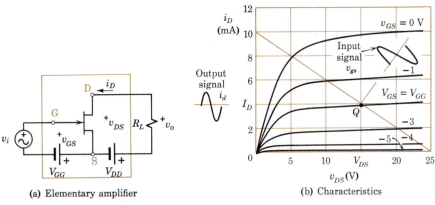

(a) Elementary amplifier

(b) Characteristics

Figure 13.5 Small-signal operation of an FET.

in their large-signal behavior. On this basis, the complicated physical device inside the colored rectangle of Fig. 13.5a is to be replaced, insofar as small signals are concerned, by a linear circuit model.

Two-Port Parameters

A diode is a two-terminal device or one-port, and its circuit model also has a single pair of terminals. Transistors are two-ports and can be represented as in Fig.

Figure 13.6 The general two-port.

13.6. In general, of the four variables identified only two are independent, and the behavior can be defined by a pair of equations such as

$$\begin{cases} v_1 = f_1(i_1, i_2) \\ v_2 = f_2(i_1, i_2) \end{cases} \quad \text{or} \quad \begin{cases} i_1 = f_3(v_1, v_2) \\ i_2 = f_4(v_1, v_2) \end{cases} \tag{13-10}$$

For linear networks, the first pair of equations would define a set of resistances as parameters and the second pair would define a set of conductances as parameters. Another acceptable formulation is

$$\begin{cases} v_1 = f_5(i_1, v_2) \\ i_2 = f_6(i_1, v_2) \end{cases} \tag{13-11}$$

For linear networks, the first equation of this pair would define a resistance and a voltage factor, and the second equation would define a current factor and a conductance. This mixture is called a set of *hybrid parameters*, and it is a particularly convenient set to use for bipolar transistors.

Evaluation of Parameters

In selecting an appropriate formulation for the field-effect transistor, we note that the input current is very small because in the JFET the gate-source junction is reverse biased and in the MOSFET the gate is insulated. Also, we note that the FET is a voltage-controlled device (Fig. 13.5b). Therefore, we write (Eq. 13-10b),

$$\begin{cases} i_G = 0 \\ i_D = f(v_{GS}, v_{DS}) \end{cases} \tag{13-12}$$

Because $i_G = 0$ at all times, only the second equation is significant. Because we are interested in small or "differential" changes in the variables, we express the total differential of the drain current as

$$di_D = \frac{\partial i_D}{\partial v_{GS}} \, dv_{GS} + \frac{\partial i_D}{\partial v_{DS}} \, dv_{DS} \tag{13-13}$$

Equation 13-13 says that a change in gate voltage dv_{GS} and a change in drain voltage dv_{DS} both contribute to a change in drain current di_D. The effect of the partial contributions is determined by the coefficients $\partial i_D/\partial v_{GS}$ and $\partial i_D/\partial v_{DS}$; these are the "partial derivatives" of the drain current with respect to gate voltage and drain voltage.

If the drain-source voltage is held constant, $dv_{DS} = 0$ and the partial derivative with respect to gate-source voltage is equal to the total derivative or

$$\frac{\partial i_D}{\partial v_{GS}} = \frac{di_D}{dv_{GS}} \bigg|_{v_{DS}=k} = g_m \tag{13-14}$$

where g_m is the *transconductance* in siemens, so called because it is the ratio of a differential current at the output to the corresponding differential voltage at the input. Typical values of g_m lie between 500 and 10,000 μS.

A second FET parameter can be defined by holding the gate-source voltage constant so that $dv_{GS} = 0$ and

$$\frac{\partial i_D}{\partial v_{DS}} = \frac{di_D}{dv_{DS}} \bigg|_{v_{GS}=k} = \frac{1}{r_d} \tag{13-15}$$

where r_d is the *dynamic drain resistance* and is just equal to the reciprocal of the slope of a line of constant v_{GS}. Typical values of r_d lie between 20 and 500 kΩ. (Instead of r_d, the *output conductance* $g_{os} = g_d = 1/r_d$ may be specified.)

For the JFET or the DE MOSFET, the drain current in the constant-current region is (Eq. 6-1)

$$i_D = I_{DSS} \left(1 - \frac{v_{GS}}{V_p}\right)^2 \tag{13-16}$$

At a dc bias voltage V_{GS}, the transconductance is

$$g_m = \frac{di_D}{dv_{GS}} = -\frac{2I_{DSS}}{V_p} \left(1 - \frac{V_{GS}}{V_p}\right)$$

The transconductance evaluated at $V_{GS} = 0$ is

$$g_{mo} = -\frac{2I_{DSS}}{V_p} \left(1 - \frac{0}{V_p}\right) = -\frac{2I_{DSS}}{V_p} \tag{13-17}$$

sometimes called y_{fs}, the *forward transadmittance*. In general,

$$g_m = g_{mo}\left(1 - \frac{V_{GS}}{V_p}\right) = g_{mo}\sqrt{\frac{I_D}{I_{DSS}}} \tag{13-18}$$

Since the manufacturer usually gives the value of g_{mo} (or y_{fs}) along with I_{DSS}, Eq. 13-18 can be used to calculate g_m for any operating value of I_D or V_{GS} (after solving for V_p from Eq. 13-17). For E MOSFETs, the data sheet usually includes a curve of g_m versus I_D.

EXAMPLE 2

Estimate the model parameters of the FET of Fig. 13.5b for operation at a Q point defined by $v_{DS} = 15$ V and $v_{GS} = -2$ V. Use Δi and Δv to approximate di and dv.

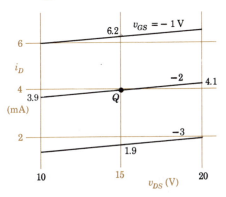

Figure 13.7 Evaluating parameters.

Assuming $V_p \cong -6$ V and $I_{DSS} \cong 10$ mA, use Eq. 13-18 to calculate the value of g_m at $V_{GS} = -2$ V and compare to the value obtained graphically.

Taking increments from the given operating point (Fig. 13.7) and reading the corresponding values yields

$$g_m = \frac{di_D}{dv_{GS}}\bigg|_{15\,V} \cong \frac{\Delta i_D}{\Delta v_{GS}}$$

$$= \frac{(6.2 - 1.9)10^{-3}}{-1 + 3} = 2150\ \mu S$$

$$r_d = \frac{1}{\dfrac{di_D}{dv_{DS}}\bigg|_{-2\,V}} \cong \frac{\Delta v_{DS}}{\Delta i_D}$$

$$= \frac{20 - 10}{(4.1 - 3.9)10^{-3}} = 50\ k\Omega$$

$$g_m = -\frac{2I_{DSS}}{V_p}\left(1 - \frac{V_{GS}}{V_p}\right) = -\frac{0.02}{-6}\left(1 - \frac{-2}{-6}\right) = 2200\ \mu S$$

approximately the same as the measured value.

Small-Signal Model

On the diagram of the elementary amplifier in Fig. 13.8a, v_{GS}, v_{DS}, and i_D are total instantaneous quantities representing small ac variations superimposed on steady or dc values. The dc values fix the quiescent point and determine the values of the FET parameters; the ac values correspond to the signals, which are of primary interest. We wish to devise a small-signal model that will hold only for ac values and that will enable us to predict the performance of the FET in an ac application such as this elementary amplifier.

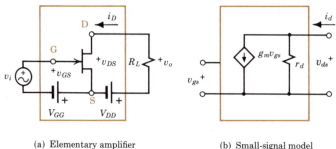

(a) Elementary amplifier   (b) Small-signal model

Figure 13.8   Derivation of a small-signal model of an FET.

For differential changes, Eq. 13-13 states that

$$di_D = \frac{\partial i_D}{\partial v_{GS}} dv_{GS} + \frac{\partial i_D}{\partial v_{DS}} dv_{DS} \tag{13-19a}$$

Replacing the differentials by small signals and using the FET parameters, Eq. 13-19a becomes

$$i_d = g_m v_{gs} + \frac{1}{r_d} v_{ds} \tag{13-19b}$$

For a functional model, we need an ac circuit that satisfies Eq. 13-19b, which says that the drain-signal current consists of two parts. In the corresponding parallel circuit, one branch contains a controlled-current source $g_m v_{gs}$ directly proportional to the input signal voltage, and the other branch carries a current v_{ds}/r_d directly proportional to output voltage. The output circuit of Fig. 13.8b is the logical result; it is the circuit model of Eq. 13.19b. The open input circuit satisfies the condition $i_g = 0$.

The functional relationship holds equally well for effective values of sinusoidal quantities (or for the corresponding phasors) and Eq. 13-19b becomes

$$I_d = g_m V_{gs} + \frac{1}{r_d} V_{ds} \tag{13-19c}$$

EXAMPLE 3

Predict the voltage gain of the amplifier of Fig. 13.8a if $R_L = 5$ kΩ.

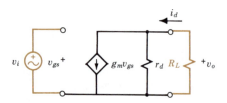

Figure 13.9   Voltage gain calculation.

The amplifier is replaced by the circuit model of Fig. 13.9 where $g_m = 2200$ μS and $r_d = 50$ kΩ. Because $r_d \gg R_L$,

$$v_o = -g_m v_i (R_L \| r_d) \cong -g_m v_i R_L$$

and the voltage gain is

$$\frac{v_o}{v_i} = -g_m R_L = -2200 \times 10^{-6} \times 5 \times 10^3 \cong -11$$

The minus sign indicates an inversion or phase reversal of the output signal.

The use of ac models in the signal analysis of FET amplifiers is illustrated in Example 3. One conclusion is that for practical values of R_L (limited by available supply voltage V_{DD}), voltage gains are small; much larger gains can be obtained with BJT amplifiers. (See Example 4.) Note that r_d typically is so large compared to practical values of R_L that reasonably good predictions can be made with a model consisting of controlled-current source $g_m v_{gs}$ alone.

High-Frequency Model

At high frequencies consideration must be given to the charging currents associated with various capacitive effects. In Fig. 13.10a, the small-signal model has been modified to take into account capacitances inherent in field-effect transistors. C_{gs} and

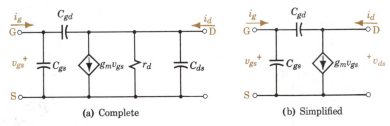

(a) Complete   (b) Simplified

Figure 13.10 High-frequency FET models.

C_{gd} represent the capacitances between gate and source and gate and drain; in a JFET (Fig. 6.1) these arise from the reverse-biased *pn* junctions and will be of the order of 1 to 10 pF. C_{ds} represents the drain-to-source capacitance and will be small, say, 1 to 5 pF. As predicted by this model, the voltage gain of an FET amplifier drops off as frequency increases. A quantitative analysis of high-frequency performance is given in Chapter 14; the simplified model of Fig. 13.10b is usually satisfactory because r_d and C_{ds} are usually negligible in comparison to the input parameters of the following stage.

BIPOLAR JUNCTION TRANSISTORS

Figure 6.16 shows the graphical analysis of an elementary large-signal amplifier. For small signals, a linear circuit model is much more convenient in both analysis and design. As with the FET, our objective is a simple model that will predict the performance of a properly biased transistor (BJT) *insofar as small ac variations are concerned*.

An Idealized Model

The *i-v* characteristics of a typical BJT are displayed in Fig. 6.15. The general relations can be expressed in terms of voltages or currents as the independent variables. Using a set of hybrid parameters, we write

$$\begin{cases} v_{BE} = f_1(i_B, v_{CE}) \\ i_C = f_2(i_B, v_{CE}) \end{cases} \tag{13-20}$$

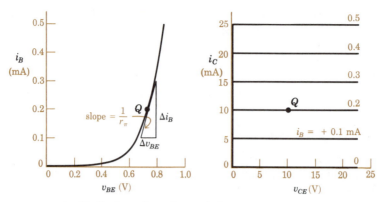

Figure 13.11 Idealized transistor characteristics.

If we limit operation to the region where v_{BE} and i_C are independent of v_{CE}, we can represent the common-emitter behavior by the idealized v-i characteristics of Fig. 13.11. The base-emitter characteristic is that of a theoretical diode and the collector characteristics indicate a constant β and negligible I_{CEO}. For operation at a quiescent point Q defined by $I_B = 0.2$ mA and $V_{CE} = 10$ V, a small change Δv_{BE} produces a small change Δi_B that, in turn, produces a larger change $\Delta i_C = \beta \Delta i_B$ (for nomenclature, see Table 12-1 on page 350). The small-signal behavior of this idealized transistor is defined by

$$\begin{cases} v_{be} = r_\pi i_b \\ i_c = \beta i_b \end{cases} \tag{13-21}$$

where

$$r_\pi = \frac{\Delta v_{BE}}{\Delta i_B} = \frac{v_{be}}{i_b} = \text{the dynamic junction resistance} \tag{13-22}$$

$$\beta = \frac{\Delta i_C}{\Delta i_B} = \frac{i_c}{i_b} = \text{the forward current-transfer ratio} \tag{13-23}$$

We see that β is a "small-signal current gain"; for most purposes, $\beta_{ac} \cong \beta_{dc}$.

To evaluate r_π we note that the base characteristic is that of a junction diode for which the theoretical relation is

$$i_B = I_s(e^{ev_{BE}/kT} - 1) \tag{13-24}$$

and the dynamic resistance is given by Eq. 13-8 with $I = I_B$. Because our biasing networks are designed to stabilize $I_C = \beta I_B$, it is convenient to express r_π in terms of I_C. At room temperature, $kT/e \cong 0.025$ V and

$$r_\pi = \frac{0.025}{I_B} = \beta \frac{25}{I_C(\text{in mA})} \ \Omega \tag{13-25}$$

This is a simple, fairly accurate method for estimating r_π in practical transistors.

On the diagram of the elementary amplifier in Fig. 13.12a, v_{BE}, i_B, and i_C are total instantaneous quantities representing small ac variations superimposed on steady or dc values. The dc values fix the quiescent point and determine the values of the BJT parameters; the ac values correspond to the signals, which are of primary interest. The

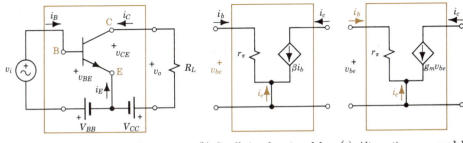

(a) Elementary amplifier   (b) Small-signal r_π-β model   (c) Alternative r_π-g_m model

Figure 13.12   Derivation of small-signal models of a BJT.

physical interpretation of the small-signal circuit of Fig. 13.12b is that a variation in base current (a signal input to the base, determined by r_π) appears in the collector circuit as a current whose magnitude is β times as great. The possibility of a current gain as well as a voltage gain is an important advantage of the common-emitter configuration for amplifiers. The use of this "r_π-β" model in predicting small-signal behavior is illustrated in Example 4.

EXAMPLE 4

An amplifier employing a transistor with $\beta = 50$ and operating at $I_C = 0.5$ mA in the common-emitter configuration is to deliver a sinusoidal signal current of 0.2 mA rms to a load resistance of 5 kΩ. Estimate the signal current gain and voltage gain.

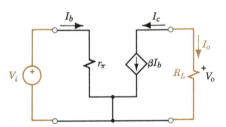

Figure 13.13   Application of the r_π-β transistor model.

For this estimate, the r_π-β model of Fig. 13.13 is satisfactory. Since $I_o = -I_c = -\beta I_b$, the signal current gain is

$$\frac{I_o}{I_b} = -\beta = -50$$

The necessary input signal current (rms) is

$$I_b = \frac{|I_o|}{\beta} = \frac{0.2 \times 10^{-3}}{50} = 4 \times 10^{-6} = 4\ \mu\text{A}$$

By Eq. 13-25 the input resistance is

$$r_\pi = \beta \frac{25}{I_C} = 50 \times \frac{25}{0.5} = 2500\ \Omega$$

$$\therefore V_i = I_b r_\pi = 4 \times 10^{-6} \times 2.5 \times 10^3 = 10\ \text{mV}$$

The magnitude of the voltage gain is

$$\frac{V_o}{V_i} = \frac{I_o R_L}{V_i} = \frac{2 \times 10^{-4} \times 5000}{10 \times 10^{-3}} = 100$$

In the r_π-β model, the source in the collector circuit is controlled by input current i_b. Alternatively, the output current can be defined in terms of input voltage v_{be}. For equivalence in Fig. 13.12c,

$$\beta i_b = \beta \frac{v_{be}}{r_\pi} = g_m v_{be} \qquad (13\text{-}26)$$

where g_m, the small-signal transconductance for the BJT, is given by

$$g_m = \frac{\beta}{r_\pi} = \beta \frac{I_B}{0.025} = 40\, I_C \qquad (13\text{-}27)$$

at room temperature.

Practice Problem 13-3

A transistor with $\beta = 50$ is operating at $I_C = 0.5$ mA in a common-emitter amplifier circuit with $R_L = 5$ kΩ. Calculate g_m, draw an appropriate small-signal model for the amplifier, and predict the voltage gain.

Answers: 20,000 μS, -100.

Where current gain as well as voltage gain is important, the common-emitter amplifier configuration possesses obvious advantages. Under certain conditions the common-base arrangement is superior and under other circumstances (discussed in Chapter 15), the common-collector configuration is advantageous.

The Hybrid-π Model

For more precise predictions of transistor behavior, more sophisticated models are necessary. In Fig. 13.14, the *base-spreading resistance* r_b takes into account the

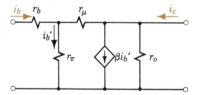

Figure 13.14 The more precise hybrid-π circuit model.

ohmic resistance of the long, thin base region. Typically, r_b varies from 50 to 150 Ω; lacking other information, a value of 100 Ω can be assumed with little error if r_π is relatively high. The resistance r_μ accounts for a very small current due to changes in collector-base voltage; this effect is negligibly small in most cases and r_μ will be ignored in our work. The resistance r_o provides for the small change in i_C with v_{CE} for given values of i_B (see Fig. 6.15).

In this model, proposed by L. J. Giacoletto, the circuit elements are arranged in a π-configuration known as the hybrid-π model.[†] It is particularly useful, when properly modified, in predicting the high-frequency behavior of transistors. Junction capacitances C_{je} and C_{jc} and capacitance C_b accounting for base-charging current can be connected in parallel with r_π and r_μ as shown in the high-frequency hybrid-π model

---

[†] It is customary to use the *internal* base-emitter voltage $v'_{be} = i'_b r_\pi$ as the control variable and $g_m v'_{be}$ as the controlled current source.

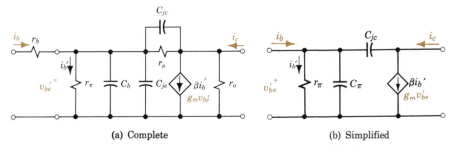

| (a) Complete | (b) Simplified |

Figure 13.15 High-frequency hybrid-π models.

of Fig. 13.15a. If r_b and r_o are neglected and r_μ ignored, and if C_b and C_{je} are combined into C_π, the simplified high-frequency model of Fig. 13.15b results. This model will be discussed in more detail in connection with amplifier performance in Chapter 14. (Note that the r_π-β model is a simple hybrid-π for moderate frequencies.)

The General h-Parameter Model

For any configuration and any small-signal application, a general linear model can be used to represent the transistor and predict its performance. Hybrid parameters are most commonly used and are most frequently supplied by manufacturers. The chief advantages of h-parameters are the ease with which they can be determined in the laboratory and the facility with which they can be handled in circuit calculations.

Choosing input current i_I and output voltage v_O as the independent variables and rewriting Eq. 13-11 in conventional transistor notation,

$$\begin{cases} v_I = f_5(i_I, v_O) \\ i_O = f_6(i_I, v_O) \end{cases} \tag{13-28}$$

For the common-emitter configuration, for example, v_I is the base-emitter voltage and i_O is the collector current.

Because of the nonlinearity of these functions, no simple model can be formulated to predict total voltages and currents. Since we are interested in small-signal behavior, let us consider differential changes in the variables. Taking the total differential, these become

$$\begin{cases} dv_I = \dfrac{\partial v_I}{\partial i_I} di_I + \dfrac{\partial v_I}{\partial v_O} dv_O \\[2mm] di_O = \dfrac{\partial i_O}{\partial i_I} di_I + \dfrac{\partial i_O}{\partial v_O} dv_O \end{cases} \tag{13-29}$$

Equation 13-29a says that a change in input current di_I and a change in output voltage dv_O both contribute to a change in input voltage dv_I. The effectiveness of the partial contributions is dependent on the coefficients $\partial v_I/\partial i_I$ and $\partial v_I/\partial v_O$; these partial derivatives of input voltage with respect to input current and output voltage and the corresponding partial derivatives of output current are the h-parameters of a small-signal circuit model.

Noting that the differential quantities di_I and dv_O correspond to small signals i_i and v_o, Eqs. 13-29 can be written as

$$
\begin{cases}
v_i = \dfrac{\partial v_I}{\partial i_I} i_i + \dfrac{\partial v_I}{\partial v_O} v_o = h_i i_i + h_r v_o \\[4mm]
i_o = \dfrac{\partial i_O}{\partial i_I} i_i + \dfrac{\partial i_O}{\partial v_O} v_o = h_f i_i + h_o v_o
\end{cases}
\qquad (13\text{-}30)
$$

where h_i = input impedance with output short-circuited (ohms),
$\quad h_f$ = forward transfer current ratio with output short-circuited,
$\quad h_r$ = reverse transfer voltage ratio with input open-circuited,
$\quad h_o$ = output admittance with input open-circuited (siemens).

The first of Eqs. 13-30 says that the input voltage is the sum of two components. This indicates a series combination of an impedance drop $h_i i_i$ and a controlled voltage source $h_r v_o$ directly proportional to the output voltage. The second equation indicates a parallel combination of a controlled current source $h_f i_i$ and an admittance current $h_o v_o$. The h-parameter model of Fig. 13.16b follows from this line of reasoning.

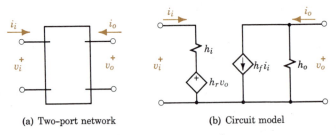

(a) Two–port network   (b) Circuit model

Figure 13.16   General h-parameter representation.

The general model is applicable to any transistor configuration. In a specific case, the parameters are identified by a second subscript b, e, or c, depending on whether the base, emitter, or collector is the common element. See Example 5.

Determination of h-Parameters

The h-parameters are small-signal values implying small ac variations about operating points defined by dc values of voltage and current. They could be determined from families of characteristic curves. In actual practice, however, the parameters are determined experimentally by measuring the ac voltages or currents that result from ac signals introduced at the appropriate locations. For example, to determine h_{fe}, a small ac voltage is applied between base and emitter, and the resulting ac currents in the base lead and the collector lead are measured (the collector circuit must be effectively short-circuited to ac). Then h_{fe} is the ratio of collector current to base current and is just equal to β. Similar measurements permit the determination of the other small-signal parameters. Note that h_{ie} is $r_\pi + r_b \cong r_\pi$.

Since transistor parameters vary widely with operating point, measurements are made under standard conditions. Usually, values are for a frequency of 1000 Hz at

EXAMPLE 5

The transistor of Example 4 has the following common-emitter h-parameters: $h_{ie}=2500\ \Omega$, $h_{re} = 4 \times 10^{-4}$, $h_{fe} = 50$, and $h_{oe} = 10\ \mu S$ as shown in Fig. 13.17.

For $R_L = 5\ k\Omega$ and $I_o = 0.2$ mA rms, estimate the current gain and voltage gain of the amplifier and compare with the results of Example 4.

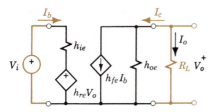

Figure 13.17 Application of general h-parameters in the CE configuration.

Applying the current-divider principle,

$$I_o = -I_c = -\frac{1/R_L}{h_{oe} + 1/R_L} h_{fe} I_b$$

The current gain is

$$\frac{I_o}{I_b} = -\frac{h_{fe}/R_L}{h_{oe} + 1/R_L} = -\frac{50/5000}{(10 + 200)10^{-6}} = -47.6$$

The required input signal is

$$I_b = I_o/(-47.6) = 200/(-47.6) = -4.2\ \mu A$$

In the input loop,

$$V_i = I_b h_{ie} + h_{re} V_o$$

$$= -4.2 \times 10^{-6} \times 2500 + 4 \times 10^{-4} \times 1$$

$$= (-10.5 + 0.4)10^{-3} = -10.1\ mV$$

Since $V_o = I_o R_L = 0.2 \times 10^{-3} \times 5 \times 10^3 = 1$ V,

$$\frac{V_o}{V_i} = \frac{1}{-0.0101} = -99$$

The simpler model introduces only small errors.

room temperature (25° C) with an emitter current of 1 mA and a collector-base voltage of 5 V. Transistor manufacturers publish typical, minimum, and maximum values of the parameters along with average static characteristics. Typical values are shown in Table 13-1. An engineer designing a transistor amplifier, for example, must arrange the circuit so that the performance of the amplifier is within specifications despite expected variations in transistor parameters. (In Chapter 15 we shall see how feedback can be used to stabilize performance.)

Table 13-1 Typical Transistor Parameters

	2N1613	2N3114	2N699B Min.	2N699B Typical	2N699B Max.	2N2222A Min.	2N2222A Max.
h_{ie} (kΩ)	2.2	1.5		2.8		2	4
h_{re} ($\times 10^{-4}$)	3.6	1.5		3.5			8
h_{fe}	55	50	35	70	100	50	300
h_{oe} (μS)	12.5	5.3		11		5	35
h_{ib} (Ω)	27	27	20	27	30		
h_{rb} ($\times 10^{-4}$)	0.7	0.25		0.5	1.25		
h_{fb}	-0.98	-0.981		-0.987			
h_{ob} (μS)	0.16	0.09	0.1	0.12	0.5		

Comparison of BJTs with FETs

The much higher values of g_m permit much higher voltage gains with the BJT as compared to the FET. On the other hand, the much higher values of input resistance for the FET permit lower "loading" of the source as compared to the BJT. The more linear transfer characteristic of the BJT permits larger signal swings without distortion; the square-law transfer characteristic of the FET is advantageous in modulation and mixing. In integrated-circuit form, the MOSFET is smaller and simpler and therefore cheaper than the BJT.

A Communication System

We now have in our repertoire of electronic devices all the essential elements of a telemetry or communication system (Fig. 13.18).

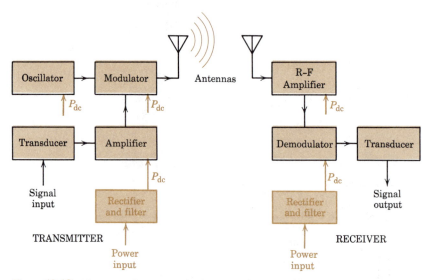

Figure 13.18 Elements of a communication system.

In a radio broadcasting system the signal input is an audible variation in air pressure on a microphone or a mechanical vibration of a phonograph needle. The transducer, a microphone or pickup, converts the acoustic vibration into an electrical signal that, when amplified, is available for modulating. The radiofrequency carrier signal is generated in the oscillator and combined with the audiofrequency signal in the modulator. Usually the modulator stage also performs amplification so that a high-level amplitude-modulated signal is fed into the transmitting antenna.

Energy is radiated from the antenna in the form of an electromagnetic wave, part of which is intercepted by the receiving antenna. The selected carrier and sidebands are amplified and fed into the demodulator. The detector output, usually amplified further, is converted by the transducer (a loudspeaker) into a replica of the original input signal. The power required may be obtained from 60-Hz ac lines, converted into dc power by rectifier-filter combinations, and supplied to each stage of the communication system.

SUMMARY

- The function performed by a diode depends on bias voltage and signal amplitude. A "small signal" is one to which a given device responds in a linear fashion.

- A modulator superimposes information on a carrier by modifying the carrier amplitude, frequency, or phase.
  An AM wave consisting of carrier and sidebands is described by

$$v_{AM} = V_C \sin \omega_C t - \frac{mV_C}{2} \cos(\omega_C + \omega_M)t + \frac{mV_C}{2} \cos(\omega_C - \omega_M)t$$

  A demodulator recovers the original signal from the modulated wave.

- The small-signal performance of complicated and highly nonlinear electronic devices can be predicted using relatively simple linear models.
  For small signals superimposed on steady values, the significant diode parameter is the dynamic resistance $r_{ac} = dv/di$, the reciprocal of the slope of the i-v characteristic at the operating point.

- To derive the small-signal circuit model of a nonlinear two-port:
  1. Write the governing equation in terms of the chosen variables.
  2. Take the total differential to obtain the equations for small variations.
  3. Note that differentials can be approximated by small signals, and write the corresponding equations for small signals.
  4. Interpret the equations in terms of an electrical circuit with appropriate parameters, including passive elements and controlled sources.

- For a field-effect transistor, the input current is negligible and the small-signal model consists of transconductance g_m and drain resistance r_d (Fig. 13.19).
  For a junction transistor, the general model is complicated; at moderate frequencies the hybrid-parameter CE model is as shown in Fig. 13.20.
  In many BJT applications, a satisfactory model consists simply of an input resistance r_π and a controlled-current source βI_b or $g_m V_{be}$ (Fig. 13.21).

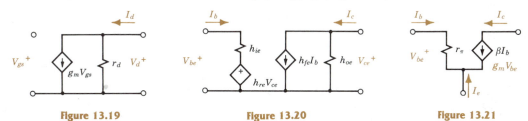

Figure 13.19   **Figure 13.20**   **Figure 13.21**

- Because individual transistors vary widely from the norm, it is customary to use parameter value ranges published by the manufacturer, minimizing the effects of parameter variations by proper circuit design.

REVIEW QUESTIONS

1. Explain the differences between V/I, $^*\Delta V/\Delta I$, dV/dI, and $\partial V/\partial I$.
2. Explain the difference between static resistance and dynamic resistance.
3. Sketch and label the spectrum of frequencies transmitted by your favorite AM station during a typical music program.
4. Explain how a nonlinear device produces amplitude modulation. Demodulation.
5. Why is a high frequency needed for a radio broadcast carrier?
6. How many telephone conversations (two-way) can be carried on a telephone cable that transmits frequencies up to 8 MHz? How many television programs?

7. Why is the ac model of an FET simpler than an equally precise model of a BJT?
8. In analytic geometry, what does $z = f(x, y)$ represent in general? Could $i_b = f(v_c, v_b)$ be similarly represented? What is the graphical interpretation of $\partial z/\partial y$?
9. Sketch a set of FET characteristics and define, graphically, the two parameters.
10. What are the criteria for selecting the quiescent point in an amplifier?
11. Does the circuit of Fig. 13.8b contain any information not in Eq. 13-19b? Why use the circuit?
12. What is meant by "The results of a graphical analysis of performance from characteristic curves lack generality"?

EXERCISES

1. A voltage $v = V_M \sin \omega_M t + V_C \sin \omega_C t$ is applied to a device whose mathematical model is $i = a_0 + a_2 v^2$.
   (a) Determine the device current.
   (b) Identify the frequency components of current.
2. The amplitude of a 740-kHz signal is varied sinusoidally from zero to $2V_C$ at a rate of 4000 times per second. Sketch the resulting modulated wave and list the frequencies of the components appearing in the transmitted wave.
3. An amplitude-modulated wave is described by the equation

   $$v = 5 \cos 2.05 \times 10^6 t + 10 \sin 2.1 \times 10^6 t - 25 \cos 2.15 \times 10^6 t$$

   Identify the carrier frequency (rad/s) and the frequencies of any modulating signals.
4. A signal is described by the equation

   $$v = (8 + 5 \sin 1000t + 2 \sin 2000t) \sin 10^5 t$$

   List the frequencies (rad/s) present in the signal and show them on a frequency spectrum. Identify the carrier and the upper and lower sidebands.
5. Channel 4, a typical TV channel, extends from 66 to 72 MHz. A ruby laser operates in the red region with a wavelength of 7500 Å. If the useful bandwidth of a laser beam is equal to 0.2% of its

carrier frequency, how many TV channels could be carried on a single laser beam?
6. Two signals $a = \sin \omega t$ and $b = 5 \sin 20 \, \omega t$ are multiplied together.
   (a) Show on a frequency spectrum all the components of the product.
   (b) If a voltage equal to the product is fed into a device for which $i_o = v + 0.2v^2$, show on a frequency spectrum all the components of the output current.
7. A signal $v = A \sin 100{,}000t + B \sin 103{,}000t$ is applied to the input of an FET whose output is defined by $i_o = av^2$.
   (a) List the frequencies of the components appearing in the output.
   (b) Show the components on a frequency spectrum.
   (c) Under what conditions would this represent demodulation, or the separation of an audible signal from a carrier?
   (d) Sketch a filter circuit to pass only the audible component to load R_L.
8. Assume that a device is "practically linear" if the distortion introduced is less than 5% of the desired component. For the MOSFET of Fig. 12.7, determine the maximum signal input for practically linear operation.
9. Derive a general expression for the dynamic resistance of a device for which $i = v + 0.01v^2$ A

and evaluate r_{ac} at $V = 10$ V. Predict $i(t)$ for $v(t) = 10 + \sin \omega t$ V by adding the responses to the two components. Why does "superposition" work here?

10. A silicon diode is characterized by $I_s = 10$ nA. Predict the dc forward resistance and the dynamic resistance at:
   (a) $T = 25°$ C and current = 0.1 mA.
   (b) $T = 25°$ C and current = 10 mA.
   (c) $T = 100°$ C and current = 10 mA.

11. Starting from the second pair of Eqs. 13-10, and assuming a linear two-port network, derive a circuit model with admittance parameters (Y_{11}, etc.).

12. Starting from the first pair of Eqs. 13-10, and assuming a nonlinear two-port device, derive a small-signal model with impedance parameters (z_{11}, etc.).

13. A semiconductor two-port is described by the equations

$$\begin{cases} i_1 = 10^{-7} v_1 \\ i_2 = 2 \times 10^{-3} v_1 + 10^{-5} v_2 \end{cases}$$

Devise, draw, and give parameter values for a circuit model of this device.

14. Represent the MOSFET of Fig. 6.4c by a small-signal model and estimate the parameters at a point defined by $v_{GS} = 0$ and $v_{DS} = 8$ V.

15. For a JFET with $g_{mo} = 4000$ μS and $V_p = -4$ V, predict g_m at:
   (a) $V_{GS} = -1$ V; (b) $I_D = 2$ mA.

16. For a DE MOSFET with $I_{DSS} = 10$ mA and $g_{mo} = 5000$ μS, predict g_m at:
   (a) $I_D = 5$ mA; (b) $V_{GS} = -1$ V.

17. A DE MOSFET with $I_{DSS} = 5$ mA, $g_{mo} = 5000$ μS and a negligibly large r_d is used in the circuit of Fig. 12.5c. If $V_{DD} = 25$ V and $R_D = 3$ kΩ, predict the voltage gain v_o/v_i.

18. The MOSFET of Exercise 16 ($r_d = 50$ kΩ) is used in the circuit of Fig. 13.8a with $R_L = 5$ kΩ and $V_{GG} = 2$ V.
   (a) Specify I_D and V_{DD} for operation at $V_{GS} = -2$ V and $V_{DS} = 8$ V.
   (b) Predict the voltage gain.

19. For the MOSFET of Exercise 14, specify the value of load resistance to provide a voltage gain of 10.

20. An E MOSFET with a transconductance of 3000 μS and a dynamic drain resistance of 100 kΩ is used in the circuit of Fig. 13.5a where $v_i = 0.5 \sin \omega t$ and $R_L = 5$ kΩ. Draw an appropriate circuit and predict output voltage v_o.

21. A semiconductor two-port is described by the equations

$$\begin{cases} i_1 = 10^{-3} v_1 \\ i_2 = 50 i_1 + 5 \times 10^{-6} v_2 \end{cases}$$

Devise, draw, and give parameters for a circuit model of this device. Define the parameters mathematically.

22. A semiconductor two-port is described by the equations

$$\begin{cases} i_1 = 10^{-6} e^{40 v_1} \\ i_2 = -0.98 i_1 + 10^{-6} v_2 \end{cases}$$

Devise and draw a small-signal model for this device at an operating point defined by $i_1 = 2$ mA and $v_2 = 10$ V.

23. A two-port electronic device is defined by the following equations

$$\begin{cases} i_1 = 0.1 v_1 + 0.1 \, dv_1/dt \\ v_2 = 5 v_1 + 20 i_2 \end{cases}$$

   (a) Devise a suitable circuit model and label the circuit elements with the appropriate values.
   (b) If this device is used as a current amplifier, describe how the current gain i_2/i_1 will vary with frequency.

24. At high frequencies a semiconductor two-port is defined by the equations:

$$\begin{cases} i_1 = 10^{-8} \, dv_1/dt - 10^{-9} \, dv_{21}/dt \\ i_2 = 2 \times 10^{-3} \, v_1 + 10^{-9} \, dv_{21}/dt \end{cases}$$

Devise, draw, and give parameter values for a circuit model of this device.

25. A silicon BJT with $\beta = 40$ and $I_{CEO} = 10$ nA is connected in the circuit of Fig. 12.15a where $V_{CC} = 20$ V, $R_B = 25$ kΩ, $V_{BB} = 3.2$ V, and $R_C = 2$ kΩ.
   (a) Write the theoretical equations for $i_B = f(v_{BE})$ and $i_C = f(i_B)$.
   (b) Sketch approximately to scale the base and collector characteristics for $0 < i_B < 0.25$ mA.
   (c) Replace the transistor by an approximate dc model based on simplification of the curves of part (b).
   (d) Redraw the circuit and use your model to predict quiescent values of I_B, I_C, and V_{CE}.
   (e) Replace the transistor by a simple small-signal model, neglect large R_B, and estimate the signal components of i_C and v_{CE} for $i_i = 0.1 \sin \omega t$ mA.

26. For the *npn* transistor of Figs. 6.11 and 12,
    (a) Write the theoretical equations for $i_E = f(v_{EB})$ and $i_C = f(i_E)$ assuming $v_{CB} > 1$ V.
    (b) Neglecting I_{CBO}, devise a small-signal model for the transistor in this configuration.

27. In the elementary amplifier of Fig. 13.12a, at the quiescent point $I_C = 1.5$ mA and $V_{CE} = 4.5$ V.
    (a) Redraw the circuit for small-signal analysis using the r_π-β model with $\beta = 60$.
    (b) Specify R_L for a voltage gain $|v_o/v_i| = 180$.
    (c) Specify the battery voltage V_{CC}.

28. The 2N3114 parameters in Table 13-1 are for $I_C = 1$ mA.
    (a) Estimate r_π, r_b, and β for $I_C = 1$ mA.
    (b) Assuming r_b and β are constant, estimate r_π and h_{ie} for $I_C = 0.2$ mA.

29. An amplifier is represented by the model of Fig. 13.14 with $r_b = 100$ Ω, $r_\pi = 4000$ Ω, $r_\mu = 20$ MΩ, $\beta = 100$, and $r_o = 200$ kΩ connected to a source v_s in series with $R_s = 900$ Ω. With $R_L = 2$ kΩ, the output voltage $v_L = -2$ V.
    (a) Draw the complete circuit and calculate i_b' and $v_{be}' = i_b' r_\pi$.
    (b) Calculate i_μ, v_s, and $A_V = v_L/v_s$.
    (c) Use the simple model of Fig. 13.15b to calculate A_V. Draw a conclusion.

30. For certain parameters, the manufacturer of the 2N699B specifies guaranteed minimum and maximum values (Table 13-1). Determine the corresponding range of values of r_π. Is high precision justifiable in calculations using these parameters?

31. A 2N3114 transistor operating at 25° C and $I_E = 1$ mA is used in an elementary amplifier to provide 2 V rms across $R_L = 5$ kΩ. The input source can be represented by V_s in series with $R_s = 1$ kΩ.
    (a) Using the general *h*-parameter model, specify V_s.
    (b) Using the r_π-β model, repeat part (a) and draw a conclusion.

32. A 2N1613 is to be used ($I_C = 1$ mA, $V_{CE} = 5$ V) in a common-emitter amplifier to provide 1.5 V rms output across a load resistance R_L. The available open-circuit signal is 10 mV from a source with an internal resistance of 500 Ω.
    (a) Draw and label the circuit with an *h*-parameter model.
    (b) Simplify the model where appropriate to permit an approximate analysis.
    (c) Specify the necessary load resistance and the collector supply voltage V_{CC}.

33. Compare the voltage gain possibilities of an FET rated at $I_{DSS} = 4$ mA and $g_{mo} = 5000$ μS and a BJT rated at $\beta = 100$ at $V_{CE} = 5$ V, both operating at $I_D = I_C = 1$ mA. Predict the actual voltage gains with $R_C = R_D = 5$ kΩ assuming the voltage source has an output resistance of 2500 Ω, and draw a conclusion.

PROBLEMS

1. When a nonlinear device is used as a detector, the important factor is the second-degree term in Eq. 13-1. Prove that an FET whose characteristic is defined by $i = av^2$ in combination with a suitable filter will recover the modulating signal from an amplitude-modulated wave.

2. The diode-capacitor circuit of Fig. 3.32 can be used as a detector for AM signals. Assume that the input voltage v is the modulated carrier of Fig. 13.2 and that time constant $R_L C$ is large compared to the period T_C but small compared to T_M.
    (a) Sketch v_L, the output signal.
    (b) If the "load," an earphone, includes an inductive reactance that is large at the carrier frequency, sketch the output current i_L.

3. A *mixer* is widely used in radio and TV receivers to obtain a signal with a frequency equal to the *difference* in the frequencies of two input signals. Because of its square-law transfer characteristics, an FET makes an ideal mixer. Demonstrate that if the sum of two signals at frequencies ω_x and ω_y is applied to the gate-source terminals of a DE MOSFET, the output current will contain a component at frequency $\omega_x - \omega_y$. What are the other components of the output and how could they be eliminated?

4. The "Darlington pair" shown in Fig. 13.22 is frequently packaged as a single component or made available as an integrated circuit. Assuming identical transistors (β, r_π) and stating any other necessary assumptions, predict the cur-

rent gain of the pair and the input and output resistances of the pair in the given circuit. Draw a conclusion about the advantages of such a component.

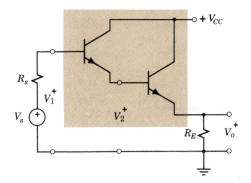

Figure 13.22

5. A BJT amplifier is terminated in a load resistance R_L. Using the general h-parameter circuit model (Fig. 13.16b), derive an expression for voltage gain v_o/v_i.

6. A BJT amplifier is terminated in a load resistance R_L.
   (a) Using the general h-parameter circuit model (Fig. 13.16b), derive an expression for input resistance v_i/i_i.

(b) Assuming $R_L = 5$ kΩ, calculate the input resistance for a 2N1613 operating under standard conditions and compare the result to simply h_{ie}.

7. In a particular transistor application (Fig. 13.23), the usual load resistance is omitted and an emitter resistance R_E is inserted. V_{BB} and V_{CC} are adjusted to bring operation to an appropriate quiescent point. Replace the transistor with an r_π-β model and derive expressions for: (a) v_E in terms of v_i (voltage gain), and (b) input resistance $r_i = v_i/i_b$. Assuming $\beta = 50$, $r_\pi = 1200$ Ω, and $R_E = 2000$ Ω, evaluate r_i and compare with the input resistance of an ordinary common-emitter amplifier (Fig. 13.12a).

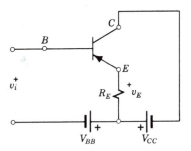

Figure 13.23

14

Small-Signal Amplifiers

RC-Coupled Amplifiers

Frequency Response

Multistage Amplifiers

To amplify the signal from a strain gage or a microphone to a power level adequate to drive a recorder or a loudspeaker usually requires several stages of amplification. If the final amplifier is a power transistor, appreciable driving current is needed and several stages of current amplification may be required. The low-level voltage or current amplification is provided by "small-signal" amplifiers that respond linearly.

In a small-signal amplifier, the input signals are small in comparison to the dc bias and the resulting output swings are small in comparison to the quiescent operating values of voltage and current. Under small-signal operation, bias is not critical and amplitude distortion is easily avoided. Although graphical analysis is possible, the linear models developed in Chapter 13 are much more convenient to use.

We have already used the linear models to predict the performance of single-stage amplifiers at moderate frequencies. Now we are going to investigate the gain of untuned amplifiers as a function of frequency and consider the effect of cascading several stages. Field-effect and bipolar transistors with *RC* coupling are commonly employed for small-signal amplification in the frequency range from a few cycles per second to several megahertz, and therefore they are emphasized in this discussion. The signals consist of sinusoids of various amplitudes and frequencies so they will be represented by phasors and the circuit parameters will be described by complex impedances. Although BJTs and FETs operate on different physical bases, their external behavior is quite similar and the same approach is used in predicting their frequency responses.

RC-COUPLED AMPLIFIERS

Each stage of a multistage amplifier consists of an electronic two-port, a biasing network, and the coupling circuits. In *RC*-coupled amplifiers, the output of each transistor is coupled to the next stage by means of a load resistor and a coupling capacitor.

Frequency-Response Curve

The voltage gain of one stage of an *RC*-coupled amplifier is shown in Fig. 14.1. The gain is relatively constant over the *midfrequency range*, but it falls off at lower

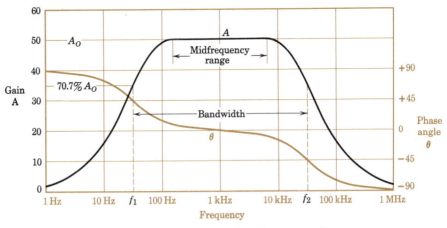

Figure 14.1 Frequency response of an *RC*-coupled single-stage amplifier.

frequencies and at higher frequencies. In simple amplifiers, the frequency-response curve is symmetric if frequency is plotted on a logarithmic scale as shown. The *bandwidth* is defined by lower and upper *cutoff frequencies* or *half-power frequencies* f_1 and f_2, just as in the frequency response of a resonant circuit. At the half-power frequencies, the response is 70.7% of the midfrequency gain, or $A_1 = A_2 = A_0/\sqrt{2}$, *and the phase angle* $\theta = \pm 45°$.

A Generalized Amplifier

In the general amplifier circuit of Fig. 14.2, the input signals come from a "source," maybe a phono pickup or a previous amplifier stage, that is characterized by an open-circuit voltage $\mathbf{V}_s$ and Thévenin equivalent impedance $\mathbf{Z}_s$. The input signals consist of sinusoids of various amplitudes and frequencies so variables are shown as effective-value phasors. This "one-stage" amplifier is characterized by

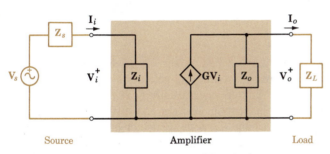

Figure 14.2 A generalized one-stage amplifier circuit.

input and output impedances $\mathbf{Z}_i$ and $\mathbf{Z}_o$ and a controlled source (controlled by input current or voltage). The amplifier gain is $\mathbf{A} = \mathbf{V}_o/\mathbf{V}_i$ where $\mathbf{A}$ is a complex function of frequency. The "load" of the amplifier may be a transducer or a following amplifier stage. Given the components of the input signal and the characteristics of the source and load, we can predict the output if we know the frequency response of the amplifier.

Typical small-signal amplifier circuits are shown in Fig. 14.3; these are "*RC*-coupled amplifiers" because of coupling capacitors C_{C1} and C_{C2} and their associated resistances. Our approach is to replace these complicated circuits by linear

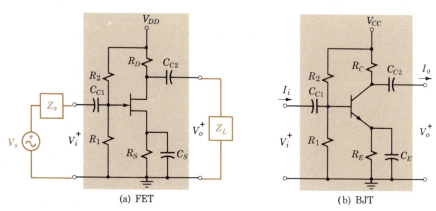

(a) FET   (b) BJT

Figure 14.3 Single-stage *RC*-coupled amplifiers.

models that respond in the same way to ac signals. We assume that the transistors are operating at proper quiescent points and the bias voltages and currents are of no concern here. Bias resistors R_S and R_E are assumed to be properly "bypassed" by C_S and C_E so they do not appear in the models of Fig. 14.4. Batteries provide constant voltages and do not allow ac voltages; therefore they are short circuits to ac signals and the upper terminal of R_D or R_C is grounded through the battery. To input signals, R_1 and R_2 are effectively in parallel, replaced by $R_1 \| R_2 = R_G$ or R_B.

The batteries are omitted. On the other hand, unintentional elements such as junction capacitances (C_j) and wiring capacitances (C_W), which do not appear on the wiring diagram, must be shown in the amplifier models. Reflecting the most common situation, the source impedance is shown as simply R_s and the load is represented by C_L and R_L in parallel.

Small-Signal Amplifier Models

The device models are those derived in Chapter 13. (See Figs. 13.10b and 13.15b.) For the FET, C_{gs} and C_{gd} are gate-source and gate-drain capacitances and g_m is the transconductance; r_d has been omitted since usually $r_d \gg R_D$. For the BJT, C_π is the effective input capacitance representing junction and diffusion capacitances, r_π is the small-signal input resistance, and C_{jc} is the collector-base junction capacitance. Mutual resistance r_μ is ignored and output resistance r_o has been omitted since $r_o \gg R_C$. Base-spreading resistance r_b has been neglected on the assumption that

$r_b \ll R_s + r_\pi$; therefore, $V'_{be} = V_{be}$. The current source can be considered to be controlled by the base current or the base-emitter voltage since $g_m V_{be} = \beta I'_b$ where g_m is the transconductance and β is the small-signal short-circuit current gain.

The determination of the values of individual elements in Fig. 14.4 is quite difficult; the specific values depend upon such factors as quiescent voltage and current,

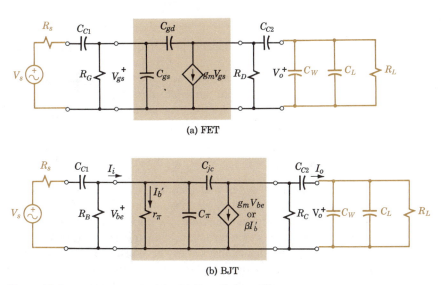

(a) FET

(b) BJT

Figure 14.4 Linear circuit models of *RC*-coupled amplifiers.

operating temperature, characteristics of the following stage, and configuration of the physical circuit. The important thing from our viewpoint here is that an actual amplifier can be replaced, insofar as small signals are concerned, by a linear circuit model that can be analyzed by methods we have already mastered. Fortunately, the models for FET and BJT are similar in form and the same analysis can be used for both.

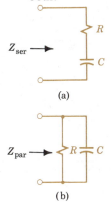

(a)

(b)

Figure 14.5 *Z* of *RC* combinations.

These linear models hold fairly well for frequencies from a few cycles per second to several megahertz. It would be possible to perform a general analysis on the circuits as they stand, but it is much more convenient, and more instructive, to consider some simplifying approximations. We recall that for series and parallel combinations of resistance R and capacitance C (Fig. 14.5),

$$Z_{ser} = \sqrt{R^2 + (1/\omega C)^2} = R\sqrt{1 + (1/\omega CR)^2} \quad (14\text{-}1)$$

$$Z_{par} = \frac{1}{\sqrt{(1/R)^2 + (\omega C)^2}} = \frac{R}{\sqrt{1 + (\omega CR)^2}} \quad (14\text{-}2)$$

We note that in Eq. 14-1 if $\omega CR \geq 10$, $Z_{ser} \cong R$ within 0.5%. Also, in Eq. 14-2, if $\omega CR \leq 0.1$, $Z_{par} \cong R$. When these criteria are met, the impedances are independent of frequency.

Practice Problem 14-1

A series circuit consists of capacitance C and resistance $R = 5$ kΩ.

(a) If the critical frequency for $Z_{ser} \cong R$ is to be 100 Hz, specify C.

(b) Calculate the error in Z at 50 Hz and at 200 Hz.

Answers: (a) 3.18 μF; (b) 2%, 0.125%.

The fact that gain is independent of frequency over the midfrequency range (Fig. 14.1) indicates that there is a range of frequencies over which the capacitive effects are negligible. This is the case in practical amplifiers (see Exercise 5), and it leads to another important conclusion. Since ωC increases with frequency, series capacitances become important only at frequencies lower than the midfrequencies and parallel capacitances become important only at frequencies higher than the midfrequencies.

FREQUENCY RESPONSE

The analysis of series and parallel RC combinations leads to the convenient conclusion that the frequency response of an RC amplifier can be divided into three regions. In the midfrequency range, the capacitances can be neglected. In the low-frequency range, the series capacitances C_C must be considered. In the high-frequency range, the parallel capacitances must be considered. Instead of trying to derive a completely general relation, we break the complex problem down into three simpler problems, solve those problems separately, and combine the results to obtain the overall frequency response.

Midfrequency Gain

In a properly designed untuned amplifier, there is a range of frequencies over which the general models of Fig. 14.4 can be replaced by the purely resistive circuits of Fig. 14.6. By making some astute assumptions, we can simplify the analysis of these circuits.

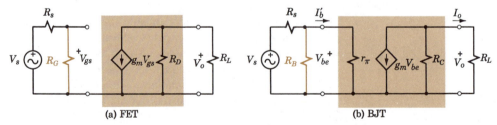

(a) FET   (b) BJT

Figure 14.6 Midfrequency models of amplifiers.

FET. Since $R_G \gg R_s$, $V_{gs} = V_s$ and R_G can be omitted in most cases. If, for convenience, we represent the parallel combination of R_D and R_L by R_o, the output voltage is

$$V_o = -g_m V_{gs}(R_D \| R_L) = -g_m V_s R_o \qquad (14\text{-}3)$$

and the midfrequency voltage gain is

$$A_{VO} = \frac{V_o}{V_s} = -g_m R_o \tag{14-4}$$

where subscripts V and O are for "voltage" and "midfrequency." The negative sign indicates a 180° phase reversal of the output signal with respect to the input.

EXAMPLE 1

A JFET with $g_m = 2000\ \mu S$, $C_{gs} = 20$ pF, and $C_{gd} = 2$ pF is used in an RC-coupled amplifier with $R_G = 1\ M\Omega$, $R_s = 2\ k\Omega$, and $R_D = 5\ k\Omega$. Signal source resistance $R_s = 1\ k\Omega$ and load resistance $R_L = 5\ k\Omega$. Coupling capacitance $C_{C1} = C_{C2} = 1\ \mu F$ and $C_S = 20\ \mu F$; the load capacitance (including wiring) $C_L = 20$ pF. Predict the voltage gain V_o/V_s at $\omega = 10^4$ rad/s ($f = 1590$ Hz).

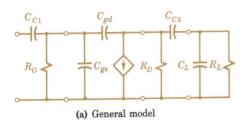

(a) General model

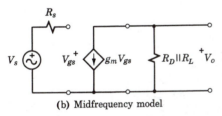

(b) Midfrequency model

Figure 14.7 Midfrequency simplification.

At the given frequency,

$$1/\omega C_S = 1/10^4 \times 20 \times 10^{-6} = 5\ \Omega$$

Therefore, $R_S = 2000\ \Omega$ is effectively shorted and the model is as shown in Fig. 14.7a.

Since C_{C1} and C_{C2} are effectively in series with R_G and R_L, 1 MΩ and 5 kΩ respectively, the reactances

$$1/\omega C_C = 1/10^4 \times 10^{-6} = 100\ \Omega$$

are negligibly small (Eq. 14-1).

With C_{C1} and C_{C2} omitted, C_L is in parallel with $R_o = R_D \| R_L = 2500\ \Omega$ and C_{gs} is in series with R_s. The reactances

$$1/\omega C_L = 1/\omega C_{gs} = 1/10^4 \times 20 \times 10^{-12} = 5\ M\Omega$$

are so large that C_L and C_{gs} can be omitted.

The reactance

$$1/\omega C_{gd} = 1/10^4 \times 2 \times 10^{-12} = 50\ M\Omega$$

is so large that a negligible current will flow in C_{gd}.

In other words, 1590 Hz is in the "midfrequency" range of this amplifier and the model of Fig. 14.7b is appropriate. The midfrequency gain (Eq. 14-4) is

$$A_{VO} = -g_m R_o = -2000 \times 10^{-6} \times 2500 = -5$$

Conclusion: Working from the simplified model, calculation of the midfrequency gain is straightforward.

BJT. For proper stability R_B should be small compared to βR_E (Eq. 12-12), whereas for negligible loading of the input circuit R_B should be large compared to r_π. Here we assume that the ideal condition $r_\pi \ll R_B \ll \beta R_E$ is achieved so the circuit is stable *and* simple to analyze. (See Exercise 6.) If R_B is neglected in Fig. 14.6b, r_π and R_s constitute a voltage divider. Where the parallel combination of R_C and R_L is represented by R_o, the output voltage is

$$V_o = -g_m V_{be}(R_C \| R_L) = -g_m \frac{r_\pi}{r_\pi + R_s} V_s R_o = -\beta \frac{V_s}{r_\pi + R_s} R_o \tag{14-5}$$

and the midfrequency voltage gain is

$$A_{VO} = \frac{V_o}{V_s} = -\frac{r_\pi}{r_\pi + R_s} g_m R_o = -\beta \frac{R_o}{r_\pi + R_s} \tag{14-6}$$

Performance of these one-stage amplifiers is influenced by the characteristics of the source (R_s) and the load (R_L). In the multistage amplifiers discussed later, the source and the load for each stage are other similar stages.

Practice Problem 14-2

A BJT with $\beta = 100$ is used in the circuit of Fig. 14.3 with $R_s = 500\ \Omega$, $R_1 = 20\ k\Omega$, $R_2 = 50\ k\Omega$, $R_C = 4\ k\Omega$, and $R_L = 4\ k\Omega$; $I_C = 2$ mA.

(a) Estimate R_B, r_π, and g_m.
(b) Making any desirable simplifying assumptions, predict the midfrequency voltage gain of the amplifier.

Answers: (a) 14.3 kΩ, 1.25 kΩ, 80 mS; (b) -114.

Low-Frequency Response

Below the midfrequencies, the susceptances of the parallel capacitances are negligibly small, but the reactances of the coupling capacitances C_C become increasingly important.

Coupling Capacitors. The general models can be reduced to the simpler low-frequency models shown in Fig. 14.8, where R_G has been retained because it is

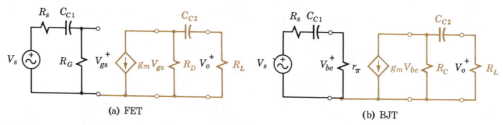

(a) FET   (b) BJT

Figure 14.8   Low-frequency models of amplifiers.

significant, but R_B has been omitted since it is large compared to r_π. With this simplification, the FET and BJT models are analogous and one analysis will serve for both.

The input and output circuits are "high-pass filters" (see Fig. 10.19) of slightly different form. The general circuit (Fig. 14.9a) can be represented by the Thévenin form as in the input circuit or by the Norton form as in the output circuit; the two forms

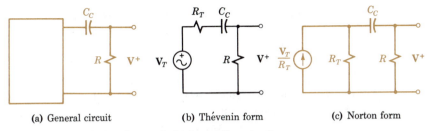

(a) General circuit   (b) Thévenin form   (c) Norton form

Figure 14.9   Equivalent forms of a high-pass filter circuit.

are equivalent. For midfrequencies where C_C is effectively shorted, the output of the Thévenin circuit is

$$\mathbf{V} = \mathbf{V}_O = \frac{R}{R + R_T}\mathbf{V}_T \tag{14-7}$$

At low frequencies where C_C is significant, the output of the Thévenin circuit is

$$\mathbf{V}_L = \frac{R}{R + R_T - j\dfrac{1}{\omega C_C}}\mathbf{V}_T = \frac{\dfrac{R}{R + R_T}\mathbf{V}_T}{1 - j\dfrac{1}{\omega C_C(R + R_T)}} = \frac{\mathbf{V}_O}{1 - j\dfrac{1}{\omega C_C(R + R_T)}} \tag{14-8}$$

after dividing through by $R + R_T$. In words, the low-frequency output $\mathbf{V}_L$ of either filter circuit is related to the midfrequency output $\mathbf{V}_O$ by a complex factor dependent on frequency and an RC product. As frequency is decreased, a larger fraction of voltage $\mathbf{V}_T$ appears across C_C and the voltage $\mathbf{V}$ at the output is reduced. The *cutoff* or *half-power frequency* where $V_L/V_O = |1/(1 - j1)| = 1/\sqrt{2}$ is defined by

$$\omega_{co} = \frac{1}{C_C(R + R_T)} \tag{14-9}$$

The behavior of the FET or BJT amplifiers in Fig. 14.8 at low frequencies can now be predicted. At a frequency defined by

$$\omega_{11} = \frac{1}{C_{C1}(R_s + R_G)} \quad\text{or}\quad \omega_{11} = \frac{1}{C_{C1}(R_s + r_\pi)} \tag{14-10}$$

the input voltage V_{gs} or V_{be} will be down to 70% of V_O and the possible amplifier gain will be correspondingly reduced. Similarly, at a frequency defined by

$$\omega_{12} = \frac{1}{C_{C2}(R_D + R_L)} \quad\text{or}\quad \omega_{12} = \frac{1}{C_{C2}(R_C + R_L)} \tag{14-11}$$

the output voltage V_o will be down to 70% of $g_m V_{gs} R_o$ or $g_m V_{be} R_o$ and the amplifier gain will be correspondingly reduced. The overall low-frequency voltage gain for the FET amplifier can be expressed as

$$\mathbf{A}_L = \mathbf{A}_O \cdot \frac{1}{1 - j\dfrac{1}{\omega C_{C1}(R_s + R_G)}} \cdot \frac{1}{1 - j\dfrac{1}{\omega C_{C2}(R_D + R_L)}} \tag{14-12}$$

or, for either FET or BJT, the *relative gain* at low frequencies is

$$\frac{\mathbf{A}_L}{\mathbf{A}_O} = \frac{1}{1 - j\omega_{11}/\omega} \cdot \frac{1}{1 - j\omega_{12}/\omega} \tag{14-13}$$

To predict the behavior of a given circuit, determine ω_{11} and ω_{12}; the higher of the two determines the cutoff frequency (unless they are very close together; see Exercise 15). To design a circuit for a specified behavior, calculate C_{C1} and C_{C2} from Eqs. 14-10 and 11, specify the larger at its calculated value, and make the smaller several times its calculated value.

EXAMPLE 2

In the circuit of Fig. 14.8b, $R_s = 1$ kΩ, $r_\pi = 2$ kΩ, $R_C = 4.5$ kΩ, $R_L = 9$ kΩ, $g_m = 60$ mS, and $C_{C1} = C_{C2} = 1$ μF. Predict the midfrequency voltage gain and the lower cutoff frequency.

At midfrequencies the capacitors are effectively shorted and

$$A_{VO} = -\frac{r_\pi}{r_\pi + R_s} g_m (R_C \| R_L)$$

$$= -\frac{2}{2 + 1} (0.06)(4500 \| 9000) = -120$$

The lower cutoff frequencies are

$$\omega_{11} = \frac{1}{C_{C1}(R_s + r_\pi)} = \frac{1}{10^{-6} \times 3000} = 333.3 \text{ rad/s}$$

$$\omega_{12} = \frac{1}{C_{C2}(R_C + R_L)} = \frac{1}{10^{-6} \times 13{,}500} = 74.1 \text{ rad/s}$$

Therefore, the lower cutoff frequency of the amplifier is

$$\omega_{12} = 333.3 \text{ rad/s} \quad \text{or} \quad f_{12} = 333.3/2\pi = 53 \text{ Hz}$$

Practice Problem 14-3

The amplifier of Example 2 is to have a lower cutoff frequency of about 265 Hz. Assuming all other circuit parameters remain unchanged, specify appropriate values of C_{C1} and C_{C2}.

Answers: 0.2 μF, 0.2 μF.

Bypass Capacitors. The foregoing analysis of low-frequency response assumes that C_S (or C_E) effectively bypasses R_S (or R_E) at the lowest frequency of interest. In practice, it turns out that C_S must be much larger than C_{C1} and C_{C2} and, in most cases, C_S is the important element in determining the lower cutoff frequency. The practical approach is to assume that the coupling capacitors are still effective at the frequencies

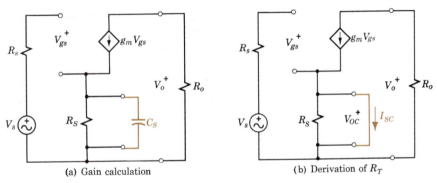

(a) Gain calculation   (b) Derivation of R_T

Figure 14.10   Low-frequency amplifier model where C_S is critical.

where C_S becomes critical. With this assumption, C_{C1} and C_{C2} are short circuits but R_S must be considered and the low-frequency FET amplifier model is as shown in Fig. 14.10a.

At low frequencies, the voltage gain is reduced because the current $g_m V_{gs}$ flowing through $\mathbf{Z}_S$ (the parallel combination of R_S and C_S) develops a voltage that subtracts from the signal voltage $\mathbf{V}_s$. In the input loop, $\mathbf{V}_{gs} = \mathbf{V}_s - g_m \mathbf{V}_{gs} \mathbf{Z}_S$, or the gate voltage is reduced to

$$\mathbf{V}_{gs} = \frac{\mathbf{V}_s}{1 + g_m \mathbf{Z}_S} \tag{14-14}$$

and the voltage gain $\mathbf{V}_o/\mathbf{V}_s$ is reduced from the midfrequency value by the factor $1 + g_m \mathbf{Z}_S$. By definition, the lower cutoff frequency occurs when $1 + g_m \mathbf{Z}_S$ has a magnitude of $\sqrt{2}$. Because $\mathbf{Z}_S$ is a complex quantity, there is no simple expression for the proper value of C_S. Instead we shall use a useful technique to determine an approximate expression.[†]

With R_S unbypassed, the gain is low. With R_S perfectly bypassed, the gain is $A_V = g_m R_o$. Therefore, let us specify C_S to be large enough so that at f_1 the reactance $1/\omega_1 C_S$ is equal to the *effective resistance* bypassed. (In other words, $\omega_{co} = \omega_1 = 1/R_{eff} C_S$.) We define the effective resistance as the Thévenin equivalent resistance looking in at the terminals of R_S across which C_S is connected. Letting $Z_S = R_S$ in Eq. 14-14,

$$V_{OC} = g_m V_{gs} R_S = g_m \frac{V_s}{1 + g_m R_S} R_S = \frac{g_m R_S}{1 + g_m R_S} V_s \tag{14-15}$$

Since $I_{SC} = g_m V_s$ with R_S shorted, the Thévenin equivalent is

$$R_T = \frac{V_{OC}}{I_{SC}} = \frac{R_S}{1 + g_m R_S} = \frac{1}{g_m + 1/R_S} = R_S \| 1/g_m \tag{14-16}$$

[†] Suggested by Malvino; a general expression is given on p. 449 of Schilling and Belove. See References at the end of Chapter 12.

or the effective resistance is the parallel combination of R_S and $1/g_m$. Therefore, the design criterion (see Example 3) is

$$\omega_1 = 2\pi f_1 = \frac{1}{C_S R_T} = \frac{g_m + 1/R_S}{C_S} \tag{14-17}$$

EXAMPLE 3

Specify C_S for a lower cutoff frequency of 20 Hz in the circuit of Fig. 14.10 where $g_m = 2000\ \mu S$ and $R_S = 2\ k\Omega$. Show that the voltage gain at 20 Hz is down to approximately 70% of its midfrequency value.

By Eq. 14-17,

$$C_S = \frac{g_m + 1/R_S}{2\pi f_1} = \frac{0.002 + 0.0005}{2\pi \times 20} \cong 20\ \mu F$$

For this value of C_S and $f = 20$ Hz,

$$g_m\,\mathbf{Z}_S = \frac{g_m}{(1/R_S) + j\omega C} = \frac{0.002}{0.0005 + j0.0025}$$

$$= 0.784\underline{/-78.7^\circ}$$

$$1 + g_m\,\mathbf{Z}_S = 1 + (0.15 - j0.77) = 1.15 - j0.77$$

$$= 1.38\underline{/33^\circ}$$

The magnitude of $1 + g_m\,\mathbf{Z}_S$ is $1.38 \cong \sqrt{2} = 1.41$; therefore, f_1 is approximately the lower cutoff frequency and Eq. 14-17 is justified.

Practice Problem 14-4

For the amplifier in Example 1, predict the lower cutoff frequency as defined by C_{C1}, C_{C2}, and C_S, and specify the lower cutoff frequency in hertz for the amplifier.

Answers: 1 rad/s, 100 rad/s, 125 rad/s, ~20 Hz.

In the BJT amplifier, C_E is the important element in determining the lower cutoff frequency. Here again an exact analysis is complicated and an approximate expression is satisfactory for our purposes. The approach employed in establishing the design criterion for C_S in the FET amplifier would work, but using a different approach may be more enlightening. As before, we assume that the coupling capacitors are still effective at the frequency where C_E becomes critical and draw the low-frequency amplifier model as shown in Fig. 14.11a.

The current $(\beta + 1)\mathbf{I}_b$ flowing through $\mathbf{Z}_E$ (the parallel combination of R_E and C_E) develops a voltage that, in effect, opposes the signal voltage $\mathbf{V}_s$. Summing the voltages

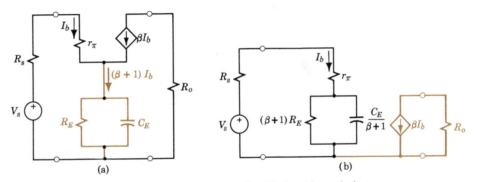

Figure 14.11   The low-frequency model where C_E is critical and its equivalent.

around the left-hand loop and solving for the current,

$$\mathbf{I}_b = \frac{\mathbf{V}_s}{R_s + r_\pi + (\beta + 1)\mathbf{Z}_E} \tag{14-18a}$$

where

$$(\beta + 1)\mathbf{Z}_E = \frac{\beta + 1}{\dfrac{1}{R_E} + j\omega C_E} = \frac{1}{\dfrac{1}{(\beta + 1)R_E} + j\omega\dfrac{C_E}{\beta + 1}} \tag{14-18b}$$

Equations 14-18a and b are satisfied by the circuit model of Fig. 14.11b. If we neglect the current through the relatively large resistance $(\beta + 1)R_E$,

$$\mathbf{I}_b = \frac{\mathbf{V}_s}{R_s + r_\pi - j\dfrac{\beta + 1}{\omega C_E}} \tag{14-19}$$

If the degrading effect of $\mathbf{Z}_E$ is to be small, the reactance of $C_E/(\beta + 1)$ must be small compared to the resistances with which it is in series. Current I_b will drop to 70.7% of its midfrequency value and the gain will drop to 70% of its midfrequency value at frequency ω_1 where

$$\frac{1}{\omega_1 C_E/(\beta + 1)} = R_s + r_\pi \tag{14-20}$$

Therefore, the design criterion is

$$\omega_1 = 2\pi f_1 = \frac{\beta + 1}{C_E(R_s + r_\pi)} \tag{14-21}$$

In practice, C_E is specified to provide the desired low-frequency response and then C_{C1} and C_{C2} are selected to be several times the values indicated by Eqs. 14-10 and 11. Note that the cutoff frequency is just a convenient measure of the low-frequency response of the amplifier and does not imply that no amplification occurs below this frequency.

EXAMPLE 4

In the circuit of Fig. 14.12, $R_s = 2$ kΩ, $R_E = 3$ kΩ, and $C_E = 50$ μF. R_1 and R_2 are very large. For $r_\pi = 1.5$ kΩ and $\beta = 50$, estimate the lower cutoff frequency and check the validity of the assumption made in deriving Eq. 14-20. For $R_C = 5$ kΩ and $R_L = 10$ kΩ, specify appropriate values for C_{C1} and C_{C2}.

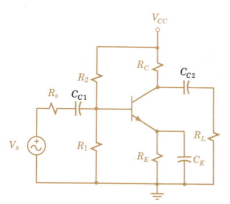

Figure 14.12 Low-frequency response.

By Eq. 14-21 (see Fig. 14.11),

$$f_1 = \frac{\omega_1}{2\pi} = \frac{\beta + 1}{2\pi C_E(R_s + r_\pi)}$$

$$= \frac{50 + 1}{2\pi \times 5 \times 10^{-5}(2 + 1.5)10^3} = 46 \text{ Hz}$$

The admittance of $(\beta + 1)R_E$ is

$$\frac{1}{(\beta + 1)R_E} = \frac{1}{51 \times 3000} \cong 6.5 \ \mu\text{S}$$

which when squared (see Eq. 14-2) is negligibly small in comparison to the square of

$$\frac{\omega_1 C_E}{\beta + 1} = \frac{1}{R_s + r_\pi} = \frac{1}{3.5 \times 10^3} = 290 \ \mu\text{S}$$

For the same value of f_1, Eqs. 14-10 and 11 indicate

$$C'_{C1} = \frac{1}{2\pi f_1(R_s + r_\pi)} = \frac{1}{2\pi \times 46(3500)} \cong 1 \ \mu\text{F}$$

Therefore, specify a value of, say, $5C'_{C1} = 5 \ \mu$F.

$$C'_{C2} = \frac{1}{2\pi f_1(R_C + R_L)} = \frac{1}{2\pi \times 46(15,000)} \cong 0.23 \ \mu\text{F}$$

Therefore, specify a value of, say, $5C'_{C2} = 1 \ \mu$F.

High-Frequency Response

Above the midfrequencies, the effects of coupling and bypass capacitances are negligibly small, but the other capacitances become increasingly important. Under these conditions, the high-frequency versions of the amplifier circuit models are as shown in Fig. 14.13, where $R_o = R_D \| R_L$ or $R_C \| R_L$ and $C'_o = C_W + C_L$ (wiring plus lead capacitance). The analysis of these analogous circuits is complicated by the presence of C_{gd} and C_{jc}. From our experience with the inverting op amp circuit, we

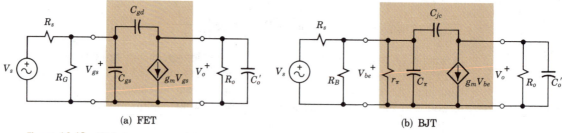

(a) FET   (b) BJT

Figure 14.13 High-frequency models of amplifiers.

know that feeding back part of the output tends to reduce the effective input and reduce the amplifier gain. At high frequencies C_{gd} and C_{jc} become low-impedance paths that limit the high-frequency response of these amplifiers. Furthermore, the circuits in Fig. 14.13 are multiloop networks with no easy way to isolate input and output circuits as we did when we analyzed low-frequency response.

But understanding the high-frequency performance of transistor amplifiers is essential in the effective application of electronics, and we do have enough circuit background to take advantage of the work of some clever engineers. As a first step, let us simplify the circuit using a technique first suggested by J. M. Miller. In Fig. 14.14, the voltage across feedback capacitance C_F causing current I_1 is

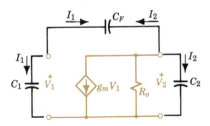

Figure 14.14 Miller capacitances.

$$V_1 - V_2 = V_1 - (-g_m R_o)V_1 = V_1(1 + g_m R_o) = V_1(1 + A) \qquad (14\text{-}22)$$

where $V_2/V_1 = -A = -g_m R_o$ is the gain of this circuit. This current could be accounted for in the input circuit by a capacitance that is $(1 + A)$ times as large under a voltage only $1/(1 + A)$ times as large, i.e., by a capacitance

$$C_1 = C_F(1 + A) = C_F(1 + g_m R_o) \qquad (14\text{-}23)$$

under a voltage V_1. Following the same reasoning, the current I_2 could be accounted for by a capacitance

$$C_2 = C_F(1 + 1/A) = C_F(1 + 1/g_m R_o) \cong C_F \qquad (14\text{-}24)$$

under a voltage V_2 since $V_2 - V_1 = V_2(1 + 1/A) \cong V_2$. When these substitutions are made and the "Miller capacitances" are combined with C_{gs} and C_o', respectively, the simplified circuit model is as shown in Fig. 14.15b where $C_{eq} = C_{gs} + C_{gd}(1 + A)$ and $C_o = C_{gd} + C_W + C_L$.

FET. In the Thévenin equivalent (Fig. 14.15a), $R_T = R_s \| R_G \cong R_s$ since $R_G \gg R_s$, in general. Now the input circuit has the form of a *low-pass filter*. At midfrequencies, $V_{gs} = V_T \cong V_s$. At high frequencies where C_{eq} is significant,

$$\mathbf{V}_{gs} = \frac{1/j\omega C_{eq}}{(R_s \| R_G) + 1/j\omega C_{eq}} \mathbf{V}_s = \frac{\mathbf{V}_s}{1 + j\omega C_{eq}(R_s \| R_G)} \qquad (14\text{-}25)$$

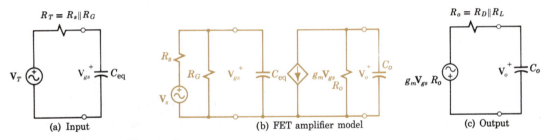

(a) Input (b) FET amplifier model (c) Output

Figure 14.15 Low-pass filter circuits.

As frequency is increased, a smaller fraction of the voltage V_s appears across C_{eq} and voltage V_{gs} is reduced. At the upper cutoff frequency defined by

$$\omega_{21} = \frac{1}{C_{eq}(R_s\|R_G)} \tag{14-26}$$

the input voltage V_{gs} will be down to 70% of V_s and the possible amplifier gain will be correspondingly reduced. Reasoning by analogy (Fig. 14.15c), we can say that at an upper cutoff frequency

$$\omega_{22} = \frac{1}{C_o(R_D\|R_L)} \tag{14-27}$$

the output voltage V_o will be down to 70% of $g_m V_{gs} R_o$ and the amplifier gain will be correspondingly reduced. The overall high-frequency gain for the FET amplifier can be expressed as

$$\mathbf{A}_H = \mathbf{A}_O \cdot \frac{1}{1 + j\omega C_{eq}(R_s\|R_G)} \cdot \frac{1}{1 + j\omega C_o(R_D\|R_L)} \tag{14-28}$$

or the relative gain at high frequencies is

$$\frac{\mathbf{A}_H}{\mathbf{A}_O} = \frac{1}{1 + j\omega/\omega_{21}} \cdot \frac{1}{1 + j\omega/\omega_{22}} \tag{14-29}$$

To predict the behavior of a given circuit, determine ω_{21} and ω_{22} from Eqs. 14-26 and 27; the lower of the two determines the upper cutoff frequency (unless they are very close together).

EXAMPLE 5

In Example 1, an FET ($g_m = 2000\ \mu S$, $C_{gs} = 20$ pF, and $C_{gd} = 2$ pF) was used in an RC-coupled amplifier with $R_s = 1$ kΩ, $R_G = 1$ MΩ, $R_D = 5$ kΩ, and $R_L = 5$ kΩ to give a midfrequency gain of $-g_m R_o = -5$. If $C_W + C_L \cong 4 + 16 = 20$ pF, predict the upper cutoff frequency.

The equivalent input and output capacitances are

$$C_{eq} = C_{gs} + C_{gd}(1 + g_m R_o) = 20 + 2(6) = 32 \text{ pF}$$
$$C_o = C_W + C_L + C_{gd} = 4 + 16 + 2 = 22 \text{ pF}$$

The equivalent input and output resistances are

$$R_s\|R_G = 1000\|1,000,000 = 999 \cong 1000\ \Omega$$
$$R_D\|R_L = 5000\|5000 = 2500\ \Omega$$

The individual cutoff frequencies are

$$\omega_{21} = 1/C_{eq}(R_s\|R_G) = 1/32 \times 10^{-12} \times 10^3$$
$$= 31.3 \text{ Mrad/s}$$
$$\omega_{22} = 1/C_o(R_D\|R_L) = 1/22 \times 10^{-12} \times 2500$$
$$= 18 \text{ Mrad/s}$$

The upper cutoff frequency of the amplifier is

$$f_2 = \omega_{22}/2\pi = 18/2\pi \cong 2.9 \text{ MHz}$$

In designing a circuit for a specified behavior, the designer does not have freedom in specifying C_{eq} and C_o since these are determined by device parameters, operating conditions, and gain. If the predicted performance is not satisfactory, one possibility is to reduce R_D, which reduces gain (and therefore C_{eq}) and increases ω_{21} and ω_{22}.

The manufacturer's data sheets usually give values that are easy to measure. The "reverse transfer capacitance" $C_{rss} = C_{gd}$; the "input capacitance with output shorted" $C_{iss} = C_{gs} + C_{gd}$. Therefore, $C_{gs} = C_{iss} - C_{rss}$. We have not considered C_{ds}, which is usually negligible. (See Table 14-1.)

Table 14-1 Typical FET Parameters

		2N4416 n-channel JFET Min/Max	3N163 p-channel E MOSFET Min/Max
Static parameters			
$V_{gs(OFF)}$	Gate-source cutoff voltage	$-2.5/-6.0$ V	
$V_{GS(th)} = V_T$	Gate-source threshold voltage		$-2/-5$ V
I_{DSS}	Saturation drain current	$5/15$ mA	
$I_{DS(ON)}$	ON drain current		$-5/-30$ mA
Small-signal parameters at 1 kHz			
$g_{fs} = g_{mo}$	Forward transconductance	$4500/7500$	$2000/4000$ μS
$g_{oss} = 1/r_d$	Output conductance	50	250 μS
$C_{rss} = C_{gd}$	Reverse transfer capacitance	0.8	0.7 pF
$C_{iss} = C_{gs} + C_{gd}$	Input capacitance	4	2.5 pF
$C_{oss} = C_{ds}$	Output capacitance	2	3 pF

BJT. Comparing Figs. 14.13a and b and reasoning by analogy, the upper cutoff frequency for a BJT amplifier is defined by

$$\omega_{21} = \frac{1}{C_{eq}(R_s \| r_\pi)} \quad \text{or} \quad \omega_{22} = \frac{1}{C_o(R_C \| R_L)} \tag{14-30}$$

where $C_{eq} = C_\pi + C_{jc}(1 + g_m R_o)$ and $C_o = C_{jc} + C_W + C_L$ (Fig. 14.15b). (More precisely, the equivalent input resistance is $R_s \| r_\pi \| R_B$, but in a well-designed amplifier $R_B \gg r_\pi$.) The overall high-frequency gain can be expressed as

$$\mathbf{A}_H = \mathbf{A}_O \cdot \frac{1}{1 + j\omega C_{eq}(R_s \| r_\pi)} \cdot \frac{1}{1 + j\omega C_o(R_C \| R_L)} \tag{14-31}$$

and the relative gain at high frequencies is

$$\frac{\mathbf{A}_H}{\mathbf{A}_O} = \frac{1}{1 + j\omega/\omega_{21}} \cdot \frac{1}{1 + j\omega/\omega_{22}} \tag{14-32}$$

The upper cutoff frequency is determined by the lower of ω_{21} and ω_{22} just as in the case of the FET amplifier.

Practice Problem 14-5

A BJT whose parameters are $g_m = 100$ mS, $r_\pi = 1$ kΩ, $C_\pi = 100$ pF, and $C_{jc} = 5$ pF is used in an amplifier where $R_s = 1$ kΩ, R_B is very large, $R_C = 6$ kΩ, $R_L = 6$ kΩ, and $C_W + C_L = 5$ pF. Predict the midfrequency voltage gain and the upper cutoff frequency.

Answers: -150, 198 kHz.

Analysis of the high-frequency response of transistors is complicated by the fact that *lumped* circuit models can only approximate effects that are, in fact, *distributed* throughout a finite region in the transistor. For example, finite times are required for signals in the form of changes in charge density to be propagated across the base of a bipolar transistor. At higher frequencies the current gain I_c/I_b begins to decrease in magnitude and a phase lag appears. In the hybrid-π model of Fig. 14.16a, these charging effects are represented by capacitance C_π.

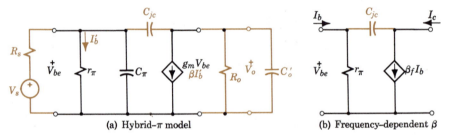

(a) Hybrid-π model   (b) Frequency–dependent β

Figure 14.16 High-frequency BJT models.

Another approach is to represent the transistor by parameters that are frequency dependent. We define β as the short-circuit current gain at low and medium frequencies and let $I_c/I_b = \beta_f$, a function of frequency. At the *beta cutoff frequency* f_β, current gain β_f is just equal to 70.7% of the low-frequency value. The significant relation between current phasors (Fig. 14.16b) is

$$\frac{\mathbf{I}_c}{\mathbf{I}_b} = \frac{\beta}{1 + j(f/f_\beta)} = \frac{\beta}{1 + j(\omega/\omega_\beta)} \tag{14-33a}$$

and the magnitude of the high-frequency current gain is

$$\beta_f = \frac{\beta}{\sqrt{1 + (f/f_\beta)^2}} \tag{14-33b}$$

where f_β may be specified by the manufacturer. At very high frequencies, Eq. 14.33b indicates that

$$\beta_f = \frac{\beta}{f/f_\beta} = \frac{\beta f_\beta}{f} \qquad (14\text{-}34)$$

or current gain is inversely proportional to frequency. Alternatively, therefore, the manufacturer may specify the frequency

$$f = f_T = \beta f_\beta \qquad (14\text{-}35)$$

at which the current gain β_f would fall to unity, or it may state the value of β_f at a specified frequency f.

To avoid working with frequency-dependent parameters, we can use constant circuit elements with the appropriate frequency response. In Fig. 14.16a, the fall-off in short-circuit current gain is accounted for by C_π alone because, when V_o is shorted, gain A in Eq. 14-23 is zero. Therefore,

$$f_T = \beta f_\beta = \frac{\beta}{2\pi C_\pi r_\pi} \qquad (14\text{-}36)$$

The determination of parameter values from manufacturer's specifications is illustrated in Example 6.

EXAMPLE 6

The manufacturer of the 2N3114 specifies $h_{ie} \cong r_\pi = 1500\ \Omega$ and $h_{fe} = \beta = 50$ at 1 kHz (appendix Fig. A8) and $\beta = 2.7$ at 20 MHz. Determine f_T, f_β, and C_π.

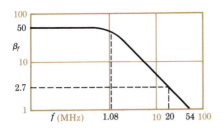

Figure 14.17 High-frequency response.

The 2.7 value is for β_f at $f = 20$ MHz. By Eq. 14-35, the current gain becomes unity at frequency $f_T = \beta f_\beta = \beta_f \times f = 2.7 \times 20$ MHz $= 54$ MHz and the beta cutoff frequency in Fig. 14.17 is

$$f_\beta = \frac{f_T}{\beta} = \frac{54\ \text{MHz}}{50} = 1.08\ \text{MHz}$$

Because the common-base emitter resistance $h_{ie} \cong r_\pi$, by Eq. 14-36,

$$C_\pi = \frac{\beta}{2\pi f_T r_\pi} = \frac{50}{2\pi \times 54 \times 10^6 \times 1500} = 98\ \text{pF}$$

Note: At high frequencies where β_f is inversely proportional to frequency, a log-log plot of $\beta_f = (\beta f_\beta) f^{-1}$ becomes a straight line with a slope of -1.

For ordinary transistors used in audiofrequency amplifiers, the decrease in β is unimportant; for special high-frequency planar transistors, the effect may be unimportant up to many megahertz. In amplifiers employing high-frequency transistors, the upper cutoff frequency may be determined by the effective shunting capacitance C_o, due to wiring and other parasitic effects, as indicated by Eq. 14-30. If the voltage gain is large, the collector-base junction capacitance C_{jc} becomes critical. (The practical value of this parameter is the common-base output capacitance designated C_{ob} or C_{cb} on the data sheet.) In practice, the engineer determines which is the limiting factor and bases the amplifier design on a careful analysis of that factor.

MULTISTAGE AMPLIFIERS

To achieve the desired voltage or current gain and the necessary frequency response, several stages of amplification may be required.

Cascading

The stages are usually connected in *cascade;* the output of one stage is connected to the input of the next. Where the preceding and following stages are similar, $R_L = r_\pi$ (Fig. 14.6b) and the midfrequency *per stage* voltage gain for the BJT is

$$A_{VO} = \frac{V_o}{V_{be}} = \frac{-g_m V_{be} R_o}{V_{be}} = -g_m R_o \tag{14-37}$$

just as for the FET. For the BJT the midfrequency *per stage* current gain is

$$A_{IO} = \frac{I_o}{I_b} = \frac{V_o/r_\pi}{V_{be}/r_\pi} = \frac{V_o}{V_{be}} = -g_m R_o = -\frac{\beta}{r_\pi} R_o \tag{14-38}$$

Since R_o is determined in part by r_π and is always less than r_π, the current gain per stage is always less than β.

By defining a single stage to include the input characteristics of the next amplifying element, the combined effect of cascaded stages is easy to predict. In Fig. 14.18,

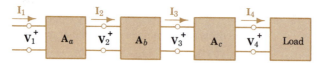

Figure 14.18   Block diagram of amplifiers in cascade.

the individual stages are represented by rectangular blocks labeled with the individual voltage gains. The overall voltage gain is

$$\mathbf{A} = \frac{V_4}{V_1} = \frac{V_2}{V_1} \frac{V_3}{V_2} \frac{V_4}{V_3} = \mathbf{A}_a \mathbf{A}_b \mathbf{A}_c \tag{14-39}$$

or

$$A \underline{/\theta} = A_a A_b A_c \underline{/\theta_a + \theta_b + \theta_c} \tag{14-40}$$

In words, the overall voltage gain is the product of the gain amplitudes and the sum of the phase shifts; both A and θ are functions of frequency. A similar statement would describe the overall current gain.

Gain in Decibels

The calculation of overall gain in multistage amplifiers is simplified by using a logarithmic unit. By definition, the power gain in *bels* (named in honor of Alexander Graham Bell) is the logarithm to the base 10 of the power ratio. A more convenient unit is the *decibel* where

$$\text{Gain in decibels} = \text{dB} = 10 \log \frac{P_2}{P_1} \qquad (14\text{-}41)$$

Since for a given resistance, power is proportional to the square of the voltage,

$$\text{Gain in decibels} = 10 \log \frac{V_2^2}{V_1^2} = 20 \log \frac{V_2}{V_1} \qquad (14\text{-}42)$$

In general, the input and output resistances of an amplifier are not equal. However, the decibel is such a convenient measure that it is used as a unit of voltage (or current) gain,[†] regardless of the associated resistances.

EXAMPLE 7

Express in decibels (dB) the relative response of a resonant circuit at the lower half-power frequency (Eq. 10-50).

Since this is a half-power frequency, the relative response in dB is

$$10 \log \frac{P_1}{P_0} = 10 \log 0.5 \cong -3 \text{ dB}$$

Alternatively, at f_1 the response is 0.707 of the resonant frequency value. Therefore, the relative gain in dB is

$$20 \log \frac{V_1}{V_0} = 20 \log 0.707 \cong -3 \text{ dB}$$

The response is "down 3 dB" at the cutoff frequencies.

The advantage of the logarithmic unit here is that when response is plotted in dB the overall response curves of a multistage amplifier can be obtained by adding the individual response curves. In Fig. 14.19, the response curves for two stages a and b are plotted in dB. From Eqs. 14-13 and 14-32 it can be deduced that for $f \ll f_1$ relative gain is proportional to frequency, and for $f \gg f_2$ relative gain is inversely

---

[†] The decibel may be used to indicate power *level* P_2 by specifying a reference level P_1, commonly taken as 1 mW into a 600-Ω resistance. A power level of 10 W would be described as $10 \log (10/0.001) = 40$ dBm or "40 dB above 1 mW."

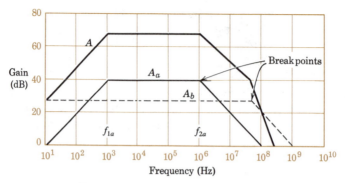

Figure 14.19 Frequency response expressed in decibels.

proportional to frequency. In the regions where gain $A = A_o(f/f_1)^{\pm 1}$, the gain in dB is $20 \log A = 20 \log A_o \pm 20 \log (f/f_1)$ or a 10 : 1 change in frequency (a "decade") results in a 20-dB change in gain. Therefore, with frequency on a log scale these relations are straight lines with slopes of 20 dB per decade, and the response of an RC-coupled amplifier can be approximated by curve A_a. (Actually, the gain is down 3 dB at the *breakpoint frequencies*.) A second amplifier with better low-frequency response is approximated by curve A_b. The overall gain of the combination is the sum of the two curves, and the straight-line approximation is easily drawn. For a single-stage amplifier with two high-frequency breakpoints, the frequency response would resemble curve A of Fig. 14.19.

Gain-Bandwidth Product

The *bandwidth* of an amplifier is a useful design criterion; for an audio amplifier we need a bandwidth of about 20 kHz whereas for a video amplifier we need a bandwidth of about 4 MHz. For an untuned amplifier the bandwidth is defined as $BW = \omega_2 - \omega_1 \cong \omega_2$ since ω_1 is small compared to ω_2. For a single FET stage (the black section in Fig. 14.20a), the gain is

$$A = \frac{V_o}{V_{gs}} = \frac{g_m}{\dfrac{1}{R_o} + j\omega C_{eq}} = \frac{g_m R_o}{1 + j\omega R_o C_{eq}} = \frac{A_O}{1 + j\omega R_o C_{eq}} \qquad (14\text{-}43)$$

Where $\omega_2 = 1/R_o C_{eq}$, if we attempt to increase bandwidth by decreasing R_o, we get a decrease in midfrequency gain. The product of gain magnitude and bandwidth is

$$A_O \omega_2 = g_m R_o \frac{1}{R_o C_{eq}} = \frac{g_m}{C_{eq}} \leq \frac{g_m}{C_{gs}} \qquad (14\text{-}44)$$

since $C_{eq} \cong C_{gs}$ if the Miller effect is not large. Because g_m and C_{gs} are device parameters:

The gain-bandwidth product is a constant for a specific FET.

For the FET of Example 5, $g_m/C_{gs} = 2 \times 10^{-3}/20 \times 10^{-12} = 10^8$ rad/s $\cong 15$ MHz. With this FET a gain of 5 permits a bandwidth of less than 3 MHz (2.9 MHz in Example 5 where $R_s = 1$ kΩ); if a bandwidth of 5 MHz is required, the maximum possible gain is $15/5 = 3$.

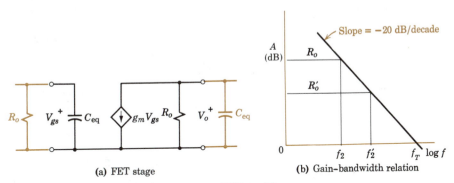

(a) FET stage   (b) Gain–bandwidth relation

Figure 14.20 Gain-bandwidth relation for an FET amplifier.

A bipolar transistor has a similar *figure of merit*, but it is usually expressed in a different way. If the high-frequency response is limited by the beta cutoff, the bandwidth is approximately f_β and the maximum possible gain is β, so the *gain-bandwidth product* is

$$A_o f_2 \le \beta f_\beta = f_T \tag{14-45}$$

The frequency f_T at which $\beta_f = 1$ typically ranges from 100 kHz for alloy junction power transistors to 10 GHz for high-frequency transistors.

EXAMPLE 8

The specifications for a certain silicon transistor indicate a minimum h_{fe} of 40 and a typical f_T of 8 MHz. Is this transistor suitable as a radiofrequency amplifier at 1 MHz?

By Eq. 14-35 at $f = 1$ MHz,

$$\beta_f = \frac{\beta f_\beta}{f} = \frac{f_T}{f} = \frac{8}{1} = 8$$

By Eq. 14-45,

$$f_\beta = \frac{f_T}{\beta} = \frac{f_T}{h_{fe}} = \frac{8 \text{ MHz}}{40} = 0.2 \text{ MHz}$$

Although some current gain is possible at 1MHz, the beta cutoff is only 0.2 MHz and another transistor should be selected.

Example 9 illustrates the complete design of a single-stage small-signal amplifier to meet frequency-response specifications.

EXAMPLE 9

Design a single-stage amplifier with a voltage gain of 34 dB, flat within 3 dB from 46 Hz to 200 KHz. Source and load resistances are $R_s = 2$ kΩ and $R_L = 10$ kΩ.
   Use a 2N3114 with $h_{fe} = \beta = 50$ (1 kHZ) and $h_{ie} = 1.5$ kΩ $= r_b + r_\pi \cong r_\pi$ at $I_C = 1$ mA and $V_{CE} = 5$ V; $C_{ob} = C_{jc} = 6$ pF and $f_T = 54$ MHz.

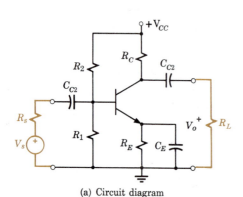

(a) Circuit diagram

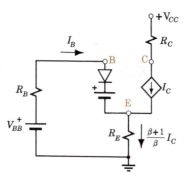

(b) Small-signal model

(c) DC bias design model

Figure 14.21 BJT amplifier design.

Replace the circuit diagram by the small-signal model of Fig. 14.21b, assuming $R_B \gg r_\pi$.
   At midfrequencies,

$$A_{VO} = \left| \frac{V_o}{V_s} \right| = \frac{\beta I_b (R_C \| R_L)}{I_b (R_s + r_\pi)} = 10^{dB/20} \cong 50$$

Solving,

$$R_o = R_C \| R_L = \frac{(R_s + r_\pi) A_{VO}}{\beta} = \frac{(2 + 1.5)50}{50} = 3.5 \text{ k}\Omega$$

Hence

$$R_C = \frac{1}{(1/R_o) - (1/R_L)} = \frac{10^3}{0.286 - 0.1} = 5.38 \text{ k}\Omega$$

Assuming high-frequency response is limited by C_{eq} with $C_\pi = 98$ pF from Example 6,

$$C_{eq} = C_\pi + (1 + \beta R_o/r_\pi) C_{jc}$$
$$= 98 + (1 + 50 \times 3.5/1.5)6 = 804 \text{ pF}$$

$$f_2 = \frac{1}{2\pi C_{eq}(R_s \| r_\pi)} = \frac{1}{2\pi \times 804 \times 10^{-12} \times 860} = 230 \text{ kHz}$$

Therefore, the high-frequency response is acceptable.
   Assuming bias is not critical (Fig. 14.21c), let

$$R_E = \frac{3}{I_E} \cong \frac{3}{I_C} = \frac{3}{0.001} = 3 \text{ k}\Omega$$

$$R_B \cong \frac{\beta R_E}{10} = \frac{50 \times 3}{10} = 15 \text{ k}\Omega$$

Then

$$V_{CC} = V_E + V_{CE} + I_C R_C = 3 + 5 + 1(5.38) \cong 13 \text{ V}$$

$$V_{BB} = R_B I_C/\beta + V_{BE} + V_E = 15 \times 1/50 + 0.7 + 3 = 4 \text{ V}$$

$$R_2 = R_B V_{CC}/V_{BB} = 15 \times 13/4 = 48.8 \rightarrow 47 \text{ k}\Omega$$

$$R_1 = R_B R_2/(R_2 - R_B) = 15 \times 47/32 = 22.03 \rightarrow 22 \text{ k}\Omega$$

For low-frequency response, use the results of Example 4 where $C_E = 50$ μF, $C_{C1} = 5$ μF, and $C_{C2} = 1$ μF.

SUMMARY

- Linear models are used to predict the performance of amplifiers in which signals are much smaller than bias values.
  The frequency-response curves of all RC-coupled amplifiers are similar in shape.
- For one stage of RC-coupled voltage (or current) amplification,

  Midfrequency gain is $\qquad A_O = -g_m R_o \text{ [FET]} \qquad \text{or} \qquad -\beta R_o/(r_\pi + R_s) \text{ [BJT]}$

  Low-frequency response is $\mathbf{A}_L = A_O \dfrac{1}{1 - j(\omega_{11}/\omega)} \cdot \dfrac{1}{1 - j(\omega_{12}/\omega)}$

  High-frequency response is $\mathbf{A}_H = A_O \dfrac{1}{1 + j(\omega/\omega_{21})} \cdot \dfrac{1}{1 + j(\omega/\omega_{22})}$

  Lower and upper cutoff frequencies ω_1 and ω_2 are determined by the critical combinations of R and C.
- For two stages of amplification in cascade, the overall gain is

  $$A \underline{/\theta} = A_1 A_2 \underline{/\theta_1 + \theta_2}$$

  Gain in dB $= 10 \log (P_2/P_1) \qquad \text{or} \qquad 20 \log (V_2/V_1).$
- The gain-bandwidth product is a measure of amplifier capability.

  $$\text{FET: } A_O \omega_2 \leq g_m/C_{gs} \qquad \text{BJT: } A_O f_2 \leq \beta f_\beta = f_T$$

REVIEW QUESTIONS

1. What determines whether a signal is "small" or "large"?
2. Distinguish between midfrequency range and bandwidth.
3. In Fig. 14.3, explain the effects of eliminating C_S and halving C_{C1}.
4. Draw a wiring diagram of one stage of an RC-coupled amplifier using an E MOSFET and draw a linear circuit model that is valid for frequencies from 20 Hz to 200 kHz.
5. For the circuit model of the preceding question, state reasonable assumptions and derive circuit models valid for low, medium, and high frequencies.
6. Define in words and in symbols the cutoff frequencies.
7. Differentiate between coupling and bypass capacitors.

8. What limits the low-frequency response of an FET? Of a BJT?
9. What limits the high-frequency response of an FET? Of a BJT?
10. Why is emphasis placed on voltage gain of an FET amplifier and current gain of a BJT amplifier?
11. Explain the Miller technique for transforming feedback circuits. Why is it useful?
12. Compare FET and BJT amplifiers on the following characteristics: input impedance, voltage gain, current gain, and linearity.
13. What is meant by "cascading"? Where is cascading used?
14. What are the advantages of the decibel as a gain unit?
15. Define the "gain-bandwidth product." Why is it useful?

EXERCISES

1. A BJT amplifier has a midfrequency gain of 200 with lower and upper cutoff frequencies at 10 Hz and 100 kHz, respectively. Draw the frequency-response curve showing gain and phase angle vs. frequency. What is the bandwidth of this amplifier?

2. In the generalized amplifier of Fig. 14.2, $Z_s = R_s = 2$ kΩ, $Z_i = R_i = 18$ kΩ, $G = 0.005$, $Z_o = R_o = 20$ kΩ.
   (a) Assuming Z_L is a pure resistance R_L, derive an expression for voltage gain V_o/V_s.
   (b) For $R_L = 20$ kΩ, evaluate the voltage gain.
   (c) Repeat part (a) assuming Z_L consists of $R_L = R_o$ in parallel with C_L.

3. In the generalized amplifier of Fig. 14.2, $Z_s \cong R_s = 10$ kΩ, Z_i consists of $R_i = 1$ MΩ in parallel with $C_i = 1$ nF, $G = 0.01\underline{/180°}$, and $Z_o \cong R_o = 10$ kΩ.
   (a) Assuming Z_L is a pure resistance R_L, specify R_L for a voltage gain $|V_o/V_s| = 50$ at $\omega = 1$ krad/s.
   (b) Predict the voltage gain (magnitude and phase angle) at $\omega = 667$ krad/s.

4. In the FET amplifier of Fig. 14.3a, Z_s and Z_L are pure resistances. Assuming that the frequency is so low that wiring capacitance and the internal capacitances of the transistor can be neglected, draw a small-signal circuit model of the amplifier.

5. A silicon transistor ($\beta = 70$, $C_\pi = 20$ pF, $C_{jc} = 5$ pF, and $r_\pi = 2.8$ kΩ) is used in the circuit of Fig. 14.3b with $R_C = 10$ kΩ, $C_{C1} = C_{C2} = 5$ μF, $R_E = 2$ kΩ, $C_E = 50$ μF, $R_1 = 10$ kΩ, and $R_2 = 100$ kΩ. It is estimated that $C_L + C_W = 50$ pF. List the assumptions made in deriving the circuit of Fig. 14.6b and check their validity at $f = 5000$ Hz.

6. If $r_\pi \cong R_s$ and $R_B \cong 10r_\pi$ are known within ± 10% in Fig. 14.6b, calculate the error made in predicting A_V by neglecting R_B. Draw a conclusion.

7. An amplifier is represented by the circuit model of Fig. 14.22. Stating any simplifying assumptions, predict the midfrequency gain.

8. An FET for which $g_m = 2$ mS, $r_d = 200$ kΩ, $C_{gs} = 10$ pF, and $C_{gd} \cong 0$ is used in the circuit of Fig. 14.23 with $R_s = 1$ kΩ, $R_1 \| R_2 = R_G = 0.1$ MΩ, $R_D = 5$ kΩ, $R_S = 2$ kΩ, R_L

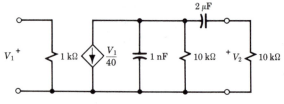

Figure 14.22

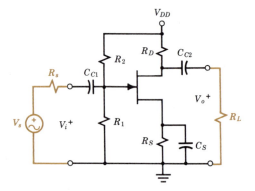

Figure 14.23

"very large," $C_{C1} = C_{C2} = 0.1$ μF, and C_S "large"; $C_W + C_L$ is estimated to be 10 pF.
   (a) Draw and label a linear circuit model valid for a wide range of frequencies, stating any useful assumptions.
   (b) Draw and label a simplified circuit model for medium frequencies, and predict the mid-frequency gain.

9. An enhancement MOSFET ($g_m = 5$ mS) is used in an RC-coupled amplifier.
   (a) Draw a circuit diagram similar to Fig. 14.23, with $R_1 = R_2 = 50$ MΩ, $R_S = 2.5$ kΩ, and $V_{DD} = 30$ V.
   (b) Draw a circuit model valid for moderate frequencies, and specify R_D for an open-circuit voltage gain $|V_o/V_i| = 40$.
   (c) For the specified R_D, with $R_s = 1$ kΩ and $R_L = 20$ kΩ connected across V_o, predict V_o/V_s.

10. A silicon transistor ($\beta \cong 100$) is used in the amplifier of Fig. 14.24, where $R_s = 5$ kΩ, $R_1 = 58$ kΩ, $R_2 = 142$ kΩ, $R_C = 8$ kΩ, and $R_E = 3.9$ kΩ. The capacitors are "large."

(a) Stating any necessary assumptions, draw a circuit appropriate for dc analysis and estimate the collector current for $V_{CC} = 10$ V.

(b) Redraw the circuit for small-signal analysis and specify resistance R_L (across V_o) for a midfrequency gain $|V_o/V_s| \cong 50$.

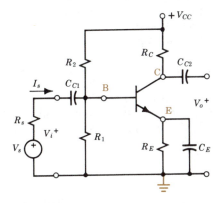

Figure 14.24

11. In the circuit of Fig. 14.24, $R_1 = 20$ kΩ, $R_2 = 80$ kΩ, $R_E = 2$ kΩ, and $V_{CC} = 10$ V. For this silicon transistor, $\beta = 100$.
    (a) Estimate I_C (by neglecting $I_B R_B$) and calculate r_π.
    (b) Specify R_C for a midfrequency voltage gain $|V_o/V_i| = 150$.
    (c) For the specified R_C and $R_s = 1$ kΩ, predict V_o/V_s.
    (d) For the specified R_C and R_s, predict V_o/V_s with $R_L = 10$ kΩ connected across V_o.

12. In the circuit of Fig. 14.24, $R_1 = 10$ kΩ, $R_2 = 90$ kΩ, $R_C = 10$ kΩ, and $I_C = 1.25$ mA. For this transistor $\beta = 50$.
    (a) Replace the transistor by an appropriate linear model and redraw the circuit to permit the prediction of the small-signal performance at moderate frequencies. (Show terminals B, C, and E on your diagram.)
    (b) Predict the midfrequency current gain I_{b2}/I_{b1} if the following stage employs a similar transistor under similar conditions.
    (c) If $R_s = 1$ kΩ, predict the overall voltage gain V_2/V_s under the conditions of part (b).

13. In the circuit of Fig. 14.24, $R_1 = 20$ kΩ, $R_2 = 60$ kΩ, $I_C = 2$ mA, and $\beta = 80$. The source is a phono pickup represented by $V_s = 10$ mV (rms) in series with $R_s = 500$ Ω. The "load" is another stage whose input resistance is

10 kΩ. Complete the design so that V_o is 1.6 V (rms) at midfrequencies.
(a) Draw an appropriate small-signal model.
(b) Specify the value of R_C.
(c) Specify values of R_E and V_{CC}.

14. The amplifier of Exercise 10 is to have a lower cutoff frequency of 20 Hz. Assuming $I_C = 0.5$ mA and C_E and C_{C2} are "large," draw an appropriate circuit model and specify C_{C1}.

15. For a certain RC-coupled amplifier, $\omega_{11} = \omega_{12} = 100$ rad/s. What is ω_1 in this case?

16. A single-stage FET RC-coupled amplifier used in a mechanical engineering experiment must have a gain at 1 Hz equal to 90% of its midfrequency gain.
    (a) What lower cutoff frequency is required?
    (b) If $g_m = 2$ mS, $R_D = 10$ kΩ, and if the input resistance to the following stage is 1 MΩ, what value of C_{C2} is required?
    (c) If the gain is halved by reducing R_D, is the required value of C_{C2} reduced?

17. For a certain n-channel JFET, the manufacturer specifies $g_m = 4000$ μS, and maximum values of $C_{iss} = 7$ pF and $C_{rss} = 3$ pF. This JFET is used in the amplifier of Fig. 14.3a with $R_s = 1$ kΩ, $R_1 = 1$ MΩ, $R_2 = 1.5$ MΩ, $C_{C1} = C_{C2} = 0.1$ μF, $R_D = 5$ kΩ, $R_S = 10$ kΩ, $C_S = 50$ μF, and $R_L = 1$ MΩ.
    (a) Draw an appropriate small-signal model and predict the midfrequency gain V_o/V_s.
    (b) Draw an appropriate small-signal model and predict the lower cutoff frequency.

18. In the amplifier of Exercise 12, $R_E = 1$ kΩ, $C_E = 50$ μF, and $C_{C1} = 5$ μF. C_{C2} is "large."
    (a) Calculate the "lower cutoff frequencies" as defined by C_E and C_{C1} separately.
    (b) What is the lower cutoff frequency for the amplifier?
    (c) Specify an appropriate value for C_{C2}, explaining your reasoning.

19. In the amplifier of Exercise 12, $R_E = 1$ kΩ.
    (a) Specify the value of C_E for a lower cutoff frequency of 50 Hz.
    (b) Specify appropriate values of C_{C1} and C_{C2}.

20. In the circuit of Fig. 14.24, $R_s = 2$ kΩ, $R_E = 2$ kΩ, $R_C = 6$ kΩ, and R_L (across V_o) $= 2$ kΩ; $C_{C1} = C_{C2} = 5$ μF and $C_E = 100$ μF. The BJT with $\beta \cong 100$ operates at $I_C = 1.25$ mA and $V_{CE} = 5$ V. Stating any necessary assumptions, predict the midfrequency response and the lower cutoff frequency. Draw and use appropriate circuit models.

21. For a certain silicon transistor in the common-emitter configuration, the small-signal current relations are:

$$\begin{cases} i_b = 0.004v_{be} + 10^{-10}\,dv_{be}/dt \\ i_c = 0.02v_{be} + 2 \times 10^{-5}v_{ce} \end{cases}$$

(a) Draw and label an appropriate circuit model.
(b) What determines the frequency response of this device?
(c) Determine the frequency at which the base signal voltage (and therefore the output current) is only 70.7% of its low-frequency value for a given input current.

22. The circuit model for an amplifier is shown in Fig. 14.22. Predict the upper and lower cutoff frequencies (rad/s).

23. For the FET amplifier of Exercise 8:
(a) Draw and label a simplified circuit model valid for low frequencies and predict ω_1 and f_1.
(b) Draw and label a simplified circuit model valid for high frequencies and predict ω_2 and f_2.
(c) For this amplifier, define "midfrequency" and sketch the frequency response curve. (Plot three key points on the A_V vs. f curve; for a log plot on ordinary graph paper let unit distance correspond to a factor of 10 in frequency.)

24. For a certain RC-coupled amplifier, $\omega_{21} = \omega_{22} = 100$ krad/s. What is ω_2 in this case?

25. For the JFET amplifier of Exercise 17, draw an appropriate small-signal model and predict the upper cutoff frequency, assuming about 4 pF of wiring capacitance in the drain circuit.

26. For the 2N699B (see Table 13-1), beta is down to unity at 70 MHz.
(a) Devise an appropriate small-signal model and evaluate the parameters.
(b) Estimate the upper cutoff frequency of the current gain in an amplifier using the 2N699B, stating any assumptions made.

27. In an amplifier using the transistor of Exercise 26, $R_s = 10$ kΩ, $R_B = R_1 \| R_2$ is negligibly large, wiring capacitance is negligibly small, $C_{jc} = 13$ pF, and $R_o = 2$ kΩ. Predict the mid-frequency voltage gain and the upper cutoff frequency.

28. An E MOSFET ($g_m = 3000$ μS, $C_{gs} = 6$ pF, $C_{gd} = 2$ pF) is used in the circuit of Fig. 14.23 with $R_s = 2$ kΩ, $R_1 = 1.25$ MΩ, $R_2 = 5$ MΩ, $R_D = 10$ kΩ, $R_S = 4$ kΩ, and $R_L = 10$ kΩ.

$C_{C1} = 0.2$ μF, C_S is "large," and C_o is negligible.
(a) Draw and label the wiring diagram.
(b) Use circuit models to specify C_{C2} for $f_1 = 20$ Hz.
(c) Use a circuit model to predict f_2.

29. A single-stage transistor amplifier has a mid-frequency current gain of 40 and a lower cutoff frequency of 60 Hz. Predict the corresponding values for an amplifier consisting of three such stages in cascade.

30. For $\mathbf{A} = 1/(1 + j\omega RC)$, explain why a log-log plot of $A(\omega)$ above $\omega_2 = 1/RC$ is a straight line of slope -20 dB per decade.

31. The input to one unit of a communication system is 5 W. Calculate the dB gain if the output is:
(a) 100 W.
(b) 20 W.
(c) 0.25 W.
(d) 0.05 W.

32. In a communication system, a microphone provides an input of 20 mV. The preamplifier provides 26 dB of voltage gain, a cable introduces 12 dB of loss, the coupling circuit introduces 6 dB of loss, and the final amplifier provides 40 dB of gain.
(a) Calculate the output voltage of each element.
(b) Calculate the overall gain in dB and the output voltage.

33. The power loss in 1 km of coaxial telephone cable is 1.5 dB.
(a) What is the power loss in 2 km of telephone cable? Three?
(b) What is the total power loss from New York to Los Angeles?
(c) What signal power input is required at New York to deliver 1 μW at Los Angeles? Would amplifiers be desirable?

34. A telephone system using cable with a power loss of 1.5 dB/km requires a minimum signal of 1 μW.
(a) If amplifiers are located 50 km apart, what amplifier output power is required?
(b) Assuming 10% overall amplifier efficiency, what total (dc) input power is required for New York to Los Angeles transmission?
(c) Compare the result of part (b) to the answer for part (c) of Exercise 33.

35. An amplifier consists of the following stages: Stage 1, midfrequency power gain = 10 dB, $f_1 = 0$ Hz, $f_2 = 3$ MHz. Stage 2, midfrequency power gain = 30 dB, $f_1 = 10$ Hz, $f_2 = 1$ MHz.

Stage 3, midfrequency power gain = 20 dB,
f_1 = 100 Hz, f_2 = 300 kHz.
  (a) Plot the individual and overall response
      curves.
  (b) If the midfrequency output is to be 2 W,
      what input power is required?
36. A transistor with a nominal β = 50 has a gain-
bandwidth product f_T = 30 MHz. Describe,
quantitatively, the high-frequency characteristics
of this transistor. What is the current gain in dB
at 30 MHz? At 2 MHz?
37. The amplifier of Exercise 22 is redesigned with
R_C and R_L reduced from 10 kΩ to 6 kΩ each to
increase the bandwidth.

  (a) Calculate the original and new values of
      midfrequency gain.
  (b) Calculate the original and new values of
      gain-bandwith product.
38. Each stage of a transistor amplifier is to provide
a current gain of approximately 20 with f_2 =
5 MHz. For a β = 80, what minimum beta cutoff frequency is required?
39. A multistage amplifier is to provide a voltage
gain of 1000 with an upper cutoff frequency of
200 kHz. Available is an FET for which g_m =
2 mS, C_{rss} = 2 pF, and C_{iss} = 30 pF. How
many stages are required?

PROBLEMS

1. A practical JFET amplifier circuit is shown in
Fig. 14.25.
  (a) Redraw the circuit for dc analysis, assume
      V_{DS} = 10 V, and estimate the bias voltage
      V_{GS} and quiescent current I_D.
  (b) Redraw the circuit for ac analysis and evalu-
      ate the parameters given that g_{mo} = 8 mS
      and I_{DSS} = 8 mA.
  (c) Predict the voltage gain.
  (d) Draw one cycle of a sinusoidal $v_i(\omega t)$ and,
      on the same graph, show the voltages with
      respect to ground at points 1, 2, 3, and 4.

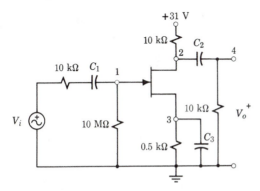

Figure 14.25

2. An elementary transistor amplifier is to be used
as a preamplifier for a "hi-fi" set. The output of
the phonograph pickup (low internal impedance)
is only 2 mV, but 0.4 V is required as the input

to the power amplifier that drives the loudspeaker. "Design" the preamplifier by specifying
a suitable transistor and collector current and the
appropriate load resistor and supply voltage.
3. Derive the criterion for C_E (Eq. 14-21) by using
the approach followed in deriving the criterion
for C_S (Eq. 14-17).
  (a) Define the condition for ω_1.
  (b) Determine the Thévenin equivalent resis-
      tance across which C_E is connected (Fig.
      14.12).
  (c) Express ω_1 in terms of the circuit parame-
      ters.
4. An FET amplifier similar to Fig. 14.3a with
R_D = 5 kΩ is built and tested. The manufacturer
specifies g_m = 4000 μS, C_{rss} = 5 pF, and
C_{iss} = 25 pF for the FET. When V_o/V_i is observed on a CRO with an input capacitance of
30 pF, the midfrequency voltage gain is observed
to be 18 and the gain is down to 12.7 at 1 MHz.
Analyze these experimental results and draw a
conclusion.
5. A small-signal wide-band amplifier is to provide
a voltage gain of 40 with a flat response from
20 Hz to approximately 100 kHz. Available is a
2N699B transistor (see Table 13-1, p. 391) with
β_f = 3.5 at 20 MHz and C_{jc} = 13 pF. The resistance of the source is R_s = 700 Ω. Stating *all*
assumptions, design the amplifier (draw the wiring diagram and specify the components and
supply voltage).

6. The E MOSFET of Fig. 6.6 is used as an amplifier with V_s in series with R_s applied from gate to ground and V_o taken between drain and ground.
   (a) Replace the MOSFET by a small-signal model and use nodal analysis to evaluate $A = |V_o/V_s|$ for $R_s = 10$ kΩ, $R_G = 1$ MΩ, $g_m = 2$ mS, and $R_D = 10$ kΩ.
   (b) Derive an expression for input resistance of the amplifier and evaluate R_i.
   (c) Replace R_G by its Miller equivalent, repeat parts (a) and (b), and draw a conclusion.

7. In the circuit of Fig. 13.8a, the input source consists of V_i in series with $R_i = 10$ kΩ and $R_L = 5$ kΩ. The JFET can be represented by the model of Fig. 13.10 with $g_m = 5$ mS, $r_d = 100$ kΩ, $C_{gs} = C_{gd} = 10$ pF, and $C_{ds} = 1$ pF.

   (a) Draw a small-signal model of the amplifier and evaluate the capacitive reactances at $f = 10$ kHz.
   (b) Redraw the circuit model neglecting any capacitances that have a small effect compared to the associated resistances and calculate the voltage gain V_o/V_i.
   (c) Recalculate the reactances at $f = 1$ MHz and redraw the amplifier circuit as in part (b). Assume $V_o = 50$ mV (rms) and calculate the necessary voltage V_{gs} across C_{gs} and the voltage across C_{gd}. Calculate the currents through C_{gs} and C_{ds} and the necessary voltage V_i (rms).
   (d) Calculate the voltage gain at 1 MHz and draw a conclusion.

15

Feedback Amplifiers

Feedback Systems

Positive and Negative Feedback Applications

Followers

A system is a combination of diverse but interacting elements integrated to achieve an overall objective. The elements may be electrical devices, assembly lines, or human beings concerned with processing materials, information, or energy. The objective may be to guide a space vehicle, to control a robot, to operate a manufacturing process, or to store and retrieve information.

The increasing complexity of man-made systems and the growing body of principles and techniques for predicting system behavior have resulted in the activity called *systems engineering*. The systems engineer takes an overall view; he or she is interested in specifying the performance of the system, creating a combination of components to meet the specifications, describing the characteristics of components and their interconnections, and evaluating system performances. Systems engineers leave the design of components and connecting links to others; their focus is on external characteristics—performance, reliability, and cost.

A *feedback* loop for returning part of the output to the input in order to improve performance is an important component in many systems. The concept of feedback is not new; it is inherent in many biological systems, and it is found in well-established social systems. The formulation of the principles of feedback is credited to H.S. Black in his studies in 1934; exploitation of the insight he provided has been widespread in engineering and electronics. In this book we have used feedback to provide precise gain in op amp circuits, to obtain oscillation in square-wave generators (Fig. 11.17), and to stabilize the operation of power transistors.

In one type of application, we use feedback to compare variables and then control an operation, as in a satellite guidance system or a voltage regulator (Fig. 11.13). A second application of feedback is to improve the performance or characteristics of a device or system. Through the intentional use of feedback, the gain of amplifiers can be made much greater, or more stable, or less sensitive to change, or less dependent on frequency. On the other hand, amplifier designers must take steps to prevent unintentional feedback from degrading the performance of their products. In this chapter we study the basic concept of feedback, learn how to reduce complicated systems to the basic form, investigate the general benefits of *positive* and *negative* feedback, and then look at a few practical applications of this important concept to amplifiers.

FEEDBACK SYSTEMS

Amplifier with Feedback

An electronic amplifier is represented by the circuit model of Fig. 15.1a. The transistors and circuit elements have been replaced by equivalent components where

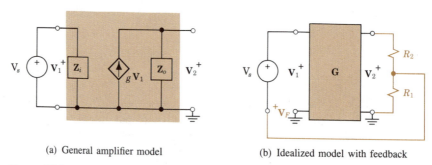

(a) General amplifier model   (b) Idealized model with feedback

Figure 15.1   An electronic amplifier with voltage feedback.

Z_i and Z_o are the input and output impedances, and g is a transconductance. Such an amplifier with feedback can be represented by the model of Fig. 15.1b. The voltage gain of the amplifier alone, called the *forward gain*, is defined by

$$G = \frac{V_2}{V_1} \tag{15-1}$$

where V_1 and V_2 are phasors and G is a complex function of frequency.

The voltage divider consisting of R_1 and R_2 provides a means of tapping off a portion V_F of the output voltage V_2 and feeding it back into the input, "closing the loop." The *feedback factor* is

$$H = \frac{V_F}{V_2} \tag{15-2}$$

In general, H is a complex quantity, although here $H = R_1/(R_1 + R_2)$ is a real number.

With a fraction of the output fed back in series with input voltage V_s, the amplifier input V_1 is the sum of the signal V_s and the feedback voltage V_F or

$$V_1 = V_s + V_F \tag{15-3}$$

The output voltage is now

$$V_2 = GV_1 = G(V_s + V_F) = G(V_s + HV_2) = GV_s + GHV_2$$

Solving,

$$G_F = \frac{V_2}{V_s} = \frac{G}{1 - GH} \tag{15-4}$$

where G_F is the gain with feedback.

Equation 15-4 is the basic equation for system gain with feedback, and by studying this equation we can predict some of the possibilities of such a system. For the moment, consider that the quantities G and H are real. If GH is very large compared to 1, $G_F = -G/GH = -1/H$, or the gain is independent of G; this is the condition we established to obtain precise amplification with an inexpensive op amp.

In general, if GH is negative so that $1 - GH$ is greater than 1, gain is reduced, but there are accompanying benefits to offset the loss in gain. If GH is positive and $1 - GH$ is between 0 and 1, gain is increased. As GH approaches 1 and $1 - GH$ approaches 0, the gain increases without limit. If we make GH precisely 1, gain is infinite and there is the possibility of output with no input; we have turned the amplifier into a *self-excited oscillator*. If GH is greater than 1 and $1 - GH$ is large and negative, we can create a *multivibrator* like the square-wave generator.

Practice Problem 15-1

In Fig. 15.1b, the amplifier gain $G = -100$. Predict the gain with feedback G_F and the type of operation for feedback factor $H = 0.1$, $H = 0.01$, $H = -0.01$, and $H = -0.005$.

Answers: -9.09, -50, infinite, -200.

Before investigating the specific behavior of feedback systems, we need a simplified notation that will permit us to disregard the details of circuits and devices and focus our attention on system aspects. Also, our approach should emphasize the paths whereby signals are transmitted.

Block Diagram Representation

Let us represent linear two-port devices by blocks labeled to indicate the functions performed and use single-line inputs and outputs labeled to indicate the signals at those points. The amplifier whose small-signal circuit model is within the colored area in Fig. 15.-1a is represented symbolically by a rectangle labeled *gain* **G** (Table

Table 15-1 Block Diagram Symbols and Defining Equations

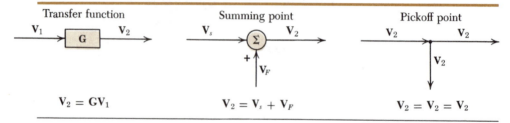

Transfer function	Summing point	Pickoff point
$V_2 = GV_1$	$V_2 = V_s + V_F$	$V_2 = V_2 = V_2$

15-1). Such a symbol can represent any two-port device for which the output signal is a known function of the input signal. In general, the ratio of output signal to input signal is called the *transfer function*.

To be represented by such a block, a device must be linear. Also, it is assumed that each device has infinite input impedance and zero output impedance so that devices may be interconnected without loading effects. In the system of Fig. 15.1b, $R_1 + R_2$ must be large with respect to Z_o so that the amplifier gain is unaffected by the feedback connection. If this amplifier is to provide the input to another device that draws an appreciable current, a bipolar transistor amplifier, for example, gain **G** must be recalculated, taking into account the finite input impedance of the next device. When these conditions are met, the blocks are said to be *unilateral*; linear unilateral blocks may be interconnected freely.

Two different types of connection are possible (Table 15-1). A *summing point* has two or more inputs and a single output. The output is the sum or difference of the inputs, depending on the signs; a plus sign is assumed if no sign is indicated. In contrast, from a *pickoff point* the signal can be transmitted undiminished in several directions. Physically this is illustrated by a voltage that can serve as input to several devices with high input impedances or by a tachometer that senses the speed transmitted by a shaft without diminishing the speed.

Using block diagram notation, we can represent the amplifier of Fig. 15.1b as in Fig. 15.2. In this case, all variables are voltages but, in general, different variables

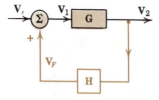

Figure 15.2 Block diagram of the basic feedback system.

may appear on the diagram. (All signals entering a summing point must have the same units.) The amplifier block represents Eq. 15-1 and the feedback block represents Eq. 15-2. The summing point fulfills Eq. 15-3, and Eq. 15-4 gives the gain of the system. This basic *closed-loop* system appears so frequently that Eq. 15-4 should be memorized.

EXAMPLE 1

Given a system described by the equations:

$$V_a = V_i + V_c \quad (1) \qquad V_b = AV_a \quad (2)$$

$$V_o = BV_b \quad (3) \qquad V_c = -CV_b \quad (4)$$

where V_i is the input and V_o is the output. Draw a labeled block diagram for the system and determine the "gain" V_o/V_i.

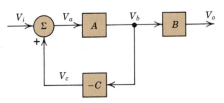

Figure 15.3 Block diagram synthesis.

Equation (1) is represented by a summing point as shown in Fig. 15.3. Equations (2) and (3) suggest cascaded blocks. Equations (4) and (1) must be satisfied simultaneously as shown.

We see that V_b is related to V_i by a basic feedback circuit. By Eq. 15-4,

$$V_b = G_F V_i = \frac{A}{1 - A(-C)} V_i = \frac{AV_i}{1 + AC}$$

The output V_o is just BV_b. Therefore, the system gain is

$$\frac{V_o}{V_i} = \frac{BV_b}{V_i} = \frac{AB}{1 + AC}$$

BLOCK DIAGRAM ALGEBRA

Analysis of the behavior of complicated systems can be simplified by following a few rules for manipulating block diagram elements. These rules are based on the definitions of block diagram symbols and could be derived each time, but it is more convenient to work from the rules.

1. *Any closed-loop system can be replaced by an equivalent open-loop system.*
2. *The gain of cascaded blocks is the product of the individual gains.*
3. *The order of summing does not affect the sum.*
4. *Shifting a summing point beyond a block of gain G_A requires the insertion of G_A in the variable added.*
5. *Shifting a pickoff point beyond a block of gain G_A requires the insertion of $1/G_A$ in the variable picked off.*

These rules and the corollaries of rules 4 and 5 are illustrated in Example 2 on p. 432.

Referring again to Example 1, if we let $G = AB$ and let $H = -C/B$, the overall gain becomes

$$G_F = \frac{V_o}{V_i} = \frac{AB}{1 - (AB)(-C/B)} = \frac{G}{1 - GH}$$

All linear feedback networks can be reduced to the basic system of Fig. 15.2 for which the gain is $\mathbf{G_F = G}/(1 - \mathbf{GH})$. The key to the behavior of a particular system is the quantity $1 - \mathbf{GH}$.

EXAMPLE 2

Determine the overall gain of the system shown in Fig. 15.4a.

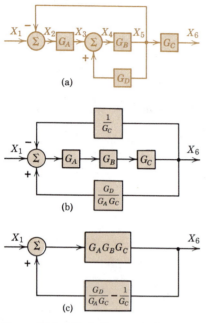

It would be possible to determine the gain by writing the equation for each element in turn and solving simultaneously. A preferable procedure is to simplify the system by block diagram algebra.

Applying rule 5, the pickoff point is moved beyond G_C and $1/G_C$ is inserted in the upper and lower loops.

Applying the corollary of rule 4 permits shifting the summing point for the lower loop before G_A and $1/G_A$ is inserted in that loop (Fig. 15.4b).

Applying rule 3, the two feedback loops are combined. Applying rule 2, the three cascaded blocks are combined. The result is in the basic form (Fig. 15.4c).

Applying rule 1, the overall gain can be written by inspection as

$$G_{61} = \frac{X_6}{X_1} = \frac{G_A G_B G_C}{1 - (G_A G_B)(G_D/G_A - 1)}$$

Figure 15.4 Block diagram algebra.

POSITIVE FEEDBACK APPLICATIONS

For **GH** positive and real but less than unity, the denominator is less than unity and the system gain is greater than the forward gain **G**. This condition of *positive feedback* was used in *regenerative receivers* to provide high gain in the days when electronic amplifiers were poor and signals were weak. Because operation with positive feedback becomes less stable as *GH* approaches unity, regenerative receivers required constant adjustment.

We have all heard the "squeal" that results when the loudspeaker of a sound system is oriented so that some of its output reaches the microphone. We now understand the basis for this annoying form of feedback. If the feedback path is just right, the value of **GH** approaches $1 + j0$, the denominator of Eq. 15.4 approaches zero, and the overall gain increases without limit. An amplifier of infinite gain can provide an output (the squeal) with no input; such a device is called an *oscillator* or signal generator and is a valuable part of many electronic systems.

Oscillation

In the amplifier of Fig. 15.1a, if Z_o consists of R_o and C_o in parallel, the output tends to decrease as the frequency increases because of the filtering effect of R_oC_o. As we saw in Chapter 14 (Eq. 14-27), the high-frequency gain can be expressed as

$$\mathbf{A} = A(j\omega) = \frac{A_O}{1 + j\omega C_o R_o} = \frac{A_O}{1 + j\omega/\omega_2}$$

where A_O is a constant representing the gain at moderate frequencies for which $\omega/\omega_2 \ll 1$, and ω_2 is the *upper cutoff frequency* at which **A** drops to 70% of A_O. At a somewhat higher frequency where $\omega = \sqrt{3}\,\omega_2$,

$$\mathbf{A}_H = \frac{A_O}{1 + j\sqrt{3}} = \frac{A_O}{2}\underline{/-\tan^{-1}\sqrt{3}} = \frac{A_O}{2}\underline{/-60°}$$

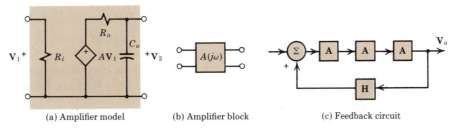

| (a) Amplifier model | (b) Amplifier block | (c) Feedback circuit |

Figure 15.5   A three-stage amplifier with feedback.

For a three-stage amplifier consisting of three similar sections (Fig. 15.5), at this frequency the overall gain is

$$\mathbf{G} = \mathbf{A}_H^3 = \left(\frac{A_O}{2}\right)^3 \underline{/3(-60°)} = \left(\frac{A_O}{2}\right)^3 \underline{/-180°} = -\left(\frac{A_O}{2}\right)^3 + j0$$

Assuming feedback is present and $\mathbf{H} = 0.008$,

$$\mathbf{GH} = -0.008\left(\frac{A_O}{2}\right)^3 + j0 = -\left(\frac{A_O}{10}\right)^3 + j0$$

Where $\mathbf{G}_F = \mathbf{G}/(1 - \mathbf{GH})$, for oscillation $\mathbf{G}_F$ must be infinite or the denominator must be zero or

$$\mathbf{GH} = -\left(\frac{A_O}{10}\right)^3 + j0 = +1 + j0 \qquad \text{and} \qquad A_O = -10$$

These results are interpreted as follows: In a three-stage amplifier with a per-stage gain of -10, if 0.8% of the output voltage is fed back to the input, oscillation is possible at a frequency for which the per-stage phase shift is 60°. In practice, there is a range of frequencies that will be amplified, but one of these will be maximized and the level of oscillation will build up until A_O is just equal to -10. In this case **G** is the frequency-dependent element; in other cases **H** may determine the frequency at which $\mathbf{GH} = 1 + j0$.

Another way of looking at the phenomenon of oscillation is to recognize that in any impulsive input, such as turning on a switch, components of energy corresponding to all frequencies are present. Also small voltages are generated by the random motion of electrons in conductors. Any such action can initiate the small signal, which builds up and up through amplification and feedback to the equilibrium level. For small signals, A_O is usually quite large (much greater than -10); as the signal level increases, the gain decreases due to nonlinearity and the amplitude of oscillation stabilizes at the equilibrium level.

The preceding illustration indicates that a circuit designed for amplification may produce oscillation if some small part of the output is coupled or fed back to the input with the proper phase relation. Opportunities for such coupling exist in the form of stray electric and magnetic fields or through mutual impedances of common power supplies. Amplifier designers use shielding and filtering to minimize coupling and reduce the possibility of oscillation.

Oscillators

When a circuit is designed to oscillate,[†] special provision is made to feed back a portion of the output signal in the correct phase and amplitude to make the **GH** product just equal to $1 + j0$. By making **GH** frequency dependent, oscillation can be obtained at a single desired frequency. Either **G** or **H** can be the frequency-dependent element of the feedback system of Fig. 15.6a.

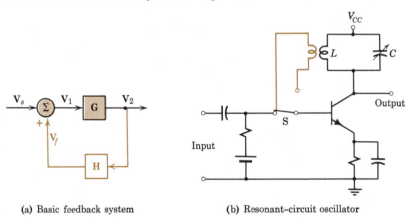

(a) Basic feedback system   (b) Resonant–circuit oscillator

Figure 15.6 Positive feedback in an oscillator circuit.

The oscillator of Fig. 15.6b is basically an amplifier with a parallel resonant circuit for a load impedance. Gain is a maximum for signals at the resonant frequency of the collector circuit. This is called a *tuned* amplifier because the frequency for maximum amplification can be selected by varying the capacitance. (Such an amplifier is used in a radio receiver to amplify the signals transmitted by a selected station while rejecting all others.) To create a "self-excited amplifier" or oscillator, switch S is

---

[†] The fact that stable operation occurs in the nonlinear region of the device characteristic complicates the design and analysis of oscillators.

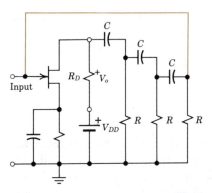

Figure 15.7 An FET phase-shift oscillator.

shifted (and the "Input" terminals are shorted) and part of the voltage across the tuned circuit is fed back into the input. In this case, **G** is a sensitive function of frequency.

For audiofrequencies, the values of L and C required for a resonant circuit are impractically large. The *phase-shift oscillator* of Fig. 15.7 avoids this requirement and is fairly common for the generation of audiofrequency signals. The output of the FET amplifier appears across R_D. The first RC section constitutes a phase-shifting network in that the voltage across R is from $0°$ to $90°$ ahead of the voltage across R_D depending on the frequency. For a particular frequency, the three RC sections produce a total phase shift of $180°$; the additional $180°$ required for oscillation is inherent in the gate-drain voltage relations in an FET. In this case, **H** is the frequency-sensitive element of the feedback system.

To provide good waveform, an oscillator should operate in the linear region of the amplifier. In the *resistance-capacitance tuned oscillator* of Fig. 15.8a, widely used as a laboratory instrument, the necessary phase reversal for positive feedback is provided by the two-stage amplifier. The input to the first stage is obtained from the voltage divider $Z_1/(Z_1 + Z_2)$ across the output of the second stage. Voltage V_{in} is in phase with V_{out} for $\omega = 1/\sqrt{R_1 R_2 C_1 C_2}$ (the principle of the Wien bridge) and oscillation occurs. In the usual case, $R_1 = R_2 = R$ and $C_1 = C_2 = C$ and the oscillator frequency is $\omega_o = 1/RC$. As the oscillations build up, the current through R_3 and

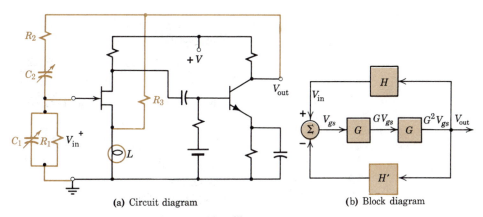

(a) Circuit diagram   (b) Block diagram

Figure 15.8 A resistance-capacitance tuned oscillator.

incandescent lamp L increases, raising the temperature of the filament. The voltage developed across this nonlinear resistance subtracts from the input voltage and limits the amplitude of the oscillations. The negative feedback loop is shown in Fig. 15.8b, where H' is a nonlinear function of output voltage.

NEGATIVE FEEDBACK APPLICATIONS

Negative feedback, where the amplitude and phase of the signal fed back are such that the overall gain of the amplifier is less than the forward gain, may be used to advantage in amplifier design. Since additional voltage or current gain can be obtained easily with transistors, the loss in gain is not important if significant advantages are obtained. Amplifier performance can be improved in the following ways:

1. Gain can be made practically independent of device parameters; gain can be made insensitive to temperature changes, aging of components, and variations from typical values.
2. Operating point can be made practically independent of device parameters, and it can be made insensitive to temperature changes or aging of components.
3. Gain can be made practically independent of reactive elements and therefore insensitive to frequency.
4. Gain can be made selective to discriminate against distortion or noise.
5. The input and output impedances of an amplifier can be improved greatly.

Stabilizing Gain

For precise instrumentation or long-term operation, amplifier performance must be predictable and stable. However, amplifier components such as transistors and IC elements vary widely in their critical parameters. By employing negative feedback and sacrificing gain (which can be made up readily), stability can be improved.

EXAMPLE 3

The midfrequency gain of a small-signal amplifier is, by Eq. 14-6,

$$G = A_{VO} = -\beta \frac{R_o}{r_\pi + R_s} \cong -100$$

In a certain mass-produced item, it is expected that random variations in parameters and the effects of temperature variations might reduce the nominal gain G_N from 100 to as low as 50.

Investigate the stabilizing effect of using two identical stages and a feedback loop with $H = -0.0099$.

For two stages, the forward gain is $G_N = (-100)^2 = 10,000$. With feedback, the nominal gain is

$$G_{FN} = \frac{G_N}{1 - G_N H} = \frac{10,000}{1 - 10^4(-0.0099)} = 100$$

If the actual gain per stage drops to $G_A = (-50)^2 = 2500$, the amplifier gain is

$$G_{FA} = \frac{G_A}{1 - G_A H} = \frac{2500}{1 - 2500(-0.0099)} = 97.1$$

With feedback (and an additional stage), the change in gain is reduced to less than 3%.

Practice Problem 15-2

Represent the noninverting amplifier circuit of Fig. 11.8 by a block diagram and evaluate G and H. Derive the expression for amplifier gain from the basic equation for a feedback system.

Answers: $G = A$, $H = R_1/(R_1 + R_F)$, $A_F \cong (R_1 + R_F)/R_1$.

Stabilizing Operation

The self-biasing circuit for decreasing the effect of changes in β or temperature on the operating point of a transistor is essentially a feedback device. As indicated in Fig. 15.9a, if I_C tends to increase, the current $I_C + I_B$ in R_E increases, raising the potential of the emitter with respect to ground. This, in turn, reduces the forward bias, reduces the base current and, therefore, limits the increase in I_C. In other words, an increase in I_C is fed back in such a way as to oppose the change that produced it.

EXAMPLE 4

Represent the response of the self-biasing circuit of Fig. 15.9a by a block diagram and derive the "stability factor" $\Delta I_C / \Delta \beta$.

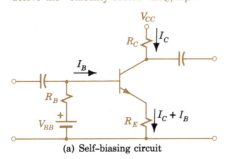

(a) Self-biasing circuit

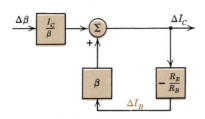

(b) Block diagram

Figure 15.9   Bias stabilization.

Where I_{CBO} is negligible (as in silicon transistors), $I_C = \beta I_B$ and

$$\Delta I_C = \Delta \beta I_B + \beta \Delta I_B = \Delta \beta (I_C/\beta) + \beta \Delta I_B \quad (15\text{-}5)$$

With fixed-current bias ($\Delta I_B = 0$), changes in β produce proportionate changes in I_C.

With self-bias (Fig. 15.9a),

$$V_B - I_B R_B - V_{BE} - (I_C + I_B)R_E = 0$$

Assuming V_{BE} is constant, for small changes this becomes

$$-\Delta I_B R_B - \Delta I_C R_E - \Delta I_B R_E = 0$$

or

$$\Delta I_B = \frac{-\Delta I_C R_E}{(R_B + R_E)} \cong \frac{-\Delta I_C R_E}{R_B} \quad (15\text{-}6)$$

where R_B is large compared to R_E.

Equations 15-5 and 6 are realized in Fig. 15.9b. For this closed-loop system $G = 1$, and

$$\frac{\Delta I_C}{\Delta \beta} = \frac{I_C/\beta}{1 + \beta R_E/R_B} \quad \text{or} \quad \frac{\Delta I_C}{I_C} = \frac{\Delta \beta/\beta}{1 + \beta R_E/R_B}$$

For $\beta R_E = 10 R_B$ (Eq. 12-18), $1 + \beta R_E/R_B = 1 + 10 = 11$ and an 11% change in β results in only a 1% change in I_C.

It can be shown (see Exercise 17) that, for small changes in G, the fractional change in overall gain with feedback is equal to the change in G divided by $1 - GH$. By definition, the *sensitivity* of gain to change is

$$\frac{dG_F/G_F}{dG/G} = \frac{1}{1 - GH} \tag{15-7}$$

In words, the sensitivity of the gain to change is reduced by the factor $1 - GH$.

Improving Frequency Response

No amplifier amplifies equally well at all frequencies; at the upper cutoff frequency, for example, the voltage gain is down to 70.7% of the midfrequency value. By sacrificing gain, however, the upper cutoff frequency can be raised. In the practical case, gain can be made insensitive to frequency over the range of interest by using negative feedback.

Using system notation, the gain $\mathbf{G}$ is frequency dependent because of storage effects in the electronic device and reactive elements in the circuit. The feedback factor $\mathbf{H}$, however, can be obtained through purely resistive elements. If $\mathbf{H}$ is obtained by means of a voltage divider tap on a precision noninductive resistor, then $\mathbf{H} = H$, a constant. If $\mathbf{G}H$ is very large at the frequency of interest, the basic gain equation becomes

$$\mathbf{G}_F = \frac{\mathbf{G}}{1 - \mathbf{G}H} \cong \frac{\mathbf{G}}{-\mathbf{G}H} = -\frac{1}{H} \tag{15-8}$$

and the gain is independent of frequency, temperature, or aging.

EXAMPLE 5

An amplifier whose high-frequency response is given by $\mathbf{G} = A_O/[1 + j(\omega/\omega_2)]$ is used in the basic feedback circuit (Fig. 15.6a). If $A_O = 1000$, $\omega_2 = 10^4$ rad/s, and $H = -0.009$, derive an expression for high-frequency response with feedback and predict the new upper cutoff frequency ω_{2F}.

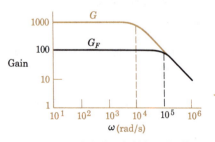

Gain

ω (rad/s)

Figure 15.10 Gain-bandwidth tradeoff.

The response is shown in Fig. 15.10. With feedback,

$$\mathbf{G}_F = \frac{\mathbf{G}}{1 - \mathbf{G}H} = \frac{\dfrac{1000}{1 + j(\omega/\omega_2)}}{1 + \dfrac{9}{1 + j(\omega/\omega_2)}}$$

$$= \frac{1000}{1 + j\left(\dfrac{\omega}{\omega_2}\right) + 9} = \frac{100}{1 + j\left(\dfrac{\omega}{10\omega_2}\right)} \tag{15-9}$$

In the midfrequency range, ω is small compared to $10\omega_2$ and

$$G_F \cong 100$$

At the new cutoff frequency, gain will be down by a factor of $\sqrt{2}$; therefore, $\omega_{2F}/10\omega_2 = 1$, or

$$\omega_{2F} = 10\omega_2 = 10(10^4) = 10^5 \text{ rad/s}$$

In Example 5, the midfrequency gain is decreased by a factor of $1 - GH$, but the bandwidth is increased by the same factor. As is usual in such cases, we trade gain for bandwidth.

Reducing Distortion

Distortion-free amplification is obtained easily at low signal levels. In the last power amplifier stage, however, efficient performance usually introduces nonlinear distortion because of the large signal swings. A common type of distortion results in the introduction of a second harmonic as shown in Fig. 12.1. (Also see Eq. 13-3.)

A nonlinear amplifier can be represented by the combination of linear blocks and rms signals shown in Fig. 15.11a with switch S open. The distortion component V_D

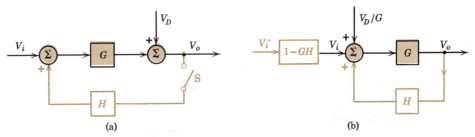

Figure 15.11   Block diagram representation of a distortion component.

is introduced in the output with no corresponding component in the input V_i. Without feedback, the output signal is

$$V_o = GV_i + V_D \tag{15-10}$$

and the distortion in percent of the fundamental is $(V_D/GV_i) \times 100$. Closing the switch S introduces feedback. In Fig. 15.11b, an equivalent distortion component V_D/G has been introduced ahead of block G. (This is allowed because with S open, $V_o = G(V_i + V_D/G) = GV_i + V_D$ as before.) Disregarding the $1 - GH$ block in color, now

$$V_o = \frac{G(V_i + V_D/G)}{1 - GH} = \frac{G}{1 - GH}V_i + \frac{V_D}{1 - GH}$$

If another stage of low-level, distortionless amplification with a gain of $1 - GH$ is added, the new output is

$$V_o = GV_i' + \frac{V_D}{1 - GH} \tag{15-11}$$

The output of the desired signal is still GV_i', but the new component of distortion has been reduced by the factor $1 - GH$. See Example 6 on p. 440.

EXAMPLE 6

The output of a power amplifier contains a 10-V, 1000-Hz fundamental and "20% second-harmonic distortion." The gain $G = 100\underline{/0°}$. Predict the effect of feedback on the distortion if $H = -0.19$.

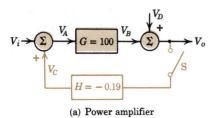

(a) Power amplifier

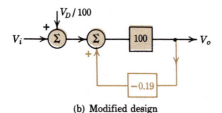

(b) Modified design

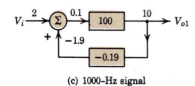

(c) 1000-Hz signal

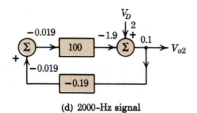

(d) 2000-Hz signal

Figure 15.12 Reducing distortion in a power amplifier by using negative feedback.

The block diagram for the system is as shown in Fig. 15.12a. Blocks G and H are linear models so the distortion component V_D is introduced as an extraneous signal not present in the input to the amplifier. By definition, 20% second-harmonic distortion indicates the presence of a 2000-Hz signal of magnitude

$$V_D = 0.20 V_{B1} = 0.20 \times 10 = 2 \text{ V}$$

To simplify the analysis, we shift the summing point for V_D in front of G and get the block diagram of Fig. 15.12b. With feedback, the gain is

$$G_F = \frac{G}{1 - GH} = \frac{100}{1 + 19} = 5$$

To retain the original power output, the input signal must be increased to

$$V_i = \frac{V_{o1}}{G_F} = \frac{10}{5} = 2 \text{ V at 1000 Hz}$$

The output is now G_F times the equivalent input or

$$V_o = G_F(V_i + V_D/100)$$

At 1000 Hz, $V_{o1} = G_F V_i = 5 \times 2 = 10$ V

At 2000 Hz, $V_{o2} = G_F V_D/100 = 5 \times 0.02 = 0.1$ V

The voltage distributions for fundamental and second-harmonic signals are shown in Figs. 15.12c and d. Since the blocks are linear, the superposition principle is applicable and the two signals can be handled separately.

The percentage of second-harmonic distortion is now

$$\frac{V_{o2}}{V_{o1}} \times 100 = \frac{0.1}{10} \times 100 = 1\% \text{ (a tolerable amount)}$$

The original amplifier required an input signal of $V_{o1}/100 = 0.1$ V; therefore, additional, distortionless, voltage gain of 2 V/0.1 V = 20 is necessary, but this is easily obtainable at these low levels.

As shown in Example 6, the relative amplitude of the distortion component compared to the desired component will be reduced by the factor $1 - GH$, which may be 10 to 100 or so.

Practice Problem 15-3

Assuming $V_i = 0$ and switch S is closed, redraw Fig. 15.12a as a block diagram with V_D as the input and V_o as the output. Calculate the gain V_o/V_D for this "distortion amplifier" and interpret the result.

Answer: $G_{FD} = 0.05$; distortion is reduced by a factor of 20.

Improving Impedance Characteristics

To minimize the "loading" effect of an electronic device such as an amplifier on a signal source, the input impedance of the device should be high. An FET with insulated gate is very good in this respect, but a BJT is not. Also, to provide efficient power transfer to the load, the output impedance of the device should be small. Neither the FET nor the BJT is optimum in this respect. In analyzing the voltage follower (p. 324), we learned that by using negative feedback, input and output impedances can be dramatically improved. A particularly useful form of feedback circuit is the FET *source follower* or its BJT counterpart, the *emitter follower*.

FOLLOWERS

In the circuit of Fig. 15.13a, the load resistor R_L is in series with the emitter and the collector is tied to V_{CC}, an ac ground. The output voltage[†] $v_o = v_E$ differs from the base voltage v_B by the base-emitter voltage $V_{BE} \cong 0.7$ V. Even for large swings of v_B, the voltage of the emitter with respect to ground "follows," hence the name *emitter follower*.

The Emitter Follower

To examine the impedance characteristics of the emitter follower at moderate frequencies, we replace the transistor by its small-signal model (Fig. 15.13b). Summing voltages around the input loop and solving for the input current,

$$i_b = \frac{v_s}{R_s + r_\pi + (\beta + 1)R_L} \tag{15-12}$$

The output voltage is

$$v_o = (\beta + 1)i_b R_L = \frac{(\beta + 1)R_L v_s}{R_s + r_\pi + (\beta + 1)R_L} \tag{15-13}$$

and the voltage gain in this feedback circuit is

$$G_F = \frac{v_o}{v_s} = \frac{(\beta + 1)R_L}{R_s + r_\pi + (\beta + 1)R_L} \cong 1 \tag{15-14}$$

[†] In this discussion we use lowercase letters to represent small signals. The relations hold equally well for effective values of sinusoidal signals.

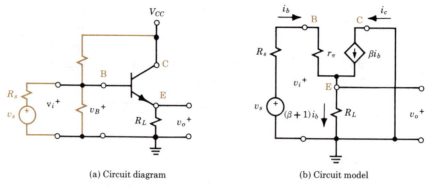

(a) Circuit diagram   (b) Circuit model

Figure 15.13 The emitter follower, a common-collector amplifier with feedback.

If β is large (50 to 150) and if R_L is of the same order of magnitude as $(R_s + r_\pi)$, then G_F approaches unity, but it can never exceed unity.

The virtue of an emitter follower is not in providing voltage gain. It does provide a current gain of $\beta + 1$ and a corresponding power gain, but its greatest usefulness is in providing impedance transformation.[†] From Fig. 15.13b,

$$v_i = i_b r_\pi + (i_b + i_c)R_L = i_b r_\pi + (\beta + 1)i_b R_L \qquad (15\text{-}15)$$

By definition, the input resistance is

$$R_i = \frac{v_i}{i_b} = r_\pi + (\beta + 1)R_L \cong (\beta + 1)R_L \qquad (15\text{-}16)$$

The effective output resistance can be evaluated from open- and short-circuit conditions. With the output terminals open-circuited,

$$v_{OC} = v_o \cong v_s$$

since the gain is approximately unity (Eq. 15-14). With the output terminals shorted (shorting R_L), $i_b = v_s/(R_s + r_\pi)$ and the short-circuit current is

$$i_{SC} = i_b + i_c = (\beta + 1)i_b = \frac{\beta + 1}{R_s + r_\pi} v_s$$

Therefore,

$$R_o = \frac{v_{OC}}{i_{SC}} = \frac{R_s + r_\pi}{\beta + 1} \qquad (15\text{-}17)$$

From Eqs. 15-16 and 15-17, it is clear that the emitter follower acts as an impedance transformer with a ratio of transformation approximately equal to $\beta + 1$. With a load resistance R_L, the resistance seen at the input terminals is $(\beta + 1)$ times as great. With the input circuit resistance $R_s + r_\pi$, the effective output resistance is only $1/(\beta + 1)$ times as large.

[†] In the equivalent circuit of the 741 (Fig. A9), transistors $Q_{1\text{-}6}$ constitute a two-stage differential amplifier (Fig. 11-20). Q_1 and Q_2 are input buffers; Q_3 and Q_4 are common-base amplifiers with active loads Q_5 and Q_6. Transistor Q_6 also serves as an inverting amplifier driving the base of Q_{16}. Transistors $Q_{8\text{-}13}$ provide bias voltages and temperature compensation. Q_{16} is an emitter follower driving a class B complementary amplifier (Fig. 12.23).

EXAMPLE 7

A 2N1893 transistor (r_π = 2.8 kΩ, β = 70) is used as an amplifier with R_s = 1.2 kΩ and R_L = 4 kΩ. Compare the input and output resistances in the common-emitter (CE) and common-collector (CC) configurations.

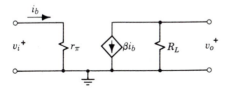

Figure 15.14 Calculation of input and output resistances.

Using the simplified circuit of Fig. 15.14, the input resistance in the common-emitter configuration is

$$R_i = \frac{v_i}{i_b} = r_\pi = 2.8 \text{ k}\Omega$$

For the common-collector circuit of Fig. 15.13a, however, the input resistance (Eq. 15-16) is

$$R_i = r_\pi + (\beta + 1)R_L = 2.8 + (70 + 1)4 = 287 \text{ k}\Omega$$

The CC input resistance is more than 100 times as great.

In the CE amplifier with R_L = 4 kΩ, R_o = 4000 Ω. For the CC amplifier,

$$R_o = \frac{R_s + r_\pi}{\beta + 1} = \frac{1200 + 2800}{70 + 1} = 56 \text{ }\Omega$$

The output resistance is reduced by a factor of 71.

The Source Follower

The common-drain amplifier or "source follower" is similar to the emitter follower in that it provides improved (lower) output impedance in return for reduced (essentially unity) gain. The input impedance, already high for an FET, can be made even higher by using the bias arrangement of Fig. 15.15 where $R_S = R_1 + R_2$. The quiescent operating point is defined by

$$V_{DS} = V_{DD} - I_D(R_1 + R_2) \tag{15-18}$$

where V_{DS} is made approximately half of V_{DD} to put operation near the midpoint of the load line. The gate-source bias,

$$V_{GS} = -I_D R_1 \tag{15-19}$$

is typically a few volts so that $R_1 \ll R_2$ and $R_S \cong R_2$.

To evaluate the performance of the source follower, we replace the FET by its small-signal model (Fig. 15.15b). In this circuit, $v_{gs} = v_i - v_o$; in feedback notation, $H = 1$ and $G = g_m R_S$. Neglecting the current in the very large R_G and any output current, $i_d = g_m v_{gs} = g_m(v_i - v_o) = v_o/R_S$. Solving,

$$v_o = \frac{g_m R_S v_i}{1 + g_m R_S} \tag{15-20}$$

and the voltage gain is

$$G_F = \frac{v_o}{v_i} = \frac{g_m R_S}{1 + g_m R_S} \tag{15-21}$$

which approaches unity in the typical case where $g_m R_S \gg 1$.

The effective output resistance can be evaluated from open- and short-circuited

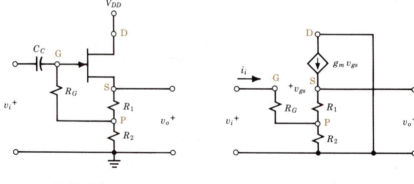

(a) Circuit diagram   (b) Small-signal model

Figure 15.15 The source follower or common-drain amplifier.

conditions. Since the open-circuit gain is approximately unity,

$$v_{OC} = v_o \cong v_i$$

With the output terminals shorted (shorting R_S),

$$i_{SC} = g_m v_{gs} = g_m(v_i - 0) = g_m v_i$$

Therefore, the effective output resistance is

$$R_o = \frac{v_{OC}}{i_{SC}} \cong \frac{v_i}{g_m v_i} \cong \frac{1}{g_m} \tag{15-22}$$

For a typical FET, the output resistance is several hundred ohms as compared to a common-source value (R_D) of several thousand ohms.

The higher input resistance resulting from the bias circuit of Fig. 15.15 can be predicted as follows. The ac input resistance is the ratio of an input voltage to the corresponding input current. Neglecting the effect of the very small current i_i on the voltage across R_2, the voltage at point P is

$$v_P = \frac{R_2}{R_1 + R_2} v_o = \frac{R_2}{R_1 + R_2} G_F v_i \tag{15-23}$$

Then the effective input resistance is

$$R_i = \frac{v_i}{i_i} = \frac{v_i}{(v_i - v_P)/R_G} = \frac{R_G}{1 - \dfrac{R_2}{R_1 + R_2} G_F} \tag{15-24}$$

If, say, $R_2 = 20R_1$ and $G_F \cong 1$, $R_i = 21R_G$; the input resistance is many megohms and the source follower presents a negligible load to the signal source. This characteristic makes it desirable as the input stage of an oscilloscope or a multistage instrumentation amplifier. (*Note:* At signal frequencies, the input capacitance $C_i \cong C_{gd}$ in parallel with R_i may determine the input impedance. See Problem 6.)

A source follower with very high input impedance followed by an emitter follower with very low output impedance provides an excellent buffer. In integrated-circuit form where active devices are cheap, sophisticated circuits provide nearly perfect buffering at low cost. (See footnote on p. 442.)

Practice Problem 15-4

Estimate the input and output resistances of the JFET amplifier in Example 1 of Chapter 14 at midfrequencies. Repeat for the same JFET used in the source follower of Fig. 15.15 with $R_G = 1$ MΩ, $R_1 = 100$ Ω, and $R_2 = 6.9$ kΩ. (*Hint:* Check G_F.)

Answers: 1 MΩ, 5000 Ω; 12.5 MΩ, 500 Ω.

SUMMARY

- A system is a combination of diverse but interacting elements integrated to achieve an overall objective.
  In a closed-loop system, a portion of the output is fed back and compared to or combined with the input. The gain of the basic closed-loop system is

$$G_F = \frac{G}{1 - GH}$$

- Complicated systems that can be analyzed into linear, unilateral, two-port elements are conveniently represented by simple diagrams consisting of blocks, summing points, and pickoff points.
  Block diagrams can be manipulated according to a set of simple rules.
- The quantity $1 - GH$ is the key to the behavior of closed-loop systems. For real values of G and H:
  If GH is positive and $0 < 1 - GH < 1$, positive feedback occurs and the system gain is greater than the forward gain.
  If $GH = 1$ and $1 - GH = 0$, the system gain increases without limit and oscillation is possible.
  If GH is negative and $1 - GH > 1$, negative feedback occurs and the system gain is less than the forward gain.
- An oscillator is a self-excited amplifier in which a portion of the output signal is fed back in the correct phase and amplitude to make $GH = 1 + j0$.
  By making G or H frequency dependent, oscillation is at a single frequency.
- The benefits of negative feedback outweigh the loss in gain:
  Gain can be made insensitive to device parameters.
  Operating point can be made insensitive to changes.
  Gain can be made insensitive to frequency.
  Noise and distortion can be discriminated against.
  Input and output impedances can be improved.
- In emitter (source) followers, the entire output is fed back into the input.
  Gain is less than unity but input and output impedances are optimized.
  The emitter follower acts as an impedance transformer with a ratio of $\beta + 1$.
  The source follower offers similar advantages.

REVIEW QUESTIONS

1. Cite an example of feedback in a biological system and describe its operation.
2. Cite an example of feedback in a social system and describe its operation.
3. Cite an example of feedback in each of the following branches of engineering: aeronautical, chemical, civil, industrial, and mechanical.
4. If the speaker of a public address system is moved farther away from the microphone, what is the effect on the possibility of a "squeal," the frequency of the squeal if it does develop, and the amplitude of the squeal?
5. Differentiate between the two general categories of feedback applications in electrical systems described in the introduction to this chapter.
6. From memory, draw and label a basic feedback system and write the general expression for overall gain.
7. What is the physical meaning for $GH = 1$? What is the ac *power* gain around the closed loop for this situation?
8. List three conditions that must be met if a system is to be represented by a block diagram.
9. Differentiate between a summing point and a pickoff point.
10. Write out the corollaries to block diagram algebra rules 4 and 5.
11. Draw a complicated system with two pickoff points and two summing points. Using the rules of block diagram algebra, reduce the system to an open-loop equivalent.
12. Define positive feedback and negative feedback. Give an example of each.
13. Explain how a multistage audio amplifier may become an oscillator at a certain frequency.
14. List five specific benefits that may result from negative feedback and indicate where they would be important.
15. List three reasons why the gain of a transistor amplifier without feedback might change in normal operation.
16. Is oscillation possible in an ordinary two-stage *RC*-coupled audio amplifier? Explain.
17. A portable high-gain amplifier tends to become oscillatory as the batteries age. It is known that the internal resistance of batteries increases with age. Explain the connection between these phenomena.
18. Explain the expression "trading gain for bandwidth."
19. Draw a block diagram to represent a "noisy" amplifier in which noise (perhaps an objectionable hum) appears in the output without a corresponding input. Suggest a means for reducing such noise.
20. Explain how an amplifier can be made insensitive to the effect of increased temperatures on transistor parameters.
21. Define the impedance characteristics of an ideal amplifier and explain your answer.
22. Why is a common-drain amplifier called a source follower?
23. Cite two virtues of the source follower.
24. Explain how an emitter follower functions as an impedance transformer. Where in a "hi-fi" set would this property be valuable?

EXERCISES

1. In Fig. 15.1b, the forward gain of the amplifier is $1000 \underline{/180°}$, $R_1 = 10$ kΩ, and $R_2 = 190$ kΩ.
   (a) For $\mathbf{V}_1 = 0.01 \underline{/0°}$ V, calculate the output voltage, the feedback factor, and the voltage fed back.
   (b) For a source voltage of $0.51 \underline{/0°}$ V, calculate $\mathbf{V}_1$ and the output voltage and the ratio of output to source voltage.
   (c) Calculate the gain with feedback from the

forward gain and the feedback factor, and compare it to the result of part (b).

2. A driver on a freeway wishes to keep his left fender two feet from the white line. Draw a labeled block diagram for the closed-loop system he uses.
3. In the block diagram of Fig. 15.16:
   (a) Replace the subloop containing A by an equivalent block A'.

(b) Combine A' with B and redraw the diagram.
(c) Determine the overall gain I_3/I_1.

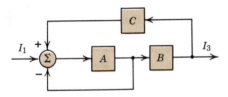

Figure 15.16

4. Reduce the block diagram of Fig. 15.16 to the basic feedback system of Fig. 15.2 by using block diagram algebra. Indicate the values of G and H. Determine the gain of the system.
5. In Fig. 15.17, $A = 10$, $B = 100$, and $C = 0.09$. Evaluate V_o in terms of V_1 and V_2. If V_2 is an undesired input, describe the operation of this system.
6. Given a system described by the equations

$$V_5 = BV_3 \qquad V_3 = AV_2$$
$$V_2 = V_1 - V_4 \qquad V_4 = CV_3$$

where V_1 is the input and V_5 is the output:
(a) Identify gain-block and summing-point relations, and draw a labeled block diagram representing this system.
(b) Determine the gain V_5/V_1.

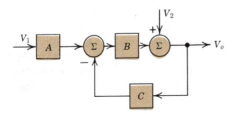

Figure 15.17

7. An amplifier with a forward gain of $100\underline{/180°}$ is provided with a feedback loop. Predict the gain with feedback and the category of feedback if the feedback factor is (a) $0.09\underline{/0°}$, (b) $0.009\underline{/180°}$, or (c) $0.01\underline{/180°}$.
8. For the amplifier circuit of Fig. 15.18 with the switches open, draw a small-signal midfrequency circuit model. Show switch S_1 on your circuit.
(a) Note the polarities of ac voltages and determine the voltage feedback factor H.
(b) Determine gain G with S_1 closed.
(c) Draw a block diagram for this circuit.

(d) Derive an expression for overall midfrequency gain from the circuit model.
(e) Repeat part (d) working from the block diagram and compare results.

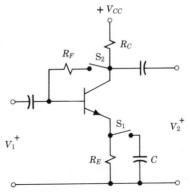

Figure 15.18

9. Analyze the amplifier of Exercise 8 with S_1 closed.
(a) Determine G with S_2 open and H with S_2 closed.
(b) From a block diagram, predict the gain with feedback.
(c) From a circuit model, predict the gain and compare to the result of part (b).
10. A feedback system is represented by Fig. 15.2 with $\mathbf{H} = 0.005\underline{/0°}$. Define the conditions necessary for oscillation.
11. A feedback system is represented by Fig. 15.16 with $\mathbf{A} = 50\underline{/90°}$ and $\mathbf{B} = 20\underline{/90°}$. Determine the value of $\mathbf{C}$ necessary for oscillation.
12. For frequencies below $\omega = 2000$ rad/s, the gain of an amplifier is described by $\mathbf{G} = 0.1\omega\underline{/180°}$. For oscillation at $\omega = 1500$ rad/s, what value of $\mathbf{H}$ is required?
13. For the circuit of Fig. 15.6b:
(a) Determine the frequency of oscillation in words and symbols.
(b) If $C = 200$ pF and $L = 25$ mH, estimate the oscillation frequency.
14. In the variable-frequency phase-shift oscillator of Fig. 15.7, $R = 1$ MΩ and $R_D = 10$ kΩ.
(a) Estimate the approximate value of C for oscillation at 20 Hz; state any simplifying assumptions made.
(b) If, instead, oscillation at 20 Hz is to be determined by a resonant circuit with $L = 1$ H, approximately what value of C is required? Is this a reasonable value for a variable capacitor?

15. An electronic amplifier with other desirable properties provides a gain G of only 5.
    (a) Design a network to provide positive feedback and an overall gain of 100.
    (b) If G increases by 2%, say, what is the new G_F? Comment on the stability of this network.

16. Draw the block diagram representation of a noninverting amplifier.
    (a) Assuming $A = 10^5$ and $R_1 = 10$ kΩ, design a negative feedback amplifier to provide an overall gain of 100.
    (b) If A increases by 2%, what is the new G_F? Compare this result to the result of Exercise 15(b).

17. Consider the basic feedback system of Fig. 15.2.
    (a) Obtain an expression for dG_F, the change in G_F corresponding to a small change in gain G.
    (b) Derive a general expression for the fractional change in overall gain with feedback (dG_F/G_F) in terms of the fractional change in forward gain (dG/G) and the values of G and H.

18. A cheap amplifier ordered from a catalog is supposed to have a voltage gain of 100, but when delivered proves to have a gain of 70. In fact, when first turned on it has a gain of only 50, which rises to 70 over the course of a minute or two. A second amplifier of the same type shows an initial gain of 90, rising to 110 during warmup.

    Design and diagram an arrangement to obtain a fairly stable gain of 100 by cascading the two amplifiers. Specify the feedback factor. Predict the percentage change in gain of your amplifier system during warmup.

19. Replace the transistor in Fig. 15.9a by an appropriate model and redraw the circuit for dc analysis with $V_{BB} = 2.9$ V, $R_B = 46$ kΩ, $R_E = 4$ kΩ, $R_C = 8$ kΩ, and $V_{CC} = 10$ V. β is around 100.
    (a) Derive a literal expression for I_C and estimate the value.
    (b) Estimate the change in I_C for a 30% increase in β.
    (c) Remove R_E, adjust the value of V_{BB} to provide the same I_B, and repeat part (b).
    (d) Explain in words the stabilizing effect of R_E.

20. The high-frequency response of an amplifier is defined by $A = A_O/(1 + j\omega/\omega_2)$ where $A_O = $ 100 and $\omega_2 = 10$ krad/s.
    (a) Plot a couple of points and sketch a graph of A/A_O versus ω on a logarithmic scale.
    (b) Assume that a feedback loop with $H = -0.04$ is provided and repeat part (a). Label this graph "With feedback."

21. The low-frequency response of an amplifier is defined by $A = A_O/(1 - j\omega_1/\omega)$ where $\omega_1 = 200$ rad/s and $A_O = 200$.
    (a) Plot a couple of points and sketch a graph of A/A_O versus ω with ω on a logarithmic scale.
    (b) Assume that a feedback loop with $H = -0.045$ is provided and repeat part (a). Label this graph "With feedback."

22. An amplifier has a midfrequency gain of 100, but the gain is only 50 at 10 Hz. Devise a feedback system and specify a real value of H so that the gain at 10 Hz is within 5% of the new midfrequency gain.

23. In Fig. 15.12a, consider V_D to be the input and V_o the output and redraw the block diagram. Calculate the gain V_o/V_D for this system and interpret the result.

24. An audiofrequency amplifier with a voltage gain of 200 has an output of 5 V, but superimposed on the output is an annoying 60-Hz hum of magnitude 0.5 V.
    (a) Devise a feedback system and specify H so that with the same 5-V output, the hum component is reduced to 0.025 V.
    (b) Draw separate block diagrams for the audio signal and the hum component, showing the actual voltages at each point in the system.

25. A class A power amplifier (see Fig. 12.21a) using a 2N3114 is to supply 70 V (rms) of signal across a 10,000-Ω load resistance.
    (a) How much ac power is supplied? To get this much power, the transistor must be driven into the nonlinear regions. It is estimated that the major element of distortion is a 7-V (rms) second harmonic. Express this as percent distortion.
    (b) The source for the amplifier is a phono pickup with an output of 6 mV and a source resistance of 1500 Ω. How many amplifier stages (operating at $I_C = 1$ mA) of what current gain will be required to drive the final amplifier?
    (c) It is suggested that the distortion can be reduced by providing feedback around the

power amplifier. Specify the feedback factor necessary to reduce the distortion to 1%.

(d) To maintain the required output voltage, how must the intermediate amplifier be modified?

26. An amplifier is represented by the general model of Fig. 14.2 with Rs in place of Zs. Reproduce the circuit and add negative voltage feedback with a feedback factor H. Derive expressions for G_F, R_{oF}, and R_{iF}, the new gain, output resistance, and input resistance. Draw a general conclusion regarding the effect of $1 - GH$.

27. An emitter follower employs a 2N699B at $I_C = 1$ mA supplied by a source for which $V_s = 2$ V and $R_s = 2$ kΩ. For an emitter resistance $R_L = 1$ kΩ, predict the output voltage and the input and output resistances.

28. A device with a high output resistance ($R_s = 50$ kΩ) is to be coupled to a load resistance R_L by

means of a 2N3114 transistor.

(a) Devise a circuit and specify R_L so that the input resistance seen by the source is approximately 50 kΩ.

(b) Predict the output resistance of the transistor under these circumstances.

29. The input to a device can be represented by $R = 100$ kΩ in parallel with $C = 100$ pF. It is fed by an FET for which $g_m = 2$ mS. Predict the upper cutoff frequency if the FET is used as:

(a) An ordinary amplifier with $R_D = 5$ kΩ.

(b) A source follower with $R_S = 5$ kΩ.

30. Modify the circuit of Fig. 15.15a so that V_i is applied (through C_C) between the gate and source terminals. Draw the small-signal model and express V_o/V_i in terms of the circuit parameters. Use this amplifier as one block in a block diagram representation of Eq. 15-21. What is the feedback factor H?

PROBLEMS

1. Given a system described by the equations

$$V_2 = V_1 - I_1 R_1 \qquad I_1 = I_2 + I_3$$

$$V_2 = I_2 R_2 = I_3 R_3$$

where V_1 is the input, I_3 is the output, and each R is a block:

(a) Draw a block diagram representing this system. (How many summing points?)

(b) Draw the corresponding electrical circuit. (What do the first two equations represent in an electrical circuit?)

(c) Calculate I_3/V_1 from the block diagram (by first reducing it to the basic system of Fig. 15.2) and then from the circuit.

2. An amplifier with a gain $G_1 = 200$ is subject to a 20% change in gain due to part replacement. Design an amplifier (in block diagram form) that provides the same overall gain but in which similar part replacement will produce a change in gain of only 0.5%. Use amplifier G_1 as part of your design.

3. In the amplifying system of Fig. 15.19, $A = -1000$ and $H = 0.03$, both independent of frequency, $R = 10$ kΩ, and $C = 1.59$ nF.

(a) Determine the upper cutoff frequency (corner frequency) of the system with switch S open.

(b) Repeat part (a) with S closed.

(c) For two identical systems with S closed in cascade, predict the overall gain and the new upper cutoff frequency. Comment on the effect of feedback.

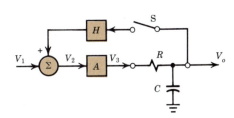

Figure 15.19

4. Ten watts of audio power is to be supplied to a 250-Ω load.

(a) What output voltage (rms) and current are required? For this voltage swing, the transistor must be driven into nonlinear regions resulting in distortion in the form of a 5-V (rms) second harmonic. Express this as "% distortion."

(b) The available phono pickup has an output of 1 mV. If the final power amplifier has a voltage gain of 50, specify the number and gain of the necessary voltage amplifiers.

(c) Show how distortion can be reduced by providing feedback around the final amplifier and specify the feedback factor to reduce the distortion to 1%. Specify the changes necessary in the voltage amplifiers to maintain the same output voltage.

5. An innovative experimenter disconnects C_E in Fig. 14.24 and finds the midfrequency voltage gain V_o/V_i greatly reduced. He expects that there may be some offsetting improvement and asks you to analyze the performance. Derive an expression for V_o/V_i and predict the effect of disconnecting C_E if $\beta = 100$, $R_C = 5$ kΩ, $R_E = 1$ kΩ, and $I_C = 1$ mA. Investigate the new input impedance and draw a conclusion.

6. A JFET for which $I_{DSS} = 5$ mA, $V_p = -3$ V, $C_{gs} = 20$ pF, and $C_{gd} = 2$ pF is used as a source follower in the circuit of Fig. 15.15a with $R_G = 5$ MΩ and $V_{DD} = 20$ V. For quiescent operation at $I_D = 2$ mA and $V_{DS} = 10$ V, design the circuit—specify R_1 and R_2. Predict the voltage gain, the output resistance, and the input impedance at 10 kHz.

7. The "hybrid" circuit shown in Fig. 15.20 provides a very high input impedance and a very low output impedance.

(a) Describe the operation of the circuit in a qualitative way.
(b) Replace the active devices by simple small-signal models and redraw the circuit for convenient analysis.
(c) For $R_1 = 10$ MΩ, $g_{m1} = 3 \times 10^{-3}$ S, $I_C = 1$ mA, $\beta_2 = 100$, and $R_3 = 3$ kΩ, predict the voltage gain V_3/V_1, R_i, and R_o.

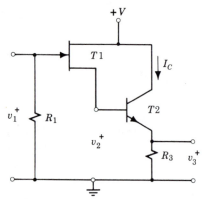

Figure 15.20

Appendix

Physical Constants

Conversion Factors

Complex Algebra

Color Code for Resistors

Preferred Resistance Values

Representative TTL IC Units

Device Characteristics

PHYSICAL CONSTANTS[†]

Charge of the electron	e	1.602×10^{-19} C
Mass of the electron (rest)	m	9.109×10^{-31} kg
Velocity of light	c	2.998×10^{8} m/s
Acceleration of gravity	g	9.807 m/s^2
Avogadro's number	N_A	6.023×10^{23} atoms/g · atom
Planck's constant	h	6.626×10^{-34} J · s
Boltzmann's constant	k	8.62×10^{-5} eV/K
Permeability of free space	μ_o	$4\pi \times 10^{-7}$ H/m
Permittivity of free space	ϵ_o	$1/36\pi \times 10^{9}$ F/m

CONVERSION FACTORS

1 inch	$= 2.54$ cm	1 hertz	$= 1$ cycle/s
1 meter	$= 39.37$ in.	1 electron volt	$= 1.60 \times 10^{-19}$ J
1 angstrom unit	$= 10^{-10}$ m	1 joule	$= 10^{7}$ ergs
1 kilogram	$= 2.205$ lb (mass)	1 tesla	$= 1$ Wb/m^2
1 newton	$= 0.2248$ lb (force)		$= 10^{4}$ gauss
1 horsepower	$= 746$ W		

[†]E.A. Mechtly, "The International System of Units: Physical Constants and Conversion Factors," NASA SP-7012, Washington, D.C., 1964.

451

COMPLEX ALGEBRA

In our numbering system, *positive* numbers correspond to distances measured along a line, starting from an origin. *Negative* numbers enable us to solve equations like $x + 2 = 0$, and they are represented in Fig. A1 by distances to the left of the origin. *Zero* is a relatively recent concept with some peculiar characteristics; for example, division by zero must be handled carefully. Numbers corresponding to distances along the line of Fig. A1 are called *real* numbers. To solve equations like $x^2 + 4 = 0$, so-called *imaginary* numbers were invented; these add a new dimension to our numbering system.

Figure A1   The system of real numbers.

Imaginary Numbers

Imaginary numbers are plotted along the *imaginary axis* of Fig. A2 so that $j2$ lies at a distance of 2 units along an axis at right angles to the *real axis*. The imaginary number $-j2$ lies along the imaginary axis but in the opposite direction from the origin. Note that all numbers are abstractions, and calling one number "real" and another "imaginary" just differentiates between their mathematical properties.

It may be helpful to think of j as an *operator*. An operator like "$\sqrt{\ }$" or "sin" means a particular mathematical operation; $\sqrt{\ }$ operating on 4 yields 2 and sin operating on 30° yields 0.5. In the same way, the minus sign $(-)$ is an operator such that $(-)$ operating on a directed line segment yields a reversal or a 180° rotation. Similarly, we define the operator j to mean a 90° counterclockwise rotation. The operator j taken twice, or $(j)(j)$, indicates a 180° rotation or a reversal, and $(j)(j)$ is equivalent to $(-)$ in effect. This concept is important because it provides an algebraic interpretation for a geometrical operation.

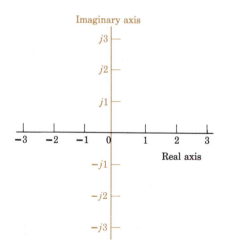

Figure A2   The complex plane.

Just as $\sin^2 \theta$ is defined to mean $(\sin \theta)^2$ and not $\sin (\sin \theta)$, so the properties of jb must be defined. It is understood that

$$j(jb) = j^2 b = -b \qquad \text{or we say} \qquad j^2 = -1$$

$$j(j^2 b) = j^3 b = -jb \qquad \text{or we say} \qquad j^3 = -j$$

$$j(j^3 b) = j^4 b = +b \qquad \text{or we say} \qquad j^4 = +1$$

The equation $j^2 = -1$ is a statement that the two *operations* are equivalent and does *not* imply that j is a number. However, and this is the beauty of the concept, in all algebraic computations imaginary numbers can be handled as if j had a numerical value of $\sqrt{-1}$.

Complex Numbers

The real and imaginary axes define the *complex plane*. The combination of a real number and an imaginary number defines a point in the complex plane and also defines a *complex number*. The complex number may be considered to be the point or the directed line segment to the point; both interpretations are useful.

The complex number **W** of magnitude M and direction θ in Fig. 10.2 on p. 278 can be expressed in rectangular form or in polar form where

$$\mathbf{W} = a + jb = M(\cos \theta + j \sin \theta) = M e^{j\theta} = M\underline{/\theta}$$

The conversion from one form to another is facilitated by Euler's theorem which states that $\cos \theta + j \sin \theta = e^{j\theta}$.

The validity of Euler's theorem is evident when series expansions are written as follows:

$$e^\theta = 1 + \theta + \frac{\theta^2}{2!} + \frac{\theta^3}{3!} + \frac{\theta^4}{4!} + \frac{\theta^5}{5!} + \cdots$$

$$\cos \theta = 1 - \frac{\theta^2}{2!} + \frac{\theta^4}{4!} - \cdots \qquad \text{and} \qquad \sin \theta = \theta - \frac{\theta^3}{3!} + \frac{\theta^5}{5!} - \frac{\theta^7}{7!} + \cdots$$

Then

$$\cos \theta + j \sin \theta = 1 + j\theta - \frac{\theta^2}{2!} - j\frac{\theta^3}{3!} + \frac{\theta^4}{4!} + j\frac{\theta^5}{5!} - \cdots = e^{j\theta}$$

Rules of Complex Algebra

Operations such as addition and subtraction, multiplication and division, raising to powers, and extracting roots are performed easily if the complex numbers are in the most convenient form. The following examples indicate the rules of complex algebra as applied to the two complex numbers

$$\mathbf{A} = a + jb = A e^{j\alpha} = A\underline{/\alpha} \qquad \text{(read: ``A at angle alpha'')}$$

and

$$\mathbf{C} = c + jd = C e^{j\gamma} = C\underline{/\gamma} \qquad \text{(read: ``C at angle gamma'')}$$

Equality

Two complex numbers are equal if and only if the real parts are equal and the imaginary parts are equal.

$$\text{If } \mathbf{A} = \mathbf{C}, \text{ then } a = c \qquad \text{and} \qquad b = d$$

EXAMPLE 1

The voltage phasors across two branches in parallel (Fig. A3) are $\mathbf{V}_1 = 2 + jy$ and $\mathbf{V}_2 = V\underline{/60°}$. Find y and V.

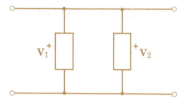

Figure A3 Phasor equality.

For branches in parallel, the voltages are equal and therefore the voltage phasors must be equal; hence

$$2 + jy = V\underline{/60°}$$

This appears to be a single equation with two unknowns, but applying the rule for equality,

$$2 = V \cos 60° \qquad \text{and} \qquad y = V \sin 60°$$

Therefore,

$$V = \frac{2}{\cos 60°} = 4$$

and

$$y = 4 \sin 60° = 2\sqrt{3}$$

Addition

Two complex numbers in rectangular form are added by adding the real parts and the imaginary parts separately.

$$\mathbf{A} + \mathbf{C} = (a + c) + j(b + d)$$

Subtraction can be considered as addition of the negative,

$$\mathbf{A} - \mathbf{C} = (a - c) + j(b - d)$$

Addition and subtraction are not convenient in the polar form.

EXAMPLE 2

In Fig. A4, $i_2(t) = 2\sqrt{2} \cos(\omega t - 90°)$ and $\mathbf{I}_1 = 3 - j4$. Find $i_3(t)$.

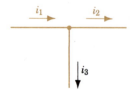

Figure A4 Phasor subtraction.

By inspection we write $\mathbf{I}_2 = 2\underline{/-90°}$ but, since the polar form is not convenient, we convert to rectangular form as $\mathbf{I}_2 = 0 - j2$. Then

$$\mathbf{I}_3 = \mathbf{I}_1 - \mathbf{I}_2 = (3 - j4) - (0 - j2)$$

$$= 3 - j4 + j2 = 3 - j2$$

In polar form,

$$\mathbf{I}_3 = \sqrt{3^2 + 2^2}\,\underline{/\arctan \tfrac{-2}{+3}} = 3.6\underline{/-33.7°}$$

$$\therefore i_3(t) = 3.6\sqrt{2} \cos(\omega t - 33.7°)$$

Multiplication

The magnitude of the product is the product of the individual magnitudes, and the angle of the product is the sum of the individual angles.

$$(\mathbf{A})(\mathbf{C}) = (A\,e^{j\alpha})(C\,e^{j\gamma}) = AC\,e^{j(\alpha+\gamma)} = AC\underline{/\alpha + \gamma}$$

Multiplication of complex numbers in polar form follows the law of exponents. If the numbers are given in rectangular form and the product is desired in rectangular form, it may be more convenient to perform the multiplication directly, employing the concept that $j = \sqrt{-1}$ and $j^2 = -1$.

$$\mathbf{(A)(C)} = (a + jb)(c + jd)$$

$$= ac + j^2bd + jbc + ajd = (ac - bd) + j(bc + ad)$$

EXAMPLE 3

In a certain alternating-current circuit, $\mathbf{V} = \mathbf{ZI}$ where

$$\mathbf{Z} = 7.07\underline{/-45°} = 5 - j5$$

and

$$\mathbf{I} = 10\underline{/+90°} = 0 + j10.$$

Find $\mathbf{V}$.

In polar form,

$$\mathbf{V} = (7.07\underline{/-45°})(10\underline{/+90°})$$

$$= 7.07 \times 10\underline{/-45° + 90°} = 70.7\underline{/+45°}$$

In rectangular form,

$$\mathbf{V} = (5 - j5)(0 + j10)$$

$$= (5 \times 0 + 5 \times 10) + j(-5 \times 0 + 5 \times 10)$$

$$= 50 + j50$$

Division

The magnitude of the quotient is the quotient of the magnitudes, and the angle of the quotient is the difference of the angles.

$$\frac{\mathbf{A}}{\mathbf{C}} = \frac{A\,e^{j\alpha}}{C\,e^{j\gamma}} = \frac{A\underline{/\alpha}}{C\underline{/\gamma}} = \frac{A}{C}\underline{/\alpha - \gamma}$$

Division of complex numbers in polar form follows the laws of exponents.

Division in rectangular form is inconvenient but possible by using the *complex conjugate* of the denominator. Given a complex number $\mathbf{C} = c + jd$, the complex conjugate of $\mathbf{C}$ is defined as $\mathbf{C}^* = c - jd$; that is, the sign of the imaginary part is reversed. (Here the asterisk means "the complex conjugate of . . . .") Multiplying the denominator by its complex conjugate *rationalizes* the denominator (i.e., converts it to a real number) and simplifies division. To preserve the value of the quotient, the numerator is multiplied by the same factor, as in Example 4.

EXAMPLE 4

In a certain alternating-current circuit, $V = ZI$ where

$$V = 130\underline{/-67.4°} = 50 - j120$$

and

$$Z = 5\underline{/53.1°} = 3 + j4$$

Find I.

In polar form,

$$I = \frac{V}{Z} = \frac{130\underline{/-67.4°}}{5\underline{/53.1°}} = 26\underline{/-120.5°}$$

In rectangular form,

$$I = \frac{V}{Z} = \frac{50 - j120}{3 + j4} \cdot \frac{3 - j4}{3 - j4}$$

$$= \frac{(150 - 480) + j(-360 - 200)}{9 + 16}$$

$$= -13.2 - j22.4 = 26\underline{/-120.5°}$$

Powers and Roots. Applying the laws of exponents to a complex number in polar form,

$$A^n = (A\,e^{j\alpha})^n = A^n\,e^{jn\alpha} = A^n\underline{/n\alpha}$$

The same result is obtained from the rule for multiplication by recognizing that raising to the nth power is equal to multiplying a number by itself n times.

Extracting the nth root of a number is equivalent to raising the number to the $1/n$th power and the same rule is applicable:

$$A^{1/n} = (A\,e^{j\alpha})^{1/n} = A^{1/n}\,e^{j\alpha/n} = A^{1/n}\underline{/\alpha/n}$$

The method for finding all n roots is illustrated in Example 5.

EXAMPLE 5

Perform the operations indicated at the right. Recalling that the number of distinct roots is equal to the order of the root (if n is an integer), locate the other roots in (b).

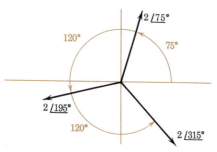

Figure A5 Roots of a complex number.

(a) $(8.66 - j5)^3 = (10\underline{/-30°})^3$

$$= 1000\underline{/-90°} = 0 - j1000$$

(b) $(-5.66 - j5.66)^{1/3} = (8\underline{/+225°})^{1/3} = 2\underline{/+75°}$

Since $8\underline{/225°} = 8\underline{/225° + N360°}$ where N is any integer, three distinct roots are obtained by assigning values of $N = 0, 1, 2$.

For $N = 1$, $(8\underline{/225° + 360°})^{1/3} = 2\underline{/75° + 120°}$

For $N = 2$, $(8\underline{/225° + 720°})^{1/3} = 2\underline{/75° + 240°}$

The three cube roots are found to be complex numbers of the same magnitude but differing in angle by $360°/3 = 120°$. (See Fig. A5.)

COLOR CODE FOR RESISTORS

Ordinary composition fixed resistors are characterized by power dissipation in watts, resistance in ohms, and tolerance in percent. They vary in size from 0.067 in. (0.1 W) to 0.318 in. (2 W) in diameter. For clarity, the resistance and tolerance values are indicated by color bands according to a standard code. (For example, bands of gray, black, orange, and silver would identify an 80×10^3 or 80-kΩ, 5%-tolerance resistor.)

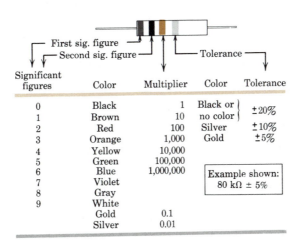

Significant figures	Color	Multiplier	Color	Tolerance
0	Black	1	Black or no color	±20%
1	Brown	10		
2	Red	100	Silver	±10%
3	Orange	1,000	Gold	±5%
4	Yellow	10,000		
5	Green	100,000		
6	Blue	1,000,000		
7	Violet			
8	Gray			
9	White			
	Gold	0.1		
	Silver	0.01		

First sig. figure —
Second sig. figure —
Tolerance —

Example shown: 80 kΩ ± 5%

A similar color scheme is used to indicate the value and tolerance of small capacitors; an extra color band indicates the temperature coefficient.

PREFERRED RESISTANCE VALUES

To provide the necessary variation in resistance while holding the number of resistors to be stocked to a reasonable minimum, the electronics industry has standardized on certain preferred values. The range in significant figures from 1.0 to 10 has been divided into 24 steps, each differing from the next by approximately 10% ($\sqrt[24]{10} = 1.10$). The preferred values, available in all multiples in ±5% tolerance resistors, are:

1.0[†]	1.5[†]	2.2[†]	3.3[†]	4.7[†]	6.8[†]
1.1	1.6	2.4	3.6	5.1	7.5
1.2[‡]	1.8[‡]	2.7[‡]	3.9[‡]	5.6[‡]	8.2[‡]
1.3	2.0	3.0	4.3	6.2	9.1

[†]Also available in ±10% and ±20% tolerance resistors.

[‡]Also available in ±10% tolerance resistors.

REPRESENTATIVE TTL IC UNITS

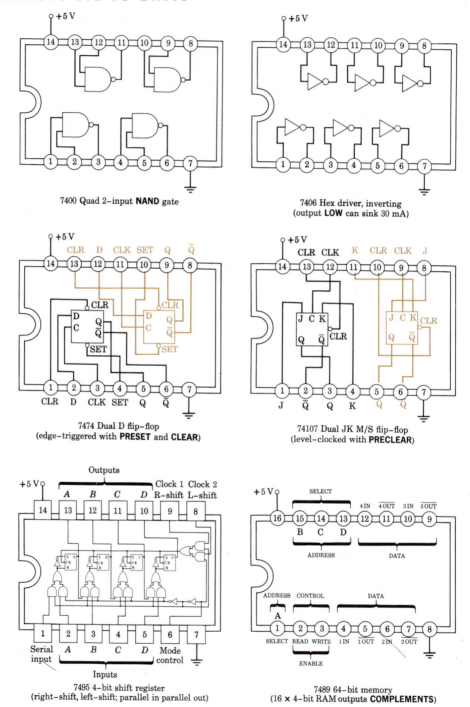

7400 Quad 2–input **NAND** gate

7406 Hex driver, inverting
(output **LOW** can sink 30 mA)

7474 Dual D flip–flop
(edge–triggered with **PRESET** and **CLEAR**)

74107 Dual JK M/S flip–flop
(level–clocked with **PRECLEAR**)

7495 4–bit shift register
(right–shift, left–shift; parallel in parallel out)

7489 64–bit memory
(16 × 4–bit RAM outputs **COMPLEMENTS**)

Figure A6 Some popular TTL integrated circuits.

DEVICE CHARACTERISTICS

2N4416 *n*-Channel Junction Field-Effect Transistor for VHF Amplifiers and Mixers

Maximum Ratings

Gate-Drain or Gate-Source voltage	−30 V
Gate current	10 mA
Total power dissipation	300 mW

Electrical Characteristics	**Min.**	**Max.**
I_{GSS} Gate reverse current		−0.1 nA
I_{GSS} (150°C)		−0.1 μA
BV_{GSS} Gate-Source breakdown voltage	−30 V	
$V_{GS(off)}$ Gate-Source cutoff voltage	−2.5	−6.0 V
I_{DSS} Saturation drain current	5	15 mA

Common-Source Parameters (V_{DS} = 15 V, V_{GS} = 0 V)		
g_{fs} Forward transconductance (= g_{mo})	4500	7500 μS
g_{fs} (400 MHz)		4000 μS
g_{os} Output conductance (= $1/r_d$)		50 μS
C_{rss} Reverse transfer capacitance (= C_{gd})		0.8 pF
C_{iss} Input capacitance (= $C_{gd} + C_{gs}$)		4 pF
C_{oss} Output capacitance (= C_{ds})		2 pF

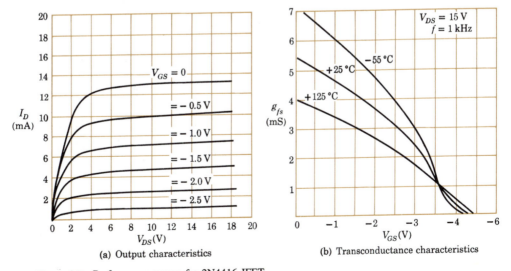

(a) Output characteristics

(b) Transconductance characteristics

Figure A7   Performance curves for 2N4416 JFET.

2N3114 *NPN* High-Voltage, Medium-Power Silicon Planar[†] Transistor

Maximum Ratings		Typical Small-Signal Characteristics $(f = 1\text{ kHz}, I_C = 1\text{ mA}, V_{CE} = 5\text{ V})$		
Operating junction temperature	200°C	h_{fe}	Small-signal current gain	50
Power dissipation: 25°C case	5 W	h_{ie}	Input resistance	1,500 Ω
25°C free air	0.8 W	h_{oe}	Output conductance	5.3 μS
V_{CBO} Collector-to-Base voltage	150 V	h_{re}	Voltage feedback ratio	1.5×10^{-4}
V_{EBO} Emitter-to-Base voltage	5 V	h_{ib}	Input resistance	27 Ω
		h_{ob}	Output conductance	0.09 μS
		h_{rb}	Voltage feedback ratio	0.25×10^{-4}

Electrical Characteristics (25°C free air)

Symbol	Characteristic	Min.	Typical	Max.	Test conditions
$V_{BE(\text{sat})}$	Base saturation voltage		0.8	0.9 V	$I_C = 50\text{ mA}$ $I_B = 5.0\text{ mA}$
$V_{CE(\text{sat})}$	Collector saturation voltage		0.3	1 V	$I_C = 50\text{ mA}$ $I_B = 5.0\text{ mA}$
I_{CBO}	Collector cutoff current		0.3	10 nA	$I_E = 0$ $V_{CB} = 100\text{ V}$
$I_{CBO}(150°C)$	Collector cutoff current		2.7	10 μA	$I_E = 0$ $V_{CB} = 100\text{ V}$
$h_{FE}(\beta_{\text{dc}})$	DC current gain	15	35		$I_C = 100\text{ }\mu\text{A}$ $V_{CE} = 10\text{ V}$
h_{fe}	20-MHz current gain	2.0	2.7		$I_C = 30\text{ mA}$ $V_{CE} = 10\text{ V}$
C_{ob}	Output capacitance		6	9 pF	$I_E = 0$ $V_{CB} = 20\text{ V}$
C_{TE}	Emitter transition capacitance		70	80 pF	$I_C = 0$ $V_{EB} = 0.5\text{ V}$

[†]Planar is a patented Fairchild process.

Figure A8 Typical characteristics of the 2N3114 transistor. (Courtesy Fairchild Semiconductor)

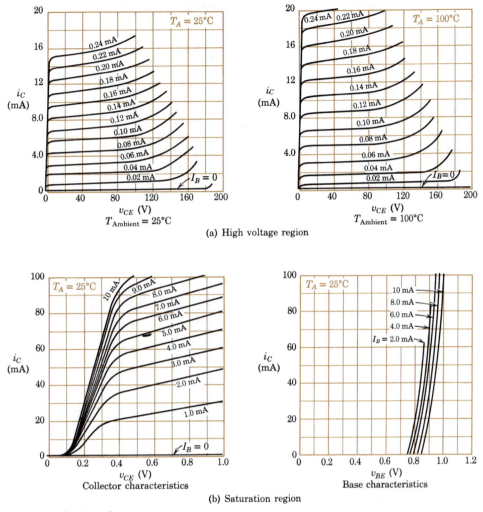

(a) High voltage region

(b) Saturation region

Figure A8   (continued)

μA741 Frequency Compensated Operational Amplifier

General Description

The μA741 is a high-performance monolithic operational amplifier constructed on a single silicon chip, using the Fairchild Planar† epitaxial process. It is intended for a wide range of analog applications. High common mode voltage range and absence of "latch-up" tendencies make the μA741 ideal for use as a voltage follower. The high gain and wide range of operating voltage provides superior performance in integrator, summing amplifier, and general feedback applications. The μA741 is short-circuit protected, has the same pin configuration as the popular μA709 operation amplifier, but requires no external components for frequency compensation. The internal 6 dB/octave roll-off insures stability in closed-loop applications.

†Planar is a patented Fairchild process.

Equivalent
Circuit

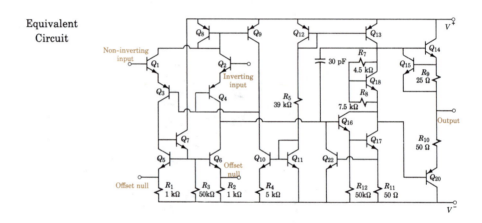

Absolute Maximum Ratings

Supply voltage	±22 V
Internal power dissipation (note 1)	500 mW
Differential input voltage	±30 V
Input voltage (note 2)	±15 V
Voltage between offset null and V^-	±0.5 V
Storage temperature range	−65°C to +150°C
Operating temperature range	−55°C to +125°C
Lead temperature (soldering, 60 s)	300°C
Output short-circuit duration (note 3)	Indefinite

Connection diagram
(Top view)

NC

Offset null 1   8 V^+

Inverting input 2   7

Output 6

Non-inverting input 3   4   5 Offset null

V^-

Note: Pin 4 connected to case

Notes

1. Rating applies for case temperatures to 125°C; derate linearly at 6.5 mW/°C for ambient temperatures above +75°C.
2. For supply voltages less than ±15 V, the absolute maximum input voltage is equal to the supply voltage.
3. Short circuit may be to ground or either supply. Rating applies to +125°C case temperature or +75°C ambient temperature.

Figure A9 Electrical characteristics and typical performance curves for the μA741. (Courtesy Fairchild Semiconductor.)

Electrical Characteristics ($V_S = \pm15$ V, $T_A = 25°C$)

Parameters (see definitions)	Conditions	Min.	Typ.	Max.	Units
Input offset voltage	$R_s \leq 10$ kΩ		1.0	5.0	mV
Input offset current			20	200	nA
Input bias current			80	500	nA
Input resistance		0.3	2.0		MΩ
Input capacitance			1.4		pF
Offset voltage adjustment range			±15		mV
Large-signal voltage gain	$\begin{cases} R_L \geq 2 \text{ k}\Omega \\ V_{out} = \pm10 \text{ V} \end{cases}$	50,000	200,000		
Output resistance			75		Ω
Output short-circuit current			25		mA
Supply current			1.7	2.8	mA
Power consumption			50	85	mW
Transient response (unity gain)	$\begin{cases} V_{in} = 20 \text{ mV} \\ R_L = 2 \text{ k}\Omega \\ C_L \leq 100 \text{ pF} \end{cases}$				
Rise time			0.3		μs
Overshoot			5.0		%
Slew rate	$R_L \geq 2$ kΩ		0.5		V/μs
CMRR	$R_s \leq 10$ kΩ	70	90.0		dB

Typical
Performance Curves

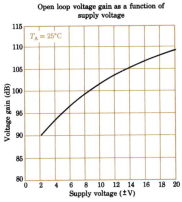

Open loop voltage gain as a function of supply voltage

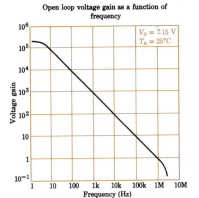

Open loop voltage gain as a function of frequency

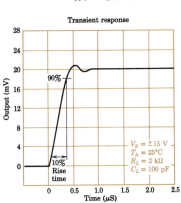

Transient response

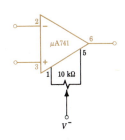

Voltage offset null circuit

Figure A9   (continued)

Answers to Selected Exercises

Chapter 1

2. 9.6 mA

5. (a) 200 kV/m

    (b) 3.2×10^{14} N, 3.5×10^{16} m/s^2

7. 0.002 C/s, 0.018 J/s

10. 0.3 A

13. (a) $RI_m \sin \omega t$ (b) $\omega L I_m \cos \omega t$

15. At $t = 1$ ms, $v_L = 25$ V, $v_C = 2.5$ V

    At $t = 3$ ms, $v_L = 0$, $v_C = 12.5$ V

18. 10 V, 0.1 W; 0 V, 0 W

21. (b) 20 W   (c) 70 W   (e) 5 J

Chapter 2

1. 1 A

4. 5 A, 6 A, 10 V

7. (a) $i_L = 4 - 2e^{-2t}$ A   (c) $w_C = 9e^{-4t}$ J

9. $v_{bd} = 15$ V

13. $v_{cb} = 10$ V

18. 2 A

22. 500 Ω

23. 4 A

26. −3 A

28. 4.33 A

31. 5 A

33. 1.5 A

35. (b) $V_T = 0.08$ V, $R_T = 10$ MΩ

38. (b) $9.1 + 0.3 \cos \omega t$ mA

41. (a) 6 mA   (b) 20 mA

Chapter 3

2. (a) $v(t) = 25.6e^{-t/2}$ V   (c) 0.28 V

4. 0.53 s

7. 16 min 23 s

9. $12e^{-200t}$ V, 1.62 V, 32.5 μA

13. (a) $40 \cos(2000t - \pi/4)$ V,

    $40 \cos(2000t + \pi/4)$ V

15. (a) $V_{av} = 15$ V, $V_{eff} = 15.81$ V,

    $I_{av} = 4.5$ A, $I_{eff} = 4.9$ A

17. (b) $AM_1 = I_1 = 2.45$ A,

    $AM_2 = I_2 = 1.73$ mA

19. 0.03015 Ω, 9994 Ω

22. (a) 19.95 kΩ

25. (a) Invert. ampl. with $R_1 = 20$ kΩ
(b) Noninvert. ampl. with $R_1 = 76.9$ kΩ
28. (a) $v_o = -2v_1 - 0.4v_2$
31. D_1 closed, D_2 open, $I = 30$ mA; 13.6 mA
35. 54 mA, 108 V, 7.2 W
37. (a) 333 μF, 5.66:1   (b) 167 μF, 5.66:1
41. (c) Clips below -2 V
43. Clips above $+2$ V and below -4 V
45. Clamps minimum at zero axis
49. (b) $v_2 = v_1 + 0.5\, V_m$
(c) $V_m = 35$ V, pk-pk VM

Chapter 4

3. $-70,330$ m/s, 2.25×10^{-21} J, -70.33 μm, 2.25×10^{-21} J
6. (a) $x = 10$ cm, $y = 0$ cm   (b) -100 V
9. (b) 8.53 μm, 0.893 ns
12. $f_e = -1.6 \times 10^{-17}$ N, $f_m = -8 \times 10^{-18}$ N, $f_g = 8.9 \times 10^{-30}$ N
15. (a) 1.875×10^7 m/s   (b) 80 V
16. (a) 0.02 cm/V   (b) 1250 V
(c) 0.0033 cm/V
19. (a) 125 Hz   (b) 2500 Hz

Chapter 5

1. 13.5 A
4. 1.83×10^9 atoms/e-h pair, 0.45 Ω-m
7. 2.04×10^{19} donors/m^3
11. (c) 5×10^{21}/m^3, 4.5×10^{10}/m^3
(d) 1 mV   (e) 245,000
14. 1.56×10^{23}/m^3
16. 39,330 A/m^2
19. (a) -0.098 V   (b) 0.98 mA
22. (a) 8 mA   (b) -20 μA   (c) 4.85 mA
23. (b) 1.8 A
27. 1.4, 1.4, 0.7, -0.3 V
30. (a) 12 V, 2 Ω (b) 5 Ω (d) 13.5 V

Chapter 6

2. 1.5 V
6. 3
8. (b) 2 mA, 10 V
11. (a) 1.2 V   (b) $i_D = (V_{DD} - v_{DS})/(R_D + R_S)$
(c) 1.2 kΩ, 14.2 V
17. From 0.04 to 2.22 mA
20. 2 mA, 283 kΩ
23. (a) 0.7 V   (b) 0.7 V, 1.4 V
(c) 0.1 mA, 3 mA, 2 V
26. 5.4 ¢

Chapter 7

3. (b) $2 < V < 3$ V, say 2.5 V
5. 16, 2^n
8. A NOR A $= \overline{A}$, $\overline{A}$ NOR $\overline{B} = A \cdot B$
11. (b) 0, 0, 15 V
(c) 0.2 mA, 10 mA, 5 V; 0.42 mA, 14 mA, ~ 1 V
(d) 15 to 1 V
14. 6.4 Ω, 480 MΩ; 25 Ω, 14.2 kΩ
17. (a) 1.7 V, < 1 V, 0, 5 V, T_2 **OFF**
(b) T_2 **ON**, 2.1 V, 1.4 V, 0.5 mA, 0.3 V
21. (a) T_1 **ON**, $V_x = 2.4$ V, $V_y = 1.0$ V, $V_F = 0.2$ V
(b) T_1 **OFF**, $V_x = 1.0$ V, $V_y = 5$ V, $V_F = 3.4$ V
24. ON, ON, 2.1, 1.4, 0.7, 0.3, > 1.4, > 1.4 V
OFF, OFF, 1.0, < 1.4, < 0.7, 4.5, 0.3, 3.3 V
28. Like 3-input **NOR** gate of Fig. 8.24b
29. $V_{out} = 10$ V, 0 V
35. (c) Divides frequency by four
37. 1000-Hz square wave

Chapter 8

1. (a) **00100, 01100, 10111, 11111**
(b) 3, 5, 10, 21
3. (a) **10100101B = 245Q = A5H**
(b) **205D = 315Q = CDH**
(c) **10000111B = 135D = 87H**
(d) **11011000B = 330Q = 216H**
7. **93D = 1011101B = 135Q**
10. **A**
14. $A \cdot B = \overline{\overline{A} + \overline{B}} = \overline{\overline{A + A} + \overline{B + B}}$
17. (c) Input mode, **Q** follows **DI**; Output disconnected
20. $f = A\overline{B} + \overline{A}B$
23. $\overline{B}\,\overline{D}$
26. (a) $D_1 D_2 D_3 =$ **CBA**   (b) $D_1 D_2 D_3 = Q_1 Q_2 Q_3$
29. (b)

CK	Q_D	Q_C	Q_B	Q_A
0	0	0	0	1
1	0	0	1	0
2	0	1	0	0
3	1	0	0	0
4	0	0	0	1

31. (b) **0000** vs. **1010** for binary; $\therefore$ must eliminate two 1s   (c) Make $CK_B = Q_A$ **AND** $\overline{Q}_D$
(d) Make $CK_D = Q_C$ **OR** (Q_A **AND** Q_D)
34. 0 to 15D, 255D
36. (c) **RS = 1, CS$_1$ = 0, CS$_2$ = 1**
(d) 6 Column Select lines.

38. (a) Let Inputs = Addresses and
Outputs = Data for 8 × 1 ROM
40. Form 2's complement of **B**
41. (b) Routes Input to selected Output line

Chapter 9

4. (a)

0100	INPUT TO A
0010	FROM PORT 2
1010	LOAD
0010	D FROM A

(c)

1101	LOAD
0000	B
0000	WITH 0000

6. (a)

CODE	INSTR
1101	LOAD A
0111	WITH
0001	0001
1000	ADD TO A
0000	B
1010	LOAD FROM A
0000	B

9. (a)

0100	INPUT TO A
0011	FROM PORT 3
1000	ADD TO A,
0001	C
0101	OUTPUT FROM A
0010	TO PORT 2

11. (a)

1101	LOAD
0010	D WITH
1100	1100
1110	LOGIC OP A
1110	AND WITH D

14. (a)

1101	LOAD
0011	E
1111	WITH 1111
1011	LOAD
0111	A FROM MEM
0001	COMPLEMENT A
1100	LOAD MEM
0111	FROM A

16. (a) Perform 5-cycle preparation routine then
execute 5-cycle loop 15 times

(c)

LOCS	CODE
0000	1101, 0111, 1111
0011	1010, 0000
0101	0111, 0000
0111	1111, 1000, 0101

19. Use: $10 = 8 + 2 = 2^3 + 2$

IN A, (1H)	INPUT N FROM PORT 1
RLCA	MULTIPLY BY 2
LD B, A	STORE 2N IN B
RLCA	MULTIPLY 2N BY 2
RLCA	MULTIPLY 4N BY 2
ADD A, B	ADD 8N + 2N = 10N
OUT (2H), A	OUTPUT 10N TO PORT 2
HALT	STOP PROCESSING

22.

	ORG 0010H
	LD B, FAH
	LD HL, 0100H
LOOP:	INC L
	LD A, (HL)
	XOR B
	JPNZ, LOOP
	LDA, L
	OUT (5H), A
	HALT
	END

24. STORE TEST PATTERN IN A
SET HL POINTER TO 1000H MINUS 1
INCR POINTER L
STORE TEST PATTERN IN MEM
COMPARE PATTERN WITH MEM
IF SAME, JUMP TO INCR L INSTR
IF NOT, DECR L POINTER TO LAST WORKING
RAM ADDR

Chapter 10

3. (b) $v = 120 \cos (50\pi t + \pi/2)$ V
(c) $v = 20 \cos (300\pi t + 1.2\pi)$
5. (a) $10 \underline{/83°}$ (b) $1.73 + j5$ (e) $17.73 - j11$
8. (a) $\pm\sqrt{2}/2 \pm j\sqrt{2}/2$
9. $20.3 \cos (\omega t + 12.37°)$ V
14. $9.24 \sqrt{2} \cos (5000t + 90°)$ V
16. $28.28 \underline{/45°}$ V, $14.14 \underline{/-45°}$ mA, $10 \underline{/0°}$ mA
18. $4 - j3 \, \Omega$
20. (a) $C_S = 2.6 \, \mu F$, $R_S = 76.9 \, \Omega$
21. (a) $R_P = 1133 \, \Omega$, $C_P = 1.48 \, \mu F$
(b) $R_P = 508 \, \Omega$, $C_P = 0.82 \, \mu F$
24. (a) $1.12 \underline{/-63.4°}$ S (b) $2.5 \underline{/0°}$ A,
$5.6 \underline{/-63.4°}$ A, $5 \underline{/-90°}$ A
27. $6.32 \underline{/71.6°}$ V, $3.16 \underline{/71.6°}$ A
29. $4 \underline{/-66°}$ A
32. $L = 15.92$ H
36. (a) 3.5 V (b) 0.18 V
38. 99 to 870 pF
40. (b) 20 (c) 250 nF, 1 H, 100 Ω
43. (a) 4.4 kΩ, 0.71 H, 90 pF in series

46. (a) Coil (25 Ω; 3.98 mH) in parallel with 15.9-nF capacitor

Chapter 11

1. (a) $R_1 = 5$ kΩ, $R_F = 1$ MΩ, inverting
4. (a) for $R_1 = 5$ kΩ, $R_F = 95$ kΩ, noninverting
8. For $R_1 = 1$ kΩ, $R = 200$ kΩ, $C = 795$ pF
10. $-10/(1 - j100/\omega)$, high-pass filter
15. (a) 10 kΩ (b) V-to-I converter, $R_1 = 1$ kΩ, 10 kΩ (c) 5.7 V, 6.0 V
19. (c) 2 sin $2\omega t + 3$ V
24. 100 MΩ, 0.1 Ω
25. For $R_i \gg (R_F + R_o)$ and $R_F \gg R_o$,
   $R_{iF} = (R_F + R_o)/(A - R_o/R_F) = 0.011\ \Omega$
27. 10 V, 2.57 kΩ
31. (c) 000, 100
33. (b) $H = 0.462$
35. $A = 520$, $A_{cm} = 49.5$, CMRR $= 10.5$
36. (b) 109

Chapter 12

3. (a) 2 mA, 20 V   (c) 0.5 sin ωt mA   (d) 5
6. (b) 5.5 kΩ, 3 kΩ
7. (a) 2 mA, 8 V   (b) 1.8 mA, 9.6 V
11. (a) -6.5 V   (b) 4.7 kΩ
13. (a) 3 kΩ, 565 kΩ (b) 0.8 mA
17. (c) 18.1 kΩ, 61.8 kΩ, 6 kΩ
19. (a) $R_E = R_C = 6$ kΩ, $R_1 = 18.3$ kΩ, $R_2 = 34.8$ kΩ
   (b) $0.46 < I_C < 0.51$ mA
21. (a) $v_{CE}(\text{sat}) = 0$, $I_{CEO} = 0$, $V_{CE} = 25$ V, $I_C = 400$ mA
   (b) 62.5 Ω, 10 Ω
23. (a) 1.5 kΩ, 3 kΩ (b) 1.5 sin ωt V
27. $N_2/N_1 = 4$, 10, $\eta \cong 50\%$
28. (a) 11.6 W, 6 W, 0.6 W, 5 W, 0
29. (b) 500 Ω (c) 10 W, 2.5 W, 25%
   (d) 7.1, 50%
32. (b) One possibility: $V_{CE} = 25$ V, $I_C = 250$ mA
   (c) 3.125 W, 3.125 W, 25%
   (d) $N_2/N_1 \cong 1.25$, 5 W, 5 W, 50%

Chapter 13

3. 2100, 50 krad/s
7. (a) 0, 3, 200, 203, 206 krad/s
   (c) If load responds to audio only.

9. $r_{ac} = 1/(1 + 0.02\ v) = 0.83\ \Omega$
   $i(t) = 11 + 1.2$ sin ωt
14. $g_m = 1$ mS, $r_d = 60$ kΩ
15. (a) 3000 μS   (b) 2000 μS
18. (b) 11.4
21. $r_i = 1$ kΩ, $\beta = 50$, $v_o = 200$ kΩ
25. (d) 0.1 mA, 4 mA, 12 V
   (e) 4 sin ωt mA, 8 sin ωt V
27. (b) 3 kΩ (c) 9 V
30. $720 < r_\pi < 3030\ \Omega$; no
33. 12.5, 100

Chapter 14

3. (a) 10 kΩ (b) $-7.4\underline{/-81°}$
7. -125
8. (a) $R_D \| R_L \| r_d = R_D$, $R_s + R_1 \| R_2 = R_1 \| R_2$
   (b) -10
11. (a) 0.65 mA, 4 kΩ (b) 6 kΩ (c) 114
   (d) 71
13. (b) 4.4 kΩ (c) 1.5 kΩ, 15 V
16. (b) 0.33 μF
18. (a) 16 Hz, 81 Hz   (b) 81 Hz
   (c) 1 μF for $R_L = r_\pi \| R_B \cong 1$ kΩ
21. (b) $C_{be} = 100$ pF   (c) 6.4 MHz
24. 64 krad/s
28. (b) 0.4 μF   (c) 2.1 MHz
31. (a) 13 dB   (d) -20 dB
34. (a) 31.6 W   (b) 31.6 kW
36. $f_2 = 0.6$ MHz, 0 dB, 23.5 dB
39. 3

Chapter 15

1. (b) $0.01\underline{/0°}$ V, $10\underline{/180°}$ V, -19.6
4. AB, $C - 1/B$, $AB/(1 + A - ABC)$
6. (b) $AB/(1 + AC)$
10. $\mathbf{G} = 1/\mathbf{H} = 200\underline{/0°}$
13. (b) 71 kHz
15. (a) $\mathbf{H} = 0.19\underline{/0°}$ (b) G_F up 65%, unstable
16. (a) $R_F = 0.991$ mΩ
   (b) G_F up 0.002%, stable
19. (b) 1%   (c) 30%
22. $H = -0.043$
24. (a) $H = -0.095$
27. 1.9 V, $R_i = 73$ kΩ, $R_o = 53\ \Omega$
29. (a) 0.32 MHz   (b) 3.2 MHz

Index

469